INTERNATIONAL SERIES OF MONOGRAPHS IN PURE AND APPLIED BIOLOGY

DIVISION: **ZOOLOGY**

GENERAL EDITOR: G. A. KERKUT

VOLUME 44

THE PHYSIOLOGY OF INSECT REPRODUCTION

THE PHYSIOLOGY OF INSECT REPRODUCTION

by

FRANZ ENGELMANN
DEPARTMENT OF ZOOLOGY
UNIVERSITY OF CALIFORNIA AT LOS ANGELES, CALIFORNIA

PERGAMON PRESS
Oxford · New York · Toronto
Sydney · Braunschweig

Pergamon Press Ltd., Headington Hill Hall, Oxford
Pergamon Press Inc., Maxwell House, Fairview Park, Elmsford, New York 10523
Pergamon of Canada Ltd., 207 Queen's Quay West, Toronto 1
Pergamon Press (Aust.) Pty. Ltd., 19a Boundary Street,
Rushcutters Bay, N.S.W. 2011, Australia
Vieweg & Sohn GmbH, Burgplatz 1, Braunschweig

First edition 1970
Library of Congress Catalog Card No. 70–114850

Printed in Great Britain by A. Wheaton & Co., Exeter

08 015559 6

CONTENTS

PREFACE

The present monograph is an outgrowth of this writer's interest in gaining for himself a comprehensive understanding of the basic phenomena governing reproductive processes in insects. During the several years of intensive reading, I became acutely aware of pertinent published results contained in a great number of scattered journals and books. Many of the original papers are not available to many scientists at various colleges, universities, and research institutes throughout the world. The need to summarize the principle findings thus became even more compelling. Topics of this monograph range from aspects of sex determination to means of control in insect societies. The book is an attempt to cover all aspects related to the propagation of the species; it is indeed a biology of the insects.

When a biologist finds a new book, he first looks through those chapters related to his own research interests and often concludes that the material is covered inadequately. Then by paging through the remainder of the book, he decides that, on the whole, the book might not be as bad as he had thought; but he is misled since he does not know all pertinent literature as thoroughly as he does that in his own research field. It is true, no author of today can deal with all aspects of biology to the same depth with the same comprehension. He naturally stresses the areas of his own research specialty a little more than the others. The reader will find that this author is no exception. I tried to extract from the vast literature the essential information on the topics treated.

This monograph in its present form would not have been possible without the invaluable help, criticisms, and discussions by some of my colleagues and friends. Various chapters were read by J. N. Belkin, C. W. C. Davis, and R. C. King and I am indebted to them for their effort and willingness to spend their time in helping to improve the manuscript. Furthermore, during the preparation of the manuscript, considerable help was received from L. Andersen, P. Girard, and J. Malamud; their efforts are gratefully acknowledged. It is my pleasure to thank Miss G. Beye and Mr. K. Pogany who prepared many of the illustrations from drafts or originals. Permission to reproduce copyrighted materials was given by the following publishers: Academic Press, New York; American Association of Advancement of Sciences, Washington; Canadian Entomological Society; Centre National de la Recherche Scientifique, Paris; Gauthier-Villars & Cie, Paris; Masson & Cie, Paris; Muséum d'Histoire Naturelle, Genève; Springer-Verlag, Berlin–Heidelberg–New York; The Thomas Say Foundation, The Entomological Society of America; Verlag für Recht und Gesellschaft AG, Basel; Verlag Paul Parey, Berlin; Wistar Institute, Philadelphia.

CHAPTER 1

THE GENITALIA

THE anatomy of the external and internal genitalia in males and females is of interest to physiologists, morphologists, and taxonomists. While the taxonomist and morphologist describe and use structural features for classification of the species and attempt to interpret the structure in terms of ontogenetic and phylogenetic origins, the physiologist can fully understand function only with a knowledge of the anatomy. Considerations of the functional anatomy of ovarian development (Chap. 5) and egg maturation as well as those on oviposition mechanisms may exemplify this. It is neither feasible nor intended to give a full account here of the genital structures in insect species of all orders. The diversity is enormous; yet certain common elements are characteristic of nearly all the species (Snodgrass, 1935, 1957; Weber, 1954; Dupuis, 1955; Tuxen, 1956). Certain aspects have been more recently reviewed by Gouin (1963), who showed how ontogenetic studies of the genitalia become quite valuable in discerning homologous structures of the species of various orders. Attempts to bring order to the confusing terminology have been made by Dupuis (1955), Tuxen (1956), and Snodgrass (1957). In the following short description of the genitalia, basically the outlines of Snodgrass (1935) and Weber (1954) have been followed.

Male internal genitalia usually consist of two testes, vasa deferentia with vesiculae seminales and accessory glands, and an unpaired ductus ejaculatorius (Fig. 1). Only in the Ephemeroptera are the original paired gonopores still present. The normally unpaired ejaculatory duct arises from an invagination of ectoderm and is consequently lined with a cuticular intima. However, the vasa deferentia can consist of both mesodermal and ectodermal elements. Some variations are found in the genital structures, such as unpaired seminal vesicles or unpaired accessory glands.

The external genitalia of a male are generally formed from the ninth sternite from an originally unpaired invagination, the lobi phallici. During further development, this anlage becomes paired and gives rise to paired mesomeri and parameri (Günther, 1961). The mesomeri generally fuse to form an unpaired intromittent organ, or phallus, which may bear an aedeagus; however, depending on the species, the parameres become phallomeres, claspers, or harpagones which are functional during copula (Snodgrass, 1957; Günther, 1961; Gouin, 1963). Parameres are considered by some investigators to be homologous to abdominal appendages similar to the gonapophyses of the female. Some recent studies on the musculature of the genitalia in a few species, however, do not support this view; these muscles are of sternal origin and are homologous to those muscles of the pregenital segments (Matsuda, 1958; Gouin, 1963); if the external genitalia were appendages, they would have no metameric muscles. This conclusion is based on embryological investigations which appear to offer reliable criteria since muscles rarely change their original positions or attachments. Unfortunately, practically no data are available on the details of either the innervation of the genital muscles or the origin of these nerves. Since the genitalia of the male arise in the ninth abdominal segment, the gonopore generally comes to lie between the ninth and tenth sternum, or occasionally, on the tenth sternum.

The female internal genitalia, consisting of paired ovaries and oviducts, a common oviduct with accessory sex glands and spermatheca, can be shown to originate from abdominal segments 7 through 9. During development ectodermal invaginations occur in all three sterna, but later these can no longer be recognized as separate parts. In Ephemeroptera,

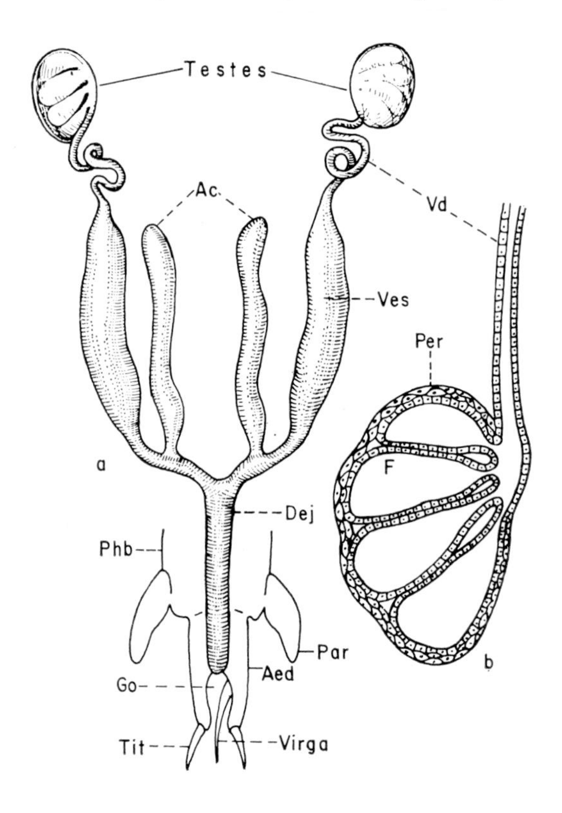

FIG. 1. Diagrammatic representation of male genitalia. (a) General structure. (b) Section through testis. Ac = accessory sex glands. Aed = aedeagus. Dej = ductus ejaculatorius. F = testis follicle. Go = gonopore. Par = paramer. Per = peritoneal sheath. Phb = phallobase. Vd = vas deferens. Ves = vesiculum seminalis. Tit = titilator. (Modified from Weber, 1954.)

paired gonopores are found on the seventh sternum and it is believed that phylogenetically this is the original type (Palmen, 1884) (Fig. 2). In the most common type, the genital opening is on the eighth sternum and joins the common oviduct (arising from the anlage of the seventh sternum) leading anteriad. The opening may be located more anteriorly, due to secondary modifications of the terminal segments. Accessory sex glands (arising from the anlage of the ninth sternum) open into the genital atrium or vagina. This type is found in the Orthoptera, Hymenoptera, Diptera, Neuroptera, and others (Heberdey, 1931; Weber, 1954; Davies, 1961). While the common oviduct has an ectodermal lining, the paired oviducts

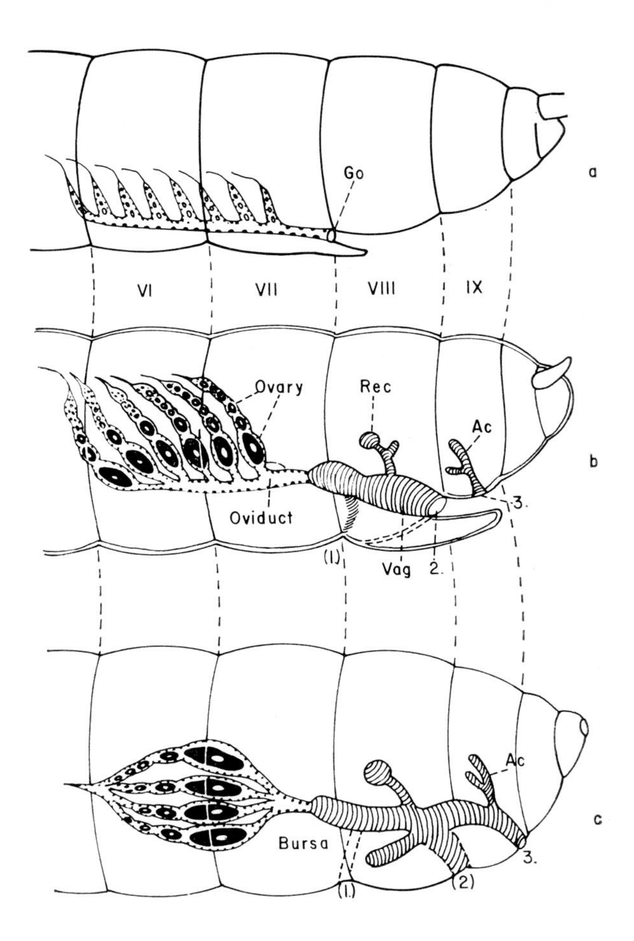

FIG. 2. Schematic representation of the types of internal female genitalia. (a) A primitive type found in Ephemeroptera. (b) Orthopteroid type; the genital opening is on the 8th segment. (c) Further development to a type found in Lepidoptera. Numbers 1, 2, 3 indicate the positions of the original genital openings. Ac = accessory sex glands. Go = gonopore. Rec = receptaculum seminis or sphermatheca. Vag = Vagina. (Modified from Weber, 1954.)

often are composed of mesodermal and ectodermal parts. Ovipositors, which are of great taxonomic value, are formed by the eighth and ninth sternum and probably in the female can be considered as abdominal appendages (Scudder, 1961a, b, 1964; Oeser, 1961; Davies, 1961).

A further modification of the female genitalia is found in the Lepidoptera. With the exception of the Microlepidoptera, all species have two genital openings (Weidner, 1934; Snodgrass, 1935). The posterior opening constitutes the actual gonopore, while the anterior one (eighth segment) forms a bursa copulatrix (Figs. 2, 3). The two openings are apparently homologous to the corresponding embryonic anlagen. As the schematic drawings show, the bursa has a connection to the common oviduct via the ductus seminalis through which the sperm migrate to reach the receptaculum seminis. All these organs are lined with a cuticular intima since they are of ectodermal origin.

Very little is known concerning the details of innervation of the female genitalia although such details could supply interesting information regarding the segmental origin of the various parts and would supplement embryological investigations. The studies on *Apis mellifera* (Ruttner, 1961) and *Leucophaea maderae* (Engelmann, 1963) generally confirm the conclusions of earlier studies with respect to the segmental origin of the genitalia. Nerves of certain ganglia innervate the corresponding genital portions. There is one noted exception in *Leucophaea*. The spermatheca in this species are innervated by branches of the seventh abdominal nerve, thus suggesting that they originate from the seventh sternum. Embryological studies in other species, however, ascribe the spermatheca to the eighth segment.

The paragenital system of the Cimicoidea, which is a peculiarity among insects, ought to be mentioned here since its functional significance has been under investigation. Males of many species of the Cimicoidea do not deposit their sperm mass into the genitalia of the female but rather penetrate a meso-ectodermal structure on the female's abdomen and deposit

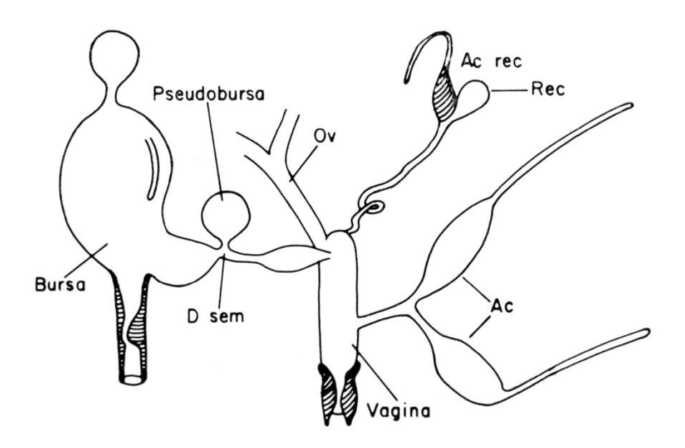

FIG. 3. Schematic representation of female genitalia of Lepidoptera with two genital openings. Ac = accessory sex glands. Ac rec = accessory glands of spermatheca. D sem = ductus seminalis which connects the bursa copulatrix with the vagina. Ov = common oviduct.

the sperm there. This organ, known as organ of Berlese or Ribaga or simply called spermalege (Fig. 41), can be located on nearly any abdominal segment but its location is species specific (Carayon, 1966). Spermatozoa migrate through the haemocoel to the conceptacula seminis where they are stored. Some species possess a structural connection from the spermalege to the conceptacula. This connection appears to be a solid core of tissue within which the spermatozoa migrate. Phylogenetically this system may have evolved from an accidental penetration of a females abdomen by the male aedeagus during copula (Carayon, 1966).

Lastly, we have to consider the structural differences of the ovaries found in the various species. Each ovary consists of a number of ovarioles which may range from 1, as in the *Coprinae* (Coleoptera)—which also have only one ovary (Heymons, 1929; Robertson, 1961)—2 as in *Glossina* (Saunders, 1961), 360 as in *Apis* (Dreischer, 1956), and up to 1200 in the army ant *Eciton* (Hagan, 1954a). Naturally, the number of ovarioles determines to some extent the reproductive capacity of a species. On the basis of structural differences, two basic types of ovarioles are distinguished: the panoistic and the meroistic (Snodgrass, 1935;

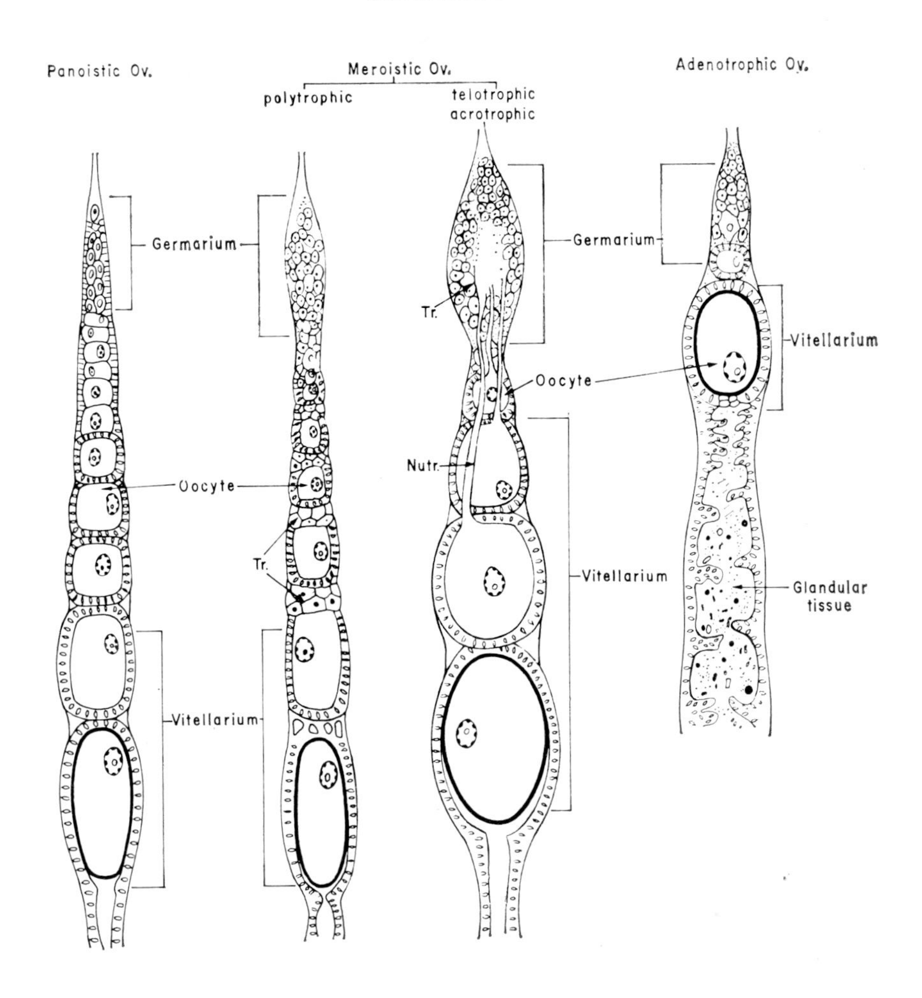

FIG. 4. The four types of ovarioles found in species of insects. The adenotrophic type is only found in *Steraspis*. Nutr = nutritive cord. Tr = trophic tissues. (Modified from Weber, 1954 and Martoja, 1964.)

Weber, 1954; Bonhag, 1958). The meriostic type can be further subdivided into polytrophic and telotrophic (acrotrophic) ovarioles (Fig. 4). In the most primitive type, the panoistic ovariole, the germarium contains only oogonia which as they migrate down the ovariole, are surrounded by mesodermal cells, thus forming follicles. This type is found in the Odonata, Ephemeroptera, Dictyoptera, Orthoptera, Plecoptera, Embioptera, Thysanoptera, and Siphanoptera. Polytrophic ovarioles are distinguished from panoistic ones in that each follicle contains a number of nurse cells together with one oocyte. Generally the nurse cells and oocyte of a follicle are derived from one oogonium (p. 45). Polytrophic ovarioles are observed in the Anoplura, Psocoptera, Dermaptera, Mecoptera, Trichoptera, Lepidoptera, Diptera, and Hymenoptera. There is some controversy as to whether the Mallophaga have panoistic or polytrophic ovarioles (Ries, 1932; Seguy, 1951).

Telotrophic ovarioles are characterized by their terminal nutritive tissue and its connection with the oocytes via a nutritive cord (Fig. 4). The Homoptera and Heteroptera typically possess ovarioles of this type. Among the Coleoptera, various authors ascribe either polytrophic or telotrophic ovarioles to the same species. The difficulty in classification of these cases may arise from the fact that the nutritive cord breaks down early in development in some species; the ovariole may then appear as a polytrophic or panoistic type. Indeed, panoistic ovarioles are reported in *Melolontha vulgaris* (Vogel, 1950) and in *Tenebrio molitor* (Huet and Lender, 1962); yet, in the latter species, Schlottman and Bonhag (1956) describe a telotrophic ovariole. Bonhag (1958) expresses doubt whether panoistic ovarioles exist in any coleopteran and suggests a reinvestigation of each report.

In one species of Coleoptera, namely *Steraspis speciosa*, a new type has been identified (Martoja, 1964). The germarium and vitellarium of this ovariole are unusually short (Fig. 4).

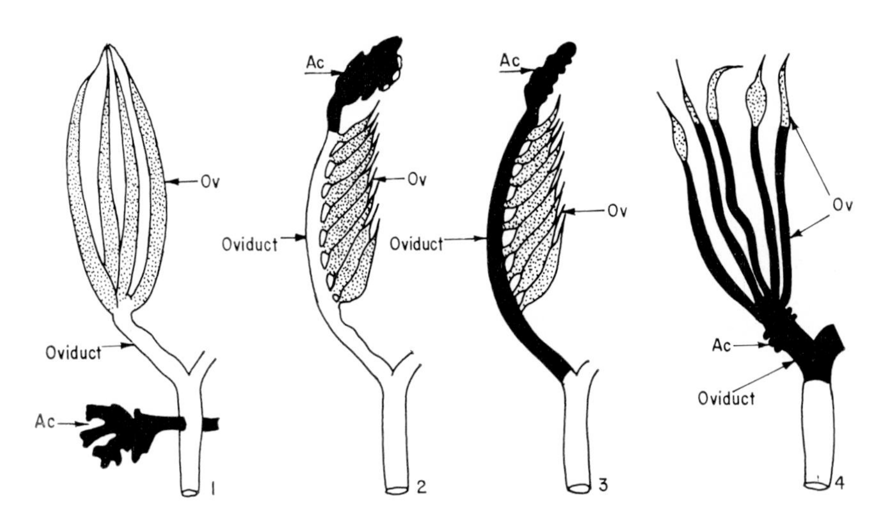

FIG. 5. Relationship of accessory sex glands to ovarioles. Glandular tissues are indicated by solid black. 1. Dictyopteran type. 2. *Acrida* type with pseudoaccessory glands. 3. *Locusta* type. 4. *Steraspis* type in which the lower portions of the ovarioles furnish secretory materials used in egg covering. Ac = Accessory sex glands. Ov = Ovariole. (Adapted from Martoja, 1964.)

Proximal to the short vitellarium, the epithelium of a long glandular section of the ovariole is thrown into folds which unfolds as an egg passes down. This unique glandular portion of the ovariole appears to secrete a covering for the egg; the material is presumably similar to that of accessory sex glands. A sphincter at the posterior end of the long ovariole closes it off from the oviduct. No nutritive cord has yet been found; Martoja, nevertheless, believed that one is dealing here with a modified telotrophic ovariole and proposed the term adenotrophic for this type. No other species is known to have a similar ovariole. A certain difficulty is encountered with this new classification, however. The distinction among the other three types is based on the location of the nutritive tissues in relation to the growing oocytes. In this new type, the oocytes apparently do not receive nutrients from the glandular portions of the ovariole. From this viewpoint, the terminology is debatable.

On the basis of the anatomical relationship between the accessory sex glands, which furnish the material for the oothecae or egg covering, and the ovarioles, Martoja (1964) distinguished four different types (Fig. 5). In the dictyopteran type, which is the most widely distributed among insects, the accessory glands are clearly associated with the common oviduct or vagina. In the next type, which is found in many Acrididae, the anterior portion of the lateral oviduct bears the accessory sex gland. In *Locusta* (type 3), the entire lateral oviduct is secretory (Lauverjat, 1964), the material of which is used in forming the plug for the egg pod. In both of these latter types no accessory sex glands are associated with the common oviduct. In type 4, which is found in *Steraspis*, materials from both the lateral oviducts and the proximal portions of the ovarioles enwrap the eggs as they are laid. The secretory portion of the ovarioles may be analogous to the secretory oviducts of *Locusta*, yet anatomically it is part of the ovariole.

CHAPTER 2

SEX DETERMINATION

THE subject of sex determination has been reviewed many times during the last few years, yet in the present context it must once again be dealt with. Emphasis will be placed, however, on the more recent findings concerning some of the details. The extensive literature on the topic is covered by Wiese (1960, 1966) and Gowen (1961) who discuss and review the older findings in detail. Additional reviews are given by Doutt (1959) and Kerr (1962). The treatment of sex determination by Goldschmidt in his *Theoretical Genetics* (1955) is very valuable and helps in gaining a proper perspective. In many papers (see reviews of White, 1954; Smith, 1960; and Ueshima, 1966) particular attention has been given to the sex chromosomes and the localization of sex determining factors or genes. It should be pointed out, however, that cytological evidence for the presence of so called sex chromosomes (e.g. *Drosophila*) is no proof of their role in sex determination. On the other hand, the absence of cytologically recognizable sex chromosomes does not negate a diplogenotypic sex determining mechanism either (e.g. *Calliphora*). In the cases of genetic determination of sex, heterozygocity in one of the sexes with respect to the sex determiners is the common pattern. Cytological heterogamety, i.e. XY or XO, may be observed in the male sex in many species of the Diptera, Neuroptera, Mecoptera, Heteroptera, Homoptera, Odonata, Coleoptera, and Orthoptera (see White, 1954). In Lepidoptera and Trichoptera, the females are known to be heterogametic; they are usually of the type XY, but the XO system is also found. Exceptions are known, as, for example, in certain dipteran species in which the females are heterogametic (Bush, 1966; Martin, 1966) rather than the males, as is commonly found. Cytological evidence exists for multiple sex chromosomes in some species of the Cimicoidae (Ueshima, 1966) as well as in many other groups of insects (White, 1954). Genetic analysis of the location of sex determiners is difficult in these cases. In addition to the cytologically detectable sex chromosomes, two lines of evidence support the hypothesis of the occurrence in many species of a homo-heterozygocity with respect to the sex determiners. Proof for heterogamety in one of the sexes is obtained by the use of genetic markers linked with the sex determiners as well as by the observation of a 1:1 sex ratio. Mendelian genetics shows that a 1:1 sex ratio in the F_1 generation can only occur if one of the sexes is heterozygotic with regard to sex determiners on either autosomes or sex chromosomes.

A. Balance Hypothesis

The idea that sex is a quantitatively variable character is contained in the early writings of Goldschmidt (1911), but it was Bridges (1916, 1921, 1922) who first provided the proof for a possible mechanistic interpretation of sex in a species of insects, namely *Drosophila*. Long before this time, however, it had been found that males of certain insect species have one "accessory chromosome"; McClung (1902) reasoned that this chromosome might bear the sex determiner. This lone chromosome is the X-chromosome in the XO system which may pair with a Y-chromosome in species where the latter is present. The concept of heterogamety in the male sex of certain species was thus established. Bridges (1916) then traced the inheritance of the X-chromosome. He found that XO-males and XXY-females can occur through non-disjunction of the homologous X-chromosomes during the maturation divisions in the oocytes. As shown by genetic markers, the X-chromosome in the XO-male is derived from the father while the XX in the XXY-female derive from the mother. Evidence was thus given that the X-chromosomes carry genes for femaleness; when these genes occur in a double dose, females are produced. In *Drosophila* (Bridges, 1916) the Y-chromosome apparently does not play a positive role in sex determination. However, since XO-males are sterile, it was suggested that the Y-chromosome may bear certain sperm viability or motility factors.

In later works, Bridges (1921, 1922, 1932) and others firmly established for *Drosophila* that it is the ratio of the number of X-chromosomes to the sets of autosomes which determines sex in this species. Triploid intersexes (3A, XX) were first found by Bridges (1921); on the basis of this observation, the balance hypothesis was founded. Table 1 gives the known possible combinations of sex chromosomes in relation to the autosomes. It is evident that progressive ploidy does not change the sex type or the ratio between X-chromosomes and autosomes. Also, the presence of Y-chromosomes, even in multiple doses, does not affect the sex. Females are produced when the ratio is 1.00 or higher, males when the ratio is 0.50 or below. The term superfemale (2A, XXX) is probably misleading: these females are poorly viable and produce only a few eggs. Intersexes are produced when an imbalance between X-chromosomes and autosomes exists, as in some triploid and tetraploid animals

TABLE 1. Balance between X-chromosomes and autosomes in *Drosophila* species and the effect on sex determination

Chromosomes			Balance X/A	Type
X	Y	A sets		
XXX		2	1.50	female (superfemale)
XXXX		3	1.30	female (triploid metafemale)
XXXX		4	1.00	female
XXX		3	1.00	female
XXX	Y	3	1.00	female
XXX	YY	3	1.00	female
XX		2	1.00	female
XX	Y	2	1.00	female
XX	YY	2	1.00	female
X		1	1.00	female
XXX		4	0.75	intersex
XX		3	0.67	intersex
XX	Y	3	0.67	intersex
X		2	0.50	male
X	Y	2	0.50	male
X	YY	2	0.50	male
X	YYY	2	0.50	male
XX		4	0.50	male
X		3	0.33	male (supermale)

Again, the presence or absence of the Y-chromosome does not influence the expression of intersexuality. These studies unequivocally show that in *Drosophila* sex determination can be explained in quantitative terms: the balance of female and male determining factors is decisive. We will see below that the factors are discrete units distributed both on the X-chromosomes and the autosomes.

The discovery of diploid intersexes in *Porthetria* (= *Lymantria*) *dispar* lead Goldschmidt to a similar view of a balance between female and male tendency genes operating in an individual (for a complete discussion, see Goldschmidt, 1955). The quantitative aspect of this hypothesis is already inherent in Goldschmidt's 1911 paper. In *Porthetria* the females are heterogametic (XY), and it is now believed that the female determining factors are outside the X-chromosomes. Originally, the location of these factors was thought to be in the cytoplasm but was later believed to be on the Y-chromosome. It is, however, possible to argue that the female determiners are distributed on the autosomes (Winge, 1937). Goldschmidt postulates that both female and male factors are present in each sex. If the male factors are present in a double dose (XX), these factors are epistatic; in a single dose, hypostatic. In the latter case, the balance is in favour of the female determiners. Goldschmidt does not accept the possibility that one factor may be dominant over the other. In any given population of *Porthetria*, the balance between male and female determiners is operational, i.e. only males and females arise. In this case sex is determined epistatically.

If, however, this normal situation is disturbed, as in crosses between races, intersexes may result. Goldschmidt explained the occurrence of these intersexes by the hypothesis that male and female determiners are of varying strength in the different races and that upon the mixture of these determiners of different races, an imbalance may be achieved. The male determiners are weak in the palaearctic region whereas, in the races of southern Japan, they are medium to strong. An individual from crosses of these races with a genetic constitution of XY may be an intersex. One can establish a graded series of the strengths of male and female determiners. In the extreme, from a cross of a race with a strong male determiner (XX) with one having a weak female determiner, all the offspring may be male regardless of their genetic constitution (XY or XX). From these observations and experiments it is difficult to argue against the hypothesis of a balance of male and female determining factors in *Porthetria*. However, any attempts to make this hypothesis applicable to other species should be undertaken cautiously. When the same techniques were used in several other species, intersexes could not be obtained. A further difficulty is met: how can one distinguish between the existence of an extremely strong female determiner which always throws the balance in favor of a female and the epistatic determination of one of the sexes? There is no means to prove either of the possibilities.

In Bridges' hypothesis of gene balance (1921, 1922), several male and female determining genes were postulated. Evidence for female genes on the X-chromosomes was then supplied by Dobzhansky and Schultz (1934) through the use of triploid intersexes in *Drosophila*. These animals were irradiated with X-rays to induce chromosome breakage and recombination with other chromosomes. Duplications of sections of the X-chromosome or additions of X-chromosome sections by this means shifted the sex type in the female direction. The longer the additions to the X-chromosome were, the more female-like the animals appeared (Dobzhansky and Schultz, 1934; Pipkin, 1940). At least ten regions of female tendency genes were identified on the X-chromosome. Their effect is additive (see reviews of Gowen, 1961, Kerr, 1962). Similar studies using triploid intersexes were conducted in the search for male-determining genes on the autosomes (see Gowen, 1961). A shift in the male direction was obtained by the addition of some regions of the 3rd chromosome; apparently no genes of male potency are located on the second chromosome (Pipkin, 1947).

Several genes on the 2nd and 3rd chromosomes in *Drosophila* affect sex by operating on the gene background (Gowen, 1961; Kerr, 1962). For instance, in *D. virilis*, a dominant gene on the second chromosome causes intersexuality in females. Another dominant gene (*Hr*) on the third chromosome of *D. melanogaster* causes a diploid female to change into a sterile type with many male characteristics (Gowen, 1942). *Hr* is an allelomorph of the recessive gene *tra* (transformer) discovered by Sturtevant (1945). When homozygous, *tra* transforms genetic females into sterile males. Combinations of *Hr* and *tra* have been reported (Gowen and Fung, 1957). The *tra/tra* individuals are phenotypically and behaviorally like males. In the early developmental stages, their testes cannot be distinguished from those of genetic males, although later the spermatocytes degenerate (Seidel, 1963). It appears that normal spermatogenesis occurs only in cells with a XY or XO genotype since, after transplantation of normal testes, the transformer males became fertile (Seidel, 1963).

In the two cases discussed under the heading balance hypothesis, namely *Drosophila* and *Porthetria*, we indeed find evidence for both male and female determining genes. The sex appears to depend on the dose or ratio of male and female factors. It remains to be seen whether this principle is operative in additional species of insects; so far, no conclusive evidence is available and further research will have to yield the desired information.

B. Epistatic Sex Determination

Under the impact of the findings in *Drosophila* and *Porthetria* (a balance between male and female determiners or genes) the prevailing tendency was to apply the same principle of sex determination to other species, to make this perhaps a universal principle. As late as 1962, Kerr writes: "The general problem of sex determination has one general solution: balance among male-tendency genes plus female-tendency genes plus environment." It now appears that this is probably the uncommon pattern. Data gathered in some species can be used alternatively to postulate an epistatic sex determination, which may apply in a greater number of species than the determination of sex by the balance between female and male determiners. Epistasis is defined as the masking of the expression of nonallelic genes. The known cases will be discussed in the following paragraphs.

Bombyx mori is heterogametic in the female sex, as is found in many other Lepidoptera. Translocation of chromosome sections or production of polyploidy in either sex by irradiation gave some insight into the sex determining mechanism (Tazima, 1944; 1964; Tanaka, 1953). It was demonstrated that W-chromosomes (presumed to be homologous to the Y-chromosomes) bear a factor with strong female potency since, even in the extreme genome of 4A + XXXY, only pure females were produced (Table 2). No intersexes were ever found. Whenever the W-chromosome was absent, a male individual resulted. It was noticed (Table 2) that the number of Z-chromosomes (homologous to the X-chromosomes) did not alter the sex; no factor for maleness was found. It should be emphasized that this finding differs from those in *Porthetria*. If one wishes to apply the balance hypothesis to *Bombyx*, one has to assume that the female tendency factor or factors are extremely strong, such that they always outweigh the hypothetical male tendency factors. The results are, however, best explained by the assumption of an epistatic sex-determining mechanism for the female sex.

In the dipteran *Megaselia scalaris*, heterogamety cannot be detected cytologically; yet, using sex linked markers, Mainx (1959, 1962, 1964a, b, 1966) and Burisch (1963) were able

TABLE 2. Chromosomes in *Bombyx mori* and their effect on sex determination

Chromosomes			Sex
W (Y)	Z (X)	Autosomes	
—	ZZ	2A	male
—	ZZZ	3A	male
W	Z	2A	female
W	ZZ	3A	female
W	ZZZ	3A	female
W	ZZZ	4A	female
WW	ZZ	4A	female

to show that sexual heterogamety does exist. A male sex determiner could be located at the end of one of the three non-homologous chromosomes. A spontaneous translocation of the sex determiner occurred between the chromosomes at a predictable frequency of 0.04–0.05 per cent. It thus appears that an epistatic sex determiner determines the male sex. According to Mainx (1962), it is not necessary to classify the chromosome which carries the male determiner as a Y- or X-chromosome since it cannot be made visible; the term should be reserved for those cases where structural differences are observable. Mainx favors the hypothesis that this is the original mechanism from which the XY system eventually evolved and later, through loss of the Y-chromosome, the XO-system.

Another case of an epistatically functioning male determiner on the Y-chromosome is found in the house fly *Musca domestica*, which has a XX–XY system (Hiroyoshi, 1964). The evidence for an epistatic male determiner was obtained as follows. With the use of marker genes, it was shown that a translocation of the male determiner from the Y-chromosome to the second chromosome—which thus becomes the functional Y-chromosome—can occur; no visible cytological evidence for the translocation can be found. In the four investigated translocation strains, no Y-chromosome as such was present; apparently the rest of the Y-chromosome had been lost. The males of two of these strains did have, however, two X-chromosomes in addition to the translocated Y-portion while in the other two strains males with XX- and X-chromosomes were found. Males with XX mated to normal females gave offspring of a 1:1 sex ratio; however, males with X furnished a sex ratio of 2 males:1 female since females of the XO constitution are apparently non-viable (Fig. 6). Since individuals without a male sex determiner always became females and since, in the presence of the male determiner, even animals with the constitution 2A, XX (normally female) were males, an epistatic male determiner must exist. No symptoms of intersexuality were detected. The X-chromosome apparently contains a factor for viability necessary in double dose because XO-females are non-viable.

Among the species of the order Diptera, some possess a cytologically detectable heterogamety (e.g. *Drosophila*); others, like the Chironomidae (Beermann, 1955) or *Culex* (Gilchrist and Haldane, 1947), do not. Even among closely related species of the Calliphoridae, one finds some with distinct heterogamety whereas, for example, in *Calliphora erythrocephala* no XY dimorphism is found. In the latter species there occur two small chromosomes which resemble those of the other related species in which one of them normally carries the male determiner. X-ray induced translocations showed that, in *Calliphora*, the male epistatic sex determiner is located on the third chromosome (Ullerich, 1963). Translocation experiments further demonstrated that the two small chromosomes do not carry the sex determiner. Until 1961 the importance of the Y-chromosome as the carrier of a sex determiner had not been shown for any insect species. Then X-ray-induced translocation of a section of the Y-chromosome onto an autosome in the calliphorid *Phormia regina* indeed proved that the Y bears the male determiner (Ullerich, 1961, 1963). Translocations between autosomes were not transmitted as sex linked, whereas those between the Y and the autosomes were transmitted exclusively by the male. In this species, X-ray treatment yielded certain exceptional individuals which allowed further insight into the mode of sex determination. Two animals were found with a constitution 2A, XXY, presumably arising through the fertilization of a normal egg by a XY spermatozoon; both of these animals were males. One can conclude that the Y-chromosome must bear the epistatically functioning male determiner. Also found were five viable XO-females which when mated with XY-males yielded one-third XX-females, one-third XO-females, and one-third XY-males; YO-males did not occur. XO-females are also known from *Lucilia cuprina dorsalis*. Intersexes could not

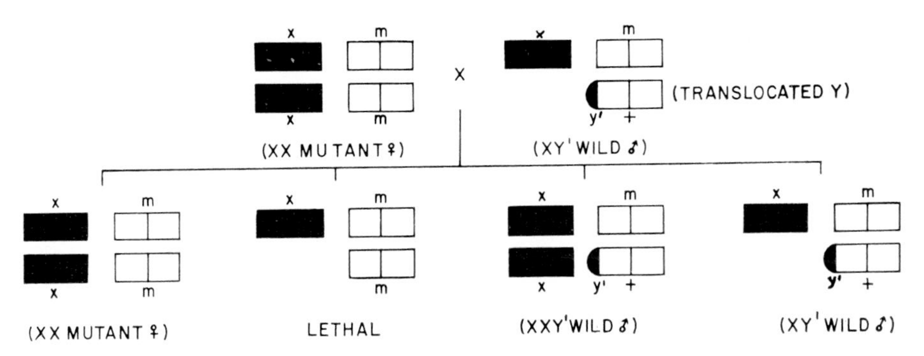

FIG. 6. Sex determination in *Musca domestica*. Only the X-chromosome and the second chromosome bearing the translocated Y-portion are indicated. Schematic representation of the origin of the aberrant sex ratio (1:2) in the F_1 generation in crosses between XX females and XY′ males. m = marker gene. (Adapted from Hiroyoshi, 1964.)

be found among the individuals of any of these constitutions. From this it is concluded that sex determination in *Phormia*, as well as in *Calliphora* and *Lucilia*, does not follow the *Drosophila* modus (Ullerich, 1963).

Results similar to those in the Calliphoridae were obtained in the analysis of the sex-determining mechanism of a Tipulid, *Pales ferruginea* (Ullerich *et al.*, 1964). Male Tipulidae are heterogametic. X-ray induced translocations produced males of the constitution XY, XYY, XXY, XYYY. Individuals with a triple Y-chromosome could arise in a mutation with a supernumerary Y through non-disjunction during spermatogenesis. Only animals with XX were females. Since intersexes occurred in none of these combinations, the Y-chromosome indeed must be the carrier of the epistatic sex determiner. Thus it is shown in a second group of Diptera that the Y-chromosome is functional in sex determination.

As these outlines may suggest, epistatic sex determination in insects appears to be a reality. Results obtained in several Diptera other than *Drosophila* are difficult to interpret in terms of a balance between male and female determiners. Can one assume that the female of these species is the neutral sex, whereas in *Bombyx* it is the male? Or do we still have to postulate both male and female determiners but assume a very weak female determiner in the dipteran species and a very weak male determiner in *Bombyx*—determiners which cannot be detected? The matter then becomes purely theoretical and is of little discussion value. What are the genetic sex-determining mechanisms in the numerous other insect species? We have hardly begun to look deeply into this fascinating field.

C. *Haploid–Diploid Sex Determination*

In 1845 Dzierzon concluded that males of the honeybee arose from unfertilized eggs since unmated queens produced exclusively drones. When the queen had been mated by a drone from another race, the drones were always of the mother's phenotype, whereas the workers carried characteristics of both parents. Later it was shown that the drones possess half the number of chromosomes of the queen and the workers, i.e. they are haploid. Sex determination in *Apis*, as well as in most Hymenoptera, thus appeared to be via a haploid–diploid mechanism. All interpretations are based, indeed, on the correlation between haploid parthenogenesis and sex. From this the question arises, of course, why the haploid nucleus of an unfertilized egg should give rise to a male. Certainly, the ratio between hypothetical male and female determiners is not changed in the haploid nucleus from that of the diploid one. In Hymenoptera a further peculiarity is found in the viability of haploid individuals; species of other orders are generally not viable as haploids. Several hypotheses have been advanced in an attempt to explain the haploid–diploid mechanism, but none is entirely satisfactory, and the discussion still persists.

Males of *Habrobracon* are also haploid (Whiting, 1935) and it was research on sex determination in this species which then yielded a plausible hypothesis. In order to accommodate the occurrence of both haploid and diploid (even though rare) males, Whiting (1943) proposed a hypothesis of multiple sex alleles. This hypothesis is based on the use of genetic markers and thus has a firm basis. According to the Whitings (1943, 1961), at least nine sex alleles—lettered xa, xb, etc.—must exist. Any heteroallelic combination will make the individual a female. Azygotes or homozygotes for any of the alleles are males. With nine sex alleles, there are theoretically nine genetically different haploid males and nine corresponding homozygous males. The probability of the combination of two of the same alleles appears to be rare, which explains the only rare occurrence of diploid males in this species. Diploid males can, when they mate, produce triploid offspring, since their spermatozoa are diploid. There are no cases known which could not be fitted to this hypothesis of multiple sex alleles at one locus.

The question then arose whether this mechanism of sex determination is also functional in other species of Hymenoptera. Cunha and Kerr (1957) believe that it is not applicable to *Apis* since no diploid males have been found in this species. They propose a balance between the genes for maleness and femaleness: those for femaleness are cumulative, while those for maleness are not. Accordingly,

$$\mathrm{M} > \mathrm{F} = ♂ \text{ and } \mathrm{FF} > \mathrm{M} = ♀.$$

This theory is difficult to test. Yet, on the other hand, if the multiple sex allele hypothesis were applicable to the honeybee one should occasionally, at least, obtain diploid drones if they are viable. Mackensen (1951) argues that the multiple sex allele hypothesis may indeed apply to *Apis*. He reasons: diploid males in this species are not viable; upon inbreeding, where the probability of obtaining diploid males becomes higher than normal, the viability of the entire colony was lowered—therefore, diploid males may exist. More direct convincing evidence for the applicability of this hypothesis was, however, obtained through the study of gynandromorphs. Gynandromorphs may have patches of diploid and haploid male and diploid female tissues, as indicated by genetic markers (Rothenbuhler, 1957; Drescher and Rothenbuhler, 1963, 1964). One can find both haploid and diploid male tissues in the eyes of these animals; diploid tissues can be identified by genetic markers as well as by the fact that they produce in some cases larger facets than the adjacent haploid areas. Apparently, diploid male tissues can survive in association with haploid male and diploid female tissues. Production of gynandromorphs can be increased by chilling the eggs soon after they have been laid. Since polyspermy is common in *Apis*, it is believed that chilling facilitates the fusion of the additional sperm nuclei which later give rise to diploid male tissues (androgenesis). In short, the studies on gynandromorphs seem to support the multiple sex allele hypothesis for the honeybee. More recently it was found that diploid drones do exist and can be reared artificially, but normally these larvae are eliminated by the workers (Woyke, 1963, 1965; Woyke and Knytel, 1966; Kerr and Nielsen, 1967).

We are still uncertain how widely among Hymenoptera this hypothesis can be applied. There is scarcely any information from other hymenopteran species which would suggest the occurrence of this sex-determining mechanism; one might, however, assume that the hypothesis applies, since the majority of the Hymenoptera do have haploid males.

Haploid males are also known in species of the Coccoidea. Both females and males arise, however, from fertilized eggs, but one chromosome set is later eliminated in the males. Haploidy in the males is thus secondarily achieved and may indirectly determine maleness. Aspects of the complex chromosome cycles will be briefly discussed below (p. 18). Haploidy in males exists also in species of the Iceryini. Here, males originate from unfertilized eggs and females from fertilized ones. The facts resemble those found in the honeybee, but no sex determining mechanisms are known for certain.

The complex life cycle of species of the paedogenetic Cecidomyidae and their peculiar mechanisms of chromosome elimination were of interest for many years. In the paedogenetic larvae only the germ line cells retain all their chromosomes during the maturation of the male or female eggs destined to become imagines. In addition, males of several species have fewer chromosomes in their somatic cells than the females (Kraczkiewicz, 1950; Geyer-Duszynska, 1959; Nicklas, 1960). Detailed studies on *Heteropeza pygmaea* then gave evidence in favor of the hypothesis that a haploid–diploid sex determining mechanism may be found among the Cecidomyidae (Hauschtek, 1962; Ulrich, 1963). Paedogenetic larvae produce daughter larvae which are paedogenetic themselves as long as culture conditions are good. However, as soon as the culture medium becomes nutritionally poor or old, males and females are produced by these larvae. Males require slightly better conditions than females do. In other words, environmental conditions ultimately influence sex determination. Male determined eggs are somewhat larger than female eggs (Fig. 7). In female eggs, one non-reductional maturation division occurs; later one of the nuclei degenerates. In the male egg, however, two maturation divisions are observed and, during meiosis, the chromosome number is reduced by half. All three polar nuclei participate in embryonic development of the male. Shortly after the maturation divisions, two somatic nuclei from the mother (how they get into the ovary is still uncertain) fuse with the egg nucleus which thus attains altogether from fifty-eight to fifty-nine chromosomes. The cleavage divisions occur thereafter in both the male and the female egg. In the further course of development, a large number of chromosomes are eliminated in both types of eggs; but, in each case, one nucleus retains the full chromosome set, seventy-seven in the female and fifty-eight in the male. These latter nuclei lie within the pole plasm of the egg and are the germ line nuclei; apparently, the pole plasm prevents the elimination of chromosomes. The somatic nuclei of the female retain ten chromosomes while those of the male retain five. Environmental factors apparently determine whether large male eggs or small female eggs are produced. The size of the egg must, in some way, regulate the elimination of chromosomes to yield either haploid males or diploid females. In other words, external factors ultimately influence the chromosomal sex determination.

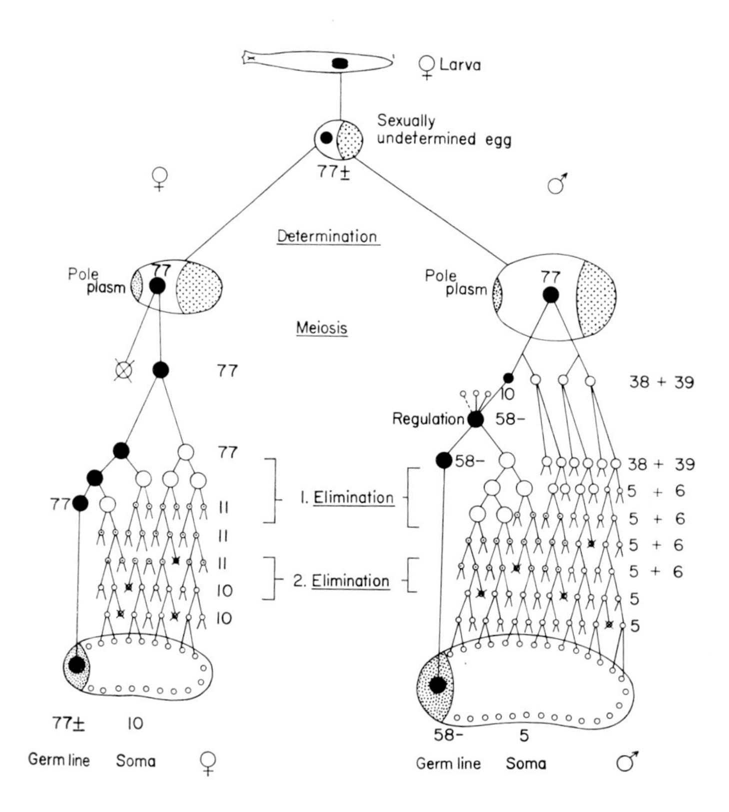

FIG. 7. Cytology of sex determination in the paedogenetic *Heteropeza pygmaea* when producing bisexual individuals. Through a sequence of differential eliminations of chromosomes in the soma and germ line, the male becomes haploid while the female remains diploid. (Adapted from Ulrich, 1963; Camentind, 1966.)

D. Other Mechanisms of Sex Determination

The Coccoidea and Sciaridae are of particular interest to the cytologist because of peculiar chromosome eliminations during development. In the present context, only aspects related to sex determination in these species will be discussed.

Some species of the dipteran *Sciara* are monogenic, i.e. a female gives rise to either male or female offspring but never a mixture of both. The cytogenetic events which lead to these unisexual broods were analyzed and summarized by Metz (1938); more recently, some of the details of this kind of sex determination have become available through the works of Crouse (1960) in *Sciara cocrophila*. In this monogenic species, both sexes develop from fertilized eggs which have three X-chromosomes and three pairs of autosomes. Two of the X-chromosomes in the zygote are derived from the sperm. Whereas oogenesis is normal and

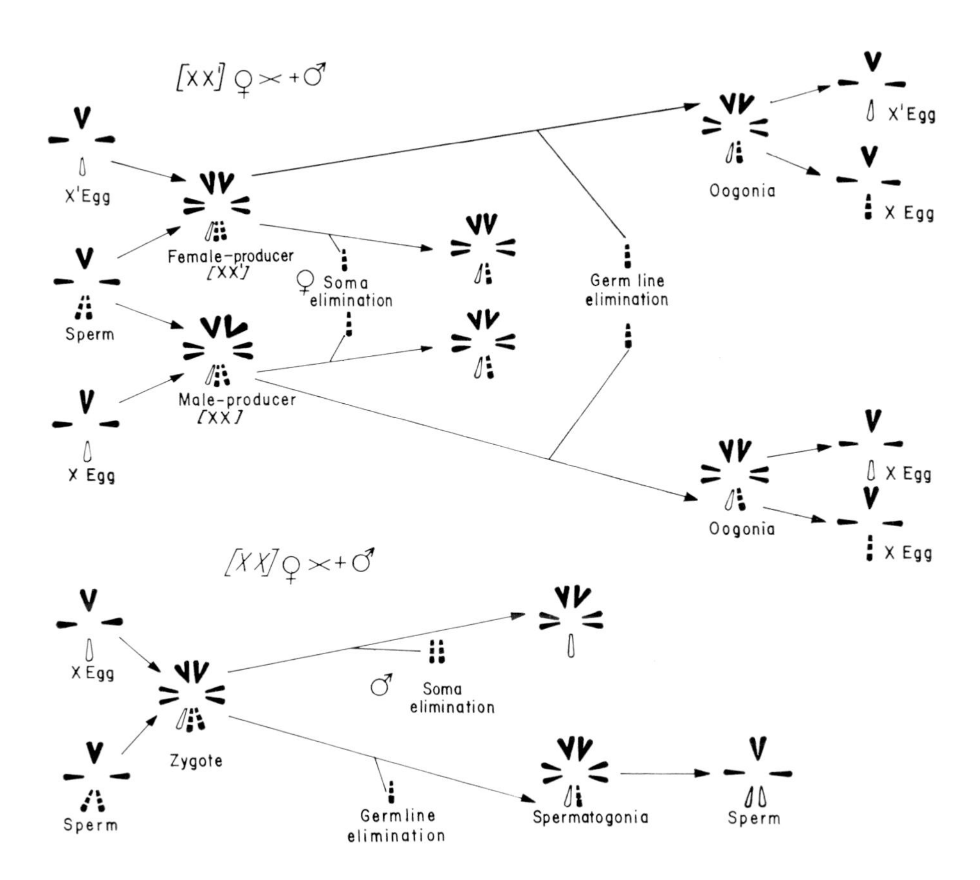

FIG. 8. Chromosome history in the monogynic *Sciara coprophila*. The X′ chromosome carries a dominant marker. The pattern of chromosome elimination in the soma and germ line during oogenesis and spermatogenesis is diagrammatically indicated. (Adapted from Crouse, 1960.)

each egg receives one X-chromosome, spermatogenesis is unusual in that one paternal X-chromosome is eliminated during the 1st meiotic division and in the 2nd meiotic division, both maternal X-chromosomes move to one pole; a bud containing only autosomes later degenerates. The single sperm thus receives a haploid set of autosomes and two X-chromosomes through non-disjunction (Fig. 8). In some of the early female embryo, one paternal X-chromosome is eliminated while, in the male embryo, both paternal X-chromosomes are eliminated. In other words, the XX/XO-system is secondarily established.

Two kinds of females, thelygenic (X′X) and arrhenogenic (XX), are known, i.e. the X′X constitution in soma of the female presumably determines that the offspring eliminate only one X, whereas the XX constitution causes the elimination of two X. X′ designates a difference from X which is cytologically not seen but whose difference can be deduced from its genetic role. X′ in Fig. 8 bears the dominant wing factor Wavy and can thus be identified in the daughters. X-ray-induced translocations of a heterochromatic portion of the X-chromosome onto an autosome (Crouse, 1960) indicated that this is the portion which determines the different mode of elimination in the two types of females. The recombination chromosome behaves like the X-chromosome: it goes through non-disjunction in spermatogenesis when derived from the mother and is eliminated from the germ line when derived from the father. Furthermore, through non-disjunction of the translocation chromosome, some eggs received 2X and others none. Consequently, the zygote may have 4X or only 2X, since the spermatozoon brings 2X. Eggs of a male producer eliminate 2X, leaving the XX genotype which is female—exceptional daughters result. The eggs without an X-chromosome eliminate one X when derived from a female producer—exceptional sons result which carry X-linked genes from the father instead of those normally obtained from the mother. These studies show that the influence of the X′-chromosome in *Sciara* is not on the sex of the offspring but rather determines the elimination of one paternal X instead of two. Provided that the somatic cells have two sex chromosomes, the gonads will develop into ovaries; if they have one X, the gonads will be testes.

Other species besides those of the Sciaridae in which monogenic reproduction is found are the dipteran *Chrysomyia albiceps* and *C. rufifacies* (Roy and Siddons, 1939; Ullerich, 1958). Thelygenic and arrhenogenic females occur in a 1:1 ratio, which suggests that this type is inherited on the basis of a homo-heterogametic principle. In analogy to the findings

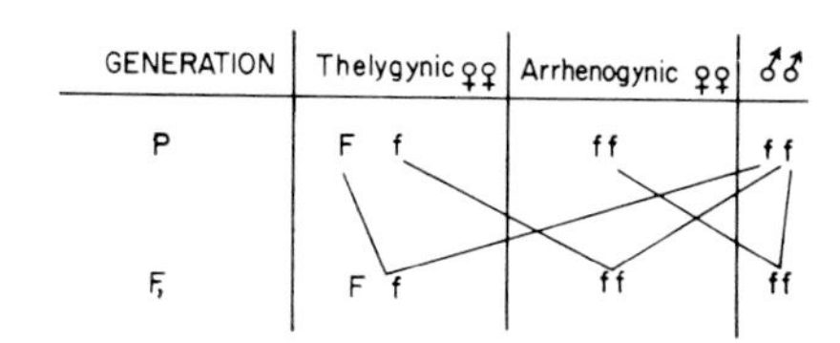

FIG. 9. Hypothetical mechanism of sex determination in the monogynic dipteran *Chrysomyia albiceps*. Two types of females exist. The thelygynic females are heterozygous for a dominant sex determiner, whereas the arrhenogynic females are homozygous for the recessive allele. (Adapted from Ullerich, 1963.)

in Sciaridae, Ullerich postulates that the thelygenic females are heterozygous for a dominant female determiner which predetermines the eggs; the arrhenogenic females are homozygous for a recessive allele just as the males (Ullerich, 1958, 1963). If this is so, the two types of females must occur in a 1:1 ratio (Fig. 9). In this system, the male has no influence on sex, but mating is essential for reproduction. No cytological differences are found in the chromosomes of the two types, and the two small sex chromosomes characteristic for many Calliphoridae are isomorphic. Ullerich was unable to localize the heterozygous mechanism even with the use of X-ray-induced translocations.

Ever since the first publication by Schrader (1921) on the chromosome system in a species of Coccoidea, cytologists have been intrigued with this object (see White, 1954). In many of these species, one complete set of chromosomes (although sometimes only one or two chromosomes) becomes heteropycnotic and is eliminated in the male. The males thus secondarily become haploid. According to the type or time of chromosome elimination during development, three categories are distinguished (Brown, 1959; Brown and McKenzi, 1962; Brown and Nur, 1964): lecanoid, Comstockiella, and diaspidid. It is reasoned that the three categories evolved in the order indicated (Brown and McKenzi, 1962). Hughes-Schrader (1948) deduced that in the process in which the males become secondarily haploid—all animals arise from diploid zygotes—it is always the paternal set of chromosomes which is eliminated. Hetero-chromatization of the chromosomes may be associated with their genetic inertness. Experimental evidence for this hypothesis was later provided (Brown and Nelson-Rees, 1961; Nelson-Rees, 1962) from the mealy-bug *Planococcus citri*. In this case, irradiation of the father caused aberrations in the heterochromatic set of chromosomes of the sons but had practically no effect on their viability, whereas the daughters invariably died. Irradiation of the mothers affected both sons and daughters. Thus, the euchromatic chromosome set derived from the father becomes heterochromatic and is genetically inert in the male.

Does the occurrence of haploidy in male coccids give evidence for a haploid–diploid sex-determining mechanism? All indications are negative. We are possibly dealing here

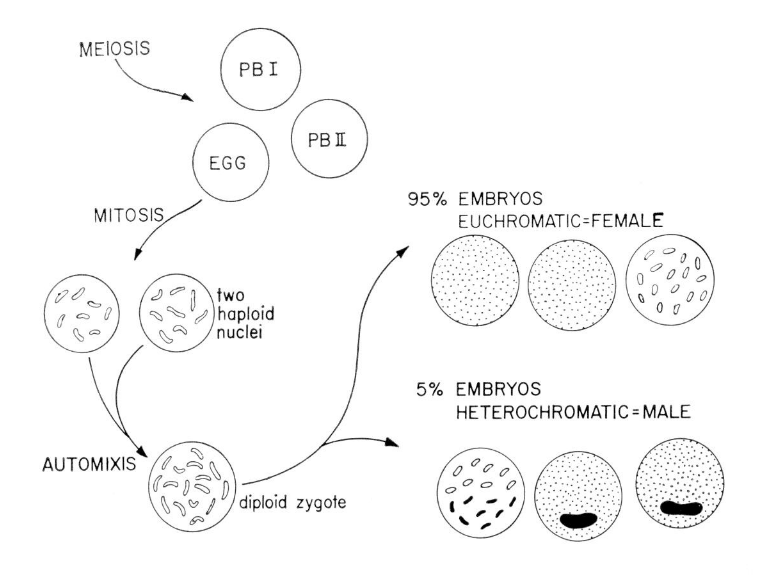

FIG. 10. Origin of female and male embryos in the parthenogenetic soft scale insect *Pulvinaria hydrangeae*. Heterochromatization is observed in about 5 per cent of the embryos; these appear to be non-viable, since no adult males are known. (According to Brown and Nur, 1964.)

with a nongenetic sex determination, with the exception of a primitive coccid, *Puto* spec., in which the male is diploid and the XX–XO system appears to be functional (Hughes-Schrader, 1944). Females of all other coccids apparently lay two types of eggs, one in which heterochromatization of some chromosomes occurs and the other in which it does not. Sex may thus be predetermined. No cytological difference can be observed in the two types of eggs, and nothing is known as to what makes the eggs physiologically different. Perhaps environmental factors induce changes in the ovarioles of the mother, which may in turn alter the physiology of the oocytes produced. It is unlikely that males have any influence on this since, even in the parthenogenetic *Pulvinaria hydrangeae*, heterochromatic chromosomes can be observed in a small portion of the eggs (Brown and Nur, 1964) (Fig. 10). Aging of the animal may be one of the factors as a correlation does exist between a higher percentage of males produced and older female. In short, we have no conclusive clue as to what determines the production of digametic egg in Coccoidea.

E. Epigenetic Factors Influencing Sex Determination

Environmental or epigenetic factors do affect the occurrence of either male or female individuals in certain species. The term epigenetic is used here as meaning mechanisms or factors outside the primary genetic sex determining mechanism influencing sex determination. A prime example may be the many species of Aphidoidea with heterogony (p. 201) in which sexual males and females arise from parthenogenetic females during certain seasons. Aphids are apparently heterogametic in the male sex (Schwartz, 1932; White, 1954). However, the actual mechanism by which day length, temperature, or foods influence chromosome behavior—and thus are involved in sex determination—are, in the end, unknown.

Females of the parasitic Ichneumonidae, particularly of the genus *Pimpla* as well as some others, apparently lay fertilized eggs into large hosts and unfertilized eggs into small ones (see Doutt, 1959). It was Chewyreuv (1913) who discovered this phenomenon. Later, this same behavior pattern was found in several additional species (Clausen, 1939; Flanders, 1939; Arthur and Wylie, 1959). Experimental proof for the female's control of release of spermatozoa for fertilization of the eggs was supplied by Brunson (1937), using *Tiphia popilliavora* (Fam. Tiphiidae). In this species, offspring grown in the third instar larvae of the Japanese beetle are female, while those grown in second instar larvae are male. Reciprocal transplantation of recently deposited eggs does not change the sex, i.e. females now arise from small second instar larvae and males from large third instar larvae. In *Nasonia vitripennis*, primarily males emerge from superparasitized hosts (Wylie, 1966), which may suggest that here also the female senses an already parasitized host and lays only unfertilized eggs. Flanders (1939) proposed that the females of some Hymenoptera, upon sensing the size of the host, control the sperm release through activation of the spermathecal glands; the secretion of these glands is essential for sperm activation. The females of the Pimplinae probably sense the size of the host with their antennae, since after the removal of the antennae they are no longer able to discriminate (Aubert, 1959, 1961): males and females then arise randomly from any sized host pupae. Basically the same mechanism holds for several additional species, even though it occasionally may break down for unknown reasons; perhaps unknown internal factors in the female interfere (Aubert and Shaumar, 1962, 1964; Shaumar, 1966).

Thelytokous females of the parasite *Ooencyrtus submetallicus* are known to produce males if the environmental temperature is raised above 29.5°C. Closely related species are arrhenotokous, and it would seem as if the normal chromosome behavior in *Ooencyrtus* breaks down at higher temperatures, in the sense that the behavior is now like that of close relatives.

As is apparent from this brief discussion, the details of how environmental or epigenetic factors influence sex determination are, in many cases, obscure. The pathways between the perception of environmental cues and the control of sex through various means are not understood. In the case of the aphids, we are able to speculate on the significance of production of sexuals in late summer, since these individuals produce forms that can survive adverse conditions. In other instances, no ecological importance can be attached to the influence of external factors on sex determination. This is particularly apparent in the case of *Ooencyrtus*, where males produced through high temperature are completely non-functional.

F. Intersexuality

The entomological literature is replete with descriptions of specimens having both male and female features. The terms hermaphrodite, gynandromorph, and intersex are sometimes applied indiscriminately to these animals. True, i.e. functional, hermaphrodites occur only in one genus (see p. 224). Gynandromorphs, i.e. mosaics in space, can accidentally arise through the simultaneous development of the zygote and additional fused sperm nuclei in the same egg. Genetic male and female tissues can thus lie side by side as shown by genetic markers as, for example, in the honey bee (Rothenbuhler *et al.*, 1952). These types as well as cases of genetically determined gynandromorphism, e.g. *Bombyx mori* (Goldschmidt and Katsuki, 1928; Katsuki, 1935), will not be dealt with further in this chapter, since their genesis contributes little to the understanding of sex determination.

This is, however, not the case with intersexes. The study of intersexes, i.e. animals in which all cells have the same genetic background but in which both male and female tissues are differentiated, has indeed advanced our understanding of sex determination in insects. Superficially, an intersex may look like a gynandromorph, and only genetic and histological analysis will determine whether one is dealing with a gynandromorph or an intersex. Environmental factors, such as temperature or parasitic infection, can influence the expression of intersexuality, as is shown in some cases (p. 20). This is even known to occur in the thelytokous stick insect *Carausius morosus*, where high temperature favors intersexuality (Bergerard, 1961).

The discovery of triploid intersexes (3A,XX) in *Drosophila melanogaster* led Bridges (1921, 1922) to the formulation of the balance theory of sex determination (p. 8). Crosses between races of the gypsy moth *Porthetria* (=*Lymantria*) *dispar* yielded intersexes, and here also the proposal was made that an imbalance between male and female determiners caused the production of intersexes. The question then arose of how some animals become more female-like while others with an identical genetic background become more male-like. A graded series from one extreme to the other could be obtained. Based on this observation, Goldschmidt (1920) formulated the hypothesis of a turning point in development. According to him, the embryo develops first as a female and then is switched in the male direction. Since the turning point may be sooner or later in development, a graded series between more female-like and more male-like individuals could be obtained. This theory, developed for *Porthetria*, was reiterated by Goldschmidt himself many times (1931, 1949, 1955).

Are there any objective means which would allow us to detect the turning point and thus verify Goldschmidt's "time law"? If the theory is indeed applicable to *Porthetria*, one would have to expect individuals which exhibit truly intermediate characteristics, particularly apparent in homologous organs of the female and male sex. The question also arises whether it applies to other species, e.g. *Solenobia triquetrella*. The situation in *Solenobia* is instructive. Intersexes of various degrees of femaleness and maleness occur in triploids of this species, which result from the fertilization of eggs from the tetraploid parthenogenetic race by males of the bisexual diploid race (Seiler, 1937). In *Solenobia* the diploid male is thought to be homogametic and has the sex determiners in a 1:1 ratio (FFMM; M $>$ F); the female is heterogametic and the sex determiners are present in the following proportion: FFM; FF $>$ M. The genesis of triploid intersexes may thus be depicted as follows (Seiler, 1937):

$$\text{FFM} + \text{FM} = \text{FFFMM}$$

Male sex determiners are thought to be located in the autosomes because both XO- and XY-females yield the same sex ratio; consequently, the Y must be inert with respect to sex determination. According to Seiler, triploidy as such results in intersexes. Goldschmidt's attempt to fit the "time law" to *Solenobia* intersexes met with opposition by Seiler and his students. Seiler rejected Goldschmidt's theory on grounds which are discussed in great detail elsewhere (Seiler, 1949, 1958) and are only briefly repeated here. The extensive studies of the homologous organs, the genitalia, which show sexual dimorphism, revealed that all cells are either completely male or completely female. The genitalia of intersexes are a mosaic of male and female cells and no intermediate areas are found. This implies that areas of either sex develop simultaneously and not (as Goldschmidt postulates) in succession. Epigenetic factors influence the expression of the sexual characteristics, but, once determined, the cells differentiate as programmed. In triploid intersexes, both sex determiners are in true equal balance and only environmental modifiers will "direct" the differentiation into either of the sexes. With this conclusion, Seiler terminated the sometimes heated discussion on intersexuality in his favor—a discussion which lasted more than 25 years. In the case of *Porthetria*, Seiler does not exclude the possibility that the time law might apply. Here, however, additional experiments are needed to provide the conclusive evidence. Nevertheless, the major contribution of Goldschmidt in this field, i.e. the hypothesis that a balance between male and female determiners determines sex in *Porthetria* (p. 10), still holds.

In further support of Seiler's interpretation of intersexuality in *Solenobia*, one additional contribution must be mentioned. In *Drosophila melanogaster* the male bears sex combs, a row of 10–13 bristles, on the 1st tarsal segment of the forelegs. Each bristle arises from one cell. As earlier studies have shown, the female also has the anlage to form bristle, but here no bristles differentiate. In triploid intersexes, the sex combs are reduced since fewer bristles differentiate, but each bristle is fully formed (Hannah-Alava and Stern, 1957). This clearly shows that a cell differentiates into either a male or a female type but never into an intermediate. The whole organ, however, appears intermediate.

In addition to the works on intersexuality mentioned so far, a number of publications ought to be reviewed in this context. These are, however, works which deal mainly with genetic modification of either sex (p. 11) and thus have no direct bearing on sex determination. The reader is therefore referred to the pertinent reviews (Gowen, 1961; Wiese, 1960, 1966).

G. Sex Ratio

Individuals of both sexes are usually produced in equal numbers, and although deviations may be observed, as a whole, the ratio between male and female individuals in a population is 1:1. Both major sex-determining mechanisms, balance between male and female sex determiners and epistatic sex determination, accomplish this ratio. In this section some, but not all, cases will be mentioned where a 1:1 sex ratio does not exist and the conditions under which this is found will be briefly analysed (see also Hamilton, 1967).

Gershenson (1928) described two strains of *Drosophila obscura* that produced a high percentage of female offspring (about 96 per cent). The factor which causes this abnormal sex ratio is transmitted by the X-chromosome and, according to Gershenson, determines that sperm containing a Y-chromosome cannot compete, consequently, nearly all-female broods are obtained. Later, it was shown that it is the males' carrying the gene sex ratio (*sr*) on their X-chromosome which causes this described effect (Sturtevant and Dobzhansky, 1936). Cytological evidence in these strains shows that normal disjunction of the X occurs during the first meiotic division, but that the Y-chromosome becomes pycnotic during the second anaphase (Novitski *et al.*, 1965). The Y-carrying sperm apparently become nonfunctional. The gene *sr* occurs in most populations. If exclusively female offspring are produced, no substantial number of progeny can arise since parthenogenesis is rare in *Drosophila*. Thus, a population in which all individuals carried the *sr* gene would decline within a short time.

A sex ratio in favor of the female can also occur in species or strains in which differential mortality of the male embryos occurs. This has been reported in several *Drosophila* species where it was also shown that the factor is transmitted by the female only. It was furthermore demonstrated that the factor can be extracted from the eggs and injected into animals of the same or other species which do not normally carry it; a skewed sex ratio is consequently induced. Further analysis revealed that this sex ratio factor is a microorganism which causes death of the male embryo (Ikeda, 1965; Leventhal, 1965; Sakaguchi *et al.*, 1965). Similarly, in a population of the bark beetle *Orthotomicus latidens*, all-female broods were found (Lanier, 1966) because male embryos did not hatch.

In the gynogenetic species like the psychid *Luffia lapidella* or the beetles of the genus *Ptinus* and *Ips* sex ratio conditions exist (p. 30). Here, exclusively female broods are produced. Gynogenesis is a form of parthenogenesis which needs the sperm only for the initiation of development.

Species which are arrhenotokous, as are many hymenopterans, often have a rather variable sex ratio. A short list of a few species of Hymenoptera has been assembled (Table 3) in order to demonstrate the phenomenon. The mated females may or may not release the stored spermatozoa during oviposition, depending on environmental conditions. Virgin females produce only males. The mode of control of sperm release which ultimately affects the sex ratio may even be genetically determined, as is suggested from studies in *Dahlbominus folginosa* and *D. fuscipennis* (Wilkes, 1964). Here, female lines with either a high or a low ratio (♀: ♂) could be selected.

TABLE 3. Sex ratio conditions in species of the Hymenoptera

Species	Sex ratio ♀ : ♂	Author
Anaphes flavipes	3:1	Anderson and Paschke, 1968
Dusmetia sangwani	7.3:1	Schuster, 1965
Hemiteles graculus	1:1	Puttler, 1963
Melittobia chalybii	20:1	Schmieder, 1933
Mormoniella vitripennis	5.5:1	Jacobi, 1939
Mormoniella vitripennis (= *brevicornis*)	7:1	Parker and Thompson, 1928
Pristiphora erichsonii	50:1	Heron, 1966
Trichogramma evanescens	6:4	Lund, 1938

An interesting observation was made in *Nasonia vitripennis*, which leads to certain speculations about the significance of varying sex ratios in this parasite. A high parasite–host ratio is accompanied by the production of more males than females. This may be the result of super-parasitism, which entails the differential mortality of female embryos as well as the deposition by the female of unfertilized eggs into already parasitized pupae of the house fly (Wylie, 1966). The ecological implication is that in the reverse situation, i.e. a low parasite–host ratio, relatively more females are produced; the propagation of the species is thus assured even though few hosts are available. The ability to control sperm release among hymenopteran species allows an economic production of males. Males can mate many females and, from this viewpoint, a 1:1 sex ratio is not essential for species propagation. The available hosts are thus "preserved" for the production of females which will further propagate the species. The population size thus fluctuates and can be adjusted to the conditions encountered (Wylie, 1966).

CHAPTER 3

PARTHENOGENESIS

PARTHENOGENESIS, the development of unfertilized eggs, is known to occur in many species of insects belonging to nearly all orders, although it has apparently not been described in Odonata and Hemiptera. The phenomenon is of interest because of the seemingly unusual mode of reproduction. Consequently, the subject has been reviewed by a number of investigators in the recent past. The older literature on the occurrence is treated by Vandel (1931), Rostand (1950), and Suomalainen (1950). The cytological aspects are excellently dealt with by White (1954, 1964) while some of the more recent findings are reviewed by Wiese (1960, 1966). Evolutionary aspects were considered by many reviewers, but the discussions by Soumalainen (1950, 1962) and White (1954, 1964) are particularly noteworthy. In the present context, it is not the intention to cover the entire literature, but rather to emphasize the important features intimately related to the propagation of the species.

A. The Phenomenon

Parthenogenesis has interested biologists as a form of reproduction as well as a mechanism of sex determination. It is particularly well known among Hymenoptera, Phasmida, and Homoptera, as the list of species (even though incomplete) illustrates (Table 4) (Soumalainen, 1950). The sex ratio, i.e. the frequency of males in a population of a haploid–diploid species, is determined by the number of eggs being fertilized. The unfertilized eggs of a species may give rise only to males, as in the majority of the Hymenoptera, or as in most other species only to females. Recently, a race of the hymenopteran *Trichogramma semifumatum* has been found which produces exclusively females, whereas unfertilized eggs of other populations give rise only to males (Stern and Bowen, 1968). A few cases are known, however, in which the progeny of a virgin female can be either male or female, as in the hymenopterans *Oecophylla longinoda* (Ledoux, 1950) and *Perga affinis* (Carne, 1962). One distinguishes, accordingly, between arrhenotokous, thelytokous, and deuterotokous parthenogenesis.

Arrhenotoky would seem to have certain disadvantages with respect to the propagation of the species since no females are produced; in certain cases, however, it may even save the population. In long-lived species, the sons can mate the females of the earlier generation which then produce daughters. An example of this is the chalcid wasp *Melittobia acasta* (Balfour-Browne, 1922). Thelytoky can be obligatory; even if males occasionally occur, they may have no function in reproduction, they may be an atavism. This is found, for example, in *Culicoides bermudensis* (Williams, 1961), *Pycnoscelus surinamensis* (Matthey, 1945), *Pristiphora erichsonii* (Heron, 1955), *Carausius morosus* (Bergerard, 1961), and in the coccid *Pulvinaria mesenbryanthemi* (Pesson, 1941), as well as in some others. Thelytokous

TABLE 4. Occurrence of parthenogenesis among insect species of various orders (this list is incomplete and does not always give credit to the earliest report)

Species	Type	Remarks	Author
Collembola			
Folsomia candida distincta	thelytoky		Green, 1964 Husson and Palevody, 1967
Isoptera			
Zootermopsis angusticollis	thelytoky		Light, 1944a
,, *nevadensis*	,,		Light, 1944a
Dictyoptera			
Brunneria borealis	thelytoky	obligatory;	White, 1948
Pycnoscelus surinamensis	,,	strains from Africa and S. America	Matthey, 1945
Several species	,,	facultative	Roth and Willis, 1956
Orthoptera			
Saga pedo	thelytoky	obligatory	Matthey, 1941
Several species of locusts and grasshoppers	,,	facultative (low viability)	Bergerard and Seuge, 1959 Uvarov, 1928 Chopard, 1948
Phasmida			
Carausius morosus	thelytoky	obligatory	Bergerard, 1961
Clitumnus extradentatus	,,		Bergerard, 1958
Sipyloides sipylus	,,		Possempès, 1958
Bacillus rossii	,,		Cappe de Baillon *et al.*, 1937
Baculum artemis	,,		,,
Carausius furcillatus	,,		,,
,, *theiseni*	,,		,,
Clonopsis gallica	,,		,,
Epibacillus lobipes	,,		,,
Leptynia hispanica	,,		,,
Ephemeroptera			
Ameletus ludens	thelytoky		Needham *et al.*, 1935
Homoptera			
Aleurochiton complanatus	arrhenotoky		Müller, 1962
Pulvinaria mesembryanthemi	deuterotoky		Pesson, 1941
Aphidina and *Coccoidea*	thelytoky, arrhenotoky		Hughes-Schrader, 1948
Adelgidae	thelytoky, arrhenotoky		Steffan, 1968 a, b
Psocoptera			
Caelicius flavidus	thelytoky		Schneider, 1955
Elipsocus hyalinus	,,		,,
Ectopsocus meridionalis	,,		,,
Liposcelis Badonnel	,,		Goss, 1954
Peripsocus subfasciatus	,,		Schneider, 1955
Pseudopsocus rostocki	,,		,,
Psocus moria	,,		,,
Psyllipsocus ramburi	,,		,,
Reuterella neglecta	,,		,,
Lepidoptera			
Bombyx mori	deuterotoky	very rare	Astaurov, 1967

Species	Type	Remarks	Author
Lepidoptera (cont.)			
Luffia ferchaultella	thelytoky		Narbel-Hofstetter, 1961
Solenobia lichenella	,,		Sauter, 1956
,, *seileri*	,,		,,
,, *triquetrella*	,,		,,
Diptera			
Cnephia mutata	thelytoky		Davies and Peterson, 1956
Culex fatigans	unknown	embryos die	Kitzmiller, 1959
Culicoides bambusicola	thelytoky		Lee, 1968
,, *bermudensis*	,,		Williams, 1961
Drosophila mangabeirai	,,		Murdy and Carson, 1959
,, *mercatorum*	,,		Henslee, 1966
Limnophyes biverticillatus	,,		Scholl, 1956
,, *virgo*	,,		,,
Lonchoptera dubia	,,		Stalker, 1956
Coleoptera			
Catapionus gracilicornis	thelytoky		Takenouchi, 1957
Adoxus vitis	,,		Jolicoeur and Popsent, 1892
Calligrapha spec.	,,		Robertson, 1966
Curculionidae	,,		Suomalainen, 1947, 1954
Listroderes costirostris	,,		Takenouchi, 1957
Otiorrhynchus cribricollis	,,		Andrewartha, 1933
Pseudocueorhinus bifasciatus	,,		Takenouchi, 1957
Xyleborus compactus	arrhenotoky		Entwistle, 1964
Micromalthus debilis	thelytoky and arrhenotoky		Scott, 1936, 1938
Hymenoptera			
Anaphes flavipes	arrhenotoky		Anderson and Paschke, 1968
Aphaereta pallipes	,,		Salkeld, 1959
Aphidius testaceipes	,,		Sekhar, 1957
Apis mellifera	,,		Dzierzon, 1845; Nachtsheim, 1913
Campoplex haywardi	,,		Leong and Oatman, 1968
Caraphractus cinctus	,,		Jackson, 1958
Cephalonomia tarsalis	,,		Powell, 1938
Coccophagoides utilis	,,		Broodryk and Doutt, 1966
Dahlbominus fuscipennis	,,		Wilkes, 1965
Diadromus pulchellus	,,		Labeyrie, 1959
Dolichovespula silvestris	,,		Montagner, 1966
Dusmetia sangwani	,,		Schuster, 1965
Habrobracon brevicornis	,,		Whiting, 1918
,, *juglandis*	,,		Whiting, 1945
Hemiteles graculus	,,		Puttler, 1963
Loxostege sticticalis	,,		Simmonds, 1947
Macrocentrus ancylivorus	,,		Flanders, 1945
Melittobia chalybii	,,		Schmieder, 1938
,, *acasta*	,,		Balfour-Browne, 1922
Microctonus aethiops	,,		Loan and Holdaway, 1961
Mormoniella vitripennis	,,		Cousin, 1933
Opius consolor	,,		Stavraki-Paulopoulou, 1966
Pachycrepoideus dubius	,,		Crandell, 1939
Paravespula germanica	,,		Montagner, 1966
,, *vulgaris*	,,		,,
Perilitus rutilus	,,		Loan and Holdaway, 1961
Pimpla instigator	,,		Chewyreuv, 1913
Plagiolepis pygmaea	,,		Passera, 1960
Praon aguti	,,		Sekhar, 1957

TABLE 4 (cont.)

Species	Type	Remarks	Author
Hymenoptera (cont.)			
Prosevania punctata	arrhenotoky		Edmunds, 1954
Spalangia nigra	,,		Parker and Thompson, 1928
,, *drosophilae*	,,		Simmonds, 1953
Stenobracon deesae	,,		Alam, 1952
Trichogramma semifumatum	,, , thelytoky		Bowen and Stern, 1966 Stern and Bowen, 1968
,, *evanescens*	,,		Lund, 1938
,, *semblidis*	,,		Salt, 1938
Aphaenogaster fulvapicea	thelytoky		Haskins and Enzmann, 1945
,, *lamellidens*	,,		,, ,,
Apis mellifera	,,		Mackensen, 1943
,, *capensis* var. *intermissa*	,,		Jack, 1916
Ceratina dallatorreana	,,		Daly, 1966
Formica polyctena	,,		Otto, 1960; Ehrhardt, 1962
Leucospis gigas	,,		Berland, 1934
Nemeritis canescens	,,		Speicher, 1937
Ooencyrtus submetallicus	,,		Wilson and Woolcock, 1960
Pristophora erichsonii	,,		Heron, 1955
Pygostelus falcatus	,,		Loan, 1961
Thrinax macula	,,		Peacock and Sanderson, 1939
Neuroterus lenticularis	arrhenotoky or thelytoky		Doncaster, 1910
,, *baccarum*	deuterotoky		Dodds, 1939
Oecophylla longinoda	,,		Ledoux, 1950
Perga affinis	,,		Crane, 1962

parthenogenesis apparently can occur in many species in which it is often not the regular mode of reproduction. Embryos and larvae of these facultative parthenogenetic species frequently have a very low viability and generally only a few animals reach a reproductive age. This is, for instance, known to be true for parthenogenetic *Culex fatigans* (Kitzmiller, 1959) and several species of grasshoppers (Uvarov, 1928) and cockroaches (Roth and Willis, 1956). In some species, parthenogenesis can be so rare that it may not be noticed under normal circumstances. In *Bombyx mori* only one adult moth was obtained by spontaneous parthenogenesis from about 100,000 unfertilized eggs (Sato, 1931; Astaurov, 1967). In this species, the incidence of parthenogenesis could be raised by HCl treatment of the eggs, by mechanical brushing, or even by heat shock. The parthenogenetic occurrence of the diploid honeybee workers or queens was debated for some time, but apparently they do indeed exist, even though seldom noticed (Mackensen, 1943). Naturally, these rare cases of thelytokous parthenogenesis do not contribute appreciably to the propagation of the species and are of only academic interest.

The occurrence of thelytokous parthenogenesis in certain species was inferred because no males had been found, but it was later established that the males were either rare, very inconspicuous, or only shortlived and had, therefore, been overlooked. In the collembolan species *Orchesella villosa* and *Sminthurides aquatica* no males were known, but Mayer (1957) eventually showed that they do exist and that fertilization is essential for embryonic and larval development.

In many species of the Aphidoidea, one to several thelytokous generations alternate with a sexual generation, i.e. these species have annual or biannual cycles known as heterogony (p.201). Fundatrices, virgo, emigrantes, exules and sexuparae reproduce parthenogenetically; the former three forms produce exclusively females, the latter one either male or female sexuals. It appears that environmental factors, such as photoperiod and temperature, influence the succession of generations, i.e. control to some extent the number of thelytokous generations. This is dramatically illustrated by the observation that *Aphis fabae* could be reared for 258 generations exclusively by thelytoky under long photoperiods (16–18 hr light) and elevated temperatures (Müller, 1954). In nature, some species may have only one thelytokous generation, while others are solely parthenogenetic, as for example, some species of the Pemphigidae. It may be a formidable task to elucidate how, in the various species, environmental stimuli finally control the cytogenetic events which lead to the production of parthenogenetic females or sexual males and females.

In some cynipids, a parthenogenetic thelytokous generation alternates with a bisexual one. For example, in *Neuroterus lenticularis*, the thelytokous females arise from fertilized eggs and lay either female or male eggs (Doncaster, 1910, 1911, 1916). What controls the rigid alternation of parthenogenetic and bisexual generations remains open to conjecture. Other sawfly species may have no sexual generations and reproduce by parthenogenesis exclusively.

Among the Cecidomyidae, some paedogenetic species can be maintained nearly indefinitely by thelytoky: 500 paedogenetic generations of *Heteropeza pygmaea* were reared in succession (Ulrich, 1963). Under proper rearing conditions, sexual forms were obtained at will; in other words, heterogony depends on environmental factors. In this species, the sexually undetermined egg of the parthenogenetic larva is pluripotent and can develop into another paedogenetic larva or into either male or a female imago. It is found that if the mother larva obtains limited food quantities and certain food qualities, her undetermined eggs will be determined to develop into sexual individuals (p. 15).

Another interesting case of paedogenetic parthenogenesis is found in the beetle *Micromalthus debilis*, a species which occurs in five different forms (Scott, 1936, 1938, 1941). Sexuals arise via arrhenotoky and thelytoky in the late summer; however, it is uncertain whether these forms are functional since they have not been observed ovipositing or mating (Scott, 1938). Males are haploid, whereas all four forms of females are diploid. Presumably, environmental factors control the type of parthenogenetic development, and a given larva can produce either a male imago or a brood of female larvae. Unfortunately, no further details about the life cycle of this species are available; it is unclear whether heterogony occurs.

Environmental factors may also influence the type of parthenogenetic offspring of workers of the ant *Oecophylla longinoda* (Ledoux, 1950). This can be inferred from the observation that males originate from large eggs, whereas winged females or workers come from small eggs in this deuterotokous species.

B. Cytology of Parthenogenesis

The various cytological mechanisms found in parthenogenetic species which in all cases result in the development of the unfertilized egg have been studied extensively over many years (see White, 1954; Suomalainen, 1962). All of the mechanisms lead eventually to a preservation of the species specific chromosome number. The distinction is made between ameiotic or apomictic and meiotic or automictic maturation of the eggs. In the apomictic type, meiosis is suppressed, no reduction of chromosomes occurs, and usually only one equational division is observed. This type is found in most of the thelytokous parthenogenetic species, for example, in the many species of Curculionidae (Suomalainen, 1940, 1954; Takenouchi, 1957), Aphidinae (Schwartz, 1932; White, 1954), Chironomidae (Scholl, 1956), and many others (Suomalainen, 1962). Occasionally two maturation divisions occur, but they are nonreductional. This is seen in *Pycnoscelus surinamensis* (Matthey, 1945, 1948) and *Carausius morosus* (Pijnacker, 1966). In meiotic parthenogenesis, the chromosome numbers are normally restored by automixis of one of the polar nuclei with the egg nucleus or by fusion of two of the polar nuclei. An example of the first case is the psychid moth *Luffia ferchaultella* (Narbel-Hofstetter, 1961, 1963) in which a recombination of the chromosomes occurs after the first anaphase. In the closely related parthenogenetic *Solenobia triquetrella*, chromosome restoration occurs in a different way. This was brought out by the extensive studies of Seiler and Schäffer (1960). Here, the derivatives of the first and second polar bodies fuse, and this nucleus gives rise to the embryo (Fig. 11). In this species two types of thelytokous heterogametic females are known, XO and XY. Figure 11 illustrates for each of the known karyotypes how one can visualize the origin of the two types of females; it is always the inner derivatives of the first and second polar body which fuse. If one were to assume that the embryo arises from the egg nucleus by a fusion of cleavage nuclei, one would have to expect both males as well as two unknown karyotypes. However, it is known that the egg nucleus later degenerates after several cleavage divisions. Interestingly, if the egg of this parthenogenetic species is fertilized, no fusion of the polar bodies takes place, i.e. their fusion appears to be prevented in this case by an unknown mechanism. This would also indicate that the sperm nucleus preferentially fuses with the egg nucleus and not with one of the polar nuclei. Similar observations could be made for both the diploid and the tetraploid races. Another case of polar body fusion which then gives rise to a parthenogenetic embryo is found in a strain of *Drosophila mangabeirai*. Here the two central polar nuclei fuse.

Fusion of the two first cleavage nuclei—which re-establishes the diploid chromosome number—is to be found in the parthenogenetic soft-scale insect *Pulvinaria hydrangeae* (Nur, 1963; Brown and Nur, 1964). In this species, about 95 per cent of the eggs give rise to female and about 5 per cent to male embryos, as judged by the chromosomal behavior (Fig. 10). What causes the heterochromatization of some chromosomes—as is characteristic for males in the 5 per cent of the eggs—remains a subject of speculation. Since all eggs are homozygous, it does not seem that sex determination depends on genetic factors residing in the chromosomes. No adult males are known, and it is assumed that the male embryo or larva of this species dies during development.

Restoration or maintenance of the diploid chromosomal number appears to be essential for most parthenogenetic species, since haploids are often nonviable. However, exceptions exist, as is known from the many hymenopteran species which produce haploid males from unfertilized eggs. Haploid males, which are characteristic of this order of insects, are called impaternate, a term which adequately describes the origin of these males (Whiting, 1945; White, 1954). Impaternate males occur also in a few species of the Iceryini and Margarodid coccids (Hughes-Schrader, 1948), Thysanoptera, and one species of Coleoptera, namely, *Micromalthus debilis* (Scott, 1936, 1941; Whiting, 1945). We are left to speculate as to the biological significance of male haploidy.

C. *Gynogenesis*

In gynogenesis, also known as pseudogamy, the sperm nucleus does not participate in development in the sense that it does not contribute to the genetic pool of the species. The

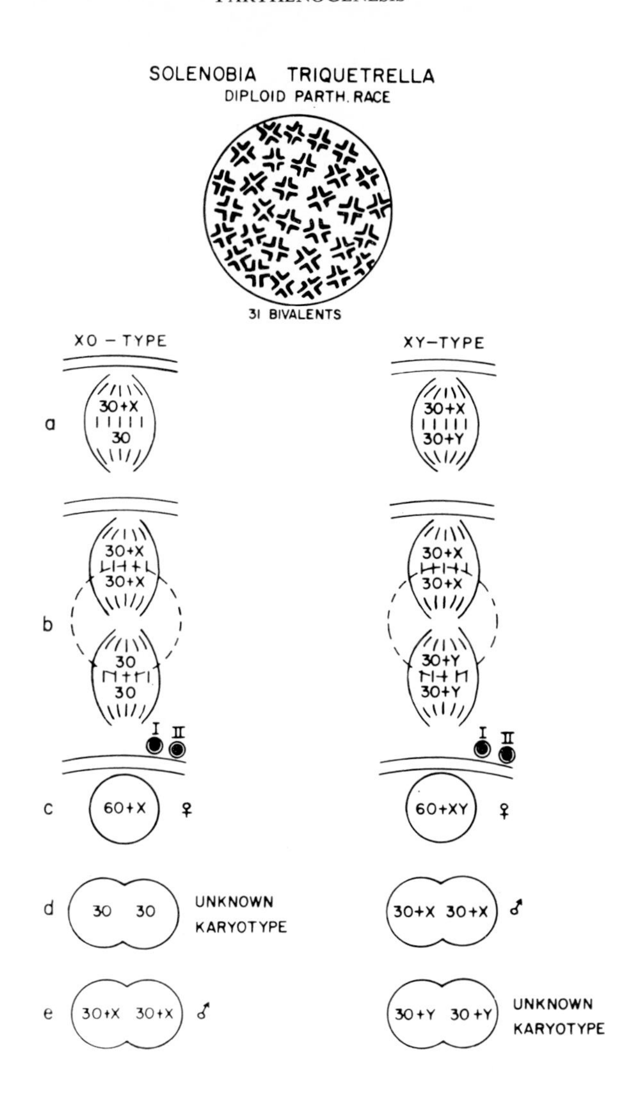

FIG. 11. Chromosome behavior and origin of the two known karyotypes in the diploid parthenogenetic race of *Solenobia triquetrella*. Males and the karyotypes in (d) and (e) are not known to exist in this race. (According to Seiler, 1960.)

function of the spermatozoon is to initiate development in the egg: in other words, the egg develops parthenogenetically. Closely related but separate species often mate. Gynogenesis is known in nematodes, amphibians and fishes (see Benazzi, 1961), and only recently it has been discovered in certain insect species. Its occurrence was first reported for a race of the psychid moth *Luffia lapidella* (Narbel-Hofstetter, 1955). The sessile sac-bearing female of both normal and pseudogamic races of this species does not lay eggs unless mated, an observation which allowed further analysis of the case. Narbel-Hofstetter found an exclusively female population which lived on trees adjacent to a bisexual population living on a wall. Only after the winged males of the bisexual race had mated with the females of the other population did the latter begin to deposit their eggs from which exclusively females hatched. Cytological observations made it clear that the spermatozoon penetrates the egg but that the pronucleus of the sperm does not fuse with the egg nucleus (Narbel—Hofstetter, 1963). It appears indeed that the sperm only triggers development.

All-female broods were observed in the ptinid beetle *Ptinus clavipes* f. *mobilis* (=*P. latro*); eggs were only fertile after these females were mated by *P. clavipes* (=*P. hirtellus*) males (Moore *et al.*, 1956; Woodroffe, 1958). As cytological studies showed, successful development depends on the presence of sperm in the egg but it has only an activating role (Moore *et al.*, 1956; Sanderson, 1960). Without the penetration of the spermatozoon, the egg nucleus does not develop beyond metaphase of the first maturation division. We are apparently dealing with another case of gynogenesis. Gynogenetic strains have been reported in several species of bark beetles of the genus *Ips* (Lanier and Oliver, 1966). As these authors reported, exclusively female broods appeared in some laboratory cultures of the bisexual forms. Since, in all cases, mating was essential and since 100 per cent hatchability was observed, one can conclude that gynogenesis may arise spontaneously in species of this genus. As has been pointed out by Lanier and Oliver (1966), sex ratio conditions alone do not allow us to conclude that gynogenesis is the normal mode of reproduction. Exclusively female broods can arise through differential mortality of the male embryos (see p. 23).

D. Polyploidy and Parthenogenesis

Peculiar to many parthenogenetic species of insects is the occurrence of polyploidy. Interesting material is provided by the several species of Curculionidae among which both bisexual and parthenogenetic races of the same species exist. Suomalainen (1940, 1947, 1950, 1954) surveyed the geographic distribution of these species in Scandinavia and the Alps and found many triploid and a few tetraploid and pentaploid species as well as races. Bisexual and parthenogenetic races generally have different geographic distributions. Bisexual forms remained in areas which were free of ice during the Würm Ice Age. One finds, however, that parthenogenetic polyploid forms now inhabit those areas which were covered by ice during this last Ice Age. It would thus appear that either the polyploid forms were better able to sustain the more adverse climatic conditions of these regions or that the polyploid species followed the retreating ice. Only speculations can be put forth as to the particular origin of polyploidy in these parthenogenetic species or races. Polyploidy apparently arose in these cases by apomictic maturation of the eggs. Observations of a similar nature suggest the same for some weevils in Japan (Takenouchi, 1957).

Polyploidy is also observed in the parthenogenetic races of the psychid moth: *Solenobia triquetrella*, *S. seileri*, and *S. lichenella* (Seiler and Puchta, 1956). *S. triquetrella* has a bisexual as well as diploid and tetraploid parthenogenetic races. This species lends itself to interesting studies on the possible origin of parthenogenesis and polyploidy. As in the Curculionidae, populations of the tetraploid race occur in isolated Alpine valleys which were freed from ice of the last Ice Age relatively late. It appears that, as the glaciers withdrew, some individuals followed and populated the isolated valleys. A tendency toward parthenogenesis can be observed even in the bisexual races of this species, i.e. in certain populations a considerable number of unfertilized eggs begin development and proceed to various degrees. Possibly some individuals can thus establish new populations which may be entirely parthenogenetic. Diploid parthenogenesis occurs via automixis of polar nuclei. Tetraploid races may have arisen via an additional fusion of cleavage nuclei. Seiler (1961,

1964) holds the opinion that tetraploid parthenogenetic forms arose from bisexual ones via diploid parthenogenetic races.

Another interesting case of polyploid parthenogenesis is the Tettigoniid *Saga pedo*. This species has a wide range of distribution in Europe from Spain, through the Rhone Valley, to the region of Vienna and Prague. The distribution is rather discontinuous, but it has been suggested that it was originally continuous (Matthey, 1941, 1946). *Saga pedo* is parthenogenetic, having sixty-eight chromosomes; this chromosomal number suggests that the species is tetraploid since a closely related species like *Saga gracilipes* has thirty-one and *S. ephippigera* thirty-three chromosomes, which is similar to other Tettigoniidae. Maturation of the egg is via apomixis.

Triploidy associated with parthenogenesis is found in the dipteran *Limnophyes virgo* (Scholl, 1956) and the chrysomelid beetle *Adoxus obscurus* (Suomalainen, 1965). Nothing can be added here with respect to the possible pathway of evolution of polyploidy. Whether parthenogenetic species of the Phasmida are polyploid has been discussed at some length by some authors (Cappe de Baillon *et al.*, 1938; Suomalainen, 1950), but since chromosome numbers are highly variable among parthenogenetic and bisexual species of Phasmida, the matter is still somewhat undecided.

Polyploidy apparently does not occur among species of Hymenoptera. This may be related to the fact that the majority of these species are arrhenotokous and that the few thelytokous species known may have arisen in more recent times. Polyploidy could not have evolved for this reason. The same may apply to parthenogenetic species of other orders in which no polyploidy is known. Admittedly, this is pure speculation and no experimental or observational evidence exists. *Solenobia triquetrella* (see above) may be the only case where we have at least some grounds for speculations on the origin of polyploidy.

E. Evolution of Parthenogenesis

A discussion of the evolutionary aspects of parthenogenesis is, in many cases, on shaky grounds, since the available material is limited; yet a few thoughts may be pertinent. Parthenogenesis most likely evolved from bisexual forms via facultative parthenogenesis. This is suggested in the finding that in many non-parthenogenetic species one frequently observes only an initial development of unfertilized eggs, which may lead eventually to complete development in some cases. Another possible evolutionary pathway may be exhibited in the species *Luffia lapidella* where a bisexual race may live next to a gynogenetic one (Narbel-Hofstetter, 1955). An egg of the gynogenetic race is capable of parthenogenetic development only after penetration of the egg by a spermatozoon. The sperm nucleus does not contribute to the genetic pool of the species, however. This species may be on its way to complete parthenogenesis, as is found in the related *L. ferchaultella*. Similarly, bisexual and parthenogenetic races of *Solenobia triquetrella* live in close proximity; the bisexual forms exhibit various degrees of initial parthenogenetic development (Seiler, 1959, 1961), thus pointing to the origin of parthenogenesis. It is also interesting to note that bisexual individuals from various locations have different potential for parthenogenetic development.

Support for the hypothesis that parthenogenesis may have arisen many times in *Solenobia* via the selection of races which are capable of developing without fertilization was given by Seiler (1963). He finds local parthenogenetic races which after previous mating, fertilize their eggs whereas others fertilize only a few or none. Some of these races behave just like the bisexual ones. However, the cytological mechanism of fusion of the pronuclei seems to have broken down in those where fertilization does not take place. The hypothesis is advanced that the first type is of relatively recent origin, whereas the latter must have evolved a long time ago. Parthenogenetic as well as bisexual females of *Solenobia* possess normal and functional external genitalia, which again supports the view that parthenogenesis evolved via bisexual forms. The same appears to apply in parthenogenetic Coleoptera (Szekessy, 1937).

Thelytoky can occur spontaneously among the Diptera, apparently by mutation. For example, no males are known in *Lonchoptera dubia* but only about 90 per cent of the eggs develop parthenogenetically (Stalker, 1956). The hypothesis is put forth that a polygenic complex favors the development of unfertilized eggs. It may have arisen in isolated populations and then spread over larger areas. Thelytokous females occur at a low percentage (0.1 per cent) in populations of *Drosophila mercatorum* (Carson, 1967), but, by repeated selection, unisexual reproduction was raised to 6.4 per cent. In this case, the character appears to be polygenic and can be transmitted also by males. Interestingly, parthenogenetic strains show sexual isolative behavior. This may indicate that the parthenogenetic strains may evolve without geographic isolation pressure (Henslee, 1966).

Evolution of male haploidy (arrhenotoky) may have occurred via the zygogenetic mode (Whiting, 1945). In arrhenotokous species the female is "allowed" to produce males "at will", as is exemplified by the many hymenopteran species in which environmental and intrinsic factors cause the female either to release sperm or to retain it in the spermatheca during oviposition. It is conceivable that thelytoky in some hymenopterans may have arisen via arrhenotoky (White, 1954). In support of this hypothesis, we may cite the thelytokous *Ooencyrtus submetallicus* which produces functionless males if reared at 29°C or above (Wilson and Woolcock, 1960); the cytogenetic mechanism for thelytoky seems to break down at elevated temperatures. It is noteworthy that closely related species are exclusively arrhenotokous (Wilson, 1962).

Parthenogenetic species have a certain advantage over bisexual ones in that their reproductive capacity is theoretically doubled. If a parthenogenetic strain arises within a population and these females are as fecund as those of the bisexual strain, the entire population may become parthenogenetic in time (Stalker, 1956). The distribution and occurrence of parthenogenetic races and species of the Curculionidae in the northern parts of Scandinavia or in the isolated Alpine valleys indicate their success in finding suitable ecological niches (Szekessy, 1937; Suomalainen, 1962). Further north, one finds even more parthenogenetic species than in the southern parts of Scandinavia. Obviously, a single female transported accidentally into an isolated area can start a new population. The same pattern may apply to the races of the psychid moth *Solenobia* found in the Alps. In this connection, it is interesting that a large number of parthenogenetic species have feeble locomotor power. As examples, one may list the several species of psychid moths, scale insects and aphids as well as the one apterous grasshopper which is parthenogenetic, *Saga pedo*. These species are handicapped in that they cannot travel great distances and, indeed, cannot extend their geographic distribution very rapidly. However, if isolated, they can start an entire new population alone.

A distinct disadvantage of parthenogenetic species is their lack of adaptability to changes in environmental conditions. In these species there is no exchange of genetic information between individuals, and new forms can only arise through mutations. The lack of adaptability must eventually lead to extinction of the species, as is perhaps exemplified by the sporadic occurrence of *Saga pedo*; this species is found today only in small isolated populations, and it is assumed that the species became extinct in intervening regions. The short outline given above shows the complexity of the phenomenon which we term parthenogenesis. It appears to have arisen many times, probably via several possible routes. However, we are far from understanding the many aspects involved and have to resort to speculations more often than is desirable.

CHAPTER 4

SEX DIFFERENTIATION

DIFFERENTIATION of both primary and secondary sex characters is undoubtedly determined by the genetic background of the individual. A vast literature dealing with aspects of sex determination, as well as its associated cytogenetics, gives ample support for this statement. The study of intersexuality particularly contributed to the understanding of sex differentiation in animals. Since it was found in vertebrates that epigenetic factors, especially sex hormones, influence sex differentiation regardless of the genetic sex, naturally, analogous possibilities were investigated in insects. In this chapter, the evidence for epigenetic sex differentiation will be dealt with in some detail, even though certain of its aspects were recently reviewed by Charniaux-Cotton (1965) and Brien (1965).

Gonads of both male and female insects are composed of germinal and mesodermal tissues. Mecznikoff (1865) recognized that germ cells originate from the posterior pole region in a dipteran egg, where they first form the so-called pole cells. These pole cells segregate during blastoderm formation and later migrate into the region of the future posterior midgut. The first experimental evidence that the pole cells are indeed the future germ cells was given by Hegner (1908, 1909, 1911) for chrysomelid beetles. He showed that cautery of the pole region or a puncturing of the eggs which resulted in loss of much of the pole plasm yielded many animals which lacked germ cells or had only few of them. Geigy (1931) irradiated the pole plasm of *Drosophila melanogaster* eggs with UV before cleavage nuclei had invaded this region; the resulting adults contained gonads which lacked germ cells. This work was later repeated, leading to a reiteration of Geigy's conclusions (Aboim, 1945). The important hypothesis emerging from these findings was that the mesodermal gonads differentiate in the absence of germ cells. However, subsequent work in several dipteran species challenged this conclusion although some authors continued to interpret their results in the sense of Geigy and Aboim. Counce and Selman (1955) treated *Drosophila* eggs with ultrasonic sound and obtained embryos without gonads, presumably because the pole cells had not migrated into the region of the posterior midgut. UV irradiation of eggs of *Drosophila* and *Lucilia cuprina* (Poulson and Waterhouse, 1960) indeed yielded some animals without gonads, but others had normal looking gonads. In cases where the complete absence of gonads was noted no germ cell was ever detectable; in animals with gonads, however, sometimes at least two or three germ cells were identified. Poulson and Waterhouse interpreted the available data to the effect that mesodermal gonad formation is induced by the germ cells, a conclusion contrary to that of Geigy (1931). One can assume with a reasonable degree of certainty that, in cases where no germ cells were found in the adult gonads, damaged pole cells had invaded the future midgut region, induced gonad formation, but then degenerated. If this interpretation is correct, the actual observations of Geigy and Aboim are valid since these authors checked only adult flies. Apparently UV irradiation does not damage either pole plasm or pole cells to the extent that these cells die immediately after the treatment; consequently, in the majority of cases gonads were induced in irradiated *Drosophila* (Geigy, 1931; Aboim, 1945; Hathaway and Selman, 1961; Jura, 1964), *Lucilia cuprina* (Poulson and Waterhouse, 1960), and *Culex pipiens* (Oelhafen, 1961). However, if total destruction of the pole plasm was achieved by cautery of the egg of *C. fatigans* and *L. sericata*, no gonads were ever found upon serial sectioning of the embryos (Davis, 1965, 1967). These results strengthen the interpretation of Poulson and Waterhouse (1960), i.e. pole cells are essential for the induction of the gonads. Experimental evidence for the role of pole cells in gonad formation is restricted exclusively to dipteran species; therefore, one must be cautious in making any generalizations. Even though the cytologically distinct pole plasm and pole cells are found in other insect species, their function may not be analogous to that of dipteran species.

Peculiar to pole plasm are the granular materials which are later found within the pole cells; this peculiarity allows these cells to be traced during their migration inwards. Evidence for the function of these granules is slim. Centrifugation of *Drosophila* eggs for 10 min at 425 g displaced these granules. It was then shown that pole cells no longer detached from the blastoderm and, furthermore, did not migrate into the region of the future midgut (Jazdowska-Zagrodzinska, 1966). This finding certainly suggests a vital role for the pole granules in both pole cell formation and gonad induction.

Male and female *Drosophila* with agamic gonads produced by UV irradiation cannot be distinguished from normal animals on the basis of their secondary sex characters (Geigy, 1931). This was taken as evidence that insects do not produce sex hormones, and it confirmed findings in other species in which gonads were extirpated or transplanted, after which no changes in sex characters were observed (Oudemans, 1899; Regen, 1910; Kopec, 1912; and others). But differentiation of sex, male and female, may be determined at different times of larval or embryonic development in different species; one certainly cannot take the evidence from a few species and draw general conclusions. Furthermore, the development of the hypothesis that sex characters may be controlled by epigenetic factors was hindered by the assumption that those factors must originate from the gonads.

The lucky coincidence of an interest in the phenomena of control of secondary sex characters and a systematic search for the causes in a suitable animal, the dimorphic beetle *Lampyris noctiluca*, led to the discovery of an androgenic hormone (Naisse, 1963a). This original contribution to insect endocrinology was based on the following experiments. After castration of third instar male larvae, many of these animals became females (Naisse, 1963a, 1965, 1966a). If testes from third or fourth instar larvae were transplanted into female larvae of similar age, the recipients developed into males, both in primary and secondary sex characters (Fig. 12). It was found that the apical mesodermal tissues of the testis follicles were prominent only during the larval instars when the testes were effective in masculinization of the female. From this, Naisse concluded that these tissues liberate an androgenic hormone. In a series of further experiments in *Lampyris* the functional mechanism was elucidated. Implantation of larval ovaries into males of the fourth instar had no effect on the male larvae but the prospective ovary was converted into a testis; normal spermatogenesis was observed in this converted ovary. Transplantations of ovaries into male pupae were ineffective, i.e. the implants were not masculinized (Naisse, 1966a). Ovaries which were transplanted into castrated male larvae did convert, but not to the same extent as those transplanted into normal male larvae; this was probably due to only a low titer of androgenic hormone remaining in the animal. Interestingly, during the phase when testes are most

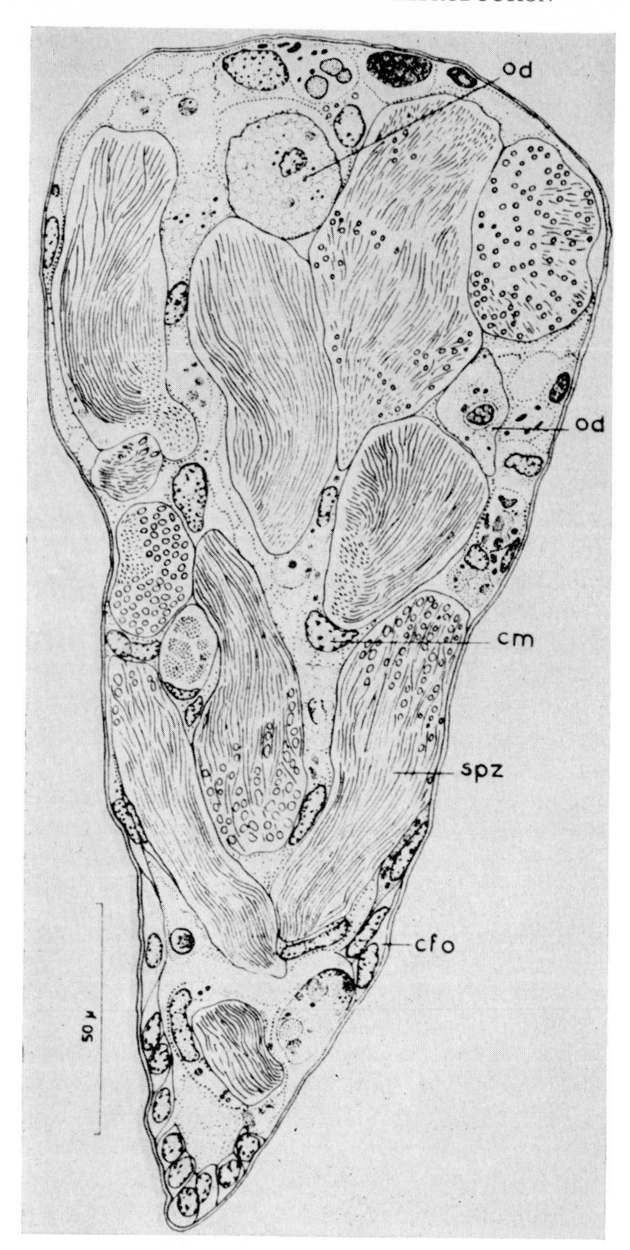

FIG. 12. Longitudinal section through an adult testis follicle of *Lampyris noctiluca* which was converted from an ovary *in situ* by the implantation of larval testes. cfo = persisting follicle cells. cm = mesodermal cell. od = degenerating oogonium. spz = spermatozoa. (After Naisse, 1966a.)

effective in influencing sex differentiation, a certain type of neurosecretory cell in the pars intercerebralis containing small secretory granules is found to be particularly active in the male (Naisse, 1963b, 1966b). From this observation the conclusion was drawn that the activity of the apical testicular cells is under the control of these neurosecretory cells. Furthermore, ablation of corpora cardiaca and corpora allata in male larvae converted these males into females (Naisse, 1966c). This latter result is difficult to interpret, however, since it is unclear whether the effect is the result of the absence of the corpora cardiaca or the corpora allata or both, whether liberation of the neurosecretory materials was inhibited or restrained by the operation, or whether operational trauma might not be responsible. Naisse interpreted her results to indicate that the corpora cardiaca and corpora allata normally stimulate the neurosecretory cells which, in turn, activate the apical tissues of the testes to produce the androgenic hormone. Irrespective of whether this interpretation applies or whether it must be modified, the available data show that the female is the "neutral sex", it differentiates without the intervention of a hormone.

How widespread the occurrence of an androgenic hormone among insects is remains to be seen. Based admittedly on insufficient evidence, we may speculate on its presence in species other than *Lampyris*. For example, some males of the membracid *Thelia bimaculata* which were parasitized during an early instar may change their external morphology (size and color) to that of a female (Kornhauser, 1919). Neither parasitized females nor parasitized older male larvae changed. Can one assume that the parasite destroyed the apical mesodermal tissues of the testis follicle and that therefore no androgenic hormone was liberated? Males and females of several species of chironomids exhibit intersexual characteristics after parasitization by Mermethidae, particularly if parasitized during an early instar (Wülker, 1961, 1962). But in this case, it is very difficult to attribute an effect of the parasite on gonadal hormones since both males and female are affected. In each case, the genetic sex was verified cytologically; the animal showed intersexuality toward the opposite sex in all instances. Is it possible that in chironomids the neutral sex might be intersexual in nature due to a balance of male- and female-determining genetic factors, and that an epigenetic factor subsequently determines the actual differentiation? A further possibility—that of parasitic starvation—cannot be the cause since, contrary to earlier findings, both sexes are affected. No satisfactory answer can be given.

Further studies on epigenetic factors affecting sex differentiation are few. Those studies which exist have dealt primarily with the effect of temperature variations. For example, in the thelytokous hymenopteran parasite *Ooencyrtus submetallicus* Wilson and Woolcock (1960) found that females exposed to high temperatures (29.4°C) produced males, those exposed to lower temperatures, thelytokous females. Intersexes (no cytological examinations were made) were obtained by alternate exposure of the females to low and high temperatures. Since the eggs could not have been affected by temperature after oviposition, one may assume either that temperature affects sex determination during maturation divisions or that sex differentiation is determined in the mature eggs at an early preembryonic time. The first of the alternatives may be the more sensible interpretation.

Another example of temperature influencing sex differentiation is the case of *Aedes stimulans*, which lives in the northern region of the United States and Canada. Exposure of larvae to temperatures of 26°C caused the male larvae to develop into females, i.e. testes became ovaries; genetic females were not affected by the temperature treatment (Anderson and Horsfall, 1963). Temperatures below 23°C had no affect on the males, and normal sex ratios were observed. The sensitive period for gonadal transformation extends over several larval instars, and testes may respond to high temperature as early as the first instar (Horsfall and Anderson, 1964; Anderson and Horsfall, 1965). Mature ova can be produced when transformed ovaries are transplanted into normal females. Transplantation of transformed ovaries is essential since males and transformed males rarely take the blood necessary for egg maturation. This experiment indicates that the transformed gonads are fully functional; however, since normally transformed males rarely feed blood and, in addition, have only partially transformed external genitalia, these animals are reproductively nonfunctional. In a further extension of these experiments genetic male gonads of this species were transplanted into closely related species and then subjected to high temperatures. Under these conditions the transplanted testes also converted into ovaries even though the host itself was nonresponsive to temperature treatment. Anderson and Horsfall (1965) interpreted this to mean that no hormonal agents are involved in the conversion and that the mechanism of feminization resides in the gonad. If, however, one postulates that the apical tissues—which may furnish the androgenic hormone in this species—cease to function at a high temperature, one could explain the feminization of the implant even in the foreign environment. It is clear that such an interpretation is hypothetical and awaits, first of all, the demonstration of an androgenic hormone in this species.

Similar reservations must apply to any interpretation of the studies on triploid intersexes of *Drosophila*. These studies showed that high temperatures (20–25°C) favor the expression of femaleness (Laugé, 1962). It is noteworthy that second instar larvae are particularly sensitive to temperature treatment (Laugé, 1964, 1966a); at 30°C, nearly complete femaleness was achieved (Laugé, 1967) if the treatment occurred during periods of segmentation of the embryo. This latter finding certainly indicates that sexual expression can be affected by temperature at a very early time in the animal's development. It may, furthermore, indicate that timing is of extreme importance in studies on sex differentiation. Sensitive periods may be different in various species.

As these short outlines illustrate, we are far from understanding the problems—cause and mechanism—which are related to sex differentiation in insects. The intricate interactions of epigenetic and genetic factors can be studied here since insects are suitable and easy to handle objects and may therefore provide the material for studies of some basic phenomena of differentiation.

CHAPTER 5

GONADAL DEVELOPMENT

A. Spermatogenesis

Mechanisms of spermatogenesis, including sperm differentiation, are rather similar in most animals, as is shown by the extensive literature dealing with this subject in both invertebrates and vertebrates. With regard to insects, the most complete review of the subject for many species was given by Depdolla in 1928. Later, spermiogenesis was reviewed by Nath (1956), the cytological aspects of meiosis by Rhoades (1961), and the premeiotic events by Hannah-Alava (1965). In the following outline I will largely follow these reviews.

In a longitudinal section of a mature testis follicle, one can follow the entire process of spermatogenesis as it proceeds from the formation of the primary spermatogonia to the primary spermatocytes, through the following first and second meiotic divisions, which give rise to the spermatids, which then differentiate to become spermatozoa (Depdolla, 1928). In the apical region of the testis follicle of the mature insect are found embryonic tissues similar to those of the embryonic and larval insects. As a rule, the apex of the follicle contains the apical cell and one or more primary spermatogonia, the germ cells. The apical cell, which is distinct from the germ cells, appears to have a nutritive role, as is suggested by its location as well as by its apparent high metabolic activity. In *Melanoplus*, RNA synthesis occurs at a high rate in this apical cell (Muckenthaler, 1964). The primary spermatogonia may undergo several mitotic divisions before they finally become spermatocytes (Hannah-Alava, 1965). Secondary and following spermatogonia divide synchronously, forming cysts ensheathed with mesodermal interstitial tissues; the results are bundles of spermatids of the same differentiation stage. The number of spermatogonial generations varies from species to species; it can be estimated from the number of spermatids contained within one cyst. For example, in *Melanoplus* seven generations are observed (Muckenthaler, 1964), but among other grasshoppers there may be from four to nine which give rise to bundles of sperm ranging in number respectively from 64 to 2048 (White, 1955).

Some discussion is found in the literature whether, in each follicle, one or a few stem cells give rise to later generations of spermatogonia or whether the later generation spermatogonia themselves can function as stem cells (Hannah-Alava, 1965). The concept of a germ line is indeed an important one. A quasi-dichotomous division of the primary germ cell, which gives rise to one primary spermatogonium and another stem-cell, appears to exist in *Drosophila melanogaster*, *Bombyx mori*, and several Acrididae (Hannah-Alava, 1965); however, in other species little is known about this phenomenon. Evidence for the stem-cell concept is principally derived by use of the brood method. In this method, X-ray irradiated adult males of *Drosophila* were allowed to mate with a series of virgin females for up to 24 days (Hannah-Alava, 1964; Puro, 1964). It was found that 6 to 8 days after irradiation sterile broods appeared, probably caused by the loss of spermatogonia which were sentisive to irradiation. Following this sterile period, clusters of identical mutations occurred at a high rate for a few days. From the size of the clusters, one can infer that, primarily, the sensitive definite or predefinite spermatogonia were affected by the X-rays. From this one probably can conclude that a stem-cell mechanism for maintenance of the germ line exists. This old concept which thus is revived can be further supported by histological evidence. From autoradiographic studies in *Melanoplus* (Muckenthaler, 1964) and *Drosophila* (Olivieri and Olivieri, 1965), it is evident that DNA synthesis occurs primarily in the spermatogonia and early spermatocytes rather than later. (RNA is synthesized both in spermatogonia and post-meiotic stages.)

Spermiogenesis, i.e. the differentiation of a spermatid to a flagellated spermatozoon, may be illustrated as follows (Fig. 13). After the second meiotic division (see Depdolla, 1928; Nath, 1956), axial filaments grow out from the centriole which is usually located at the posterior end of the nucleus. Usually, an axial filament consists of nine double hollow fibers arranged in a circle and two central singlets (Nath, 1956). At first the mitochondrion is found near the centriole, but later it is stretched into the tail alongside the axial filaments and is then known as the Nebenkern. An acrosome is found anterior to the nucleus in the majority of the species studied. This acrosome arises from the Golgi apparatus (Nath, 1956), first forming an acroblast which then separates and lodges anteriorly, where it undergoes further differentiation (Kaye, 1962). The function of the acrosome is presumably to aid in the penetration of the eggs by the sperm; it may furnish enzymes which dissolve the egg membranes. However, this explanation may not be sufficient for all species since in several species of tiger beetles (Nath, 1956), among them *Cicindela belfragei* (Gassner and Breland, 1967), no acrosome could be identified with certainty. Also, in several Thysanura, the acrosome is located in the neck region of the sperm posterior to the nucleus (Nath, 1956; Nath *et al.*, 1960; Bawa, 1964).

Several additional aberrations of the typical spermatozoon can be observed in some species. For instance, in species of the Thysanura it appears that centriole and acrosome have exchanged places, a centriole may be found at the anterior end of the nucleus (Nath, 1956; Nath *et al.*, 1960; Bawa, 1964). In these species the axial filament arises consequently from the centriole anterior to the nucleus. In the tiger beetles the centriole is also located anteriorly (Nath, 1956; Gassner and Breland, 1967). In some species the axial filaments may show aberrations which may or may not be of functional significance; only histological evidence is presently available. For instance, in the mature sperm of the cricket *Acheta domestica*, the axial filaments are solid rather than hollow as is normally found (Kaye, 1964). An abnormal arrangement of the axial filaments is also found in the fungus gnat *Sciara coprophila*. In this species, an average of seventy doublets (in contrast to the usual nine) are observed throughout the length of the tail; they are arranged in a ring for most of the length, but posteriorly they form spirals (Makielski, 1966; Phillips, 1966a, b). The usual axial filaments grow out from the centriole in this species just as in others. As for the function of these unique axial filaments, nothing unusual can be said. Their close association with the mitochondrial Nebenkern, as in other species, may suggest their role in energy transfer and mobility of the sperm tail. This is, however, only inferred from the structural arrangements.

Complete spermatogenesis takes place during larval and pupal life in many species having a short adult life span (Depdolla, 1928; Sado 1963; Tazima, 1964) (Fig. 14). For example, no further spermatogonial divisions occur in 4th instar larvae of the gypsy moth, *Porthetria dispar*, and meiosis takes place at this time (Rule *et al.*, 1965). In *Rhyacionia buoliana* mature spermatozoa are found shortly after pupation (Shen and Berryman, 1967). The same

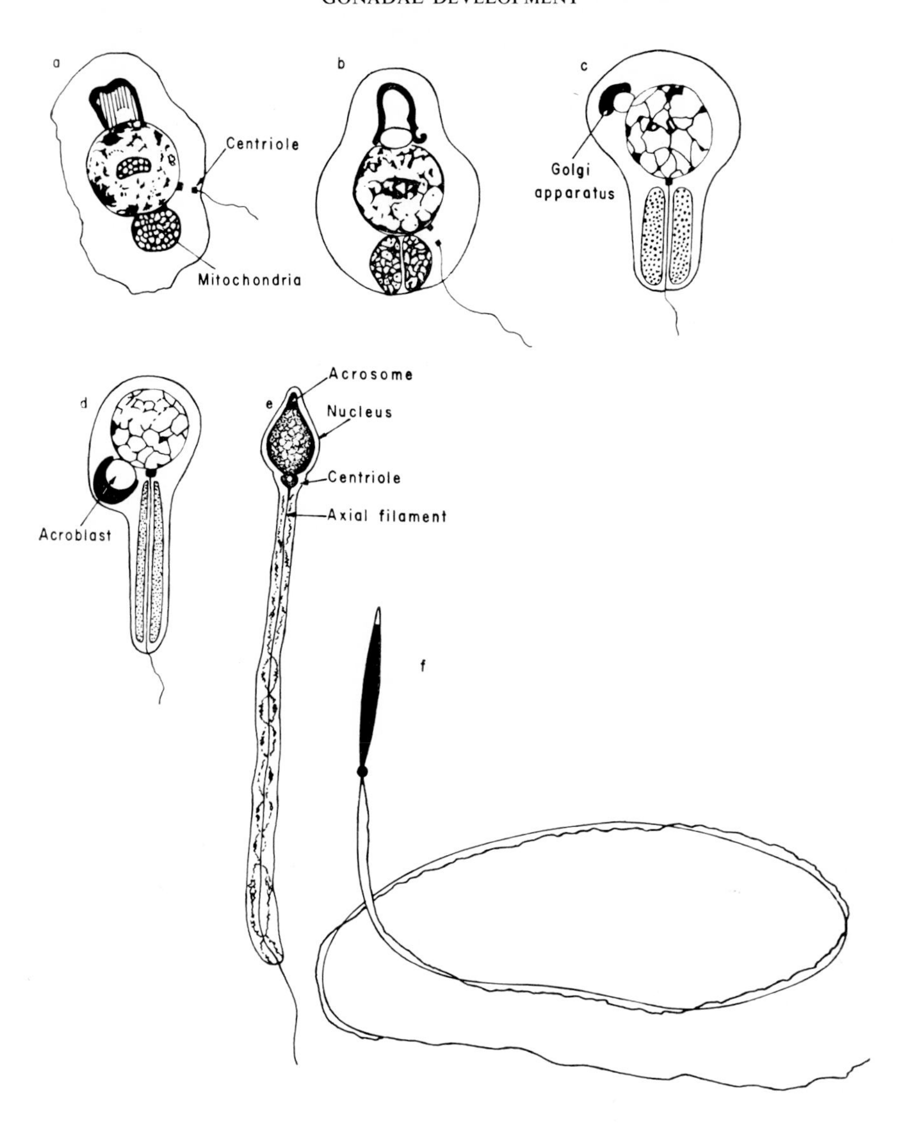

FIG. 13. Schematic representation of the sequence of sperm differentiation. (Modified from Depdolla, 1928.)

observations can be made for many additional species, but these quotations should suffice to illustrate the point. Quite in contrast to this, in species with a long adult life span, the full process of spermatogenesis may take place after emergence. Both brooding procedures and labeling techniques gave convincing evidence for this. X-ray treatment of adult *Drosophila* males which were then reared with normal females resulted in sterile broods 6 to 8 days later, presumably because spermatogonia had been affected. Both subsequent broods containing mutations and later normal broods were reared, thus indicating that new spermatogonia had been formed by the stem-cells (Hannah-Alava, 1964, 1965). From this, one can also conclude that spermatogenesis takes about 10 days in *Drosophila.* A more direct method, certainly, is the labeling of primary spermatogonia with a follow up of the distribution of the label during the following days. This was done, for instance, in *Anthonomus grandis* males which were injected 5 days after emergence with H^3-thymidine. The label appeared immediately in the spermatogonia and in the early spermatocytes (Chang and Riemann, 1967). Approximately 10 days later the label was found in the mature spermatozoa. In adult *Melanoplus*, complete spermatogenesis takes about 28 days, as shown by similar techniques (Muckenthaler, 1964). In this species seven successive spermatogonial generations occur within 8 to 9 days, meioses then take about 9 to 10 days and spermiogenesis another 10 days.

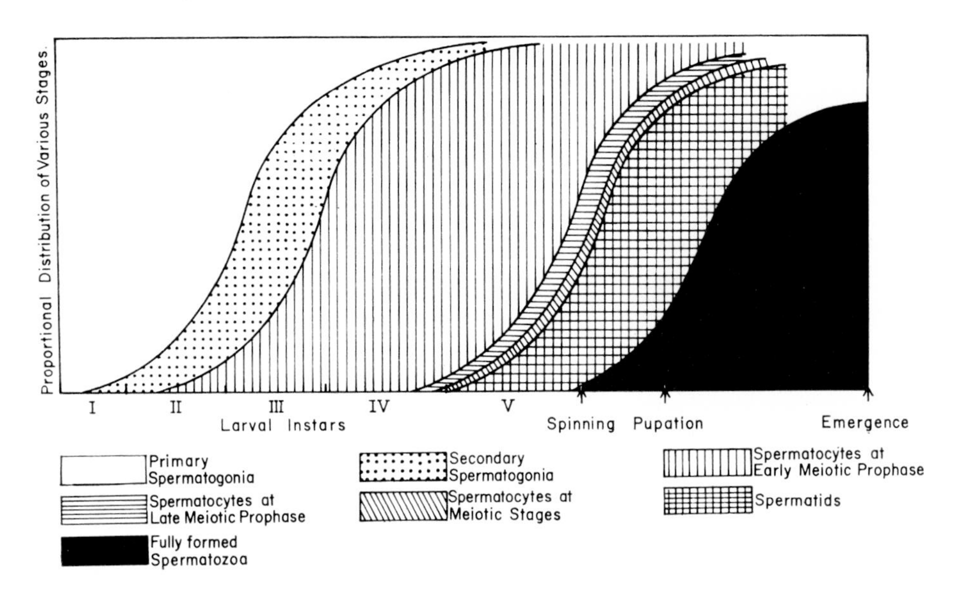

FIG. 14. Schematic illustration of spermatogenesis in *Bombyx mori* in relation to the developmental stages. Note that fully formed spermatozoa are already present at pupation. (Modified from Sado, 1963.)

Evidence for a hormonal control of spermatogenesis analogous to the control mechanisms for oogenesis in females has been sought by some investigators. However, as the above citations concerning timing and time sequences in various species may suggest, an endocrine control pattern either may differ in different species or may not exist at all. In analogy to the findings in females, the corpora allata may be implicated, but spermatogenesis (including spermiogenesis) in many species occurs during the pupal stages (Depdolla, 1928), a time when the corpora allata are inactive. Furthermore, after allatectomy in *Schistocerca gregaria*, no effect on spermatogenesis was noticed (Cantacuzène, 1967). Deprivation of food, which inactivates the corpora allata in adult *Lucilia cuprina* (Mackerras, 1933), *Phormia regina* (Cowan, 1932), and *Scatophaga stercoraria* (Foster, 1967), to mention only a few examples, did not affect spermatogenesis. The molting hormone ecdysone likewise is probably not involved since spermatogenesis was observed in testes of larval *Porthetria dispar* (=*Lymantria dispar*) transplanted into adults (Kopeć, 1912). Similarly, in *Drosophila*, larval cysts transplanted into adult female abdomina underwent complete sperm differentiation autonomously (Garcia-Bellido, 1964a; Hadorn *et al.*, 1964). The possibility, however, exists that hormones act indirectly on the process of spermatogenesis rather than directly. This is shown, for example, in *Hyalophora cecropia* where cysts cultured *in vitro* undergo differentiation when blood from developing pupae—which presumably contains certain stimulatory factors—is added to the medium, but not when blood from diapausing pupae is used (Schmidt and Williams, 1953). Addition of ecdysone to the medium did not enhance spermatogenesis, however; some other component of the blood from developing pupae appears to be responsible for the promotion of spermatogenesis (Bowers, 1966, personal communication). This same interpretation might also apply to results obtained in transplanted *Drosophila* testes (Garcia-Bellido, 1964a, b); either starvation or brain surgery of the adult host female affected both cyst formation and sperm differentiation in this case. None the less, it cannot be ruled out that neurosecretion or other hormones normally influence spermatogenesis. As these two reports on *Hyalophora* and *Drosophila* show, our knowledge of control of spermatogenesis is rather fragmentary; at this time we have no concrete evidence for or against direct hormonal involvement in any species.

B. Oogenesis

1. OOCYTE DIFFERENTIATION

For many years a number of students of insect reproduction have been studying the cytological events in the ovarioles during oogenesis. Even though the effort has been great, our knowledge of the details is still rather scanty, with the possible exception of differentiation in the polytrophic ovariole of *Drosophila* due to the intensive efforts of R. C. King and associates. Ovarioles of the three types—panoistic, polytrophic, and telotrophic—differ to some extent in the process of oocyte differentiation but the differences are apparently only of minor significance. In all three types of ovarioles, oocyte differentiation takes place in the germarium, i.e. the anterior part of the ovariole. This germarium contains oogonia, primary oocytes, nurse cells (in polytrophic and telotrophic), prefollicular, and follicular epithelial cells (Snodgrass, 1935).

In the panoistic type, follicles are formed as the oocytes move down the ovariole with mesodermal tissues growing around the oocytes. Various zones of differentiation are recognized. For instance, in *Periplaneta americana* (Bonhag, 1959; Anderson, 1964) the zone next to the terminal filament is characterized by the presence of primary oocytes in prophase of the first meiotic division; further downward, the oocytes, though still in prophase, contain high concentrations of RNA. Up to this zone, oocytes are not yet arranged in a row. Then in the next zone toward the vitellarium, the oocytes are lined up in a row and giant lampbrush chromosomes appear in the nuclei. Further posteriorly, yolk deposition begins. It is believed that in an ovariole of the panoistic type all germ cells differentiate into oocytes. Apparently, nothing is known about the presence of stem-cells which would give rise to an oogonium and a stem cell with each division.

In the germarium of the telotrophic ovariole, such as in *Oncopeltus fasciatus* (Wick and Bonhag, 1955) and *Tenebrio molitor* (Schlottman and Bonhag, 1956), the prefollicular tissues differentiate into the follicular epithelium, while the oogonia give rise to oocytes and apical trophic tissues. The trophic tissues remain connected to the oocytes for some time via a nutritive cord and, histologically, one can demonstrate that materials do indeed migrate down the cord. In *Oncopeltus*, the trophic tissues appear to be a syncytium, whereas in *Tenebrio*, distinct cell boundaries are observed. During the pupal period of *Tenebrio*, i.e.

before yolk deposition had begun, a flow of RNA rich material was observed along the trophic cord (Lender and Laverdure, 1964), an observation which suggests the functional significance of the cord in the previtellogenetic stages.

Polytrophic ovarioles are identified by the presence of a number of trophocytes or nurse cells together with an oocyte, surrounded by an envelope of mesodermal cells thus forming a follicle. The number of trophocytes per follicle differs among the various species (Bonhag, 1958). For example, there may be only one as in the earwigs, three as in *Panorpa communis* (Ramanurty, 1964), seven as in *Anopheles maculipennis* (Nicholson, 1921), *Bombyx mori* (Colombo, 1957), and *Hyalophora cecropia* (King and Aggarwal, 1965), or fifteen as in *Calliphora erythrocephala*, *Musca domestica* (Bier, 1963a), *Drosophila melanogaster* (Brown and King, 1964), and others. In several species—for example, *Anopheles* (Nicholson, 1921), *Drosophila* (King *et al.*, 1956), and the gall fly *Cynips folii* (Krainska, 1961)—it has been shown that nurse cells and oocytes arise from the same oogonium through mitotic divisions. Differentiation of the primary oogonia into nurse cells and oocyte has been studied intensively in *Drosophila* (see King, 1964a; Koch *et al.*, 1967). Presumably, whatever applies to *Drosophila* also applies in principle to many other species with polytrophic ovarioles, although each species will have to be studied in detail to be certain. This word of caution is strengthened by a recent study of the paedogenetic larvae of the Diptera *Heteropeza pygmaea* where, according to Panelius (1968), the trophocytes are of somatic origin and are not derivatives of the oogonium. In *Drosophila*, the best known case, the apical region of each germarium contains one or a few stem-line oogonia each of which during pupal and early adult life (King *et al.*, 1968) gives rise to a cystoblast and another stemline oogonium (Fig. 15). Evidence for this is both histological and experimental (Brown and King, 1964). Irradiation with X-rays induced clusters of identical viable mutations apparently because the mutant stem-cell furnished a succession of mutant oocytes. As the cystoblast passes down the germarium, it divides four times, giving rise to a cyst of sixteen cells, fifteen of which will become nurse cells. Finally, mesodermal cells grow around and in between the cysts, forming the follicular epithelium; these cysts pinch off from the germarium and thus become a follicle (Koch *et al.*, 1967). The formation of the cyst in *Drosophila* could be followed in detailed microscopic studies. All sixteen cystocytes are descendents of the cytoblast and are interconnected by fifteen canals, each surrounded by a protein-rich ring. These connecting rings form around the mitotic spindle and are characteristic only of the cystocytes. No ring canal is found between the stem-line oogonium and the cystoblast. Ring canals had been known from cysts of polytrophic ovarioles of several other species (see Brown and King, 1964). Already in 1912 they had been described for several species of Carabidae by Kern and in the beetle *Dineutes nigrior* had been termed intercellular strands (Hegner and Russel, 1916); for the same structures as observed in *Drosophila* by Meyer (1961) and in *Calliphora* and *Musca* by Bier (1963a), the term fusome was preferred.

From these studies arose the question how the oocyte is determined from among the cystocytes. No leads are available from any species other than *Drosophila*. Here it was found that the oocyte is always one of the two cystocytes which arise from the first division of the cystoblast. These two cystocytes are distinguished from the others in that they have four ring canals (Fig. 16) and are the richest in the pericanalicular materials which form around all the ring canals after mitosis (Koch and King, 1966; Koch *et al.*, 1967; Koch and King, 1968). Only these two cells enter meiotic prophase and they are the only ones which form synaptonemal complexes during this period (Smith and King, 1969). Synaptonemal complexes are also found in *Aedes aegypti* pro-oocytes and, in a modified form, in the nurse cells

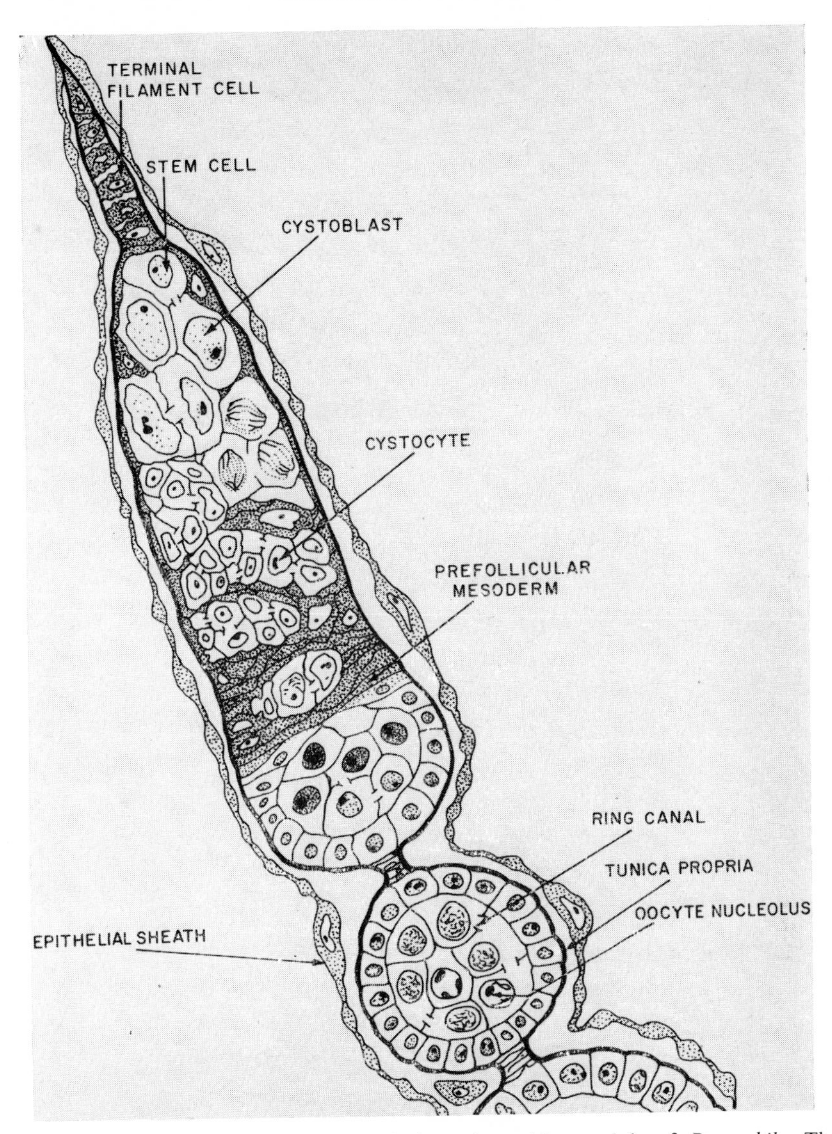

FIG. 15. Anterior portion (germarium) of the polytrophic ovariole of *Drosophila*. The sequence of stages in differentiation proceeds from anterior to posterior. The stem cell continuously gives rise to cystoblasts. (From Koch *et al*., 1967.)

of this species (Roth, 1966). In *Drosophila*, the first cystocytes always take a unique position as is illustrated in Fig. 16. Later in differentiation, one of the pro-oocytes becomes endopolyploid like the other nurse cells and from then on, functions as a nurse cell. The mechanism by which one of the pro-oocytes becomes determined as the actual oocyte remains somewhat obscure. It is found, however, that the definite oocyte has seven times more plasmalemma in contact with the follicle cells than does the other pro-oocyte, a finding which may be related to the determination of the oocyte (Koch and King, 1968). A characteristic feature of the nurse cells of polytrophic ovarioles is their polyploidy, which arises through

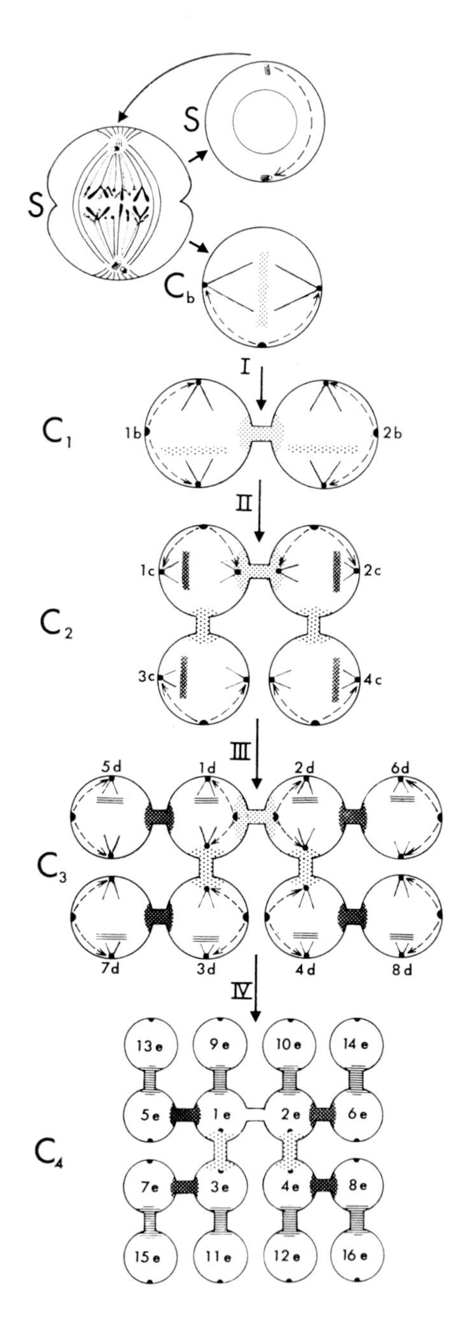

FIG. 16. Model illustrating the steps which lead to the production of a cluster of sixteen interconnected cystocytes in *Drosophila*. The stem cell divides into another stem cell and a cystoblast (C_b). In a series of four divisions, the oocyte and fifteen nurse cells are differentiated. The ring canals of each successive division are shaded differently. (From Koch *et al.*, 1967.)

endomitosis. The ploidy can be determined by labeling with H^3-thymidine as was done, for instance, in *Drosophila* (King and Burnett, 1959; Jacob and Sirlin, 1959), in *Cochliomyia hominivorax* (LaChance and Burns, 1963), and in *Panorpa* (Ramamurty, 1964). The ploidy may be $2^9 = 512n$ or higher. Nurse cells are rich in RNA which appears to flow into the oocyte through the ring canals before and during early vitellogenesis. In this connection, the existence of ring canals is indeed of great significance since they allow an unrestricted flow of material between the cells.

2. VITELLOGENESIS

The study of vitellogenesis, i.e. the formation and incorporation of yolk into the growing oocytes, has become one of the most dynamic research fields in insect physiology in recent years. This is illustrated by the fact that, although Bonhag had reviewed the material in 1958, another review soon became necessary (Telfer, 1965). The use of the electron microscope and labeling techniques has greatly advanced our knowledge about mechanisms of both protein uptake and its synthesis by the ovarian and extraovarian tissues.

The majority of investigations were concerned with the formation of proteinaceous yolk in the oocyte, yet a few recent findings concerning other components deserve mention here. Engels and Drescher (1964) showed that labeled carbohydrates are found in the oocyte within 3 min of application of H^3-glucose in *Apis mellifera*. The rapid uptake of the label suggests that it is the oocyte itself which is involved in carbohydrate synthesis rather than the other suspected tissues. Furthermore, simultaneous application of H^3-histidine in *Apis* (Engels, 1966) and *Musca* (Engels and Bier, 1967) at various stages of vitellogenesis revealed that carbohydrate incorporation occurred primarily during the later stages of oocyte maturation as protein yolk synthesis gradually ceased. It appeared that protein and glycogen synthesis did not occur simultaneously (Engels, 1966; Engels and Bier, 1967). This finding may not apply to other species, however.

Little is known on the secretion of lipids into the growing oocytes. Mature eggs of many species contain large amounts of lipids but relatively small quantities of carbohydrates. It is suspected that the Golgi elements may play a role in lipid synthesis. However, the nurse cells also seem to contribute to the lipid pool of the oocyte, as seen in *Drosophila* (Falk and King, 1964). Control mechanisms for lipid yolk formation in the oocytes are not well understood in any species. It is, however, suggested that the hormone of the corpora allata is involved in some way, at least in the cockroach *Leucophaea* (Gilbert, 1967).

As might be expected from the structural differences of ovarioles among the different species, several sites and mechanisms of yolk synthesis and incorporation are found: oocyte, follicular epithelium, nurse cells and extraovarian tissues participate. In any one species all of these tissues may be functional in yolk deposition. In some of the species one or the other may be dominant, but in none of the species do we reliably know about the relative importance of the various routes in yolk deposition.

In the following discussion, the role of the various tissues in the formation of proteinaceous yolk will be dealt with. Several observations have suggested the possibility that the oocyte and the germ nucleus participate in yolk formation in some species. For example, follicles of panoistic ovarioles do not obtain nutrients from special nutritive tissues except, perhaps, from the follicular epithelium. It is interesting to note that oocyte nuclei of this type of ovariole contain giant lampbrush chromosomes, which are never found in either telotrophic or polytrophic ovarioles (see Telfer, 1965). These oocytes also contain several nucleoli. The lampbrush chromosomes (Fig. 17) and the nucleoli appear to be involved in a high rate of RNA synthesis—as is exemplified by *Locusta migratoria*—thus suggesting their involvement in yolk protein synthesis (Kunz, 1966; Bier *et al.*, 1967). Within a few minutes after the injection of H^3-uridine in *Blattella germanica* (Zalokar, 1960) and *Gryllus bimaculatus* (Favard Séréno and Durand, 1963), label appeared in the oocyte nucleoli again suggesting that the oocyte nucleus makes nucleolar RNA. These examples are indeed suggestive of the participation of the germ nucleus in protein synthesis but do not constitute conclusive proof.

Similarly, nucleolar buddings and their transfer to the ooplasm must be cautiously interpreted. These nucleolar extrusions were observed in certain saw-flies (Gresson, 1929), *Periplaneta* (Bonhag, 1959; Anderson, 1964), and *Blatta orientalis* (Gresson and Threadgold,

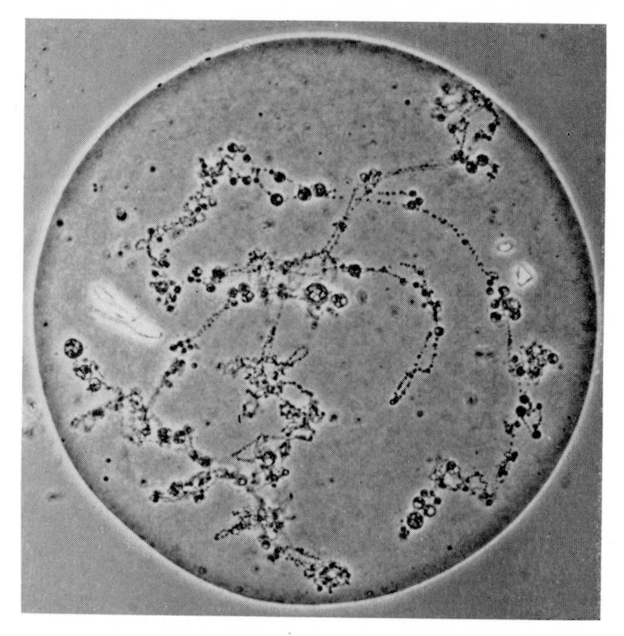

FIG. 17. Isolated unfixed nucleus of the oocyte of *Locusta migratoria*. Loop-like structures appear on the chromosomes causing them to appear similar to amphibian lampbrush chromosomes. It can be shown that H^3-uridine is incorporated particularly into the loops. Phase contrast. (Courtesy of W. Kunz.)

1962). Interestingly, the nucleolar buds of *Periplaneta* and *Blatta* are rich in RNA; they apparently do not give rise to yolk proteins *per se*, but it may not be unreasonable to assume that this RNA is associated with yolk protein synthesis.

Further evidence for protein synthesis in the oocyte comes from recent studies in *Drosophila*. Here, as in some other species of Diptera, most of the ribosomal RNA in the oocytes is derived from the nurse cells (see below). During the major growth phase of the oocyte, yolk spheres which are not membrane bound grow in close association with the rough surfaced endoplasmic reticulum (Cummings, 1968). Since these yolk spheres are not membrane bound in contrast to those of the preceding growth phase, one can conclude that they are not derivatives of pinocytotic vesicles. This finding probably indicates that proteins are synthesized on RNA from the nurse cells since no evidence has been obtained for nucleolar RNA synthesis in the germ nucleus.

Two lines of information point to the fact that, in Blattaria, yolk is not exclusively synthesized by the oocyte. First of all, immunological techniques show that in *Leucophaea maderae* several of the blood serum proteins are incorporated into the maturing oocytes (Engelmann, 1966; Engelmann and Penney, 1966). These proteins are synthesized elsewhere and pass through the ovarian sheaths and follicular epithelium as intact molecules. Secondly, and supporting the finding just mentioned, electronmicrographs of the yolk-depositing oocyte in *Periplaneta* show an excessive development of oolemma foldings (Anderson, 1964) which seem to be pinched off later, giving rise to the growing yolk spheres. Evidence from other species—which is given below—supports the view that the small vesicles pinched off from the oolemma contain blood proteins. What portion of the oocyte proteins is made up by the blood proteins and what portion by the porteins synthesized by either the follicular epithelium or the germ nucleus is uncertain in these species.

Probably the first demonstration of the presence of extraovarian proteins in the oocytes of insects was given by Wigglesworth (1943), using a number of blood-sucking Hemiptera. The eggs of *Rhodnius prolixus*, *Triatoma infestans*, *T. brasiliensis*, and *Cimex lecturlarius* contained hemoglobin, which could only be derived from the host blood. Then, convincing immunological evidence for the uptake of extraovarian proteins during egg maturation was brought forth by Telfer (1954) in *Hyalophora cecropia*. He found that all blood proteins had their counterpart in the mature oocytes, a fact which suggests that these proteins are taken up as intact antigens (Telfer, 1954, 1960). Along with the uptake of the blood proteins, injected foreign proteins can also be incorporated into the oocytes. Telfer showed that during oocyte maturation the blood protein concentration, particularly that of the specific female protein, drops. It is interesting to note that this specific female protein is selectively accumulated in the oocytes and reaches a concentration 20 times higher than in the hemolymph: it is transferred to the oocytes against a concentration gradient. The mechanism for the selective uptake of proteins by the oocytes is unknown and remains speculative. Is it that the oocyte membranes specifically absorb this protein, or does the follicular epithelium actively select for certain proteins? Do the oocyte or follicle membranes act as a filter for specific proteins? For a discussion of these aspects, see King and Aggarwal (1965). In *Hyalophora*, as the relative concentration of the female protein in the blood drops, the rate of growth slows down (Fig. 18). This observation suggests an important role of the female protein in egg maturation processes (Telfer and Rutberg, 1960). This hypothesis is, furthermore, supported by the fact that egg maturation is slow and that eggs do not grow to normal size in ovaries that have been transplanted into males, presumably because the concentration of the female protein in males is very low. However, if female blood was injected into the males simultaneously with the implantation of the ovaries, the oocytes grew to normal size (Stay, 1965). Since the early demonstration of blood protein uptake by oocytes in the *Cecropia* silkmoth, the same phenomenon has been shown conclusively by immunological techniques in the hemipteran *Rhodnius* (Coles, 1965a), the blattarian *Leucophaea* (Engelmann and Penney, 1966), and the dipteran *Sarcophaga bullata* (Wilkens, 1967a, b). Further histological evidence points to the possibility that this is true for many other species as well.

What is the exact route of entry for blood proteins into the growing oocytes? Histologically, one finds that during oocyte growth in *H. cecropia* (King and Aggarwal, 1965), intercellular spaces open up between the follicular epithelium. Sections treated with fluorescein-labeled antibodies clearly and elegantly show how blood proteins are first taken up through these gaps; they are next to be found in the brush border, and then later in the yolk globules of the oocytes (Telfer, 1961; Telfer and Melius, 1963). Thus, it appears that a major portion of the proteins does not pass through the follicular epithelium. In the cortex of the oocytes of *H. cecropia*, the oolemma is thrown into innumerable folds and pits (Fig. 19) which may contain ferritin after injection of the animal with these molecules. Small vesicles in the cortex have the same content as the pits, indicating that these spheres are pinched-off oolemma pits (Stay, 1965; King and Aggarwal, 1965). It is thus obvious that the oocytes obtain extraovarian proteins which may be both secreted elsewhere by the animal and derived from foods via this process of pinocytosis. Pinocytosis in an insect oocyte was first shown by Roth and Porter (1962, 1964) in *Aedes aegypti* (Fig. 20). They also showed that the small pinched-off vesicles appear to coalesce further inside the oocyte, thus forming the

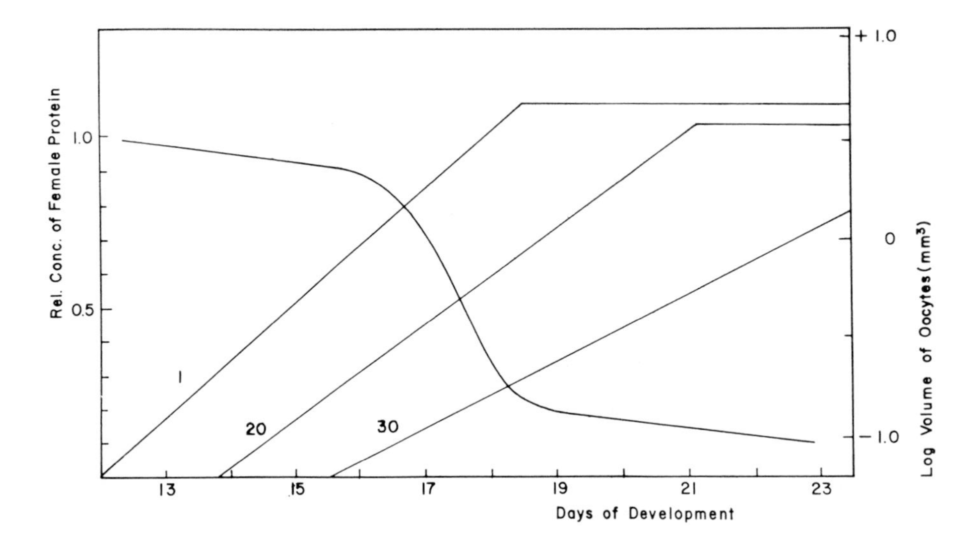

Fig. 18. Changes in the hemolymph concentration of the female protein in relation to the rate of oocyte maturation in *Hyalophora cecropia*. Growth curves of the 1st, 20th, and 30th oocytes. (Modified from Telfer and Rutberg, 1960.)

large yolk globules. Pinocytosis was also observed in the maturing oocytes of *Periplaneta* and *Rhodnius* (Anderson, 1964) as well as *Lygaeus Kalmii* (Kessel and Beams, 1963). In this context, it may be suggested that the permeability of the follicle to vital dyes in *Blattella* (Iwanoff and Mestcherskaja, 1935) and *H. cecropia* (Telfer and Anderson, 1968) is related to pinocytotic processes.

From these findings, one may speculate just how much the pinocytosis of extraovarian proteins contributes to the maturation of the oocytes. Interestingly, it has been found in *Aedes* that follicle and nurse cells do not contain the cytological machinery usually associated with protein synthesis, such as an extensive rough-surface endoplasmic reticulum and a large Golgi apparatus (Roth and Porter, 1964). Numerous ribosomes, however, are found. These observations may suggest, at least for *Aedes*, that proteins are not synthesized by the ovarian follicle cells. This may not be so in other species since in *Calliphora* (Bier, 1962) and *Musca* (Bier, 1963a) (Fig. 21) H^3-histidine is found in epithelial cells within a short time after injection and later in the ooplasm. This indeed suggests participation of the epithelium in yolk formation. Once the label reaches the cortex of the oocyte, one is, of course, uncertain whether the label is within extraovarian or ovarian proteins. In these species, however, it appears that follicle cells secrete proteins prior to the formation of the chorion. The same seems to apply for *Rhodnius* (Vanderberg, 1963).

As shown by labeling techniques, follicular epithelial cells of a variety of species are rich in RNA. This is demonstrated, for example, in the orthopteran *Gryllus bimaculatus* (Favard-Séréno and Durand, 1963), the dipterans *Calliphora* and *Musca* (Bier, 1963a), and the mecopteran *Panorpa communis* (Ramamurty, 1963). It is possible that the high rate of RNA synthesis in these cells is associated with yolk protein synthesis, particularly when this can be shown during the early phases of egg maturation. On the other hand, it has also been

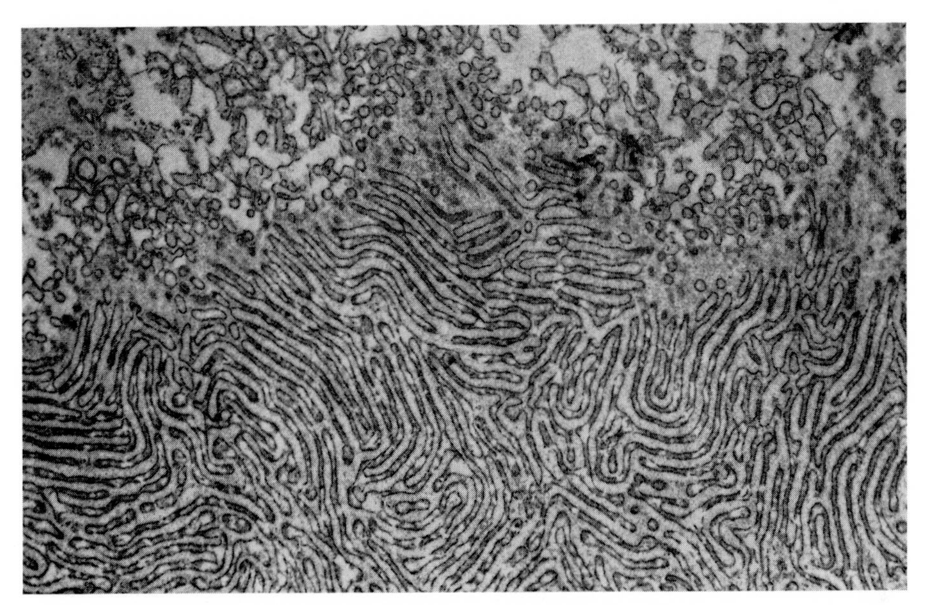

FIG. 19. Tangential section of the surface of the oocyte in *Hyalophora cecropia*. The oolemma is thrown into innumerable folds which are functional in the pinocytotic uptake of blood proteins. (Courtesy of B. Stay.)

shown on the basis of electronmicrographs that follicle cells, at least in several Diptera (King and Koch, 1963; Cummings, 1968), are associated with the formation of the vitelline membrane. This membrane is later thrown into innumerable folds. Is it then possible that follicle cells have a dual function in certain species? Certainly, at the present time, we cannot separate the function of follicle cells in membrane formation from that in the production of yolk precursors. Until we have data from many more species, the different species cannot be compared in this respect; it has been shown repeatedly that the various species of insects have rather diverse mechanisms governing egg growth.

The role of the nutritive tissues in polytrophic and telotrophic ovarioles during egg maturation undoubtedly is complex and may differ in different species. In telotrophic ovarioles of *Oncopeltus* (Bonhag, 1955) and *Rhodnius* (Vanderberg, 1963), for example, a flow of RNA and other materials is observed in the nutritive cord, particularly during the early phases of egg maturation. This nutritive cord breaks down in the later stages of maturation; therefore, at this point the germinal nutritive tissues can no longer play a role.

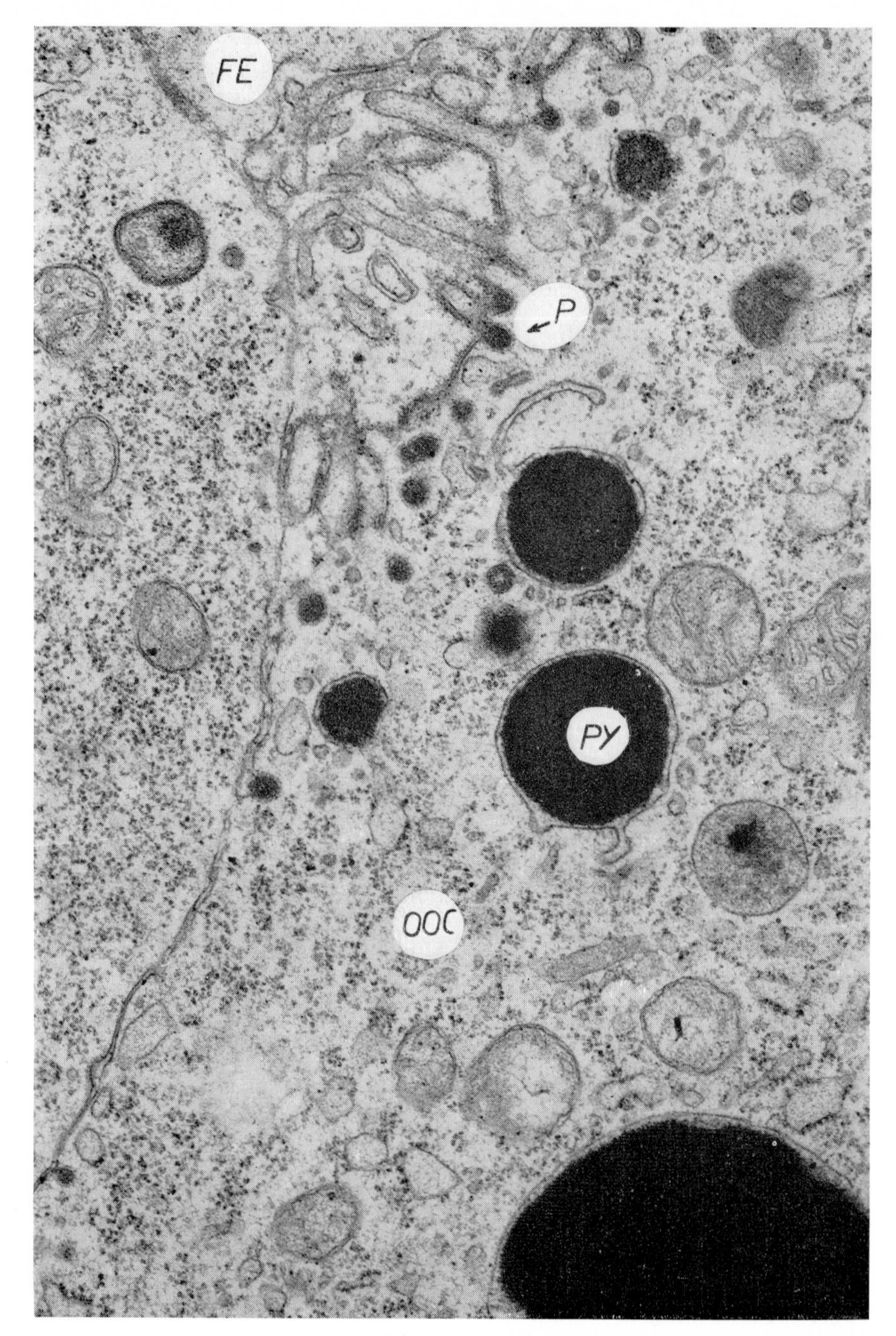

FIG. 20. Section through the interface between follicular epithelium (FE) and oocyte (OOC) in *Aedes aegypti*. Pits (P) along the oolemma are filled with dense material similar to that in the protein yolk globule (PY). (Courtesy of T. F. Roth.)

The bulk of the oocyte proteins is apparently derived from both follicular epithelium and extraovarian tissues in these species. As is apparent from these and other reports, nutritive tissues seem to supply mostly RNA to the oocytes both in telotrophic and in polytrophic ovarioles. For example, in the ovariole of *Panorpa*, the nurse cells contain large amounts of RNA which reach the oocyte via the ring canal; but, as yolk deposition begins, the nurse cells break down; they do not supply yolk proteins (Ramamurty, 1963, 1964). In several Carabidae and Hymenoptera likewise, the nurse cells do not seem to supply protein yolk to the oocytes; rather a flow of RNA is seen going into the oocytes from the nurse cells throughout the growth period of the egg (Bier, 1965). On the other hand, dipteran oocytes apparently

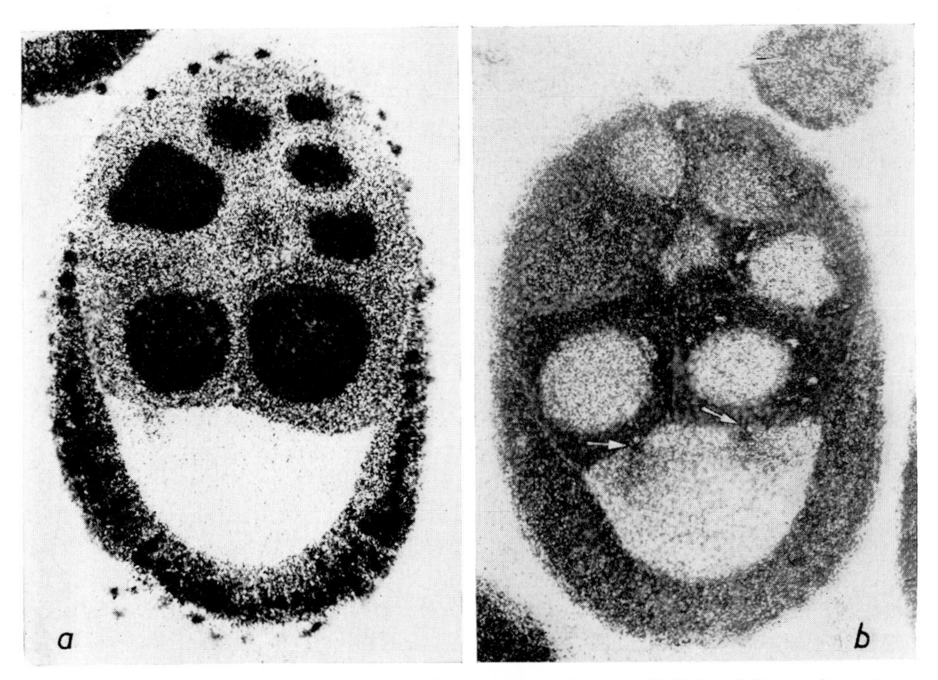

FIG. 21. Autoradiography of sections through stage three of ovarian follicles of *Musca domestica*. (a) Incubation with H^3-cytidine for 1 hr; the nuclei of the nurse cells and the follicular epithelium are heavily labeled. (b) Incubation with H^3-cytidine for 5 hr; the cytoplasm of the nurse cells, particularly that of the proximal two, is now labeled. Labeled RNA enters into the oocyte through the ring canals (arrows). (Courtesy of K. Bier.)

derive most of their RNA as well as some proteins and lipids from their nurse cells. In *Drosophila*, for example, precursors of both RNA and proteins are derived from the nurse cells (Hsu, 1952, 1953; Jacob and Sirlin, 1959; King and Falk, 1960; King and Mills, 1962). In an elegant series of experiments on *Musca* and *Calliphora*, Bier (1963a, b, 1964a) showed that a directed flow of RNA proceeds from the more anterior nurse cells through the posterior ones and into the oocytes (Fig. 21). The labeled RNA was first seen in the nuclei of the nurse cells, then in the cytoplasm; later, a distinct stream of it entered the oocyte. Bier (1963a) also showed in *Musca*, that, simultaneously with the RNA, some proteins from the nurse cells enter the oocytes. Electronmicroscopically, it could be shown in *Drosophila* that a stream of ribosomes enters the oocyte during the early growth phase (Cummings, 1968). Ribosomes enter primarily through the ring canal, yet it has also been shown that the cell membranes between the nurse cells and the oocytes have relatively large pores through which nutritive materials could enter the oocyte.

The collective information now available for Diptera shows that oocyte maturation may be facilitated by a number of different mechanisms, all or some of which may be functional in any one species; one pathway may be dominant over the others. In none of the species explored were all these possibilities investigated; therefore, no generalization can be made. Extraovarian proteins have been found in the oocytes of *Sarcophaga* (Wilkens, 1967). Pinocytosis in the oocytes has been observed in *Aedes* (Roth and Porter, 1962, 1964) which may indicate uptake of extraovarian proteins in this species too. RNA and proteins seem to be secreted into the oocyte by the follicular epithelium of *Calliphora* and *Musca* (Bier, 1963a). The RNA probably can be associated with protein synthesis within the oocyte. In *Drosophila* (Cummings, 1968), *Musca*, and *Calliphora* (Bier, 1963a, b, 1964a) most of the RNA from the nurse cells seems to be involved in protein synthesis in the oocyte. This conjecture is substantiated by another observation in a sterile mutant of *Drosophila*, su^2-*Hw*. In this mutant, eggs do not mature beyond stage eight or nine (beginning of yolk deposition) (Klug *et al.*, 1968); the cessation of growth is associated with an abnormal clumping of the chromosomes in the nurse cells. Based on tracer techniques, the hypothesis is advanced that ribosomal biogenesis in this mutant is impaired (instable ribosomal RNA) and that consequently little RNA enters the oocyte where yolk synthesis normally occurs (Klug, 1968). As has been mentioned repeatedly, the nurse cells in polytrophic ovarioles break down at certain stages of oocyte maturation. These breakdown products are incorporated into the oocyte and thus constitute an additional source of oocyte proteins and RNA. As we have just seen, egg maturation processes may be rather diverse even among species of one order, the Diptera; these differences may be even more extreme in species of different orders.

Once the oocyte is fully grown, the follicular epithelium secretes the egg chorion. It is generally agreed that, in all the species studied, the chorion which can be a complex array of layers (see Beament, 1946), is formed by this epithelium. Another egg membrane, the vitelline membrane, is secreted during the early phase of the oocyte growth period. Early studies in *Culex pipiens* (Nath, 1924) and several grasshoppers (Slifer, 1937) caused belief that this membrane is secreted by the oocyte itself. However, as is mentioned above, it has now become apparent in several dipteran species that it is actually secreted by the follicle cells (King and Koch, 1963; King, 1964b).

After the mature egg is ovulated, the follicular epithelium breaks down and degenerates. The degenerating tissues form a plug which stays in the ovariole for some time. This plug usually attains a reddish or yellowish color and was, therefore, termed the corpus luteum by Stein (1847). It ought not to be compared with the corpus luteum of vertebrate ovaries. In some species of the Acrididae, a variable number of eggs which have grown to various submature sizes degenerate while eggs in the neighboring ovarioles continue to mature. These eggs also form so-called corpora lutea (Phipps, 1949, 1966). Oosorption in *Nasonia vitripennis* is associated with aminopeptidase and esterase production in the follicular epithelial cells (King and Richards, 1968). Depending on the species, the corpora lutea disappear from the ovarioles more or less rapidly, although, in some species, one can recognize these plugs even after several eggs have been ovulated. Parous mosquitoes can be reliably separated from the nulliparous animals because corpora lutea stay within the ovarioles for some time (see Detinova, 1968).

CHAPTER 6

MATING

A. Mating Behavior

A vast entomological literature covers the various aspects of courtship in insects, and it is a formidable task to extract the essentials of today's knowledge. Innumerable reports mention details of mating behavior and copula, but in relatively few cases is a full analysis available. It has been shown that tactile, visual, auditory, and chemical senses are employed in the recognition of the sexes. Elaborate behavioral patterns are found in some species whereas, at the other extreme, some animals exhibit no or very little courtship; in these latter cases, the male may mount the female without any prelude. No or little courtship is found particularly among the Coleoptera and Hemiptera (e.g. Hewitt, 1906; Butovitsch, 1939; Kullenberg, 1947). Since the male has to recognize the female, it is possible that subtle behavioral features are perceived by the male. However, males of some species, e.g. *Sarcophaga Meigen*, approach anything that moves, provided it is of a certain size (Thomas, 1950). The mate is only recognized when the male tries to copulate.

Research on mating behavior has concentrated over many years on the types which involved sound emission and perception. Consequently, a particularly large body of literature is available on this aspect. In more recent years, insect species which emit attractants perceived by the opposite sex became rewarding research objects. As the number of reports increase, it appears that sex attractants (pheromones) are to be found in very many species. This is the most common means of sex recognition over long and short distances. Research on other types of mating behavior does not appear to have attracted a large number of physiologists and ethologists. Nevertheless, many fascinating cases have become known. Insects employ many different means to overcome the "behavioral barriers to mating". Female "coyness" is overcome by the persistent courting of the male in some of the species; in others, males may be self-stimulated by their own song. In some cases, appeasement of the female is achieved through extensive courtship by the male. The study of courtship in insects involves the application of many divergent means, i.e. for true understanding, we have to deal with the whole animal as well as with single sensory units. The integrating role of the CNS in perception and response to various stimuli has been studied. The intricate complexity of the animal's behavior is best indicated by the observation that animals are in a labile status and orient to environmental cues only when the intrinsic physiological status has reached a certain condition (Kühn, 1919).

1. COURTSHIP IN APTERYGOTA

As has been pointed out by Schaller (1953, 1954, 1965) and his students, indirect spermatophore transfer among Apterygota appears to link lower Insecta to other Arthropoda behaviorly. Species among the spider-like arthropods and the millipedes exhibit, in ascending phylogenetic order, increasing complexity in behavior concerned with sperm transfer, just as do primitive insects. Elaborate courtship dances, for instance, in scorpions, pedipalps, and chilopods are associated with sperm transfer just as in some apterygotes. A rather primitive behavior pattern is found among the Collembola in *Orchesella villosa* and *Tomocerus vulgaris* (Schaller, 1953; Mayer, 1957). Here the male places stalked spermatophores—sometimes up to 300—on the substratum. The male will deposit a number of spermatophores even if there are no females in the neighborhood. When the female then, by chance, finds a spermatophore, she stands over it, exudes a drop of liquid from the vulva

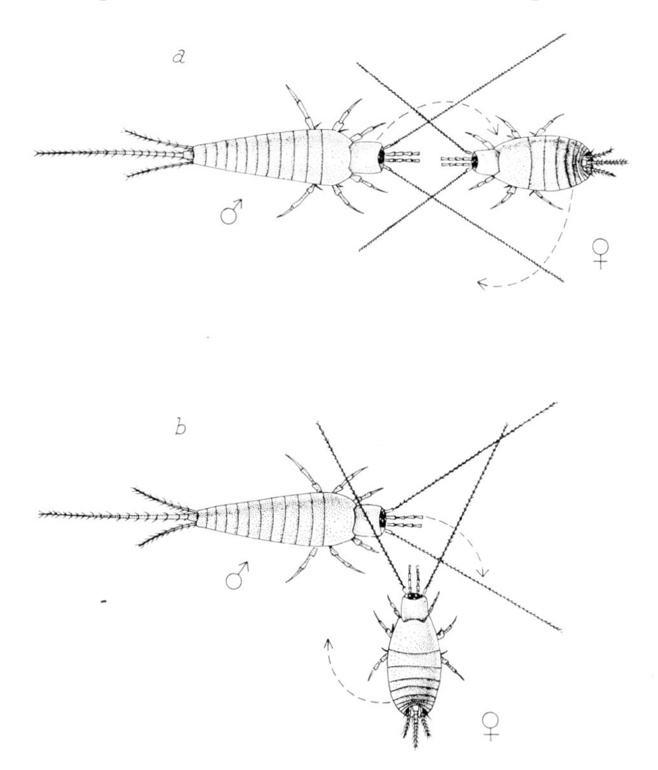

FIG. 22. Two successive phases in the courtship behavior of *Machilis germanica* during the spinning of the thread onto which the spermatophores will later be attached. The male guides the female, causing her to move as the arrows indicate. (From Sturm, 1955.)

which breaks the sperm drop on the stalk; the free spermatozoa apparently migrate into the vulva. This is done without the attendance of the male. A similar situation has been observed in the dipluran *Campodea remyi*, but here the spermatophores are only deposited when a female is present (Schaller, 1954). A further step towards an assured pick-up of the sperm is found in the collembolan *Podura aquatica* (Schliewa and Schaller, 1963) where distinctive courtship behavior occurs. The male, after depositing the spermatophores, pushes the female towards them; she will indeed pick up one or two, after which the pair separates.

Only if both partners are active during courtship does a proper sperm transfer take place in the thysanurans *Thermobia domestica* (Spencer, 1930; Sweetman, 1938; Sahrage, 1953) and *Lepisma saccharina* (Sturm, 1956). In the latter species, the male spins fine threads between the substrate and a vertical structure, deposits the spermatophores on the floor, and then guides the female underneath the threads to pick up the spermatophores. Only tactile stimuli appear to be involved in this elaborate courtship which may last for 30 minutes or longer. Similarly intriguing is the courtship of *Machilis germanica* (Figs. 22, 23) (Sturm, 1955). The courting male first drums on the female with his mandibular palpae. If she is ready to mate, she will lift her abdomen and move towards the male. Then the male spins a thread onto which he deposits three or four spermatophores. While holding this thread, he forces the female towards the spermatophores which she then picks up with her ovipositors. A female may mate several times at daily intervals. It was noticed that females with large eggs in their ovaries responded faster than others to the courting male; this may suggest that, to some extent, the internal status regulates the behavior of the female.

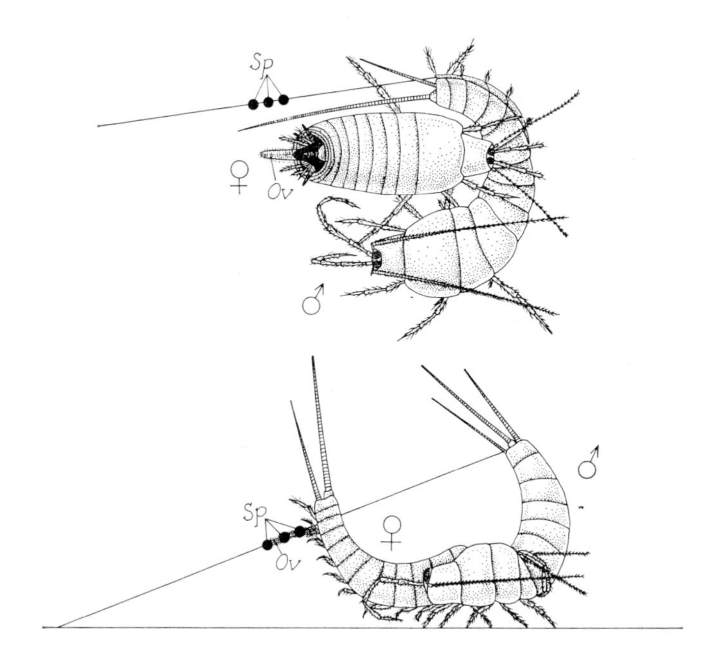

FIG. 23. Advanced courtship in *Machilis germanica*. After the male has placed a few spermatophores on the thread he guides the female, causing her to pick them up with her ovipositor. (From Sturm, 1955.)

Among Apterygota, the symphypleonous collembolan *Sminthurides aquaticus* exhibits probably the highest stage in the evolution of indirect sperm transfer. With his antennae the male of the species grasps those of the female. The heavier female lifts him, may even jump, and carries him around for 2 to 3 days (Falkenhan, 1932; Mayer, 1957). The male may lower his abdomen, deposit a spermatophore, and draw the female over it. After a few days, the male loosens his grip and walks off (Mayer, 1957).

As has been pointed out by Schaller (1965), Apterygota which exhibit indirect sperm transfer are bound to humid environments since, otherwise, the relatively unprotected sperm droplet would dry up within a short time. Species in which the sexes do not engage in mutual stimulation and courtship activities also deposit many more spermatophores than those which have a definite courtship dance; furthermore, these species live in rather dense populations. These features seem to be adaptations to insure that at least some spermatophores are found by the females.

2. VISUAL STIMULI AND SEX RECOGNITION

Many diurnal insect species, as can be expected, use mainly visual cues in finding their mates. Important visual stimuli for attraction of the sexes are movement, color, and form, as can be demonstrated by experiments employing dummies. These stimuli may vary in relative importance according to the species. Indiscriminate attraction to a moving object may be corrected by close range contact chemoreception, as has been shown in some butterflies, or by tactile stimuli, as in Odonata.

The male of the greyling *Eumenis semele*, which approaches flying objects, follows butterfly dummies made of cardboard, particularly when they are grey and less so when they are white (Tinbergen *et al.*, 1942). The response to a dummy on a fishing pole is better if characteristic dancing movements rather than smooth ones are made (Fig. 24). This very short description of behavior in *Eumenis* contains nearly all the essential features to be found in other butterfly species, i.e. attraction to characteristic movements and color. As in *Eumenis*, the male of the nymphalid *Argynnis paphia* follows the female in a characteristic flight pattern (Magnus, 1950), as do *Heliconius erato hydara* (Crane, 1955), *Hypolimnas misippus* (Stride, 1956, 1958a, b), *Papilio dardanus*, *P. deneodocus* (Stride, 1958b), and *Pieris brassicae* (David and Gardiner, 1961).

As important as the movements of the female are her wing color and color pattern, as has been shown in all cases examined. This is particularly apparent in populations of *Pieris napi* and *P. bryoniae*, which occasionally occur in the same habitat. Males of *napi* clearly prefer their white females over both the yellow morph and the females of *bryoniae* which do not have so much white in the wing color. Dummies painted with zinc-white were effective in attracting the *napi* males (Peterson *et al.*, 1951; Peterson and Tenow, 1954). Strangely enough, males of *bryoniae* prefer females of *napi*. Since matings between these latter species were seldom observed, additional isolating mechanisms are probably at work. For *Heliconius erato hydara*, flapping colored wings—the color being close to the natural color-elicited courtship flight (Crane, 1955); colors like green, blue, violet, or black rarely elicited a response from the male. In a series of publications, Magnus (1953, 1955, 1958a, b) showed that in the fritillary butterfly *Argynnis paphia*, dummies painted in the natural colors (golden-yellow or orange) attracted males from a distance; moving wings were more effective than those at rest. Even moving striped cylinders were approached by the males, indicating that the form of the moving object is of minor importance. At close range, however, the dummy must have the odor of the female in order to stimulate continued courtship behavior in the male. Similar observations have been made for *Lemenitis camilla camilla* (Lederer, 1960). It is interesting that in *Hypolimnas misippus* the color of the wings of the courting males elicited the courtship response of the female; males with colorless (scales removed) or painted-over wings were less successful than males with natural wing colors (Stride, 1958b). Obviously, not only does the male respond to color and movements, but the female can also actively respond in courtship behavior.

Males of several species pursue a moving object irrespective of its shape, although there are qualitative differences for the various shapes. For *Eumenis semele*, circles appear to be more attractive than the shape of a butterfly, and dummies larger than natural size are approached more often than are the latter (Tinbergen *et al.*, 1942). The males apparently receive "over-optimal stimuli" from larger dummies. Quite similar observations have been made in

Argynnis (Magnus, 1953, 1955, 1958a, b). Here, both pure colors (golden-yellow, orange) and dummies larger than natural size function as overoptimal stimuli. In *Argynnis*, a graded response to the larger dummies could be obtained (Fig. 25) (Magnus, 1958b).

Just as has been shown in several butterflies, males of *Musca domestica* (Kloboucek, 1913; Vogel, 1954, 1957) and *Sarcophaga carnaria* (Vogel, 1958) also follow flying objects, and will even try to mount ink dots or nail heads. After mounting a female, the male normally moves foreward, "kisses" the female on the head, and then slides back and copulates. However, if the male has mounted a dummy, this full behavioral sequence of movements does not occur. As in *Eumenis* and *Argynnis*, *Musca* and *Sarcophaga* approach oversized dummies more frequently than they do normal sized objects (Vogel, 1957, 1958). Also in these species, the males can apparently receive overoptimal visual stimuli.

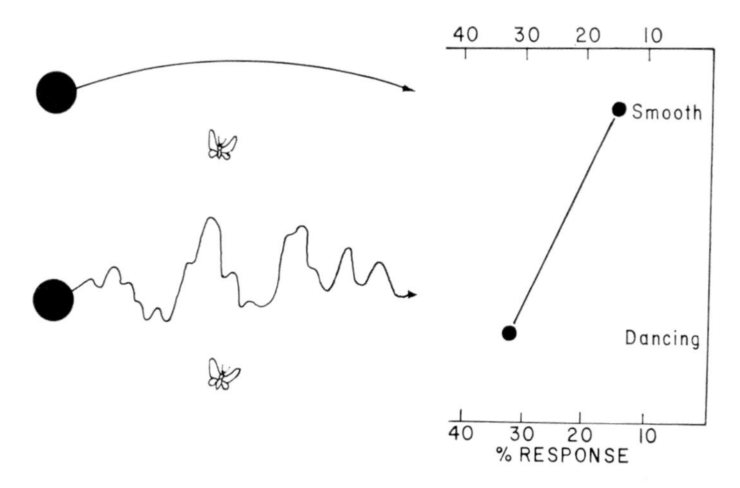

FIG. 24. Response of the male grayling, *Eumenis semele*, to dancing dummies. The male more often pursued those which made irregular dancing motions, the natural flight pattern of the females. (Modified from Tinbergen, 1951.)

As to the physiological significance of overoptimal stimuli, we can only speculate. Since visual acuity in insects is relatively poor—i.e. they are unable to recognize small objects at more than a few meters—large objects can be seen at greater distances. In nature, so-called overoptimal stimuli appear to play no role in the finding of a mate, with one possible exception as is explained below. Males of some species of the Diptera *Empididae* spin balloons which they offer to the courted females during the courtship flight. In certain species, the balloon, which serves as nuptial gift (p. 83), contains prey. These species are diurnal and dance in the sunlight. In many of these, the females appear to approach the dancing males. It may be conjectured that a male carrying a large balloon makes himself more obvious to the female, but whether or not this interpretation is correct has yet to be proven experimentally. Since certain butterflies and Diptera are attracted by overoptimal stimuli, the conclusion that a balloon-carrying male represents an "overoptimal" stimulus may not be far from reality. Here overoptimal stimulation may indeed have a physiological meaning.

In mantids, apparently the males can visually recognize the females at some distance; antennal amputation did not impair the approach to the female (Rau and Rau, 1913). As the male of *Mantis religiosa* recognizes a female, he virtually "freezes", then slowly approaches her (Roeder, 1935); but, as soon as the female moves, he freezes again. This behavior may continue for some time. When the male is close enough, he then jumps onto the back of the female. Some reports mention that the female eats the male during copula; but, as has been stated by Roeder, the male is usually not attacked if he manages to get into the mating position. However, if the female does capture the male, she bites his head off first; severance of the ventral nerve cord posterior to the suboesophageal ganglion then causes the male to make vigorous copulatory movements in searching for the female genitalia (Roeder, 1935; Roeder *et al.*, 1960). Invariably this results in actual copulation.

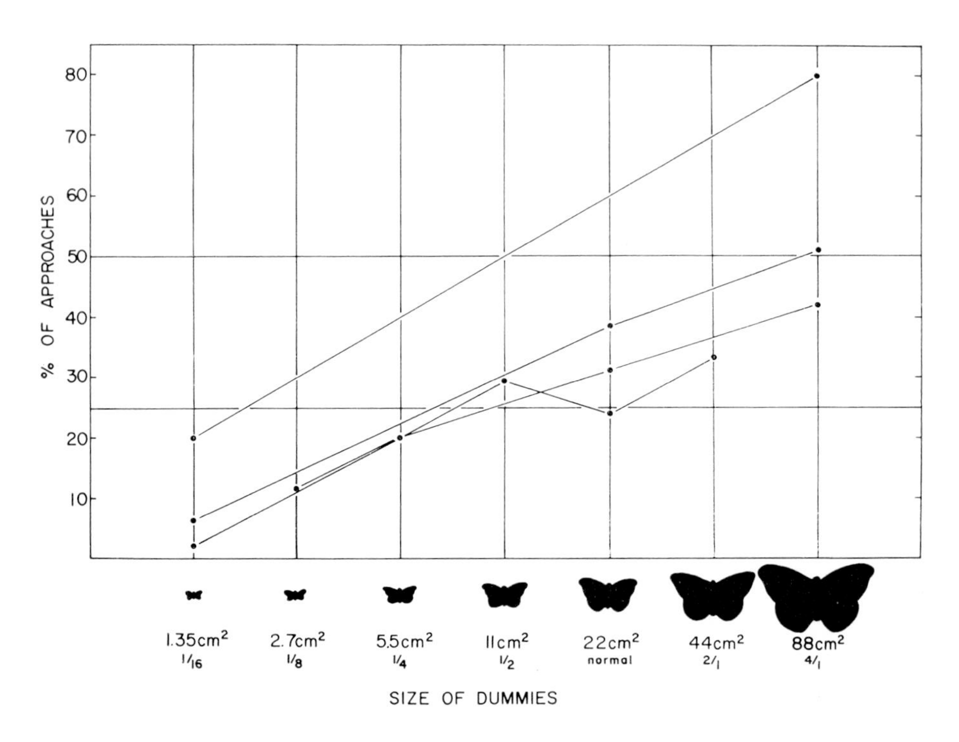

FIG. 25. Approaches of the male fritillary butterfly, *Argynnis paphia*, to non-moving colored dummies of different sizes. In the graph the results of four separate series of observations are given. Note the overoptimal stimulation by larger than natural-sized dummies. (Adapted from Magnus, 1958.)

In Odonata, a typical diurnal species, movement and species coloration are of prime importance in the recognition of the sexes. Distance recognition is visual; but, as soon as contact is made, tactile stimuli are essential for further courtship display. Males of many species establish a territory near the breeding sites (e.g. Jacobs, 1955; Corbet, 1963a) and, if other males intrude, they are attacked. However, intruding females are courted and copulation may follow. Females are presumably recognized as such by their coloration, which in many species is different from that of the male.

To illustrate this point, Buchholtz (1951) used dummies on fishing poles; she found that *Calopteryx splendens* males primarily responded to dummies which had transparent wings with some green in them. Fully painted wings did not elicit courtship except if painted blue-green or yellow-green—colors which are close to that of the female of the species. Courtship in *Calopteryx* consists of a series of characteristic behavior responses from both sexes. Mutual recognition is essential for mating. Certain color patterns on the wings and also the transparency of the wings are important in both females and males. Wing transparency is, for example, essential in order to elicit "proper" courtship in other *Calopteryx* species from Turkey and Syria as well (Buchholtz, 1955). Furthermore, properly painted dummies elicited courtship in *Platycnemis pennipes* and *P. dealbata* (Buchholtz, 1956). It should be pointed out, however, that visual recognition in these latter species appears to be of importance only until the male has grasped the female. Thereafter, removal of the female's head, wings, and even abdomen did not impair the male's further behavior. Color dimorphism of the sexes guides the males of *Orthetrum albistylum speciosum* (Ito, 1960), *Calopteryx maculatum* (Johnson, 1962a, b), and *Erythemis simplicicollis* (Andrew, 1966) in finding the females.

Another fascinating aspect of courtship in insects also involves visual recognition. Probably for as long as man has existed on this earth, he has been intrigued by light-producing animals. This may apply particularly to those species which can turn their lantern on and off. After much speculation over a long time span, it was finally shown that many of the nocturnal Lampiridae use their light-producing organs either to signal their presence or to respond to the light flashes of their mates. For example, the male of *Photinus pyralis* emits flashes at definite intervals while flying over herbage. The female, sitting on grass stems, responds by emitting flashes when the male is at a distance of 5 to 6 meters (McDermott, 1911; Mast, 1912). The male then turns towards the flashing female and some alternate flashing may be observed. Artificial flashes, but not continuous light, attract males just as do the flashes of the females (McDermott, 1911; Mast, 1912). Clearly, distance recognition of the other sex is visual in this species. Males generally do not respond to male flashes; and, as was later confinrmed by Buck (1937), unless the female responds within a definite interval after the male has flashed, he does not react and approach the female. This applies not only to members of the genus *Photinus* (Mast, 1912; Hess, 1920; Buck 1937, 1948) but also to those of *Photuris* and *Pyractomena* (Harvey, 1952; Lloyd, 1964). In some species, for example *Pyractomena dispersa*, the male and female flashes consist of several pulses (Lloyd, 1964). This female does not respond to only a single pulse per flash. In other words, both timing and flash pattern aid in species recognition. Once male and female are close to each other, different stimuli (tactile, contact-chemoreceptory) govern further courtship. Males of the *Pteroptyx* spec., a species which lives in the swamplands south of Bangkok, gather in large numbers in mangrove trees and flash synchronously; it appears that certain trees serve as congregational trees (Buck and Buck, 1966, 1968). It is tempting to assume that synchronous flashing may serve to attract females of the species and thus facilitate the meeting of the sexes. However, no actual observations have been made to verify this speculation.

In another species, *Lampyris noctiluca*, females glow for prolonged periods (Elmhirst, 1912; Schwalb, 1961), as do females of *Phausis spendidula* (Schwalb, 1961). Males flying in the area close in on the glowing females and drop to the ground. Then males of either of these two species find their way on foot to the female. Chemosensors probably guide the male at this close range. Some interesting aspects have been reported and also, in part, experimentally approached by Schwalb (1961). Females of different glowing species have different light patterns which seem to be recognized by the searching males. Light dummies which had the *Lampyris* pattern were the only ones attractive to *Lampyris* males. On the other hand, *Phausis* males were attracted by a variety of light patterns, including that of

Lampyris. Obviously, *Lampyris* males can discriminate better than *Phausis* males. These brief outlines show that in flashing species it is the interval and pulse frequency of the flashes which are important for recognition, whereas in glowing species it may be the light pattern which guides the male.

As has already been stated, visual sex recognition generally operates only over distances of a few meters due to the relatively poor visual acuity of insects. This applies to both diurnal and nocturnal species. Interestingly, visual attraction is usually found in species which live and fly in open areas, like waters (Odonata) or meadows and woods with little underbrush (butterflies); dense vegetation would certainly limit distant recognition by visual cues.

3. AUDITORY STIMULI IN COURTSHIP BEHAVIOR

Emission and perception of sound appear to be superior to vision for distant recognition since vegetation is no obstacle. This means of recognition becomes particularly effective in species which live low on the ground in grasslands. It is not surprising, therefore, that grasshoppers are among the insects which most effectively use sound to attract the sex partner. Entomologists' delight in the variety of insect songs has resulted in a large body of literature. Audition in insects has been excellently treated from several angles during recent years by several researchers active in this field (Ossiannilsson, 1949; Jacobs, 1953; Frings and Frings, 1958; Alexander, 1960, 1961; Haskell, 1961; Dumortier, 1963).

Sound production in insects is most often related to courtship and reproductive behavior; only this aspect will be considered within the present context. Sounds of different quality are emitted by several species of Orthoptera, depending on the environmental and internal situation. In a given species, there may be normal or calling, courtship, and copulation songs. The song may consist of phrases of species specific pulses or of only one pulse, each pulse having its own structure (Fig. 26). In other species, as in some cicadas, sound serves to aggregate the species, both males and females, and thus facilitates encounters between the sexes. In the establishment of territories and dominance hierarchies among crickets, song also seems to play an important role (Alexander, 1961).

I will first turn to the function of song in orientation and courtship because it is the best understood type of song. As early as 1855, Lespés observed that females of *Liogryllus campestris* orient towards the chirping male and approach him. Similar early observations were made in several species of Orthoptera (Hancock, 1905; Jensen, 1909; Houghton, 1909; Regen, 1909, 1910; Fulton, 1915). The first experimental proof, however, that females of *Gryllus campestris* phonotactically find the males was given by Regen (1912) who could attract females with a Galton whistle provided the proper frequency was used. The female does orient towards a hidden microphone which emits the male song (Regen, 1913). Thus, vision and smell as senses used in orientation cannot play a role; even the removal of her antennae did not hinder the female (Regen, 1923). After destruction of the tympanum in a female (*Gryllus*), she no longer approached the male. Likewise, in females of *Gryllulus domesticus* destruction of the tympanum resulted in the partial cessation of response to the courting song of the male; in 27 per cent of the cases females still approached the males (Haskell, 1953). The additional removal of the cerci almost completely abolished her response, indicating that the tympanum and the cerci are hearing organs in this species.

The actual range at which *Gryllus* females respond to the male song is approximately 10 m, far less than the range at which the song is audible to the human ear. This applies, likewise, to *Pholidoptera linerea*, whose range is about 20 m (Baier, 1930), and *Tettigona cantans*, whose females approach the male from distances as far as 30 m (Dumortier, 1963). The attraction of females to males or to loudspeakers emitting the male calling song has also been demonstrated in several additional species, e.g. *Metriopa* (Zippelius, 1949), *Chorthippus bicolor* and *C. biguttulus* (Weih, 1951), *Oecanthus pellucens* (Busnel and Busnel, 1954), and *Acheta pennsylvanicus* (Alexander, 1960).

The calling songs of orthopteran species can be simple successions of oscillations of the sound producing apparatus resulting in trills as in species of the Oecanthinae, or they can be rather complex arrays of pulses, each one in itself consisting of variable length and sound frequency (Alexander, 1960). The tettingoniid *Amblycorpha uhleri* has perhaps the most complex calling song known. Between these extremes are found many variations.

Species	Song Type	Formalised pulse structure (not to scale)	Pulse repetition frequency in c/s	Song characteristics to human ear
Chorthippus brunneus	NORMAL		—	Single pulse at irregular intervals
	COURTSHIP		5—8	Series repeated clicks
	COPULATION		5	Series repeated clicks
Chorthippus parallelus	NORMAL		5	Phrases of 3–15 pulses repeated at varying intervals
	COURTSHIP		—	Single pulse of sound

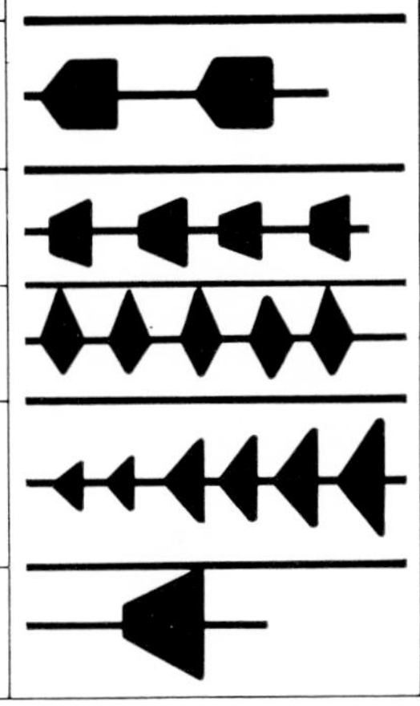

FIG. 26. Diagrammatic representation of the structure of various types of songs in two species of *Chorthippus*. (Modified from Haskell, 1957.)

Sound production by male insects serves as a means of sex recognition and attraction not only in Orthoptera but also in other species where, however, few studies are available. In one example, the water bug *Corixa striata*, the female responds to the male's stridulation by circling on the water surface. Upon perception of the water ripples, the male approaches the source of the water disturbance (he will orient towards anything which makes water ripples) and thus finds the female (Schaller, 1951). In two additional Corixiadae, *Sigara striata* and *Callicorixa praeusta*, similar observations have been made (Finke, 1968). In both of these latter species, the males may stridulate in chorus. Males of the fruit flies *Dacus tryoni* and *D. cacuminatus* strike their wings against a row of bristles on the 3rd abdominal tergite (Monro, 1953), thus producing a faint flute-like sound which attracts the females (Myers, 1952).

As we have seen, the males produce sound to attract their females. However, in some members of the Acrididae, as well as other insects, the females also produce sound. Shortly after the males of *Chorthippus bicolor* or *C. biguttulus* have sung, the responsive female answers with her own song (Weih, 1951). Then the male sings again, and an alternating singing follows while the animals approach each other. Similarly, mutual attraction by sound is found in several other species of *Chorthippus* (Haskell, 1958), *Gomphocerus rufus* (Jacobs, 1950; Loher and Huber, 1966), *Omocestus viridulus*, and *Stenobothrus lineatus* (Haskell, 1958). Females usually begin to sing only after they have been stimulated by the singing male (Jacobs, 1953), and the response of the female obviously further stimulates the male (Alexander, 1960). In *Microcentrum rhombifolium*, however, the female initiates the calling song which then stimulates the male (Alexander, 1960).

Interestingly enough, males of some silent grasshopper species, such as *Locusta migratoria* (Mika, 1959) and *Schistocerca gregaria* (Loher, 1959), may make silent vibratory movements with their hind femora, as if they were singing. Whether this movement is part of the sexual behavior is not entirely clear. It may even be only a visual stimulus to the partner. At this point, one might speculate that sound production in insects evolved from a silent-vibration of the hindlegs or wings (Alexander, 1960).

Mutual attraction by song emitted by both male and female has been found in the homopteran *Calligypora lugubris* (Strübing, 1958) which emits a very faint song hardly audible to the human ear. The beetles *Conotrachelus nenuphar* are also mutually attracted by sound (Mampe and Neunzig, 1966).

So far, we have considered only the so-called normal or calling song. As the term given to this type of song indicates, it serves primarily to signal the presence of the opposite sex and to stimulate the mutual approach of the sexes. Often, after visual or tactile perception in crickets and grasshoppers, for example, this type of song changes to the courtship song (Haskell, 1961; Alexander, 1962). Courtship songs may help overcome the female's resistance in *Chorthippus* spec. (Perdeck, 1957), as well as in other species (see Alexander, 1960), indicated by an initial rejection of the male which after some time, turns into acceptance. In *Oecanthus nigricornis*, if the female attempts to leave after the transfer of the spermatophore, the male trills; this appears to quiet her down (Houghton, 1909). Or, in *Chorthippus bicolor*, the song of the male acts as a kind of self-stimulation (Loher and Broughton, 1954); the song becomes more and more intense as the male approaches the female.

Another aspect of sound emission in insects ought to be considered, namely, the observation that in some cases the males seem to stimulate each other. This is exemplified in the following few cases. Males of the hemipteran *Sehirus bicolor* are rarely heard singing when kept in isolation, yet in groups they do sing frequently (Haskell, 1957a). Several species of cicadas are known to sing in chorus. Sometimes within a few seconds, out of complete silence many animals begin to sing, a phenomenon which can only be explained by the assumption that the animals were mutually stimulated. Chorus singing appears to attract other individuals of the same species, thus facilitating the meeting of the sexes (Poulton, 1921). The chorus song appears to signal the species, and it was observed that it is structurally different in the cicadas *Magicicada septendecim* and *M. cassinii*; in other words, song causes the aggregation of only a given species (Alexander and Moore, 1958). The function of song in the congregating cicadas contrasts, to some extent, with that in the Orthoptera, as these outlines may indicate. In Orthoptera, song serves to space the males and attract the females, whereas in cicadas both sexes are attracted to already grouped individuals (Alexander, 1957).

The sound of the wing beat of the female in some mosquito species is known to attract the males (Kahn *et al.*, 1945). *Aedes aegypti* males can even be attracted to a tuning fork or a loudspeaker emitting a tone of a given frequency (Roth, 1948; Roth and Willis, 1952a). The males were no longer attracted, however, if the flagellum of the antennae was removed, thus indicating that the antennae are the sound perceiving organs. The antennae appear to vibrate with the frequency of the sound emitted by the female (Tischner and Schief, 1955); this then activates the Johnston's organ. It has also been shown that the frequency of the sound of females is approximately 150 c/sec lower than that of the males; males are not attracted to sounds of other males and generally do not approach another male (Tischner and Schief, 1955). In the field, males of *Anopheles albimances* were attracted on a large scale by sound; and sound traps were used, therefore, in an attempt to eradicate the species (Kahn and Offenhauser, 1949).

Males of many species of mosquitoes form swarms near so-called swarm markers, such as trees, open roads, or rocks; many other species do not form swarms. Among the latter species are the crabhole mosquito *Deinocerites cancer* (Downes, 1966). *Opifex fuscus* (Haeger and Provost, 1965), *Atrichopogon pollinivorus* (Downes, 1955b), and several *Clunio* species (Caspers, 1961; Hashimota, 1965). For the swarming species, it has often been assumed that the male swarms attract the females, which are then approached with mating ensuing. It has indeed been observed that some males will leave the swarm and seize approaching females. It was pointed out, however, by Nielsen and Haeger (1960) that the swarming of males is not essential to elicit copulatory behavior. According to their view, females apparently are not attracted to the swarm but only pass by incidentally. Copulation can occur while the females are at rest. Interestingly enough, males of several species of simuliids are visually attracted to flying females or, for that matter, to any flying object passing by the swarm (Wenk, 1965b). All of these species have poorly developed Johnston's organs.

As will be apparent in the chapter on control mechanisms involved in mating behavior (cf. p. 85ff.), only a little is known—and then only in few species—about the nervous control of the animal's behavior, or more specifically, of acoustic behavior. Naturally, sound emission and perception are controlled by the CNS, as is exemplified by males of *Gryllus campestris*. Here, the male normally emits the calling song when a spermatophore is ready to be deposited. If, however, the ventral nerve cord was transected, the male no longer courted (Huber, 1955). Apparently, after transection, the information that a spermatophore is ready no longer reaches the brain where sound emission is controlled. A mated female of *Gomphocerus rufus* will not accept the courting males for some time; but, if the spermathecae are denervated, she becomes ready to mate irrespective of the presence of a spermatophore (Loher and Huber, 1966). The CNS is the integrating and controlling system in acoustic behavior, as has been shown in several additional species of grasshoppers. Upon sound stimulation, recordings from the tympanal nerves in males or females during various reproductive stages in the species of *Chorthippus parallelus*, *C. brunneus*, *Stenobothrus lineatus*, and *Omocestus viridulus* showed no differences in patterns of spike sequences (Haskell, 1956). Depending on the reproductive status females do, however, react differently to sound stimuli. It is therefore clear that information received by the tympana and other receptors (chemo-, tactile, or proprioceptors) is integrated in the CNS: the animal behaves accordingly.

4. PHEROMONES

Communication by chemical means among animals, particularly insects, has aroused the interest of entomologists and physiologists in recent years. The term pheromone was created to replace the somewhat unsatisfactory and self-contradictory word ectohormone (Karlson and Butenandt, 1959; Karlson and Lüscher, 1959). According to the definition, pheromones are chemicals which are secreted to the outside of an individual and which act on another animal of the same species. Olfactorily acting substances function as chemical releasers in the terminology of the ethologists, whereas many pheromones, which are taken in orally, are primers, acting on the endocrine system or in developmental processes (Wilson and Bossert, 1963). However, no clear-cut distinction can be made between the two types of pheromones because, as will be seen, some of the olfactory pheromones may act as primers too. Nevertheless, a rough grouping is of some use.

In contrast to visual and auditory stimuli, chemical stimuli are not perceived by the recipient immediately upon release; a certain delay is inherent in the message transfer. On the other hand, chemicals act over greater distances than do either visual or auditory stimuli. They are probably the most widely used means of sex recognition and attraction in insects, as the steadily growing lists of species in which they had been found illustrates. Several reviews of the subject have appeared in recent years, and the reader is referred to these for a nearly complete list of species which have been investigated in this respect (Wilson and Bossert, 1963; Butler, 1967a; Jacobson, 1965). In the present discussion, emphasis will be placed on the biological implications of pheromone emission and perception.

(a) *Olfactorily acting pheromones*

Probably the first suggestion of sex attraction in insects as mediated by chemicals was given by von Siebold in 1837 for a female butterfly. It has subsequently been found that sex attractants are emitted by either females or males of over 200 species (Jacobson, 1965). The list of species which emit a pheromone has been rapidly growing since 1965. Females which emit a sex attractant have been found among the many moths—classical examples being *Bombyx mori* (Kellog, 1907) and *Porthetria dispar* (Collins and Potts, 1932; Schedl, 1936)—some Dictyoptera, Orthoptera, Coleoptera, Hymenoptera, Diptera, and one species of Homoptera (Tashiro and Chambers, 1967). Some males of species in these orders, as well as of some Mecoptera and one neuropteran species, also emit sex attractants. Cases where the males attract the females are, however, not as frequent as vice versa. In a few species, both males and females emit chemical sex attractants. This has, for example, been shown both in *Galleria mellonella* (Vöhringer, 1934) and in *Ceratitis capitata* (Féron, 1959).

Upon perception of the pheromone, the insect typically shows signs of alertness, most often begins to wave its antennae in slow motion, and increases its locomotor activity. In *Tenebrio molitor*, a sequence of behavioral activities culminates in the male's mounting of the female (Valentine, 1931). At first the antennae are moved in feeble oscillation. The animals then make jerky leg movements and vibrate their palpi; this is followed by energetic antennal movements and a commencement of walking. Similar motions are found in both male and female, but the male is generally the more active partner. Since the above described movements begin before the animals have seen each other or have had contact, one may assume that here both sexes emit an odoriferous sex attractant (Valentine, 1931). In analogous fashion, stimulated *Bombyx* males, as well as males of several other noctuid moths, at first clean their antennae and lift their heads, vibrate their wings in small amplitudes, then, with large amplitudes, run in circles or zig-zags, and finally make oriented runs (Schwinck, 1954, 1955a). Raised antennae and wing vibrations as a prelude to flight can be taken as signs of pheromone stimulation in *Galleria mellonella* (Barth, 1937a; Röller *et al.*, 1963), *Plodia interpunctella* and *Ephestia kühniella* (Schwinck, 1953), and *Trichoplusia ni* (Shorey *et al.*, 1964; Shorey, 1964). Males of *Neodiprion pratti pratti* (Bobb, 1964) and females of *Anthomonus grandis* (Keller *et al.*, 1964) raise their antennae high when they approach the partner. The behavior of cockroaches is similarly quite characteristic of pheromone stimula-

Fig. 27. Courtship in *Leucophaea maderae.* (a) The male raises his wings while he encircles or pushes the female. (b) The female climbs up the male's back, nibbling on his terga. By doing so, she puts herself into the suitable mating position. (c) Immediately after genital contact is made, the female swings around by 180°. The pair stay in this position.

tion. In the males, rapid antennal oscillation signals the perception of the female. This is followed by increased locomotor activity and by searching movements which are accompanied by wing raising and a rocking of the body (Roth and Willis, 1952b, 1954; Wharton *et al.*, 1954a, b) (Fig. 27).

These observations certainly suggests that the antennae bear the chemoreceptors functioning in pheromone perception. Indeed, recordings from an isolated male antenna of *Bombyx mori* which was exposed to the female sex lure showed a typical change in potential, the EAG (Schneider and Hecker, 1956) (Fig. 28). No EAG could be recorded in this species from the female antennae. The amplitude of the typical slow potential change after pheromone exposure was reportedly larger when the antenna was stimulated with high concentrations of the attractant than with low ones (Schneider, 1962). In a later study using various

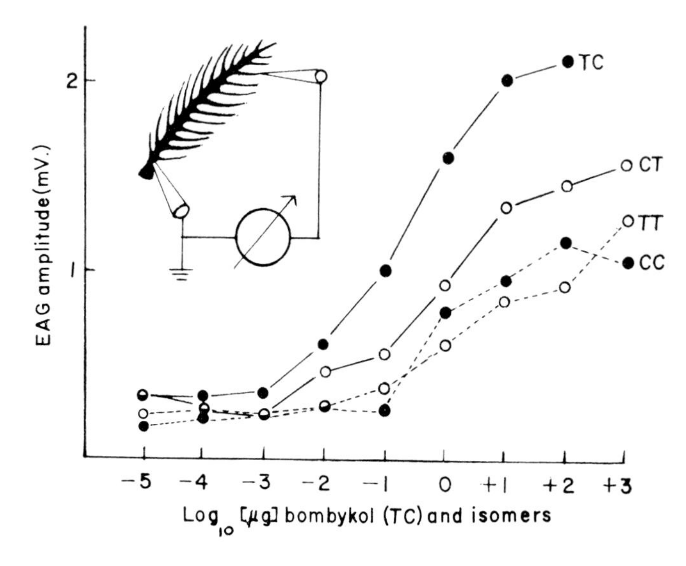

FIG. 28. Recording scheme and electroantennogram (EAG) of males of *Bombyx mori* exposed to various concentrations of Bombykol and its isomers. The EAG amplitudes are the mean amplitudes from between 30 and 200 recordings. (From Schneider, 1966.)

isomers of the synthetic pheromone Bombykol (Schneider, 1966; Schneider *et al.*, 1967), it was shown that the *trans–cis* isomer was the most potent chemical, eliciting an EAG at a concentration of 10^{-3} μg/ml of solvent, which is equivalent to approximately 10^7 molecules per ml of air. This figure is considerably higher than that given in earlier publications from similar observations. It was also found that the other possible isomers are far less potent than the *trans–cis* one, both in the electrophysiological (Fig. 28) as well as in the behavioral studies. Behaviorally, the intact male responds to the pheromone at concentrations of about 2×10^2 molecules per ml of air. An EAG similar to that of *Bombyx* could be obtained in the gypsy moth *Porthetria dispar* (Stürkow, 1965) and in *Periplaneta americana* (Boeckh *et al.*, 1963) upon exposure to the sex lure. In the latter species a typical EAG is also obtained from the female antennae after exposure to her own sex attractant, which is in contrast to the findings in *Bombyx*. Since the female is not attracted to other females, however, additional factors must determine the behavioral response. Recently, in drones of the honeybee separate receptors for the queen substance and the odor from the Nassanov glands have been found on the antennae; these receptors respond with different spike patterns upon exposure to the two pheromones (Kaissling and Renner, 1968). The receptors for the queen substance also respond to caproic acid, but considerably higher concentrations are needed to elicit a response.

That information from the individual chemoreceptors concerning the perception of a sex attractant is summated and integrated in the central nervous system can be deduced from results obtained after partial amputation of antennae in *Bombyx*: as more and more of the antenna was removed, the animal became less and less sensitive (Schwinck, 1955a). Similar observations have been made in several cockroach species (Roth and Barth, 1967). Antennectomy did not, however, completely eliminate mating in *Leucophaea maderae*, presumably because there are also sense organs on the maxillary and mandibular palpi which are functional during mating behavior (Roth and Barth, 1967). In some cockroaches, e.g. *Leucophaea*, distant perception—in which antennae often play a major role—is of relatively minor importance to mating. It is contact chemoreception which is the principle means of sex recognition; therefore, only when practically all chemosensors had been removed did mating fail to occur.

For many years it was believed that insects have an extremely sensitive chemosensory apparatus because individuals could be attracted over great distances. For example, *Porthetria* males travelled up to 16 km to the source of the sex lure, 10–15 virgin females located in a trap (Collins and Potts, 1932). The range of other species does not seem to be great. Only about 8 per cent of marked males of the carpenterworm moth, *Prionoxystus robiniae*, found a baited trap after they had been released about 1.6 km away; while 30 per cent were recaptured of those which were released about 0.8 km from the trap (Solomon and Morris, 1966). Many reports on sex attractants mention that the animals fly upwind: this is only natural, for the attractant is only found downwind of the source. From the origin of the sex attractant is emitted a tunnel-shaped odor trail, along which the attracted individuals fly (Wilson and Bossert, 1963). Wind velocity influences the shape and length of the odor trail. The question then arises how animals are then stimulated to fly upwind over prolonged periods of time. It was conjectured that perception of change in the intensity of the odor as the animal approaches the source stimulates its continued flight. Such a mechanism would demand high sensitivity to minor changes in odor concentration, and it is doubtful that insects have such high acuity.

An alternative mechanism has been proposed by Schwinck (1954) for *Bombyx* and several noctuid moths. *Bombyx* males, when stimulated by the female pheromone, make unoriented runs in zig-zags or circles (Fig. 29) which may continue for prolonged periods in still air as long as the pheromone is present. Eventually the male may run into or literally bump into the female. From this, one may deduce that the pheromone stimulates the male to increased locomotor activity, which may be interpreted as searching. If, however, an air current is passed over the female toward the male, he soon makes orientated movements towards the female (Fig. 29). Odor-free air currents do not elicit a searching activity. It can thus be concluded that the stimulated males are anemotactically attracted to the source of the sex pheromone. This could explain the attraction of individuals over many kilometers in several species of moths, like *Porthetria* and *Orgyia antiqua* (Schwinck, 1954) or *Trichoplusia ni* (Shorey, 1964), as well as other species responsive to pheromones. Travel along an odor gradient apparently does not occur.

Odoriferous glands of varying structure are found in both males and females. They can be

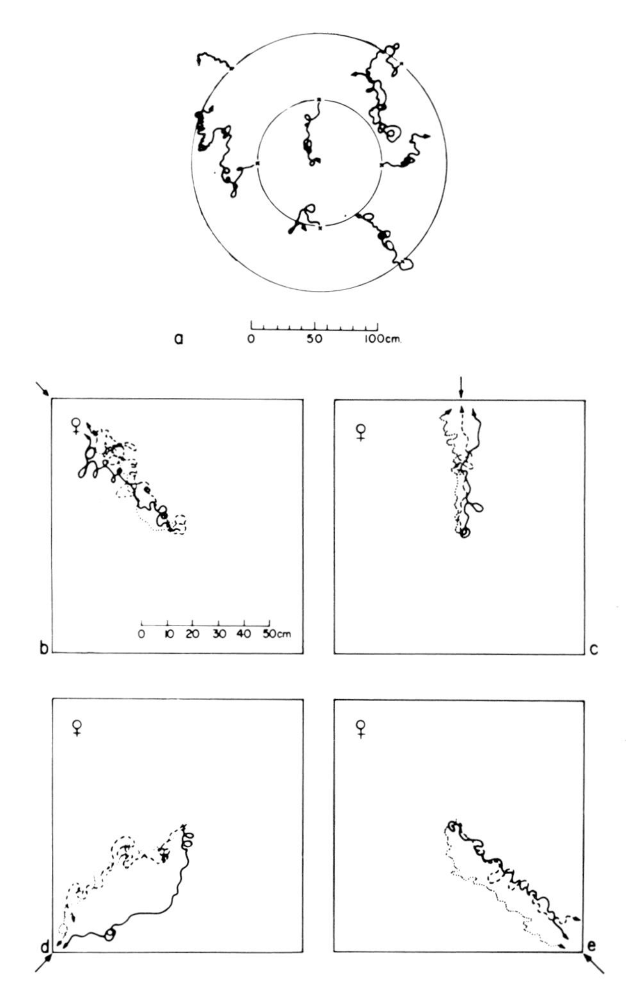

FIG. 29. Anemotactic response of pheromone stimulated males of *Bombyx mori*. (a) Walking paths of stimulated animals without a directed air current (b, c, d, e). Walking paths of stimulated males when an air current was passed over them. Note that the males will even walk away from the pheromone source under these conditions. (Adapted from Schwinck, 1954.)

located on the body, wings, and legs. Because of their variable structures, and presumably also because the emitted scent can in some cases be perceived by humans, the scent glands have interested entomologists in the past. This is apparent from the vast literature (see Richards, 1927; Jacobson, 1965) which is largely descriptive and often contains speculations on the function of those glands. The presence of glandular tissues in the epidermis or the perception of an odor by the investigator does not necessarily prove the presence of an attractant for the opposite sex of the species. Many females of the noctuid moths have scent organs near the tip of the abdomen in the form of saddle-like scent fields as in *Phaera leucephala*, extrudable scent rings as in *Pteroctoma palpinum* or *Colocasia coryli*, dorsal scent folds as in *Porthetria similis* or *Dasychyra pudibunda*, and intersegmental scent sacs as in *Saturnia pavonia* and others (Urbahn, 1913). These scent organs are generally modified

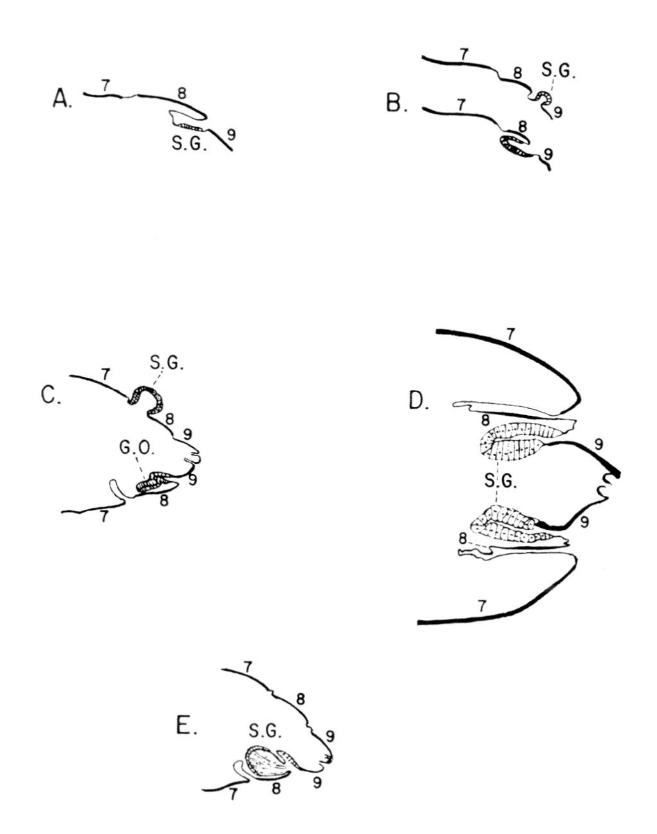

FIG. 30. Location of abdominal intersegmental scent glands in various species of Lepidoptera. (A) Dorsal scent field in *Phalera buccephala*. (B) Extrudable dorsal scent gland in *Cucullia verbasci*. (C) Dorsal and ventral scent glands in *Argynnis adippe* (G.O. = glandula odorifera). (D) Extrudable scent rings in *Plodia interpunctella* or *Heliothis virescens*. (E) Ventral scent sac in *Aglia tau*. (Adapted from Urbahn and Barth.)

intersegmental membranes between the 8th and 9th or 7th and 8th abdominal segments, and may occur dorsally, ventrally, or as rings (Urbahn, 1913; Götz, 1951; Jefferson *et al.*, 1968) (Fig. 30). Extrusion of scent folds, such as the sacculi laterales of *Bombyx* exposes the actual odoriferous surface, making a quick and efficient liberation of the odor possible. Electronmicrographs of the scent cells of these sacculi laterales show that the cell membranes adjacent to the cuticle are thrown into many folds, thus increasing the surface of the cell 60- to 70-fold; the occurrence of these folds may be related to the liberation of the pheromone (Steinbrecht, 1964a; Steinbrecht and Schneider, 1964). The appearance of the folds during late pupal life is correlated with the beginning of pheromone secretion in the cells (Steinbrecht, 1964b). In contrast to those of *Bombyx*, the apices of the scent cells of *Trichoplusia ni* possess distinct villi; each villus has a diameter of about 500–700 Å and is 2–3 μ long (Miller *et al.*, 1967).

Scent glands or scent scales are located on the wings of both male and female *Galleria mellonella* (Barth, 1937a, b; Khalifa, 1950b). Fluttering of the wings presumably aids in dissipating the scent. It is known in this species that the males and females attract each other by emitting a pheromone; but, since mating can occur even after both wings have been removed (Khalifa, 1950b), additional means of sex recognition must exist. This may also be true in the case of the honey bee queen (*Apis mellifera*), who can still be found and mated by the drones even after her mandibular glands which liberate the sex attractant have been removed (Morse *et al.*, 1962).

No glandular tissues which are definitely responsible for the production of a pheromone have as yet been found in the females of several species of cockroaches. In all of those species which do emit pheromones, the hypodermis over the entire body may perhaps secrete the attractant (Wharton and Wharton, 1957; Barth, 1962). Also, in the female mealworm beetle, *Tenebrio molitor*, glands have so far not been identified, yet the female definitely attracts the male chemically (Tschinkel *et al.*, 1967), and a male scent, the origin of which is likewise uncertain, may cause the females to gather in the male's vicinity (August, 1967). This account of scent glands in females is admittedly sketchy; but, from the physiological viewpoint, nothing essentially new could be added by listing all these species in which odor glands have been described.

In the male, odoriferous glands the secretion of which presumably attracts females are found particularly among Lepidoptera. For example, numerous scent scales which emit an odor that supposedly attracts the female are located on the hind legs of the male in *Hepialus hectus* (Deegener, 1902). The location of these scent scales on club-like enlarged hind tibiae is probably unique, and it is unfortunate that no physiological studies have been conducted or reported. The males of the lepidopteran *Opsiphanes invirae isagoras* bear scent scales on the pleura of the 4th and 5th segment, on the tergites of segments 2 to 6, and on the hind wings (Barth, 1952), but again there have been no reports of experiments which might clarify the function of these various scent glands. This is true for several additional lepidopteran species. The males of several Blattaria have tergal glands on an abdominal segment, the particular segment varying according to the species. For example, in *Supella supellectilium* males, these glands are found on the 7th abdominal tergum (Roth, 1952), and in *Leucophaea maderae* on the 2nd. It has frequently been assumed that, in cockroaches, these glands are analogous to the metanotal glands of, for example, *Oceanthus* spec., because, during normal courtship, the female climbs on the back of the male and seemingly feeds on them. However, several other species of cockroaches, e.g. *Periplaneta americana*, do not possess tergal glands; yet the same behavior is observed in the female: she nibbles on the terga of the male and puts herself into the proper mating position. If, in *Leucophaea*, the tergal glands were removed and female terga were grafted to the same place, normal mating behavior was not eliminated in either the male or the female roaches. These operated males successfully copulated several times in succession (Engelmann, unpublished). Thus, tergal glands in the male are apparently not essential for mating.

The male of the dipteran *Ceratitis capitata* exposes glandular tissues by extruding the anus; the females are then attracted (Lhoste and Roche, 1960). Males of *Danaus gilippus berenice* possess a pair of hair pencils at the end of the abdomen, which presumably emit a pheromone (Brower *et al.*, 1965). This has been deduced from observations of courting pairs: the male, in pursuing the female, induces her to land by brushing her anterior end with his hair pencils. Brower *et al.* believe that the pheromone of the male calms the female and induces her to land. Interestingly, a chemical which could be extracted from the hair pencils of another Danaid butterfly, *Lycorea ceres ceres*, has remarkable similarity to the pheromone of the silkmoth *Bombyx mori* (Meinwald *et al.*, 1966). However, no test for its biological activity has been reported so far; therefore, one is left with mere speculation.

Scent brushes have been reported in several additional male Lepidoptera, such as *Phlogophora meticulosa*, *Leucania conigera*, and *L. impura* (Aplin and Birch, 1968).

Lastly, the interesting behavior of the Australian scorpion flies *Harpobittacus australis* and *H. nigriceps* deserves mentioning. The males of these species take a calling position similar to that of some Lepidoptera, extruding two scent vesicles from between the 6th and 7th, and 7th and 8th segments (Fig. 31) (Bornemissza, 1966a, b). Females are attracted to males which have taken this position. The male then offers prey to the female for food during the ensuing copula (see p. 84).

It has long been held that sex attractants are species specific, serving an important role in interspecific isolation. From an evolutionary and behavioral point of view, this is the most

FIG. 31. Extrudable dorsal scent glands in males of the scorpion fly *Harpobittacus australis*. It is observed that females approach only those males which extrude their scent glands. (Courtesy of G. F. Bornemissza.)

probable explanation for their occurrence; if sex attractants were not species specific, a given species would be "led astray" so often, and the chances of mating would be reduced in sympatric species. In *Trichoplusia ni* the release of the species' own pheromone in high concentrations (1×10^{-10} g/l of air) disoriented the males; they then, after a short time, were unable to find their females (Shorey *et al.*, 1967). In the field, pheromone concentrations are normally rather low, and probably only rarely are males of another species attracted.

Generally, traps baited with a given species catch only the males of the same species, as shown in *Cacoecia muriana* (Franz, 1940), and several species of the Tortricidae (Roelofs and Feng, 1968). However, this does not always hold, particularly under laboratory conditions. Attractants prepared from *Porthetria monacha* females attracted 97 per cent males of *P. monacha* but also 3 per cent of *P. dispar* (Görnitz, 1949). The few *dispar* males might have flown into the trap accidentally. Extracts of *P. dispar*, however, lured males of both *dispar* and *monacha* (Görnitz, 1949; Schwinck, 1955b), although, in cage experiments, mating between the two species has not been observed. Males of *Plodia interpunctella* were attracted to females of *Ephestia kühniella* and vice versa (Schwinck, 1953). Whether the sex attractant of the two species is similar or even identical has not yet been shown. Males of *Trichoplusia ni* are attracted to *Autographa californica* female extracts and males of *Heliothis zea* react to extracts of *H. virescens* and vice versa; mating is attempted but is unsuccessful in these cases (Shorey *et al.*, 1965). It was noted that in the latter cases, the species' own pheromone was always more potent than that of another species. *Trichoplusia* extracts are attractive to several other noctuid moths (Berger and Canerday, 1968). No species specificity was noted in electrophysiological studies on several species of moths (Schneider, 1962). A typical EAG could be elicited with extracts from a variety of species in several cases. It was, however,

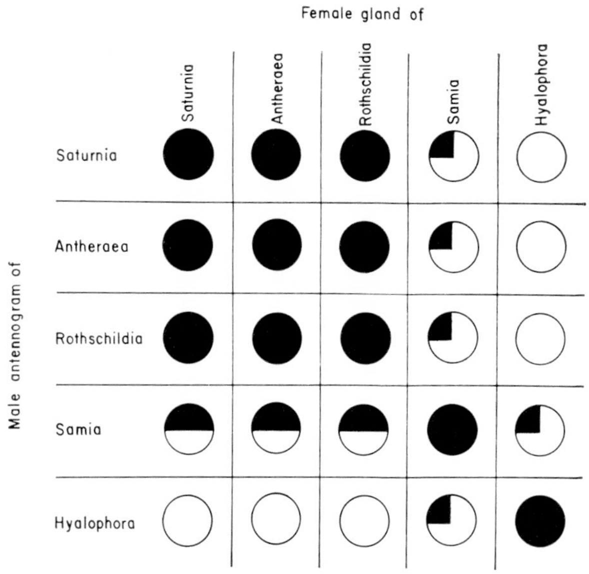

FIG. 32. The reaction (EAG) of males of species of the same Saturniid subfamily to the sex lure of their own and other related females. Full black discs represent the maximal response to the glands of the same species. (Adapted from Schneider, 1966.)

noted that a decreasing response was obtained the less related the two species were (Fig. 32) (Schneider, 1966). Not only did the pheromone receptors respond to the foreign attractant but also the intact animal showed a behavioral response. A further example for interspecific attraction can be given here. Recently, it has been shown that drones of *Apis mellifera* are attracted to ethanol extracts from queens of *A. cerana* subspec. *indica* and *A. florea* (Butler *et al.*, 1967). In drones of *A. mellifera*, an EAG was obtained during exposure to the species' own queen substance and to the mandibular gland secretion from *A. cerana* (Ruttner and Kaissling, 1968). Gas chromatography of these extracts revealed that they probably contained 9-oxodecenoic acid, the sex attractant of *A. mellifera*. From this it appears that in areas where these species coexist interspecific attraction of the males to the other females may occur; however, interspecific mating is not possible because of anatomical differences. Generally these two species are also allopatric.

From these findings one may question the assumption that sex attractants function as isolating means between closely related species. Yet, as actual field observations show, interspecific attraction and mating is rare. Other factors such as differing flight times during the day or different reproductive seasons can account for the rare interspecific attraction in sympatric species (cf. p. 96).

Extracts from females of several species of click beetles, *Limonius californicus*, *Agriotes ferrugineipennis*, *Ctenicera sylvatica*, and *C. destructor*, attract only the respective males; here true species specificity of the lure appears to exist (Lilly and McGinnis, 1965).

The most fascinating aspects of pheromone emission and perception, at least from a biological point of view, are the control mechanisms which govern production, release and perception of the pheromone. These aspects of control and integrated behavior will be discussed below (cf. p. 94).

Attempts to extract, isolate, and purify sex attractants have been made in several species, but to date only a few pheromones have been successfully identified and synthesized. Along with any extraction procedures a bioassay for the detection of the pheromone has to be developed. A few of the typical assays will be mentioned here. As long ago as 1893, females of *Porthetria dispar* were used in New England for the attraction of males in the field. Later, instead of living virgins, extracts were used in large-scale experiments by Collins and Potts (1932). In *Bombyx mori* the males show a sequence of behavioral responses after exposure to the sex lure; a sequence which apparently demands an increase in pheromone concentration. It can be broken off at any point when the concentration is not high enough to elicit the next step (Schwinck, 1954, 1958). This behavioral sequence can be used to estimate the amount of pheromone present. All males of *Trichoplusia ni* responded to ether extracts equivalent to one-tenth of an abdominal tip of a female, but only about 20 per cent responded to 10^{-3} of that extract (Shorey *et al.*, 1964). In an olfactometer (a Y-tube), 1.08 μg of the synthetic sex attractant on filter paper was needed to elicit a response from 50 per cent of the males of this species (Toba *et al.*, 1968). A log-response curve was obtained from the response of *Tenebrio molitor* males to ethanol extracts from the females (Tschinkel *et al.*, 1967). In another species, the sawfly *Diprion similis*, a bioassay was conducted in the field by using extracts prepared from virgins. In this case about 0.004 μg of the purified extract attracted males from a distance of 30–70 m.

The chemical identification of a pheromone demands rigorous procedure and testing since minute impurities may contribute to either its activity or its inactivity. Only synthesis and subsequent biological assay of the compound provides final proof for its identity. The first identified olfactory sex attractant was that of the female silkmoth *Bombyx mori*; it was isolated from 500,000 scent glands. In 1959 Butenandt *et al.*, after more than 20 years of intensive work, reported the identification of the pheromone which they named Bombykol. It is an alcohol of the formula:

$$CH_3(CH_2)_2CH = CHCH = CH(CH_2)_8CH_2OH$$

Butenandt and Hecker (1961) synthesized the four isomers of bombykol and found that in the bioassay only the *trans–cis* isomer had an activity similar to that of the natural compound. The other isomers were considerably less potent. From this Butenandt and Hecker concluded that the natural sex attractant is 10-*trans*-12-*cis*-hexadecadienol. This conclusion has since been confirmed (Eiter *et al.*, 1967). Soon after this success, Jacobson *et al.* (1960, 1961) reported the identification of the female lure of the gypsy moth *Porthetria dispar*. The natural pheromone was named "Gyptol" and has the formula:

$$CH_3(CH_2)_5\underset{\underset{\underset{O}{\|}}{OCCH_3}}{\underset{|}{C}}HCH_2CH = CH(CH_2)_5CH_2OH$$

Again, it is an alcohol, with a remarkable similarity to Bombykol. Synthetic pheromone homologues have been made available and tested in a bioassay. One of these homologues, gyplure, which has two additional CH_2 groups after the double bond, is as attractive as the natural compound (Jacobson and Jones, 1962). This last report crowned the effort of 30 years of research on the gypsy moth sex attractant which began with Collins and Potts in 1932 (Jacobson and Beroza, 1963). Repetition of the synthesis of both gyptol and gyplure in a different laboratory challenged the reports of Jacobson *et al.* Both compounds, when synthesized according to Jacobson, were biologically inactive (Eiter *et al.*, 1967). Possibly minute impurities in the original material accounted for its biological activity.

Gary (1962) and Butler and Fairey (1964) showed that one component which had been isolated earlier from the mandibular glands of the honeybee queens, and which functions in the control of queen rearing by the workers, has sex attractant qualities. Drones are attracted to dummies impregnated with this substance. It is the fatty acid 9-oxodec-*trans*-2-enoic acid:

$$CH_3CO(CH_2)_5CH = CHCOOH$$

One of the additional sex attractants isolated and synthesized recently is that of the cabbage looper, *Trichoplusia ni* (Berger, 1966); it was found to be the alcohol *cis*-7-dodecen-1-ol acetate.

$$CH_3CO_2(CH_2)_6\ CH = CH(CH_2)_3CH_3$$

Its congeners were synthesized but found to be practically inactive (Green *et al.*, 1967). *Pectinophora gossypiella* females emit 10-propyl-*trans*-5,9-tridecadienyl acetate which was given the common name "Propylure" (Jones *et al.*, 1966):

$$\begin{array}{l} CH_3CH_2CH_2 \\ \qquad\quad | \\ \qquad\quad C = CH(CH_2)_2CH = CH(CH_2)_4O\underset{\underset{O}{\|}}{C}CH_3 \\ \qquad\quad | \\ CH_3CH_2CH_2 \end{array}$$

However, the chemical identity of this pheromone has again been questioned by Eiter *et al.* (1967). The carpet beetle, *Attagenus megatoma*, produces *trans*-3,*cis*-5-tetradecadienoic acid (Silverstein *et al.*, 1967):

$$CH_3(CH_2)_7\overset{\overset{H}{|}}{C} = \overset{\overset{H}{|}}{C} - \overset{\overset{H}{|}}{C} = \underset{\underset{H}{|}}{C}CH_2COOCH_3$$

The fall armyworm moth, *Spodoptera frugiperda*, secretes *cis*-9-tetradecen-1-ol acetate (Sekul and Sparks, 1967; Warthen, 1968) and the attractant of the sugar beet wireworm, *Limonius californicus*, has been identified as valeric acid (Jacobson *et al.*, 1968). It has been known for a long time that valeric acid attracts the males of this species, and it indeed appears to be the natural attractant. The attractant of another moth, the leaf roller moth *Argyrotaenia velutinana*, was also isolated, synthesized, and identified as *cis*-11-tetradecenyl-acetate (Roelofs and Arn, 1968). The *trans* isomer of this compound is not attractive to males even in amounts of 100 μg whereas the *cis* isomer attracts them in amounts of 0.1 μg both in the laboratory and in field assays. The frass of several species of bark beetles is attractive to both the male and the female. The attractant of one of these species, *Dendroctonus brevicomis*, has been identified and the trivial name "Brevicomin" was given to it (Silverstein *et al.*, 1968). This compound is attractive in 1 μg quantities and is produced by virgin females. Attractants of other bark beetles have also been isolated.

The female sex attractant of *Periplaneta americana* was isolated by Jacobson *et al.* (1962) and identified as 2,2-dimethyl-3-isopropylidene-cyclopropyl proprionate. Some controversy arose regarding the identification of this substance as the actual sex attractant (Wharton *et al.*, 1963) and it was later shown that the synthetic compound had no attractiveness (Jacobson and Beroza, 1965). The chemical nature of the female pheromone of *Periplaneta americana*, therefore, remains uncertain. Undoubtedly, many pheromones will be chemically identified in the near future. Pure or synthetic pheromones may play an increasing role in surveying the field of harmful pest insects. Thus, research on the chemical nature of these compounds is of more than purely academic interest. However, their function in attempts to eradicate pest insects is of uncertain value since many attempts have failed to yield significant results.

(b) *Contact chemoreception and gustation*

A variety of species are known to offer some kind of secretory product to the partner during courtship or copula. These substances may serve as aphrodisiacs or may pacify the receiver (Butler, 1967b). Glandular structures which may serve this function have been described in butterflies as well as in other insects (Richards, 1927), but experimental evidence is rarely available.

Hancock (1905) described metanotal glands in males of the cricket *Oecanthus fasciatus* (so-called Hancock's glands) on which the female feeds during the prelude to copula. It appeared that the female was attracted to this secretion and, while feeding, put herself into the proper mating position. The alluring secretion functions only in short distance or contact perception; in *Oecanthus*, long-distance recognition and attraction occur by sound emission and perception. These observations were essentially confirmed in *O. fasciatus* (Jensen, 1909) and extended to *O. nigricornis* (Houghton, 1909; Fulton, 1915) and *O. pellucens* (von Engelhardt, 1914; Hohorst, 1936). In *Oecanthus* species, the availability of this secretory material seems essential for normal courtship, because, if the glands were covered, the female was not attracted to mount (Hohorst, 1936), and copula did not take place. In these species, it has also been shown that, in order to continue courtship, the male must receive some kind of stimulus, probably tactile, while the female mounts; if the female does not feed, he chases her away (Hohorst, 1936).

Alluring glands analogous to Hancock's glands have been described in several other orthopteran species in all of which the female feeds on the secretion during courtship. In *Isophya acuminata* two rows of hairs are found on the 1st abdominal tergum which are connected to glands inside (von Engelhardt, 1915). These structures are probably analogous to the metanotal glands of Oecanthidae. This may also be true in *Hapithus agitator* (Alexander and Brown, 1963) or in *Homeogryllus reticulatus* and *H. tessellatus*, two relatively primitive crickets from the Ivory Coast (Leroy, 1964). In another cricket, *Discoptila fragosoi*, which occurs in caves and cellars of the Crimea, Boldyrev (1928b) saw that during copula the female licks a thick fluid secreted on the underside of the tegmina. Yet another extremely interesting case was also reported by Boldyrev (1928a): the female of the Tettigoniid *Bradyporus multi-tuberculatus* mounts the male as in other orthopteran species and feeds on a brownish fluid exuded from pronotal slits on the male. The female seems to rub the ridges bearing these slits until the fluid appears. The brownish fluid may be hemolymph which is also exuded from the slits upon rough handling. Can we speculate that the exuded blood functions analogously to the alluring glands of the other species?

The females of many orthopteran species not known to have such glands mount the male in a manner similar to that of females of species which do possess alluring glands. For example, this has been shown in *Nemobius silvestris*, where the females palpate and lick the abdomen or perhaps chew the wings of the males during mounting (Gerhardt, 1913; Richards, 1952; Gabbutt, 1954). Perhaps, in this case, some substances enticing to the female are secreted by the body surface. In this species, Richards (1952) and Gabbutt (1954) could not find the spines on the hind tibiae of the males on which the females presumably feed during courtship (Fulton, 1931). Licking of the male dorsum by the mounting female has been reported in *Locusta caudata* and *Ephippiger limbata* (Gerhardt, 1914) and may occur in many additional species. Rather unfortunately, there have been practically no reported experiments concerning the alluring glands in orthopterans. Only from observations on the species mentioned can we assume that alluring glands play an important role in courtship.

The functional significance of the occurrence of alluring glands may be to keep the female quiet during the transfer of the spermatophore. As was pointed out by Fulton (1915), the female may be kept on the back of the male longer than she would otherwise stay; this would serve the function of restraining her from eating the spermatophore before the spermatozoa have been emptied. This view has been adopted by many investigators and may particularly apply to the many orthopteran species which do eat the spermatophore after transfer. On the other hand, this view could be challenged on the grounds that the spermatophore often is eaten only long after the female has dismounted or, as in cockroaches, never. Since hardly any experiments are known which involve the removal of the alluring glands, one is left with the interpretation that the feeding or nibbling on the secretion pacifies the female.

The function of tergal glands in a variety of cockroach species is still a matter of debate, since in hardly any case does conclusive experimental evidence show that the females are attracted either long or short distances or that feeding on the glands is essential for normal mating. It is uncertain whether the female is actually attracted to the alluring glands at all or whether she merely feeds on them when she encounters these structures. While in some species, e.g. *Blattella germanica* (Roth and Willis, 1952b) and *Leucophaea maderae* (Roth and Willis, 1954), the female, just as in *Oecanthus*, normally feeds on the tergal glands during courtship, other species have no tergal glands, but the female still exhibits a similar characteristic behavior. Complete removal of the tergal glands in *Leucophaea* and replacement of the terga with female cuticle did not impair normal courtship and mating in either males or females (Engelmann, unpublished). As this result indicates, one should be cautious with sweeping conclusions based on uncontrolled observations (Roth, 1965; Butler, 1967a).

A secretion which may serve as an aphrodisiac is offered by males of several species belonging to the Malacodermata. These beetles possess glandular structures which were first described on the elytral tips of *Axinotarsus pulicarius* (Evers, 1948). Later, it was observed that similar glands, called excitators, are located on the frons of *Malachius bipustulatus* and are offered to the females during courtship (Evers, 1956), although actual induced mating was not seen. Other species of the Malacodermata may have excitators or alluring glands not only on the head or elytra, but also on the antennae, tibiae, palpae, or thorax (Matthes, 1959, 1960, 1962a). No distant attraction has been observed in any of these species; the sexes always meet by chance (Matthes, 1959, 1962a). Once the sexes have recognized each other by antennal contact, the male may turn around by 180°, offering the excitators of his elytra to the female, or may turn his head towards the female, offering the head or antennal excitators (Matthes, 1959, 1962a, b). The females of *Anthocomus fasciatus*, *A. coccinus*, *Malachius marginellus*, and *Axinotarsus pulicarius* were seen to bite into the elytral glands of the males. If the female attempted to leave, the male pursued her and again offered his alluring secretion; while the female fed, the male stayed quiet. Finally after repeated offerings of the alluring glands, the male may mount and copula takes place. A series of photographs illustrates the typical sequence of behavior (Fig. 33). One may conjecture that feeding on the secretion of the alluring glands may help to calm the female and to enhance her readiness to accept the male.

Nothing is known about the chemical nature of aphrodisiacs in any species. In one instance, the honeybee *Apis mellifera*, it is at least plausible that the sex attractant of the queen may also serve as an aphrodisiac (Butler, 1967b). Apparently, only after the drone has come into actual contact with the sex attractant does he mount.

5. TACTILE STIMULI

From the numerous descriptions of mating behavior in many insect species, it is obvious that tactile stimuli play an essential part in the successful completion of courtship. There is, of course, often no way to distinguish between contact chemosensory perception and involvement of tactile stimuli since both mechanoreceptors and chemoreceptors are found on the antennae, palpi and tarsi, the organs with which the animals explore their partners. A few examples, however, may unequivocally illustrate the role of mechanoreception during courting in either the male or the female.

For instance, it has been observed by Gerhardt (1913) that a male of *Liogryllus campestris* in the "excited" phase of courting may extrude a spermatophore upon being touched on the abdomen; normally, the female mounts the male and may thus elicit the delivery of a spermatophore. Stroking the abdomen of a male of *Gryllus campestris* with a brush caused him to make copulatory movements (Huber, 1955). In this and the related species, *Acheta pennsylvanicus*, the male antennates the female after mating; this apparently keeps her quiet and may prevent her from eating the spermatophore before the spermatozoa have been delivered (Alexander, 1961). The same observations have been made in *Gryllus domesticus* (Khalifa, 1950a).

In many species of cockroaches, the male raises his wings during courtship, which appears almost like an invitation to the courted female to climb to his back. Invariably, a receptive female will slowly climb up the dorsum of the male, continuously licking his terga and finally reaching a position which allows intromission (Roth and Willis, 1952b, 1954; Roth and Barth, 1967). The behavior of a responsive female cockroach resembles that of a tree cricket during the prelude to copula. Interestingly, if an excited male is touched on the abdomen, he may make violent copulatory movements, even extrude his penis in an attempt to copulate. Apparently, purely tactile stimuli are part of the normal sequence of courtship movements, particularly during the final phase.

FIG. 33. Courtship behavior of two species of Malachiidae. (a) *Axinotarsus pulicarius*. The male and female meet frontally. After antennal contact is made, he turns and offers his elytral excitators (b) into which the female bites. (c) *Malachius bipustulatus*. The male offers his frontal excitators on which the female feeds. (Courtesy of D. Matthes.)

During copula or immediately before, males of many species stroke the female with their antennae, seemingly quieting and reducing the mobility of the female. This has been shown, for example, in the hog louse *Haematosiphon suis* (Florence, 1921), in *Cassida viridis* (Engel, 1935), *Antestia* spec. (Fiedler, 1950), and *Saperda populnea* and other Laminiidae (Funke, 1957). Palpation of the female by the male in several species of *Pyrota* appears to quiet her down and to allow mating (Selander, 1964), but only receptive females will be quieted. Restless females of *Androchirus erythropus* are pacified during copula by a stroking of the head, thorax, and elytra (Campbell, 1966).

Tactile stimuli of various qualities may be essential in courting, as can be shown in some additional examples. Males of longhorned beetles lick, palpate, or tap the female's back during mating. In *Strangalia maculata*, the male even bites into the antennae of the female and pulls her head backward in a violent movement before the actual copula is accomplished (Michelsen, 1963, 1965). Biting is an essential part of courtship in this species; if biting is not permitted, mating does not take place.

In various species, a stimulus-response sequence is found during courtship; however, very few details are known in most cases. A sequence of mutual activities can be broken off if the next stimulus does not follow, as can be seen in courtship patterns of various kinds. This also applies to the types of courtship where largely tactile stimuli are involved, as, for instance, in *Tipula oleracea* (Stich, 1963). In this species, at the beginning of courtship the male grasps the leg of the female; this behavior has a certain filtering effect in the ethological sense since, if this is not permitted, courting by the male does not continue. Quite certainly, the peculiar grasping of the female's leg involves primarily tactile stimuli. In several plecopteran species, such as *Capnia bifrons*, *Isoperla grammatica*, *Dinocras cephalotes*, and *Perla marginata*, both the male and the female drum on the substrate with their abdomen during courtship. Generally, the male drums first, which action is followed by the female's response (Rupprecht, 1968). The vibration of the substrate is perceived by the subgenual organs of the legs and elicits a directed response, i.e. the animals orient towards each other. Solely mechanoreception serves in orientation of the animals.

6. NUPTIAL GIFTS

One of the most fascinating aspects of the mating behavior of insects, but one little studied physiologically, is the exchange of prey or food before or during copula. The phenomenon of food presentation by males has been observed among several species of Diptera, in one hemipteran, and in two genera of Mecoptera. It is amazing that the observations have been largely descriptive and that essentially no experimental studies have been published. It would be a rewarding endeavor to analyze the behavior of nuptial gift exchange on a comparative basis since several species known to exhibit this behavior are easily available in many parts of the world. The case of the Empididae, members of which occur both in North America and Europe, will be treated first. In 1877, Osten Sacken saw a *Hilara* spec. that carried a balloon made of silk-like threads which was twice the size of the insect itself. Soon thereafter, naturalists became interested and reported similar observations for several species including *Hilara satrix* (Handlirsch, 1889), *H. sartor* (Mik, 1894), *Empis poplitea* (Aldrich and Turley, 1899), and *Hilara maura* (Eltringham, 1928). It soon became apparent that, in each case, it was the male who was carrying a balloon which often contained prey. During copula, the balloon is transferred to the female and she often eats the prey. It also became apparent that some species of the same group of Diptera do not make balloons but rather offer the prey to the mate without a balloon. In others, the balloon is made but contains no prey. It is tempting to speculate on the evolution of balloon-making (Poulton, 1913; Kessel, 1955) in Empididae and, in so doing, I shall largely follow the discussions of the quoted authorities. The first step may be the simple exchange of food (prey). This is found, for instance, in *Empis livida*, *E. tessellata*, *E. opaca* (Hamm, 1908, 1909a, b) and *Pachymeria femorata* (Hamm, 1909b). Next, a balloon is made and the prey enclosed in it is eaten by the female during copula. Species where this behavior is found are *Empis poplitea* (Aldrich and Turley, 1899), *E. borealis* (Howlett, 1907), *E. bullifera* (Kessel and Kessel, 1951), and *Hilara wheeleri* (Kessel, 1959). From this stage, balloon making seems to have become some sort of ritual. Even when the prey within the balloon is inedible, the balloon is transferred just as though there were still edible prey within. The balloon of *Empimorpha geneatis* contains a prey too small to be of any food value (Kessel and Karabinos, 1947). Those of *Hilara maura* (Eltringham, 1928) and of *H. pilosa* (Trehen, 1965) contain inedible objects such as seeds, pieces of wood, etc. Probably the highest degrees in evolution of balloon making toward a completely ritualized transfer are to be seen in *Hilara sartrix* (Handlirsch, 1889), *H. sartor* (Mik, 1894), and *H. granditarsis* (Melander, 1940). All of these species make a balloon which contains no prey. As has been pointed out elsewhere (cf. p. 60), balloon carrying may serve as an "advertisement" for the male,

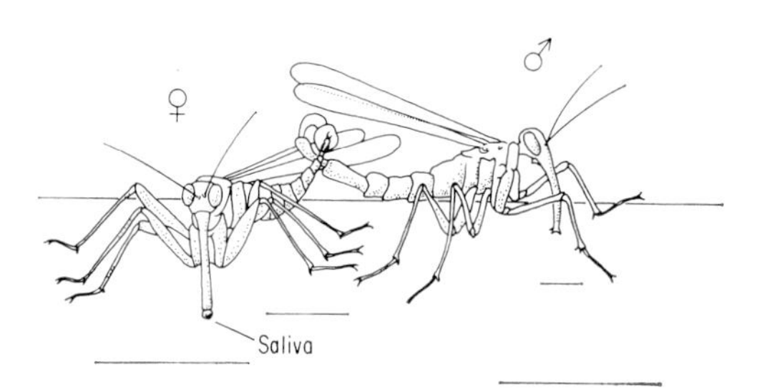

FIG. 34. Position of male and female *Panorpa communis* during copula. The female feeds on the drop of saliva during copula. (From Mercier, 1914.)

because the large object can be seen well by females, which were indeed observed approaching the males. The balloon may serve as an "overoptimal stimulus". Such a thought is strengthened by the finding that *Platyperina pacifica*, which does not make balloons, flies about with its unusually long legs hanging down (Kessel and Kessel, 1961) and thus creates the impression of a large insect.

The males of the mecopteran *Harpobittacus nigriceps* and *H. australis* offer prey on which the females feed during copula (Bornemissza, 1964). Possibly the females become acquiescent through such a feeding; this may be essential for the period of sperm transfer. This could also be true for the several species in which the male produces a liquid droplet (saliva) on which the female feeds during copula, as, for instance, in the fly *Rivella boscii* or in *Panorpa* spec., among them *communis* (Fig. 34) (Mercier, 1914; Steiner, 1930). As a matter of fact, males of *Panorpa* mate only when the salivary glands are well developed, i.e. approximately on the 3rd day after emergence (Mercier, 1920).

Among Hemiptera, one species, *Stilbocoris natalensis*, has been found in which the male offers the female a seed as a nuptial gift. In an elaborate ritual, the male transfers the seed which he has injected with saliva to the courted female (Fig. 35) (Carayon, 1964). The injected saliva apparently predigests part of the seed. Males which do not have any seeds

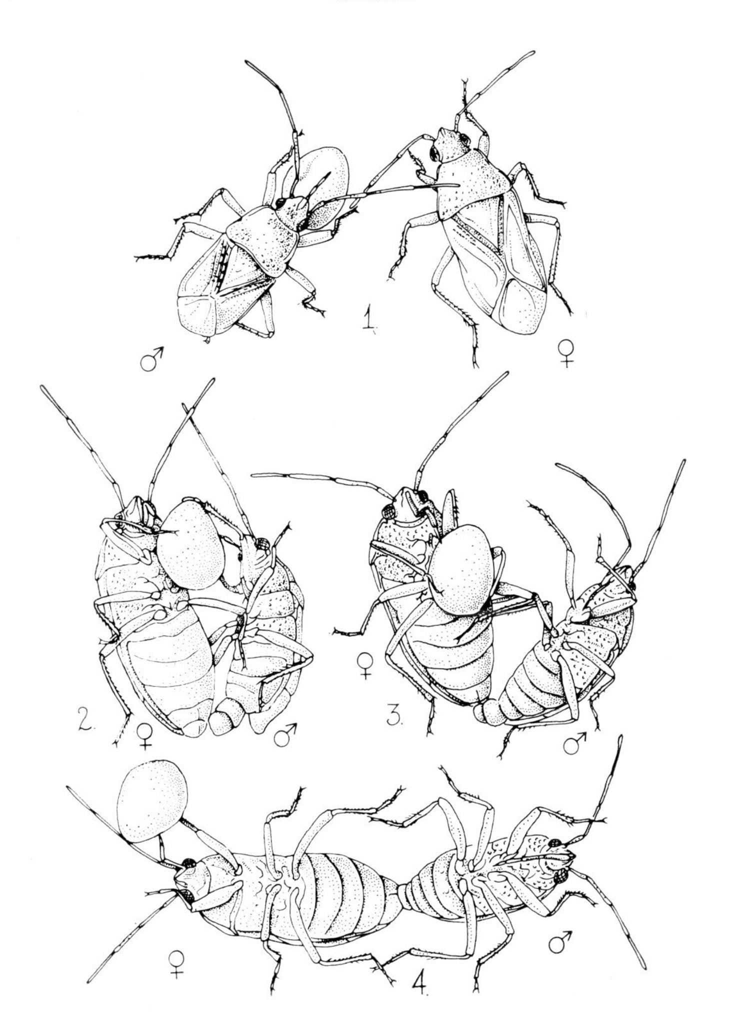

FIG. 35. The offering of a nuptial gift by the male as a prelude to copula in *Stilbocoris natalensis*. The male injects saliva into the seed prior to transferring it to the female. Males without a seed are unsuccessful in courting. (From Carayon, 1964.)

available, but which nevertheless court, are unsuccessful, and no copula is observed. This last-mentioned report gives a definite clue as to the function of the nuptial gift exchange.

7. CONTROL OF MATING BEHAVIOR

Age and the frequency of mating in either sex may be related to the animal's reproductive status, i.e. in females, the period of maturation of the oocytes and the depletion of sperm, and, in males, the availability of sperm or spermatophores. Very unfortunately, in the majority of species information on the control of mating behavior is scanty; often only circumstantial evidence can be found. In the following discussion, the available information will be presented.

Females, particularly those of many Hymenoptera and Lepidoptera, often mate as soon as they emerged, sometimes only a few minutes afterwards. In the extreme, the female is mated even before she has emerged from the pupal skin, as in the dipteran *Opifex fuscus* (Kirk, 1922; Edwards, 1926) or in species of the Clunioninae (Oka, 1930, Tokunaga, 1935). The female of the cockroach *Diploptera punctata* is mated before her wings are fully expanded and hardened (Roth and Willis, 1955c). In none of these cases is anything known about the control mechanisms; it appears that mating is not governed by the status of the reproductive organs. It is, however, important to note that in many of these species except *Diploptera*, eggs have matured by the time of emergence, and also many of these animals are rather short-lived and do not take any food as adults.

Some kind of maturation process—a postemergence maturation—before the animals mate may be found in animals which mate only some time after emergence. For example, male crickets may display characteristic courtship behaviour and mate only when a spermatophore is present in the genitalia or when the seminal vesicles are filled with semen (not necessarily with sperm) (Haskell, 1961; Dumortier, 1963). Apparently, sensory information from the genitalia may cause the initiation of courting since, as in *Gryllus campestris*, after severance of the ventral nerve cord, no courting was observed, presumably because the information did not "get through" to the brain (Huber, 1955). Spermatophore production is unrelated to the presence of spermatozoa in crickets and grasshoppers, as is shown by the fact that castration had no effect on courtship in male *Schistocerca gregaria* (Husain and Mathur, 1945; Odhiambo, 1966b), *Omocestus viridulus*, *Chorthippus brunneus*, and *C. parallelus* (Haskell, 1960). Also, castrated males of *Drosophila melanogaster* (Geigy, 1931), *Musca domestica* (La Breque *et al*., 1962), and *Scatophaga stercoraria* (Foster, 1967) mated as readily as normal ones. These few reports indicate that mating behavior is not governed by gonadal development in males.

In most males of the mosquito species, mating is delayed until the hypopygium is turned by about 180°, which is completed within the 1st day after emergence. Turning of the hypopygium may be termed a postemergence maturation process. The phenomenon was discovered in *Anopheles* by Christophers (1915) and later found to exist in many species (see Feuerborn, 1922; Clements, 1963). As soon as the inversion is complete, the males approach the females.

Considerably more attention has been given to control mechanisms of mating behavior in the female than in the male. Females of many species mate only some time after emergence. Obviously, some kind of maturation process precedes the time of acceptance of the male's courtship. Prior to acceptance, there is a period of active refusal of the courting males. This is found particularly often in species where the female plays an essential role in courtship. For example, young females of several species of *Pieris* refuse the courting males by curving their abdomen upward, an action which does not permit copulation (Peterson and Tenow, 1954; David and Gardiner, 1961), while *Hypolimnas misippus* females simply close their wings (Stride, 1958b), and unreceptive females of *Heliconius erato hydara* close and open their wings, behavior which causes the male to leave (Crane, 1955). Unreceptive females of the damselflies *Calopteryx splendens* (Buchholtz, 1951), *C. maculatum* (Johnson, 1962a), and *Platycnemis pennipes* (Buchholtz, 1956) frequently spread their wings, which action drives the courting males off. A female locust or grasshopper usually kicks with her hindlegs if a courting male is very close and she is unwilling to mate (Renner, 1952; Mika, 1959; and others). Similar evasive behavior is found in many more species of several orders. Females which are unreceptive will not only actively refuse the male but will also remain passive to long-distance sound, and to optical and pheromonal stimuli.

What are the internal conditions which control this kind of behavior? One may postulate that perhaps the ovarian maturation processes or ovulation interfere, i.e. the female is ready to accept the male only at certain times. In a number of species, as will be shown below, this is indeed the case. However, as has been shown by Regen (1909), removal of the ovaries from a female *Gryllus campestris* did not eliminate her responsiveness to the courting males. The same has been observed in *Drosophila melanogaster* (Geigy, 1931), *Euthystira brachyptera* (Renner, 1952), *Locusta migratoria* (Mika, 1959), *Leucophaea maderae* (Engelmann, 1960a), and undoubtedly in many other species. However, some kind of control is exerted by the gonads in some species, though the mode is rather difficult to determine. *Euthystira brachyptera* females normally drop the spermatophore after mating, but castrated females do not; as a consequence, the castrated mated females stay passive and no longer respond to courting males (Renner, 1952). No explanation can be given for this retention of the spermatophore by the castrated females.

One report involving another grasshopper, *Chorthippus parallelus*, shows that somehow the ovaries may influence the female's responsiveness to the male (Haskell, 1960). Castration of the females was directly followed by a loss of responsiveness to courting males within 2 days. Blood transfusions from normal females restored her responsiveness for 1 or 2 days. This would indicate that some blood-borne factor originating in the ovaries affects mating behavior. Are we dealing here with an insect sex hormone, an agent which many investigators have been searching for unsuccessfully in many species for so long?

If, as has just been made plausible, the ovaries of some species liberate a hormone which controls mating behavior, one could further assume that this agent is liberated only at certain stages of egg maturation. In a search for evidence of this, no conclusive case became apparent other than the one given above. Many species, such as *Anabrus simplex* (Gillette, 1904), *Nomadacris septemfasciata* (Burnett, 1951), or *Blaberus craniifer* and *Byrsotria fumigata* (Roth and Stay, 1962b)—to mention only a few—mate at any time during their ovarian cycle. In other species, e.g. *Leucophaea maderae* (Engelmann, 1960b), or *Nauphoeta cinerea* (Roth, 1964a), the females mate usually only when the oocytes are still small; this, however, as will be shown below, does not necessarily indicate that the ovaries control mating behavior. Furthermore, mature gonads seem to be a prerequisite for mating in females of the dragonfly *Anax imperator* (Corbet, 1957), a species which stays away from ponds and waters during the maturation period; thus the females do not encounter courting males. Also in *Machilis germanica* the female will accept the courting male only if she has maturing oocytes in the ovaries (Sturm, 1955). These two additional reports again give no direct evidence for gonadal control of mating behavior in the female.

Since egg maturation in many insect species is controlled by the corpus allatum hormone, one may perhaps search for endocrine involvement. Environmental factors and food supply which control in many species endocrine activities may indirectly or directly be involved. This may be inferred from the reports that both starved *Dysdercus sidae* females (Ballard and Evans, 1928) and *Calliphora erythrocephala* fed sugar only (Strangeways-Dixon, 1961a) do not mate. Also, females of *Haematosiphon inodorus* (Lee, 1955) mate more often if fed than if starved. To be certain, starved females of many species mate as readily as normally fed ones do.

An incidental observation on *Leucophaea* females then gave the first clear evidence for any species that the hormone liberated by the corpora allata is involved in control of the female's receptiveness (Engelmann, 1960a, b). In cockroaches the female takes an active part in courtship, and mutual stimulation through several steps leads to copula (Roth and

Willis, 1952b, 1954; Barth, 1964). The female in many species, including *Leucophaea*, climbs onto the dorsum of the male, nibbles on the terga and thus comes into a position which enables the male to interlock the genitalia (Fig. 27). In other words, unless the female actively responds in this manner, mating does not take place. Forceful takeover of the female by the male has never been observed, but sometimes even an unreceptive female may by chance walk onto the back of a courting male, particularly under crowded conditions. In this species allatectomy performed either in the last instar nymphs or newly emerged adults made the females "reluctant" to respond to the courting males. About 30 per cent of the allatectomized females mated after some time when kept in small containers, and up to 50 per cent eventually mated when kept under crowded conditions (Engelmann and Barth, 1968). During the observation of a *Leucophaea* colony composed of normal and allatectomized individuals one can actually see that the operated females are sluggish in response to male courting, whereas normal females of the same age, when brought into the colony, almost immediately accept the courting males. Clearly, the corpus allatum hormone affects the behavior of the female through the central nervous system. These observations, which have been confirmed in collaboration with Barth, are in contrast to those of Roth and Barth (1964) on the same species. In their colonies, 100 per cent of the allatectomized females mated. No explanation for this controversy is apparent. Mating behavior mediated by pheromones as controlled by the corpus allatum hormone is discussed below.

In analogy to the situation in *Leucophaea*, copulatory behavior involving sound production and perception in *Gomphocerus rufus* is under corpus allatum control (Loher, 1962). Operated females do not respond to the songs of the males by emitting their own song as they normally would, and they behave defensively towards the male if he comes into accidental contact. Here again, the corpus allatum hormone appears to influence the central nervous system, which controls sound emission as well as the full range of behavior responses.

Control of mating behavior by the corpus allatum hormone in another grasshopper, *Euthystira brachyptera*, is far less complete than in the cases just mentioned. Allatectomized females of this species still mated but do not become ready to accept the male until somewhat later than usual (Müller, 1965a, b). Additional implantation of active corpora allata or treatment with farnesol accelerated the first acceptance. Müller concluded from these findings that the first acceptance of the male is, to some extent, under the control of the corpus allatum hormone. Does the hormone hasten a central nervous system maturation process which normally takes several days? One additional case remains to be discussed in which the hormone from the corpus allatum may play a role in mating behavior and which may be analogous to the findings in *Euthystira*. Implantation of active corpora allata in female *Drosophila melanogaster* pupae caused the resulting adult females to mate on the 1st day after emergence, whereas controls usually mated on the 2nd day (Manning, 1966, 1967). Non-protein fed females did not mate, indicating that in those animals the corpus allatum was inactive. The same dependence of male acceptance on corpus allatum activity seems to be present in *Lucilia cuprina* since, if eggs did not mature, mating was not observed (Barton Browne, 1958). Even though there was no direct observation of corpus allatum activity in the latter case, one can infer from the data given that the corpus allatum is involved. The same may hold for the eye gnat *Hippelates collusor* (Adams and Mulla, 1968).

Absolutely no influence of the corpora allata on mating behavior was found in *Galleria mellonella* (Röller *et al.*, 1963) or similarly in *Diploptera punctata* (Engelmann, 1960a). Allatectomy in *Gryllus bimaculatus* (Roussel, 1967) did not affect the female's normal mating behavior either. From the few reports regarding crickets, grasshoppers, cockroaches, and some Diptera, it appears that the corpus allatum fully controls mating behavior in some species, in others only to some extent, while in some there is no known influence of the corpus allatum hormone. Hormonal involvement is perhaps only one of the many possible control means which have evolved among insects. In the majority of species, we just do not know the underlying causes for the animal's behavior.

Many insects species mate only once in their lifetime, or rarely twice, whereas some mate very frequently, irrespective of whether the sperm reserves are depleted or not. Storage of spermatozoa is possibly one of the factors which influence repeated mating in some species. On the other hand, the capacity for storage of spermatozoa is remarkable in some other species. Social insects, such as the honeybee queens, supposedly store sperm for several years. Overwintering *Podisa maculiventris* (Couturier, 1938), *Dytiscus marginalis* (Blunck, 1912), and *Orthacanthacris aegyptia* (Grassé, 1922) mate in fall but only lay their eggs the following spring (with no further mating). Associated with single mating in a lifetime is the refusal of courting males by mated females. What are the underlying causes for this reaction? There is evidence in *Musca domestica* that the secretion of the male's copulatory duct which is transferred during mating causes the female to refuse other males (Riemann *et al.*, 1967). Implantation of the copulatory duct but not the testes into virgin females resulted in non-acceptance of courting males. A dose-response curve for virgins injected with extracts was established (Adams and Nelson, 1968). Similarly, in *Aedes aegypti*, the secretory product of the male accessory sex glands, which is normally transferred during copula, prevents further insemination (Craig, 1967). In the latter case, the substance is non-dialysable, heat labile, and precipitable with ammonium sulfate; it is probably a protein. Nothing is known about the mechanisms that make females of many cockroach species unreceptive to courting males after mating. For example, females of *Leucophaea* and *Diploptera* mate only once during an egg maturation period (Engelmann, 1960b). They will accept another male after parturition even though there are still sufficient spermatozoa in the spermathecae to fertilize at least another two batches of eggs (Engelmann, unpublished). Is it the distention of the bursa copulatrix which makes the female unresponsive to courting males shortly after a successful copula? It appears that females of *Leucophaea* (Engelmann, unpublished) or those of *Nauphoeta cinerea* (Roth, 1962, 1964a) mate repeatedly if the spermatophore is removed shortly after mating; also after severance of the ventral nerve cord, females would accept several males in rapid succession. In the latter case, the female obviously did not get the information that mating had already taken place (Roth, 1962).

Among the insects best studied with respect to courtship and mating are several species of crickets and grasshoppers. Many of these species exhibit a sequence of changes in behavioral patterns after emergence. For example, a *Euthystria brachyptera* female is at first unresponsive to a courting male, then may passively accept him, and later may actively follow his calling song (Renner, 1952). After copula, the female shows a passive acceptance, but shortly before ovulation she refuses the male. Only if no further mating has taken place for some time will the female become actively responsive again after several batches of eggs have been laid. This latter change in behavior seems to be correlated with the gradual depletion of spermatozoa from the spermathecae. Several mechanisms underlie the changes in behavior of the female. Factors such as depletion of spermatozoa or of eggs within the oviducts affect the behavior in opposite directions, i.e. acceptance or refusal. Refusal or passive resistance (also called secondary defense) of courting males after a copula is likewise found in other grasshoppers. Busnel *et al.* (1956) report its occurrence in *Ephippiger bitterensis*, Haskell

(1958) in *Chorthippus parallelus* and *C. brunneus*, and Loher and Huber (1964) in *Gomphocerus rufus*. Information from the genitalia (filling of the spermathecae) apparently makes the female non-responsive to the male's courting. The information appears to travel via the ventral nerve cord anteriorly, as has been suggested by results from the aforementioned experiments. Removal of the spermatophore shortly after mating and before the sperm mass has been transferred into the spermatheca in *E. bitterensis* (Busnel *et al.*, 1956) or *Gomphocerus* (Loher, 1966a, b) caused females to accept courting males shortly thereafter. In the latter species, denervation of the spermatheca or severance of the ventral nerve cord resulted in a permanent readiness to respond to male courting in spite of the presence of a spermatophore; repeated matings took place, behavior similar to that of *Nauphoeta* females after a similar operation (Roth, 1962). The mechanism involved in the control of secondary defense is obviously different from that of newly emerged females of *Gomphocerus* which exhibit a so-called primary defense. During this period of primary defense, the corpora allata are inactive; and it was shown that, for acceptance of the male, the corpora allata must be active; this is the case a few days afterwards (Loher, 1962, 1966a; Loher and Huber, 1966). As has been shown for *Gomphocerus*, both endocrine and nervous systems ultimately control the various phases of the animal's behavior. Furthermore, prolonged isolation of *Gomphocerus* males (Loher and Huber, 1964) or *Gryllus campestris* females (Hörmann-Heck, 1957) resulted in an overt readiness of these animals to respond to courting males. Clearly, the physiological status of the female determines her responsiveness. The same probably applies to many other species. In grasshoppers and crickets, sound signals are an integral part of courtship behavior, but they become effective only when the level of motivation has reached a certain height. "Motivation" may describe the physiological status governed by the endocrine and nervous systems of the animal.

Emission and perception of pheromones used as a means of communication between individuals in courtship are probably controlled by the animal's physiological status, just as are the other types of communication. This is best illustrated as follows. It has long been known that virgins of many species are more attractive to males than are mated females, i.e. they presumably liberate more pheromones than do mated animals. In field studies, for example, virgins of the gypsy moth *Porthetria* were used (Collins and Potts, 1932). *Clysia ambiguella* and *Polychrosis botrana* females attract their males throughout their life while virgin, but not when mated (Götz, 1939). Similarly, virgins of *Periplaneta* produce sex attractants at a high rate. When the females are mated, production of the attractants declines rapidly within 18 hr and falls nearly to zero within 2 weeks (Wharton and Wharton, 1957) (Fig. 36). Virgins of species such as the pine shoot moth *Rhyacionia buoliana* (Pointing, 1961) and the cucumber beetle *Diabrotica balteata* (Cuthbert and Reid, 1964), as well as many others stayed attractive if kept virgin; but, soon after mating, they lost their attractiveness. Loss of attractiveness may not be linked to cessation of pheromone production, rather the control of release may be affected by mating. Only a few reports are available for information on this aspect. For instance, no decrease in extractable pheromone of *Trichoplusia ni* females was noticed after mating (Shorey and Gaston, 1965c). The same situation has been reported in additional noctuid moths (Shorey *et al.*, 1968). On the other hand, mated *Bombyx mori* females produced less extractable pheromone than did virgins. Pheromone production was, however, not entirely eliminated in mated females (Steinbrecht, 1964b). In other words, mating influences the production of the sex attractant, at least in *Bombyx* and in *Periplaneta*, as far as we know. Certainly, many moths no longer extrude their scent glands after mating, i.e. they do not take the typical calling position. In the latter cases, either the filling of the bursa with a spermatophore or the presence of sperm in the spermathecea seems to influence the female's behavior. We know no details about the actual mechanisms by which mating controls the pheromone production or release in any species which emits pheromone.

Animals often or continuously exposed to pheromonal stimulation seem to adapt to the odor; their threshold for the perception and response to it may rise. This has been shown, for example, in *Periplaneta* males whose response declined in successive tests (Wharton *et al.*, 1954a, b). A similar observation was made for *Trichoplusia* males which ceased to respond after previous exposure, even though they had no access to females (Shorey and Gaston, 1964). Adaptation to pheromones or odors in general is probably widespread. An interesting example, which may have some bearing on the interpretation of the results from behavioral studies in a variety of species, should be mentioned in this context. As was pointed out earlier (cf. p. 87), females of *Leucophaea* invariably accept the courting males

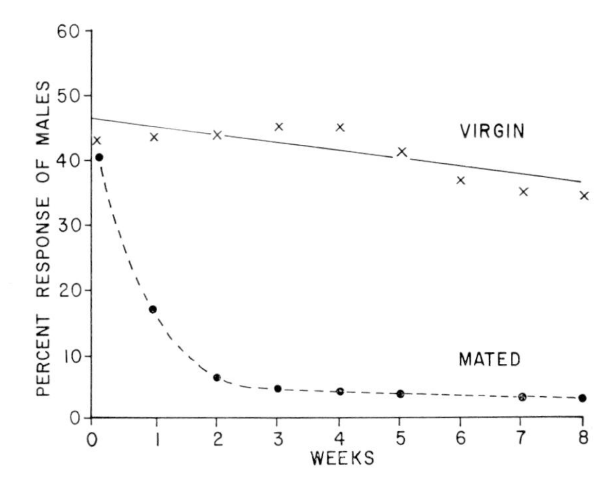

FIG. 36. The effect of mating on the weekly production of sex attractant in *Periplaneta americana* as tested by the filter-paper method. The test began with 114 virgins and 25 mated females. (Adapted from Wharton and Wharton, 1957.)

when their oldest oocytes have just begun to deposit some yolk, i.e. the corpora allata are in their initial activity phase. This applies in the strict sense only to animals which, irrespective of age, have had continuous access to males (Engelmann, 1960b). If, however, the females were kept isolated for some time, it was observed that they might accept males even when their ovaries contained large or nearly mature eggs (Engelmann, unpublished). One can interpret these results to mean that, in the latter cases, the females were not adapted to the male pheromone and, as a consequence, the threshold for their acceptance was lower than if they had been kept with males all the time.

Endocrine control of production of a sex attractant has been postulated so far for one insect species, the cockroach *Byrsotria fumigata* (Barth, 1961, 1962, 1965, 1968; Roth and Barth, 1964). In this species, only a small proportion of allatectomized females mated, and it was shown that only 14 per cent of these operated females produced a sex attractant (Barth, 1962), as tested by the filter-paper method (Wharton and Wharton, 1957; Barth 1961). Pregnant females of this ovoviviparous cockroach, the corpora allata of which were inactive, did not produce the pheromone. Reimplantation of corpora allata into previously allatectomized females restored pheromone production to some extent, but still many individuals did not produce the attractant. Perhaps in some cases the implants did not take. It is, furthermore, interesting to note that *Byrsotria* females produce pheromone throughout the activity phase of the corpora allata, i.e. during the entire egg maturation period, but do not accept courting males once they have been mated (Barth, 1962). This certainly shows that other aspects of reproductive physiology have to be considered for complete understanding of the behavior of the species. Barth (1962) and Roth and Barth (1964) interpret their results to the effect that a limited amount of pheromone is produced by the animals most of the time, but that the corpora allata greatly enhance the rate of production, thus stimulating courting of the male. According to these authors, the corpus allatum hormone in *Byrsotria* does not affect the behavior of the female directly; the central nervous system is not influenced. This contrasts to the situation in *Leucophaea* (Engelmann, 1960a). For *Byrsotria* one would have to assume that the female would be passive during normal courtship; one actually observes, however, that she plays an active part in courtship—the female has to accept the courting male. Probably, the corpus allatum hormone both stimulates pheromone production and affects the central nervous system, thus changing the female's behavior. In *Byrsotria* it has also been shown now that farnesylmethylether, a compound with corpus allatum hormone activity, stimulates pheromone production of allatectomized females, as tested by the filter-paper method (Emmerich and Barth, 1968); interestingly, no dose response (0.1–0.8 μl of a 10 per cent solution per animal tested) was observed.

How widespread among insects is the action of the corpus allatum hormone on pheromone production? Certainly, no influence of the hormone on pheromone production could be found in the *Cecropia* silkmoth, since even isolated abdomina were attractive to males (Williams, 1952). Similarly, allatectomized females of *Galleria* (Röller *et al.*, 1963), *Bombyx* (Steinbrecht, 1964b), and *Antheraea pernyi* (Barth, 1965) produced sex attractants. All of these latter species are short-lived and one may speculate that other control mechanisms are effective here (Barth, 1965). However, observations in the cockroach *Periplaneta* demand a different interpretation. As was demonstrated, mated females produce very little pheromone (Fig. 36) whereas virgins readily elicit a response in the male (Wharton and Wharton, 1957). In both virgins and mated females the corpora allata are active, as is shown by the production of eggs. One must therefore assume that, in this long-lived species, pheromone production is not under corpus allatum control. No clue as to the actual control mechanism is yet available.

Pheromone production or release may be influenced by a variety of different means in the various species. Of great interest in this connection is the finding that the female of the moth *Antheraea polyphemus* only releases her pheromone if she is in close proximity to oak trees, an observation made a long time ago by many entomologists. It is now known that *trans*-2-hexenal, a volatile compound found in oak leaves, stimulates the female to release her sex attractant. Interestingly, leaves of several trees which also can serve as food plants contain the same chemical, but, in these cases the action of *trans*-2-hexenal is blocked by other substances (Riddiford, 1967; Riddiford and Williams, 1967). How widespread this kind of exogenous control of pheromone release may be among insects can only be speculated upon. Possibly the observation that *Heliothis zea* females are mated at a high percentage only in the presence of certain plants may be explained in a similar manner (Snow and Callahan, 1967). For another moth, *Acrolepia assectella*, it has been shown that males are attracted to their females only when the latter are in the vicinity of the pear tree, an observation which leads one to suspect that here also the females release their pheromone only when close to their host plant (Rahn, 1968).

8. ENVIRONMENTAL FACTORS INFLUENCING MATING BEHAVIOR

The majority of insect species characteristically mate at a certain time in the normal day–night rhythm. Diurnal species, such as *Oncopeltus fasciatus* (Caldwell and Dingle, 1965) or *Lucilia sericata* (Cousin, 1929), to mention just two species, mate nearly any time during the daylight. Most moths mate during the night hours, but a few species, e.g. the carpenterworm moth *Prionoxystus robiniae* (Solomon and Morris, 1966), mate in bright daylight. Often, mating in certain moth species occurs during a given time period of the night (see below). Rather interesting is the report that the flies *Dacus tryoni* and *D. cacuminatus* mate only in twilight (Myers, 1952). In these species fading daylight is essential for courtship, as is shown by the absence of mating if the transition from light to dark is made abruptly. This phenomenon is probably more widespread than is presently known and may occur in a variety of species from different orders, particularly those active in twilight. The lepidopteran *Paramyelois transitella* appears to be another example which exhibits a dependence on a twilight

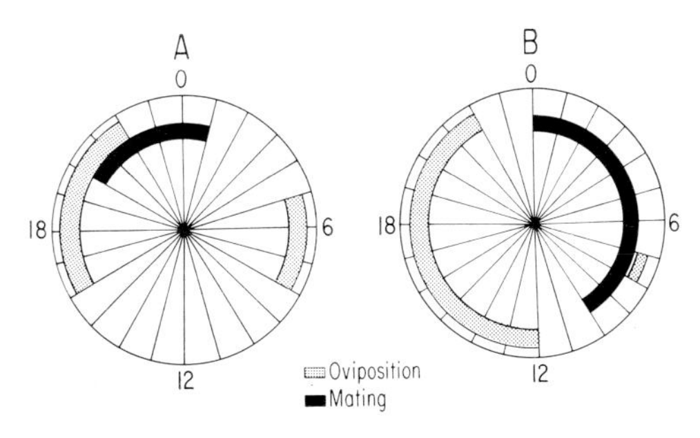

FIG. 37. Diurnal rhythm of mating and oviposition in the moths *Polychrosis botrana* (A) and *Clysia ambiguella* (B). The record was obtained by the usage of a rotating baited sticky board to which there was only a narrow access. (Adapted from Götz, 1941b.)

period for mating (Goodwin and Madsen, 1964). If there is no twilight, mating behavior is seriously disturbed.

Many species have circadian rhythms for a variety of activities, such as locomotion, feeding, or mating. These activities often occur simultaneously at certain periods during the 24-hr day. Early observations on pheromone-emitting species showed that either the males are more active at a certain time or that the females emit the sex attractant only then. Götz (1941a, b) baited a sticky trap which turned once in 24 hours and to which the males had access only through a narrow slit, thus enabling the author to determine the hour when the males were caught. He showed that *Clysia ambiguella* males were mainly caught between 2 and 6 a.m., whereas those of *Polychrosis botrana* were caught mostly between 9 and 11 p.m. (Fig. 37). These observations suggest that, during these hours, the males are active and the females emit their pheromone. The males of the cabbage looper *Trichoplusia* respond to pheromone stimulation best during dark periods (Shorey and Gaston, 1965a). Males of this species are inhibited in their response by light of as low intensity as 0.3 lux (Shorey and Gaston, 1964). The females presumably release the sex attractant during the night hours. Since no correlation was found between time of day at which the pheromone was extracted and quantity of pheromone (Shorey and Gaston, 1965c), it is the release mechanism which is affected by the day–night rhythm. Obviously, in evaluation of these behavioral observations, one must keep in mind all aspects of both male and female activity as well as environmental influences which may dominate intrinsic phenomena.

As in pheromone-mediated mating behavior, many sound-producing insects sing and mate during certain periods of the light–dark cycle. *Ephippiger ephippiger* sings and courts only during the night (Dumortier *et al.*, 1957); in continuous dark this circadian rhythm of singing persists for at least 4 days. Light inhibits courting. *Gryllus bimaculatus* courts and mates only during the early evening (Royer, 1966). Undoubtedly, a list of species which exhibit similar behavioral activity patterns could be assembled, but these two examples may suffice here.

In 1930, Oka reported that the chironomid *Clunio pacificus* always emerges around new-moon. This short-lived species of Clunioniae as well as others of the same genus mate soon after emergence and then die within a few hours. It was then found that several *Clunio* species from Japan emerge and mate likewise always at the spring low tides (Tokunaga, 1935). Later, Caspers (1951) showed for a European species, *Clunio marinus*, that it is not the low tide which synchronizes the emergence, but rather the lunar phase. Both in the laboratory, as well as in the field, this species emerged and swarmed around full or new-moon. The same species kept under a 12/12 hr light/dark cycle did not show lunar periodicity, but if, in addition, artificial moonlight was supplied for a few days, the peak emergence of the animals was 15 and 30 days later (Fig. 38) (Neumann, 1963). From the appearance of the two peaks, Neumann postulated and later confirmed (1965, 1966a, b) that different populations have different responses to moonlight, i.e. they have either a 15- or 30-day emergence interval. The picture may be even more complex. For example, certain populations from Helgoland apparently do not respond to a moonlight cycle in the laboratory; yet, in nature, a clear emergence rhythm is observed. This leaves the alternative that tidal rhythms are the environmental cues for emergence and mating in this case (Neumann, 1966b). Another species, *Clunio takahashi*, likewise emerges with tidal rhythms, but not with lunar periodicity (Hashimoto, 1965). Furthermore, a population of *Clunio marinus* from the Baltic Sea, which does not experience any tidal rhythm and which lives permanently submerged, has no lunar emergence peaks at all (Neumann, 1966b). It should be interesting to work out the underlying control mechanisms, genetical and endocrinological, which govern the biology of the various *Clunio* populations. Molting involves the activation of cerebral neurosecretory cells and the prothoracic glands; before metamorphosis, the corpora allata presumably are inactivated. Moonlight apparently synchronizes these events in many individuals, but different populations respond to different cues.

The unique observation on mating behavior in Dynastid beetles from Venezuela deserves mentioning (Beebe, 1944). Rainfall elicited mating behavior from males of *Megasoma elephas* and *Strategus alveus*. Mating behavior could also be evoked by artificially drenching the terrarium in which these species were living. This is certainly a peculiar instance of behavior induced by a change in environment. No further details are known about this at present.

9. MATING BEHAVIOR AS AN ISOLATING MECHANISM

Just as morphological differences between closely related species prevent mating and thus in essence function to screen out the species which are genetically incompatible, so also may courtship behavior serve the same "purpose". Mechanical prevention of mating may be more archaic than the use of subtle behavioral means. As has been frequently illustrated, ethological barriers to mating are found in a variety of species, both among the vertebrates and the invertebrates (Mayr, 1963). In a search for examples in insects one finds many cases which well illustrate the principle involved. Here, only a few cases will be mentioned.

A variety of means, ranging from acoustical, olfactory, and visual to mechanoreceptory, which are employed by the various species, may play a decisive role in the determination of the further courtship behavior of either sex: The sensory input will affect the response of the recipient animal. Manning (1966b) remains unconvinced that behavior may be a means of

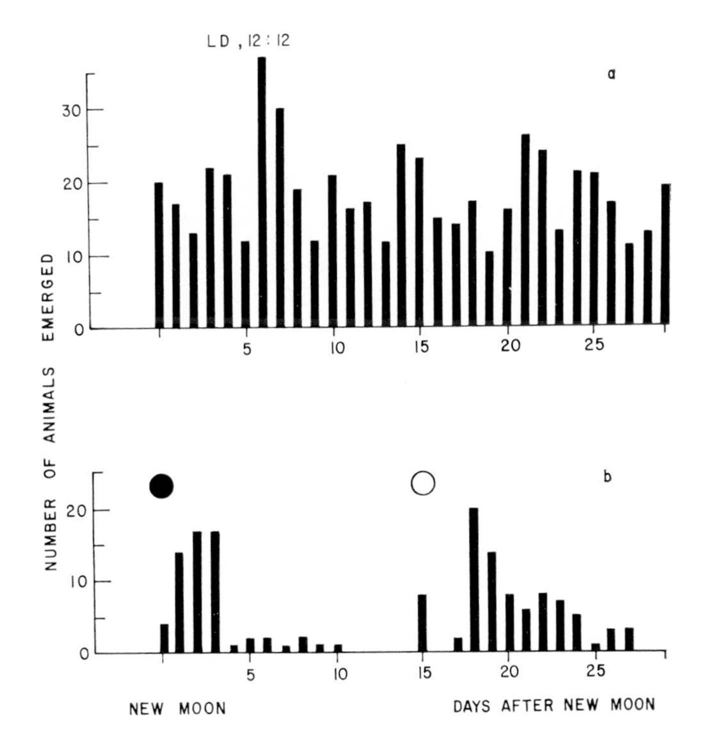

FIG. 38. Daily emergence and mating of *Clunio marinus* under artificial day/night (12:12) conditions (a) and under the natural light regime (b). The moon phases are indicated. (Adapted from Neumann, 1963.)

discrimination, but if we define behavior as the response to the sum of sensory input we cannot escape the fact that behavior indeed serves to discriminate. The study of a single variable may not provide us with a meaningful answer since the animal in nature reacts to a complex of independently acting mechanisms (Wilson and Bossert, 1963); this will become more apparent in the cases discussed below. Furthermore, laboratory studies necessarily introduce an artificial environment (temperature, humidity, space, vegetation, etc.) which may be a significant factor in interpreting the results obtained.

Closely related sympatric species may be reproductively isolated because of their different reproductive seasons, and, in this case, we do not have to invoke behavioral differences. This is well demonstrated by the two crickets, *Acheta veletis* and *A. pennsylvanicus*, whose song patterns are hardly different from each other but whose active reproductive periods are

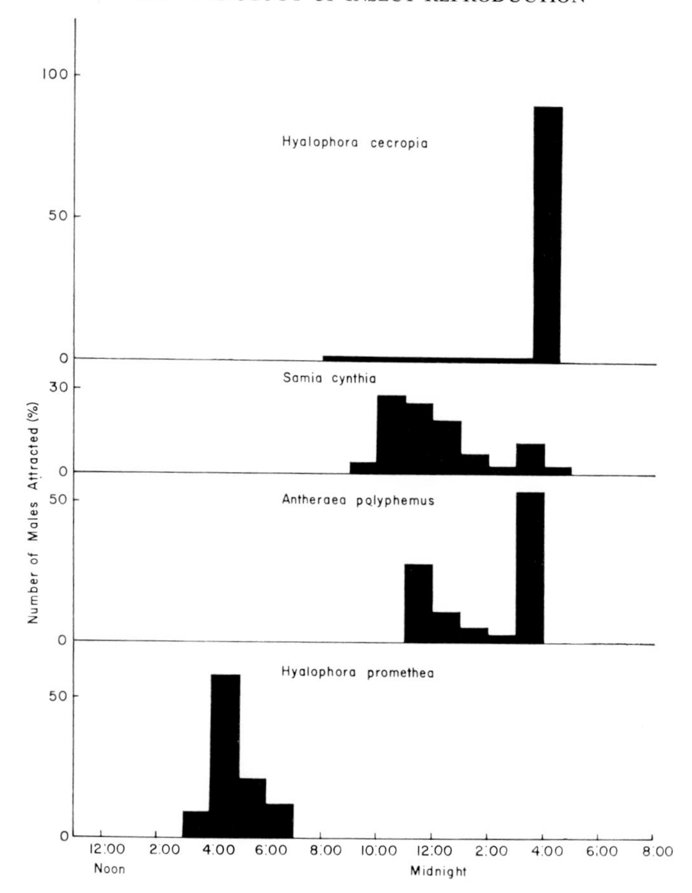

FIG. 39. Time of day when males of four species of Saturniids are attracted to their females. Note that even though a male may be attracted to a female of another species (see Fig. 32), their daily flying time differs greatly, a fact which eliminates interspecific attraction. (Based on data given by Rau and Rau, 1929.)

at different times of the year (Alexander, 1961). Also the sympatrically occurring Saturniids *Hyalophora cecropia* and *Samia cynthia* are to some extent seasonally separated (Rau and Rau, 1929), although males of either of these two species may be attracted to the sex lure of the other species; however, the threshold for pheromonal attraction by the other species is higher than that for their own species (Fig. 32) (Rau and Rau, 1929; Schneider, 1966). Additional means of sexual isolation in the sympatric species *Hyalophora cecropia* and *Samia cynthia* and other Saturniids is found in their different circadian flight times (Fig. 39) (Rau and Rau, 1929). Thus even though the chemosensory apparatus of the one species responds to the pheromone of the other species, several behavioral differences can be recognized: seasonal and diurnal flight times and a high threshold to the foreign pheromone. It has been noticed that the more closely related species are, the better does the chemosensory apparatus of one respond to the pheromone of the others (Schneider, 1966).

Relatively few experiments have been performed to find the isolating behavioral mechanisms in insects. Nevertheless, some of the known cases are significant. The crickets

Nemobius fasciatus socius and *N. fasciatus timucelus* exhibit hardly any morphological differences; they even mate and produce viable offspring in captivity (Fulton, 1933). In the field, the two subspecies do not attract each other, apparently because their song patterns are different. Interestingly, hybrids sing an intermediate type of song which is never found in nature; this clearly illustrates that mating normally does not take place between the species (Fulton, 1933). In the two sympatrically living species of grasshoppers, *Chorthippus brunneus* and *C. biguttulus*, heterospecific matings can occur since they have no morphologically isolating mechanisms. Yet it is known that only about 15 per cent of the animals mate heterospecifically. Interspecific mating in these species is prevented because of an ethological barrier, i.e. the song patterns of the species are different (Perdeck, 1957); consequently, species isolation in the field is effective. Similar isolating means have even been found between strains of the species *Acheta domesticus* (Ghouri and McFarlane, 1957). Also populations of *Ephippiger bitterensis* from different geographic locations have different song patterns, ranging from one to five strokes per pulse (Busnel, 1963). It is not quite clear whether these differences function in the isolation of the populations since, in locations between the extremes, the song pattern is in between the two. It appears that some interbreeding occurs.

Closely related species may employ acoustical or pheromonal means for isolation; but, as can be expected, visual stimuli also may serve the same goal. This has been shown in the butterflies *Pieris napi* and *P. bryoniae* whose natural habitats overlap to some extent. Interspecific matings are anatomically possible, but hybrids are rarely found. The males of *P. napi* are more strongly attracted to their own white females than to the somewhat darker *bryoniae* females (Peterson and Tenow, 1954). Differences in color patterns or movements may account for subspecific isolation of several Calopterigidae (Buchholtz, 1955), as well as of other Odonata (Johnson, 1962b).

A combination of the various behavioral cues used in species isolation is found in many *Drosophila* species. As has been studied by geneticists, *Drosophila* exhibits a sequence of activities which lead to mating: the male first approaches almost any moving object (visual), flicks or shortly vibrates his wing or wings, taps the female with his forelegs (tactile), circles around her, vibrates the wings (auditory), licks her genitalia (chemical), and then mounts. A great many qualitative differences in any of the various activities are found among the many species and strains of *Drosophila*. These differences account for either rejection or acceptance of the courting male. Courtship may be broken off by either sex at any point of the sequence, apparently because the necessary stimuli for continuation are not provided by the partner. Because of the males' promiscuity, the success of courtship is most often the choice of the female (Brown, 1964; Ehrman, 1964). The divergence of the sexual behavior seems to parallel morphological differences in the various species (Spieth, 1947, 1949, 1951, 1952; Spieth and Hsu, 1950; Brown, 1965). From the biological point of view, i.e. preservation of the species and success in mating, the elaborate courtship behavior in *Drosophila* species, with the possibility of recognizing a foreign species by several means, seems to "make sense". Hybrids are frequently non-viable; in other words, sperm is not "wasted". The quantitative nature of the various stimulus response sequences is exemplified by at least one case. The frequency of wing vibration of the male, as also the wing movement, is altered after experimental removal of various areas of the wing. It has been shown (Ewing, 1964) that success of mating was possibly correlated with the size of the wing in *D. melanogaster*; wingless males were least successful (Bastock, 1956; Ewing, 1964). The females of several species use their antennae to recognize the male and accept his courting (Mayr, 1950; Bastock, 1956). Later, it was shown that the antennae serve in perception of sound produced through wing vibrations by the males (Waldron, 1964); since the arista functions as a sail, the twisting of the funiculus activates the Johnston's organ (Bennet-Clark and Ewing, 1967; Manning, 1967b). The frequency of wing vibration is different in different species, and since it now seems demonstrated that sound is perceived by the female, discrimination of the species is possible. However, this is quite clearly not the only means of species recognition or isolation, as has been mentioned above.

B. Insemination

Internal fertilization, making possible an economical use of spermatozoa, is found throughout the class Insecta. Since the majority of adult insects are terrestrial, internal fertilization became imperative as a protective means; secondarily aquatic insect species still exhibit this means of sperm transfer. A variety of mechanisms, all leading to the same result, are found in various species belonging to different orders. As has been pointed out repeatedly (see Ghilarov, 1958; Hinton, 1964), transfer of sperm via a spermatophore is considered to be a primitive feature and is itself a step in the evolution from external to internal deposition. Males of higher insect species, e.g. Mecoptera and Diptera, generally transfer free spermatozoa directly into the female. The ability to form a spermatophore may have been lost independently, however, in members of several groups of insects.

1. TRANSFER OF SPERMATOZOA VIA SPERMATOPHORE

In several species of Apterygota, it has been shown that sperm droplets are deposited on the substratum by the male and then picked up by the female. This has been reported, for example, in *Thermobia domestica* (Spencer, 1930; Sweetman, 1938; Sahrage, 1953). We owe much to F. Schaller and his students for our present-day knowledge of sperm transfer in primitive insects and for the speculations and discussions on the evolution of insemination processes. Sperm droplets are deposited in *Thermobia* and *Lepisma*; it appears that this mode is probably the most primitive mechanism of sperm transfer and was to be found in ancestral types of insects. A further evolvement of this pattern can probably be seen in *Machilis germanica*, in which the male attaches the sperm droplets to fine threads that were spun during the initial phases of the courtship dance (Sturm, 1955). In the Collembola *Orchesella villosa* (Schaller, 1953; Mayer, 1957), *Campodea remyi* (Schaller, 1954), and *Podura aquatica* (Schliwa and Schaller, 1963), the sperm is then placed on top of a tiny stalk approximately 0.25 mm in height and later picked up by the female during courtship. In the latter cases we may definitely speak of spermatophores. A humid environment which prevents quick drying up of unprotected sperm droplets is essential for the successful transfer of sperm in all of the Apterygota studied.

True spermatophores deposited into the female genitalia have been described, and their occurrence in the different orders of insects has been reported by several investigators (Boldyrev, 1927; Ghilarov, 1958; Hinton, 1964; Khalifa, 1949b). They have been found in Orthoptera, Dictyoptera, Mantoidea, Dermaptera, Psocoptera, Neuroptera, Lepidoptera and some Hemiptera, Trichoptera, Coleoptera, Hymenoptera as well as in a few Diptera. From some of the reports, it is not clear whether failure to find a spermatophore stems from the fact that it had already been digested or disposed of by the time the animal was checked or whether the species studied indeed does not make a spermatophore. It has also been pointed out by Khalifa (1949b) that even among related species of insects, some species form a true spermatophore while others do not. Sperm transfer without the use of a spermatophore can occur even in species which normally form one (George and Howard, 1968).

Rather extensive observations regarding spermatophore production have been made in Orthopteroid insects. According to Khalifa (1949b), a certain evolutionary trend can be recognized within this group. In Gryllidae and Tettigonidae, the spermatophore is built before copula, then placed into the vulva of the female, and is often still visible from the outside after mating (Fischer, 1853; Lespés, 1855; Houghton, 1909; Gerhardt, 1913, 1914, 1921; Fulton, 1915; Hohorst, 1936; Gabbutt, 1954; and many others). A complicated spermatophore is formed during copula and then placed deep into the bursa copulatrix in Blattaria (Willis *et al.*, 1958) and Mantoidea (Binet, 1931). In Acrididae, the intromittent organ penetrates deeply into the spermathecal duct, pressing a simple tubular spermatophore into it (Iwanowa, 1926; Mika, 1959; Gregory, 1965).

Precise information on how the spermatophore is built is available for only a few species. For example, in *Oecanthus pellucens* where the spermatophore is formed prior to mating (Hohorst, 1936), the accessory glands which line the male genital atrium first secrete the main spermatophore body. The sperm mass is placed into the center of the atrium and thus fills the center of the spermatophore as its walls thicken (Fig. 40). In the cockchafer *Melolontha melolontha*, the spermatophore is formed within the aedeagus during mating and, as the phallus withdraws, it stays in the bursa (Landa, 1960). It is formed in this species by a mixing of the secretion of the accessory sex glands and that of the ejaculatory duct (Landa, 1961). In several species of Trichoptera, as for example, in *Sericostoma personatum*, the proteinaceous product of the two male accessory glands is poured into the bursa, mixes there, and a thick mass results; then the sperm is injected into the bursa, forming a more or less spherical sperm sac (Khalifa, 1949b). Contrary to the general belief that Diptera do not form a spermatophore (Khalifa, 1949b; Hinton, 1964), its occurrence was demonstrated in at least a few species. For instance, the male of the chironomid *Glyptotendipes paripes* transfers a preformed spermatophore during the 3–5 sec of copula (Nielsen, 1959). A spermatozoa-filled spermatophore is found in *Simulium salopiense* (Davies, 1965) and in *Boophthora erythrocephala* (Wenk, 1965a, b), as well as in some other Simuliids. In *S. salopiense*, the male fills the intergenitalic cavity of the mating pair with a viscous mass into which the sperm mass is injected. The viscous mass is thus hollowed out to contain the spermatozoa (Davies, 1965).

As these few quotations suggest, the mode of spermatophore production by the male can be quite variable: it may be formed within the male genitalia, either before or during mating, or secretions may be poured into the female genitalia. Little is known about the chemical composition of spermatophores in insects. The few reports which deal with this subject generally agree that they are largely made up of proteins (Khalifa, 1949a; Gregory, 1965). The only spermatophore definitely known to contain chitin is that of *Heliothis zea* (Callahan, 1958). An interesting finding is reported by Roth and Dateo (1964) for *Blattella germanica* and six additional cockroach species. Spermatophores of these species are covered with a layer of uric acid. Those of thirty-three other species did not contain uric acid. Obviously, males of some cockroach species can eliminate uric acid in this manner. Whether a functional significance can be attached to this mechanism remains to be shown.

The loss of the spermatophore from the mated females seems to be essential since in most cases it would block the passage of the eggs during ovulation. Characteristically, many Lepidoptera which have a separate copulatory orifice do not extrude the spermatophores. Females of the Gryllidae, e.g. *Gryllus*, *Nemobius*, and *Oecanthus*, generally drop or remove

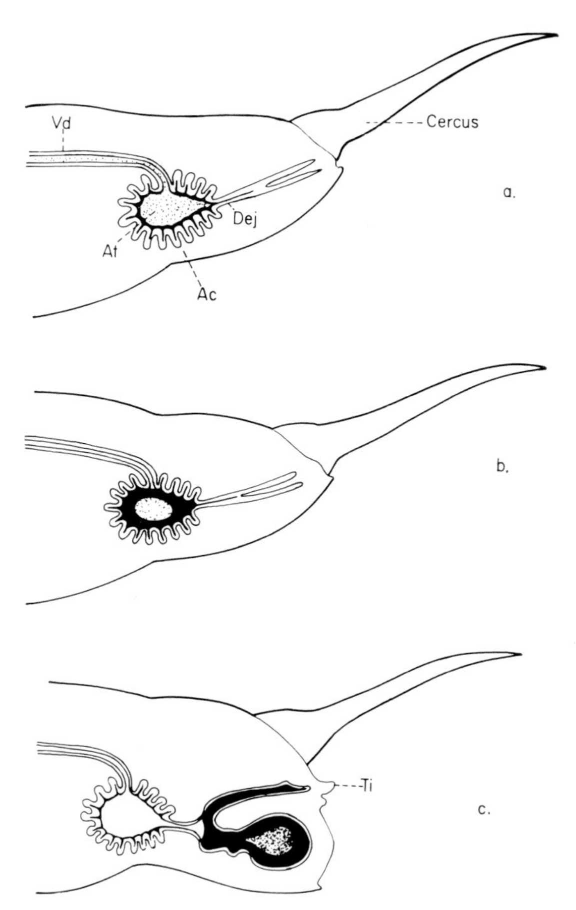

FIG. 40. Formation of a spermatophore in the atrium (At) of the male cricket *Oecanthus pellucens*. After completion (a, b), it is pressed into the ductus ejaculatorius (Dej.). (Adapted from Hohorst, 1936.)

the spermatophore within minutes to hours after mating and then eat it. This behavior also has been reported in the Neuroptera *Osmylus*, *Chrysops*, *Sisyra fuscata* (Withycombe, 1922), *Sialis lutaria* (DuBois and Geigy, 1935), and *Nothochrysa californica* (Toshi, 1965). Reportedly, the spermatophore is extruded by the female *Blattella germanica* within 1 day (Willis *et al.*, 1958), *Eurycotis floridana* in 4 days, *Diploptera punctata* and *Leucophaea maderae* within 5 to 8 days (Engelmann, 1960b) and *Pimpla instigator* (Khalifa, 1949b) after 16 to 20 hr. Extrusion of the spermatophore in *Leucophaea* and *Diploptera* occurs only after it has been partially digested by the secretion from one of the accessory glands (Engelmann, 1960b). Secretory activity of the accessory sex glands is under the control of the corpora allata; consequently, after allatectomy, the spermatophore stays in the bursa in the majority of females. A comparable function of the accessory sex glands and their control has not been demonstrated in other species, but one may surmise that analogous mechanisms are also operative in other insect species.

In the grasshopper *Gomphocerus rufus*, the spermatophore is partially digested within the spermathecal duct and the remaining parts extruded. If extrusion was prevented by ligation, the entire spermatophore was digested and resorbed (Loher and Huber, 1966). Complete digestion of the spermatophore by the female occurs normally in *Melolontha* (Landa, 1960) as well as in *Galleria mellonella* (Khalifa, 1950b). In these species, no remains are ever extruded. In *Galleria* the secretion which digests the spermatophore has proteolytic properties. Fibrin introduced into the bursa was digested completely. Undoubtedly, spermatophores are removed by digestion in many additional insect species. For example, the protein mass injected by the male into the bursa of the trichopteran *Anabolia nervosa* becomes amorphous and the spermatophore collapses and disappears within a few days (Khalifa, 1949b). Since the precise processes involved in spermatophore removal are not known in the majority of species, any further discussion of the subject would be pure speculation.

Males of the butterflies deposit a spermatophore into the bursa of the female, but in some species they deliver a structure which is clearly separate from it. This so-called sphragis, also known as spermatophragma, is secreted by the male accessory sex glands. It has been found in *Parnassius apollo* (Eltringham, 1925; Petersen, 1929) and *Euphydryas editha*, as well as in species of *Argynnis* and *Acraea* (Labine, 1964). The sphragis supposedly functions as a plug and prevents multiple insemination; it has, however, been found that some females have a second spermatophore in the bursa in spite of the presence of a sphragis (Petersen, 1929; Labine, 1964).

Lastly, the unique mode of sperm transfer via a spermatophore in Odonata must be considered. The familiar occurrence of dragonflies in tandem over ponds and lakes represents part of the mating flight (Wesenberg-Lund, 1913). The male, either before or during flight, places his spermatophore into a sperm pouch on the 2nd and 3rd (Fraser, 1939) or 3rd and 4th (Brinck, 1962) abdominal sterna. While the male holds the female with his anal forceps, the female bends her abdomen forward to pick up the spermatophore with her external genitalia. It appears that this bizarre mode of sperm transfer must have evolved from a more typical one since primary genital openings are present on the 8th abdominal segment and typical genitalia are found in both sexes. The male still has a vestigal penis, which, however, does not function as an intromittent organ. How this unique mechanism has evolved must indeed remain speculative (Fraser, 1939; Brinck, 1962).

2. DIRECT SPERM TRANSFER

A variety of insects of several orders do not transfer sperm via a spermatophore, i.e. sperm is introduced directly into the female genitalia. Among Hemiptera, Neuroptera, Trichoptera, Coleoptera, and Hymenoptera, there are species in which no spermatophore has ever been found (Hinton, 1964). None of the Mecoptera and few of the Diptera make spermatophores. Since spermatozoa are not contained in a case, it is obvious that the sperm mass must be placed deep into the female in order to assure a minimum of loss. This has indeed been found in *Trichocera annulata* and *Tipula paludosa*, where sperm is injected deep into the spermathecae (Neumann, 1958). In *Drosophila melanogaster*, nearly all spermatozoa were found within the spermathecae at the termination of a copula which had lasted about 8 min (Garcia-Bellido, 1964c); this indicates that the sperm mass was probably deposited close to the spermathecae. Free spermatozoa have been found in the bursa copulatrix of many Diptera shortly after mating. In several species of Trichoptera, free spermatozoa are deposited into the bursa (Khalifa, 1949b).

The escape of free spermatozoa from the female genitalia seems to be prevented in some mosquitoes by the deposition of a "mating plug" by the male during copula. This structure was first found in *Anopheles gambiae*, *A. funestus*, and *A. longipalpis* (Gillies, 1956), and later described in additional mosquitoes (Lum, 1961; Hamon, 1963). The mating plug should not be confused with a true spermatophore as is found in Simuliidae. Lum (1961) showed that the yellowish secretion of the male accessory sex glands is emptied into the female atrium and forms a yellowish plug there. Apparently it does not have a sperm-containing lumen. Within 24 to 36 hr after mating, the mating plug is dissolved, and the bursa shrinks to its normal size. One may speculate on the phylogenetic origin of this uniquely occurring mating plug. Is it the remains of once occurring spermatophores in all the dipteran species or has it evolved as an entirely new mechanism?

Honeybees do not produce a spermatophore, and the sperm seems to be injected into the bursa. During copula the everted endophallus (bulbus) appears to penetrate into the vaginal orifice; then the bulbus is torn off, occasionally accompanied by an explosive sound, and remains with the queen. This so-called mating sign (which is a mixture of semen, bulbus, and secretory material ejaculated after copula) stays with the queen for variable length of times (Snodgrass, 1956; Ruttner, 1956a, b). A homecoming queen may still carry the mating sign, but often it has been removed before she reaches the hive. This strange mechanism of sperm transfer in honeybees, which involves the sacrifice of the drone, allows the transport of sperm into the spermatheca without much loss of material. It might also prevent any immediate 2nd mating which could result in dislodging of the previously injected spermatozoa. It has been found, however, that a queen may mate several times during one extended mating flight.

3. HAEMOCOELIC INSEMINATION

Extragenital insemination, i.e. injection of spermatozoa through the cuticle, occurs in two groups of insects: the Cimicoidea and the Strepsiptera. For the Cimicoidea, the matter has been excellently reviewed twice in recent years, and little can be added to the experts' reports (Hinton, 1964; Carayon, 1966b). Nevertheless, the subject should be viewed in the present context.

In 1896 Ribaga reported the finding of a peculiar structure on the 4th abdominal sternite of the female bed bug. Berlese (1898) confirmed this and extended the knowledge without actually discovering its true function. In 1913 Patton and Cragg, and later Hase (1918), discovered its role as a copulatory organ and gave an accurate description of copulation and insemination. Spermatozoa are injected through this so-called organ of Ribaga, or spermalege (Carayon, 1966b), even though genital openings exist at the 7th and 8th segments. From the site of sperm injection, the sperm pass through the haemocoel to the oviducts (Cragg, 1920) and penetrate into the conceptacula seminis (sperm reservoirs) within about 12 hr after mating (Davis, 1956, 1964). In *Cimex lectularius*, the seminal conceptacles are not homologous to the spermathecae of other insect species. They are mesodermal structures annexed to the oviducts (Carayon, 1966). From there the spermatozoa migrate within the oviductal walls (Abraham, 1934) towards the ovarioles and accumulate in the syncytial bodies (Davis, 1964); fertilization occurs in the ovarioles (Fig. 41). The lateral oviducts possess an intra-epithelial network, called "spermodes", through which the spermatozoa actually migrate (Carayon, 1966), i.e. they do not penetrate the oviductal cells. If spermatozoa were artificially injected into the spermalege or haemocoel, they did not migrate toward the seminal conceptacle; however, if semen was injected with the sperm, they did (Davis, 1965a). Spermatozoa could also be activated by 0.07 M sodium citrate. Stimulation of egg maturation via the activation of the corpora allata is initiated by the spermatozoa from the upper part of the lateral oviducts (see p. 161).

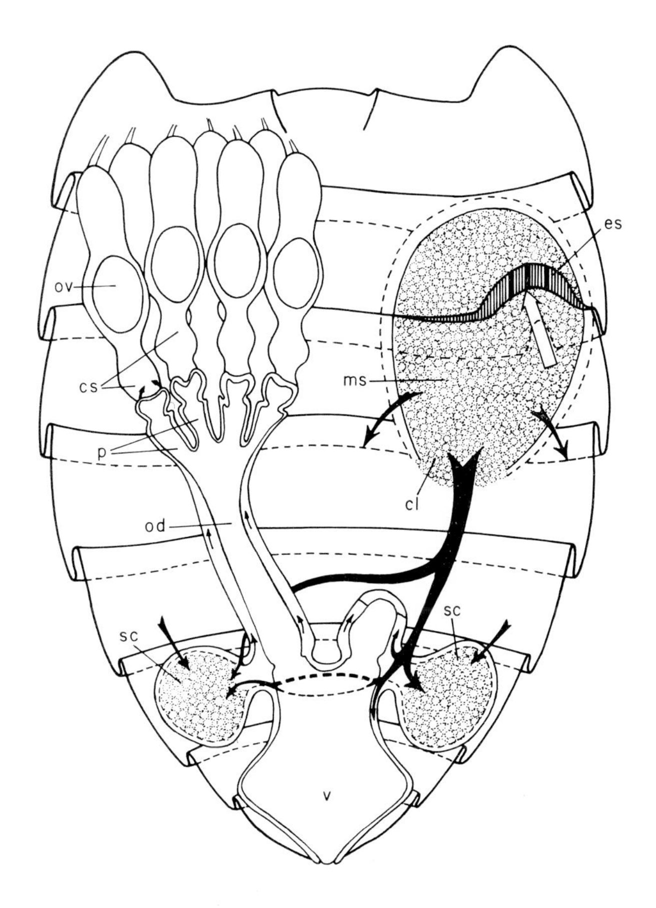

FIG. 41. Diagram of the paragenital system and the process of insemination in Cimicidae. Arrows indicate the pathway of the spermatozoa from the spermalege to the corpora seminales at the base of the ovarioles. cl = conductor lobe found in some species. cs = corpora seminales. es = ectospermalege. ms = mesospermalege. od = oviduct. ov = oocyte. p = pedicel. sc = conceptaculum seminalis. v = vagina. (From Carayon, 1966; copyrighted with The Thomas Say Foundation, The Entomological Society of America.)

Haemocoelic or extra-genital insemination is found in four families of Cimicoidea: namely, Anthocoridae, Cimicidae, Polyctenidae, and Plokiophelidae. Members of these families have been studied by Carayon for several years and the results were summarized in the recent monograph of Cimicidae (1966b). Based on these studies, Carayon (1966a, b) and Hinton (1964) speculate on the possible evolution of haemocoelic insemination. In some species belonging to the genera *Prostemma* and *Alloeorhynchus* (family Nabidae), sperm is injected into the haemocoel through the roof of the vagina, whereas in others the sperm may either be injected into the lumen of the oviducts or into a pouch-like organ of the genitalia; the latter organ may be analogous to the spermalege (Carayon, 1952a, b, c). In *Alloeorhynchus vinulus*, insemination occurs either through perforation of the vagina or integument (Carayon, 1966a). In these species it appears that haemocoelic insemination involves only an accidental integumental penetration. A further step forward is probably found in *Primicimex cavernis*, where the integument may be penetrated at almost any place on the abdomen, as judged by numerous scars in older females; no spermalege is present in this species (Carayon, 1954). A typical spermalege is to be found in species of the genus *Xylocoris*, as well as others (Carayon, 1952d, 1953). It may be located at almost any abdominal segment. The highest stage in perfection may be found in *Stricticimex brevispinosus* (Carayon, 1959) and some other species (Carayon, 1966b). In these cases, true haemocoelic insemination does not occur, but insemination takes place via a spermalege. The spermalege has, however, a more or less solid tissue connection to the oviducts. Spermatozoa migrate within this conductor cord to the genitalia.

What may be the evolutionary advantage of haemocoelic insemination? It was noticed by many investigators of Cimicoidea that sperm masses are digested within the haemocoel of the female; and, following this observation, it was presumed that the female can survive for long periods without food, using the large masses of semen as nutrients (Hinton, 1964). It is, however, doubtful whether injected sperm masses are indeed sufficient to bridge long starvation periods and allow survival of the population since the nutrient reserves of the population (males and females) are thereby not increased. Furthermore, females which do not feed, only rarely will mature a few eggs after mating (Davis, 1964). Bloodsucking Hemiptera are well adapted to survive extended periods without food. If the theory of nutritional value of semen does not hold, haemocoelic insemination in Cimicoidea appears to be only an interesting oddity, a deviation from the main line of insect development.

Little other than the mere occurrence of haemocoelic insemination is known in Strepsiptera. This is probably due to the fact that the life histories of many of the parasitic Strepsiptera are incompletely understood. The larviform viviparous female lives permanently within the host and is not visible except during a brief period of a few days. At the end of the larval development, the female penetrates the cuticle of the host with its cephalothorax but does not emerge from it. This was first seen by Perkins (1918) and later confirmed by Hofeneder (1923); the males copulate with this protruding front part of the Stylops. The winged males, which for a long time were not even known, live for only a few hours, during which time they have to find their mate. Males generally do not leave the brood from which they emerged. Hughes-Schrader (1924) saw that in *Acroschismus wheeleri* the spermatozoa pass along the brood canal and its ducts into the haemocoel. Similarly, extra-genital insemination was found in the strepsipteran *Corioxenos antestia*, a parasite of the coffee bug *Antestia* spp. (Kirkpatrick, 1937), *Halictophagus tettigometrae* (Silvestri, 1941), several species of *Mengenilla* (Silvestri, 1942), and *Elenchus tenuicornis* (Baumert, 1959). From these reports one can probably assume that haemocoelic insemination occurs in all of the parasitic Strepsiptera. Within a few minutes of copula, spermatozoa are found within the haemocoel of the Stylops female, which would suggest that the aedeagus penetrates via the brood canal into the coelom (Lauterbach, 1953, 1954). Since eggs develop free-floating in the coelom of the female, the free spermatozoa have ready access to them. Spermatozoa were found in the coelom of the female up to several weeks after mating.

4. THE INSEMINATION REACTION

In 1946 Patterson reported a rapid and copious secretion of fluid into the vagina of several *Drosophila* species shortly after copula. This was found both in intra- and interspecific matings. The material, which is secreted by the vagina itself, greatly distends it; but within 6 to 9 hr the vagina returns to its normal condition. However, in interspecific mating this secretory material often hardens and forms a solid plug which remains in the vagina for several days, often causing damage to the tissues (Patterson, 1946, 1947). In homogamic mating, the reaction was similar for all species examined. In these cases, the contents of the swollen vagina, which consists of excess spermatozoa and the reaction material, were expelled in the form of whitish droplets.

In heterogamic matings, the reaction generally proceeds faster and the spermatozoa often move more slowly to the entrance of the ventral receptacle. They often form a tangled mass near the opening of the seminal receptacle, thus retarding the entrance of the spermatozoa. In many cases they do, however, enter the receptacle and are then located at its distal end (Patterson, 1947). The consistency of the reaction mass depends on the cross, indicating that the semen is responsible for the degree of reaction. The reaction occurred even if sterile males were mated with normal females, suggesting that it is not the sperm which causes the secretion by the vaginal lining (Patterson, 1947); presumably it is the material from the paragonia (Wheeler, 1947). Patterson and Wheeler favor the idea that possibly foreign proteins cause this reaction in the vagina. Among the seventy-eight species examined by Wheeler and Patterson, seventeen did not exhibit the insemination reaction although it was found within the others in various degrees. Interestingly enough, hybrid males of *D. pavani* and *D. gaucha* mated to non-hybrid females elicited a stronger reaction than did pure heterogamic males (Koref-Santibañez, 1964).

Can we speculate on the physiological meaning of the insemination reaction? It is possible that the plug within the vagina may prevent immediate loss of spermatozoa in intraspecific matings. In interspecific matings, on the other hand, filling of the seminal receptacle is frequently prevented and thus potentially leaves room for spermatozoa from subsequent matings. However, in these cases the damage caused by the hardened plug and its prolonged presence apparently prevents further copulations with the species' own male. In other words, it is difficult to visualize any advantage to the species. Furthermore, both the species which show the reaction and those which do not propagate successfully.

5. PASSAGE OF SPERMATOZOA WITHIN THE FEMALE

Once spermatozoa have been deposited in the female reproductive tract, they have to move into the storage place, the spermatheca. Mobility of spermatozoa has been observed *in vitro* in many species, but, as was pointed out by Hinton (1964), little is known of the actual transport mechanisms *in vivo* for most species. The observed *in vitro* mobility of spermatozoa gives no evidence for directed movement within the reproductive tract. The questions indeed arise whether the sperm mass actively moves into the spermatheca, and how much the movements of the spermathecal ducts help in the transport. From the limited information in a few species, we must assume that a variety of mechanisms are at work. Our basic knowledge of the matter will be presented.

In species which have a spermatophore, spermatozoa may escape from it in several ways. For instance, in the house cricket where the spermatophore remains in the vulva for only a few minutes, the sperm mass must be transferred into the bursa within a short time period. Khalifa (1949a) found that the spermatophore in this species contains a pressure body and an evacuation fluid. The evacuation fluid is taken up by the pressure body, which swells and forces the spermatozoa into the female genital tract. In many Lepidoptera, the hard spermatophore is broken open by a chitinous structure in the bursa, the lamina dentata or signum (Weidner, 1934; Hewer, 1934). Some species may have several signa. The sperm mass is then released and moves through the sometimes long and coiled ductus seminalis into the spermatheca. There is little doubt that in Lepidoptera the secretion of the bursa copulatrix also, partially or totally, digests the spermatophore and thus may aid in the release of spermatozoa. Rhythmic contraction of the genitalia in *Porthetria dispar* may be responsible for passive sperm movement (Klatt, 1920). Chemotactile attraction of spermatozoa to the spermatheca has been claimed, but actually there is little concrete evidence for this, and the assumption still awaits rigorous experimental testing. A secretion from the bursa may be responsible for the activation of spermatozoa in *Galleria mellonella* (Khalifa, 1950b) since mixing the sperm with male secretion did not activate them. On the other hand, the sperm fluid in *Zygaena* (Hewer, 1934) and some component of the spermatophore in *Melolontha* (Landa, 1960) activate the spermatozoa. However, whether activation of the sperm can alone account for its movement into the spermatheca has not been shown. Within 5 minutes after mating, the spermatheca are filled in *Drosophila melanogaster*; Noñidez (1920) claims that some secretion from the seminal receptacle causes activation and rapid migration of the spermatozoa. Only activated spermatozoa ascend the non-contractile spermatheca in *Aedes aegypti*, in which species it was also shown that sperm from ejaculates reach the storage space better and faster than sperm from the seminal vesicles (Jones and Wheeler, 1965a, b); this again might indicate that some component of the semen activates the spermatozoa. Dead sperm were not transported into the spermatheca in this species.

In contrast to those reports, Ruttner (1956a, b) believes that in the honeybee queen spermatozoa are at least in part passively transported into the spermatheca by the contractions of the spermathecal duct and abdomen. The sperm are practically squeezed through it. Similarly, rhythmic movements of the bursa, oviducts, and spermathecal ducts are responsible for sperm movement in *Rhodnius prolixus* (Davey, 1958). The secretion of the opaque male accessory sex gland, which is normally transferred together with the spermatophore, causes rhythmic contractions of the female's genitalia. After removal of this gland, a normal spermatophore is transferred, but the sperm does not leave the bursa, and no contraction of the female genitalia is induced. On the other hand, dead spermatozoa could be transported into the spermatheca provided that a normal spermatophore (including the secretion of the opaque gland) was transferred. This report on *Rhodnius* is probably still the only one which clearly shows that the secretion from the male causes the female to transport the spermatozoa within her reproductive tract. One could surmise that similar mechanisms are operative in other insect species, but unequivocal experiments still have to be conducted in each case.

A unique mechanism of sperm transport in the female tract is found in the Coccid *Aspidiotus nerii* (Krassilstschik, 1893) and *A. ostreaeformis* (Pesson, 1950). In these viviparous species, large cells seem to proliferate in the lower parts of the oviducts; these cells which contain several spermatozoa then migrate up the lateral oviducts into the pedicel. How those cells take up the spermatozoa is obscure. They certainly do not serve as storage organs for the sperm, since spermatozoa are found within the spermatheca.

CHAPTER 7

FACTORS THAT AFFECT EGG PRODUCTION AND FECUNDITY

IT IS apparent from the many reports on insect reproduction that a variety of factors, both external and internal to the animal, influences total egg production. Nutrition, which has been shown to affect the total egg output in several ways, is probably the most important single factor in the majority of insect species. Other factors in addition to feeding, e.g. mating, are known to influence egg laying in many species. Moreover, light, temperature, and humidity may not only interfere directly with egg production but may also have an effect on feeding and mating activities, which in turn influence egg production. Numerous reports on the number of eggs laid by various species, particularly those injurious to agricultural products and those of medical interest, are found in the entomological literature. However, records of egg laying or total egg output do not necessarily give a true picture of the actual reproductive potential of the species under study since mature eggs may often not be deposited for various reasons. For example, unmated females of many species which mature a full complement of eggs lay only a portion of them. In other species mating primarily causes an increase in egg production and secondarily induces oviposition. In other words, the effect of mating on total egg output can be twofold. Furthermore, the proper substratum, or the host in parasitic insects, may not be available for oviposition; many eggs may therefore be retained in the ovaries or genitalia and later resorbed. Since these various aspects often have not been considered separately, it is frequently not possible to determine from the literature the reproductive capacity of a given species.

Aging accounts for a decrease in the rate of egg production in many species, i.e. egg output slowly declines as the female ages. In *Aedes aegypti*, for example, it has been shown that successive clusters contain fewer eggs. The females laid approximately 15 per cent fewer eggs in successive batches (Putnam and Shannon, 1934), presumably because of a steady increase in the number of degenerating follicles in the ovaries. A decrease in egg output with age is independent of food availability, substratum for oviposition, or chances to mate with young males, as can be seen from observations on *Drosophila melanogaster* having the genotype lgl/Cy (Fig. 42) (Hadorn and Zeller, 1943).

In Table 5 the total number of eggs laid by various insect species is listed. This list of data was accumulated while scanning the literature on insect reproduction. If known, the number of eggs deposited under optimal conditions was chosen. Where data on egg laying in both mated and virgin females of bisexual species were available, only those for mated females are quoted. This table shows the great variability in egg production even among species of the same genus or family. Extremely high fecundity is found in social insects, such as army ants and bees; the total egg production by one female in these species can only be estimated. The total reproductive capacity of queens of certain termite species is difficult to assess since at

TABLE 5. Number of eggs laid by insect species (mostly laboratory observations)*

Species	Number of eggs laid average	maximum	Remarks
Thysanura			
Folsomia candida	168		
Thermobia domestica	62		
Orthoptera			
Acheta configuratus	667		
,, *domesticus*	1060		
,, *sigillatus*	864		
Anacridium aegypticum	300		
Camnula pellucida	179	305	
Euthystira brachyptera	116		
Gryllus campestris	1000		
,, *domestica*	2636		
Locusta migratoria	497		
Melanoplus bilituratus	252		
,, *bivittatus*	355		
,, *differentialis*	292		
,, *sanguinipes*	329		
Nomadacris septemfasciata	600		
Oecanthus nigricornis	61	165	
Schistocerca gregaria	697		
Trimerotropis pallidipennis	225	955	
Phasmida			
Bacillus rossii	398		
Parasosibia parva	350		
Phobaeticus sinetyi	137		
Dermaptera			
Euborellia annulipes	160		
Prolabia arachidis	76		
Dictyoptera			
Blatta orientalis	161		
Blattella germanica	220		
Diploptera punctata	30–40		
Periplaneta americana	938		
,, *australasiae*	600		
Psocoptera			
Ectopsocus meridionalis	155	195	
Hemiptera			
Alydus pilosulus	46	133	
Antestia lineaticollis	144		
,, spec.	148		
Anthocoris confusum	120		
,, *nemorum*	73		
,, *sarothammi*	99		
Antillocoris minutus	94	158	
Calidea dregii	150–200		
Carpilis consimilis	70	147	
Cimex lectularius		541	
Delochilocoris umbrosus	123	152	
Drymus unus	131	250	

* In this table authors are not listed because the reference list would have become rather cumbersome.

Species	Number of eggs laid average	maximum	Remarks
Hemiptera (cont.)			
Dysdercus fasciatus	300		
„ *sidae*	496		
Emblethis vicarius	181	218	
Eremocoris ferns	198	291	
Eurygaster integriceps	85	300	
Euschistus conspersus	225	640	
Geocoris punctipes	178	469	
Haematosiphon inodorus	33	96	
Hesperocimex sonorensis	343		
Ligyrocoris carius	133	151	
„ *diffusus*	166	325	
„ *silvestris*	272	475	
Megalonotus sabulicolus	167	129	
Megalotomus quinquespinosus	237	615	
Murgantia histrionica	51		
Nysius huttoni	204		
Pachybrachius albocinctus	140	231	
„ *basalis*	108	210	
Perigenes constrictus	268	330	
Peritrechus fraternus	193	386	
Plinthisus americanus	67	97	
Podisus maculiventris	550	1097	
Pseudocnemodus canadensis	154	280	
Ptochiomera nodosa	131	230	
Rhodnius pictipes	377	513	
Scolopostethus diffidens	132	184	
„ *thomsoni*	104–250		
Stynocoris rusticus	98	128	
Trapezonotus arenarius	97	141	
Triatoma sanguisuga	711	1166	
Zelus exsanguis	406		
Zeridoneus castalis	164	231	
Homoptera			
Deltocephalus sonorus	27	50	
Empoasca fabae		226	
Icerya purchasi	1250		
Lacifer laeca	325		
Lecanium kanoensis	1200	1700	
Lipaphis erysimi	30–60		
Phenacaspis pinifoliae	40		
Pineus pineoides	4–6	8	hiemosistens
Planococcus citri	577		
Pseudaulacaspis pentagona	150–200		
Pseudococcus obscurus	400		
Psylla nucatoides	463		
„ *pyricola*	664		
Stenocranus minutus	385		
Therioaphis maculata	90		
Neuroptera			
Agulla bractea	460	792	
Chrysopa californica	716		
„ *carnea*	477	1022	
„ *oculata*		617	
Nothochrysa californica	52		
Mecoptera			
Panorpa communis	150		

TABLE 5 (cont.)

Species	Number of eggs laid average	maximum	Remarks
Lepidoptera			
Achroia grisella	300		
Acleris variana	63		
Acrolepia assectella	233		
Agrotis orthogonia	235	564	
,, *segetum*		1053	
Arctia caja	1150		
Bombyx mori	750		
Bupalus piniarius	175		
Cacoecia murinana	105		
Carpocapsa pomonella	180		
Choristoneura fumiferana	160		
Chorizagrotis auxiliaris	1000–2500		
Colias philodice eurytheme	715	1172	
Cosmotriche potatoria		269	
Dasychira pudibunda	434		
Dendrolinus pini	254		
Diacrisia strigatula	626		
Elasmopalpus lignosellus	126		
Ephestia kühniella	422		
Euproctis chrysorrhoea	470		
Fumea crassiorella	100–200		
Galleria mellonella	1550	1900	
Gastropacha quercifolia	600		
Hyalophora cecropia	281	390	
Laothoe populi	112		
Lasiocampa callunae		238	
,, *pini*	88–330		
Lymantria monacha	130–170	447	
Malacosoma neustria	279		
Oncopera intricata	589		
Pammene juliana	180	336	
Panaxia dominulla	165		
Panolis flammea	193		
,, *piniperda*	150–200		
Platynota flavedana	284		
Plodia interpunctella	160–300		
Plutella maculipennis	177		
Porthetria dispar	556	800	
Prionoxystus robiniae	720	1008	
Pseudoletia unipuncta	1451	1756	
Pyrausta nubilalis	533		
Rhyacionia buoliana	116		
Scrobipalpa ocellatella	208		
Sphinx pinastri	200		
Tineola biselliella		200	
Tinolius eburneigutta	552		
Trichoplusia ni	604		
Tryphaena pronuba	1371	1457	
Diptera			
Anastrepha ludens	1769	4164	
Anopheles maculipennis	450	1145	
Bactrocera cucurbitae	246	687	
Boophthora erythrocephala	186		
Ceratitis capitata	280	568	
,, ,,		625	
Chironomus plumosus	1676		
Chortophila brassicae	160		

Species	Number of eggs laid average	maximum	Remarks
Diptera (cont.)			
Chrysomyia rufifacies	377		
Cochliomyia hominivorax	231		
Dacus dorsalia	1100		
,, *oleae*	276		
Dasyneura leguminicola	96		
Drino munda	19	36	
Drosophila melanogaster	1532		wild-type
,, ,,	912		inbred
,, ,,	1800		lgl/Cy
Glossina palpalis	9	14	
Hippobosca variegata	4.5		
Hylemya antiqua	150		
Lampetra equestris	83		
Megarhinus brevipalpis	85		
Melophagus ovinus	12		
Miastor metraloas		47	
Musca autumnalis	67	230	
,, *domestica*	920	1278	
,, *domestica vicina*	267	700	
Mycophila nikoleii		54	
,, *speyeri*	23	44	
Pegomyia betae	75		
Phytobia maculosa	367	1404	
Rhagoletis pomonella	395		
Stegomyia fasciata	750		
Syrphus corollae	404		
,, *ribesii*	143	336	
Tekomyia populi		86	
Wyeomyia smithii	28		
Hymenoptera			
Apanteles angaleti	120	142	
,, *ruficrus*	216	490	
Apis mellifera	*ca.* 120,000		per year
Caeloides brunneri	21		
Campoplex haywardi	87		
Caraphractus cinctus		153	
Coccophagoides utilis	26		
Diadromus pulchellus	78		
Dibrachoides dynastes		122	
Dusmetia sangwani	35		
Eciton hamatum	*ca.* 28,000		per brood
,, ,,	*ca.* 200,000		per year
Euchalcidia caryobori	88	181	
Leptomastix dactylopii	392 ± 60		
Melittobia acasta	878	1217	mated
,, ,,	35	92	virgin
Meteorus loxostegi	200–240		
Mormoniella vitripennis	607	685	
Neodiprion rugifrons	90–130		
Opius consolor	45		
Pygostolus falcatus		46	
Scambus buolianae	500		
Spalangia drosophilae	143		
Syntomosphyrum albiclavus		195	
Trichogramma evanescens	66.1 ± 2.5		fed sugar
	81.5		fed honey

TABLE 5 (cont.)

Species	Number of eggs laid average	maximum	Remarks
Coleoptera			
Acanthoscelides obtectus	121		
Adalia bipunctata	1535 ± 158		
Agrilus viridi	8		
Agriotes lineatus	165		
,, *obscurus*	225		
Aleochara taeniata	324	639	
Anthonomus grandis	180	283	
Anthrenus verbasci	66		
Aphodius howitti	50		
,, *tasmaniae*	55		
Araecerus fasciculatus	79	125	
Bruchus obtectus	67		
,, *quadrimaculatus*	130	196	
Calandra granaria	191	362	
Calligrapha alni	422		
,, *apicalis*	168		
,, *multipunctata*	359		
,, *scalaris*	167	258	
,, *vicius*	79		
,, *virginea*	301		
Calosobruchus analis	77		
,, *chinensis*	53		
,, *maculatus*	97		
Caryedon gonagra	175	207	
Chilomenes sexmaculata		2384	
Chrysomela varians	45		
Coccinella septempunctata	814	3765	
,, *undecimpunctata aegyptiaca*	746		
Conoderus vespertinus	541		
Ctenicera aeripennis	300		
,, *destructor*	355	638	
Dermestes vulpinus	300–400		
Ernobius mollis	20	66	
Galerucella viburni	250	465	
Gastrophysa polygoni	2500	2930	
Haltica ampelophaga	500	524	
Laemophloes minutus	242 ± 28		
Lampyris noctiluca	60–90	198	
Lasioderma serricorne	76	93	
Leptinotarsa decemlineata	1027		isolated
,, ,,	321		grouped
Melolontha melolontha	24	50	
Nebria brevicollis	31 ± 8		
Oncideres texana	175	207	
Oncopera intricata	589		
Orina vittigera	50	57	
Otiorrhynchus cribricollis	51		
Phausis splendidula	60–90	147	
Phyllopertha horticola	14	30	
Pityogenes chaleographus	10–26		
Prionoplus reticularis	400		
Rhipidius quadriceps	1800	2200	
Rhizopertha dominica		415	
Rhynchophorus palmarium	245	718	
Sitodrepa pauicea	57	72	
Sitona crinitus	519	3000	
,, *cylindricollis*	1700		

Species	Number of eggs laid average	maximum	Remarks
Coleoptera (cont.)			
Sitona decipiens	348		
,, *flavescens*	396		
,, *hispidula*	719	1200	
,, *lineata*	1800	4500	
,, *lividipes*	1168		
,, *regensteinensis*	400		
,, *ulcifrons*	529		
Tenebrio molitor	83	178	
Tribolium confusum	971	1582	
,, *destructor*	1200		
,, *ferrugineum*	518		
Trogoderma anthrenoides	54		
,, *parabile*	77		
,, *versicolor*	95–120		
Xyloborus compactus	20		
Zabrotes subfasciatus	39		
Strepsiptera			
Corioxenos antestiae	3500	3720	
Elenchinus chlorionae	1200–1500		
Mengenilla quaesita	2000		
Stylops spec.	1000–2000		

their peak of productivity they may lay an egg every 2 sec, and it is uncertain how long this period can last. The other extreme is observed among viviparous flies or the oviparous generation of aphids; some of the latter species lay only a single egg at the end of the reproductive season. The number of eggs laid by a species is certainly in part associated with the number of ovarioles per ovary. This is well illustrated by some species of the Coleoptera. At the one extreme, over 200 ovarioles are found in the *Meloe* species while at the other extreme, in the *Coprinae*, only one ovariole and one ovary are found (Robertson, 1961). In the latter case, one observes a certain relationship between the extremely low number of ovarioles, which means a very low reproductive potential, and the behavior of the female: these species exhibit a certain degree of parental care.

It is doubtful whether the compiled data are representative of egg laying in natural environments since, in most cases, they were collected from laboratory colonies; animals reared in the laboratory usually live under better conditions, particularly with respect to food availability and temperature, than are to be found in the field. Nevertheless, the figures give a good indication of the reproductive potential of many species and thus may be of value.

A. *Nutritional Requirements*

Both the quantity and quality of food influence total egg production and egg laying of a given species. The effect of foods on egg production results from the balance between a multiplicity of factors, such as species specific requirements, temperature, metabolism, food constituents, and quantity. Nutrition in the larval stages often significantly influences fecundity of the resulting adults, particularly in species where the adult females do not feed. In other cases, specific factors supplied with the normal food, such as hormones of the host on which the animals feed or specific vitamin and mineral requirements, affect the total egg output. The majority of studies have been concerned with food requirements of immature stages; as a consequence, we have relatively little detailed knowledge about adult food needs. Experiments which attempt to associate certain nutrient factors with specific developmental processes are often inconclusive since materials needed in trace amounts may be carried over from the egg and larval stages into the adult. Long-term experiments, which would be essential to determine such correlations, have rarely been carried out. Moreover, microorganisms, which are frequently found in the intestine and fat bodies of the insects, may supply essential vitamins; therefore, axenic conditions would be required in order to observe the actual nutritional requirements of the animal. The subject of nutritional requirements for reproduction has been reviewed in some detail by Johansson (1964); nevertheless, it seems necessary to deal with the matter once again in the present context.

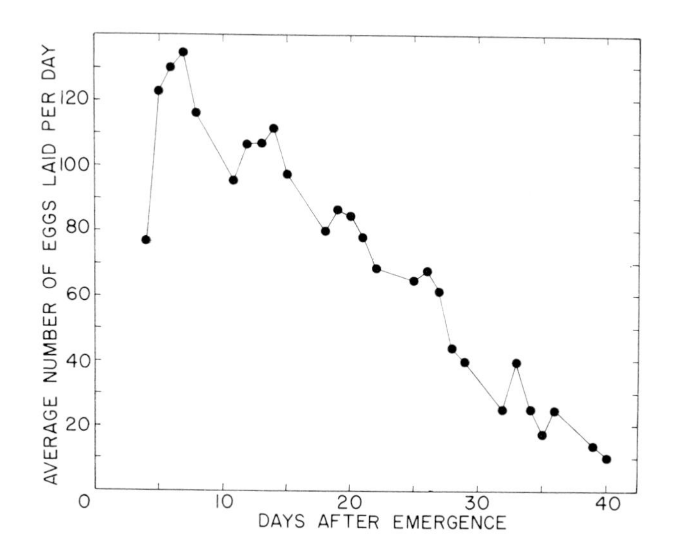

FIG. 42. Average daily number of eggs laid by six *Drosophila melanogaster* females of the genotype lgl/Cy. The decrease with age in number of eggs laid is typical for many species. (Modified from Hadorn and Zeller, 1943.)

1. FOOD QUALITY

The effect of different kinds of food on reproductive processes has been reported for a variety of phytophagous, carnivorous, and omnivorous insect species. In many cases the normal food seemed superior to other foods in promoting egg production. This may be due to superior nutritional value or it may be caused by the animals' having eaten greater quantities of the normal food than of the other foods tested. It is often difficult to decide which of these possibilities applies since most reports do not distinguish between quality and quantity. Therefore, reported differences in the effect of food qualities may have been essentially reports on differences in food quantities consumed.

While it is known for a variety of insect species that feeding is essential for egg maturation and for increased egg production, not much else is known about specific food requirements for most of these species. The few known cases will be treated in the subsequent paragraphs.

Tribolium confusum laid 521 eggs when fed on whole wheat flour, 333 on bran, and only

187 on white flour (Good, 1933); obviously whole wheat flour was the most nutritious material. The same was found for the flour beetle, *Tribolium ferrugineum* (Good, 1933). However, in neither case could the specific component in whole wheat flour which best promoted production of eggs be determined. The beetle *Epitrix tuberis* laid more eggs if fed on potato, its normal food plant, than if fed on tomato, bean, or march elder (Hill, 1946). *Solanum edinese* and *Solanum tuberosum* were far superior to other solanaceous food plants for the potato beetle, *Leptinotarsa decemlineata* (Trouvelot and Grison, 1935). The beetles lay eggs even when fed on lettuce, but their egg output is rather poor (Hsiao and Fraenkel, 1968). Again, cotton seeds, the natural food for *Dysdercus superstitiosus*, allowed a higher rate of egg production than did several other seeds (Geering and Coaker, 1960). Differences in the nutritional value of various food plants were also shown for *Melanoplus bilituratus* (Pickford, 1958) and for several Middle European grasshoppers (Kaufmann, 1965).

Green food plants are important for maximal egg production in the locusts *Dociostaurus maroccanus* (Merton, 1959) and *Schistocerca gregaria* (Cavanagh, 1963). The essential factor is not definitely known in either of these cases, except perhaps for *Schistocerca*, in which the addition of gibberellin A_3 to old leaves markedly increased egg production. In senescent leaves, only small quantities of gibberellins seem to be present (Ellis *et al.*, 1965). For the Colorado beetle, old leaves were of lesser nutritive value than young leaves, i.e. the beetles laid significantly fewer eggs if fed on old leaves (Grison, 1944, 1948). *Choristoneura fumiferana* laid twice as many eggs when fed on new leaves than when fed on old ones (Blais, 1952). Similarly, old leaves were of poor nutritive value for *Heliopeltis theivora* (DeJong, 1938). Second and third generations of the beetle *Haltica ampelophaga* (Picard, 1926) and the silkmoth *Bombyx mori* (Yamashita *et al.*, 1961), living during the summer and early fall, laid fewer eggs than did females of the first generation of the year. This could possibly be caused by the fact that later generation females had to feed on older food plants, which perhaps contained fewer nutrients than did those younger plants available in early summer. Again, it is not certain whether the animals of the later generations ate as much as did those of the first generation; possibly the younger leaves are more palatable.

Species which do not feed as adults but whose larvae have been reared on different foods lay more eggs in the instances where the larvae have fed on more nutritious foods. For example, females of *Trogoderma versicolor* reared on wholemeal or oatmeal as larvae laid on the average 105 eggs, whereas they averaged 83 on maize and only 67 on wheat (Norris, 1936). *Porthetria* (=*Lymantria*) *dispar* reared on *Quercus pedunculata* as larvae laid 586 eggs, whereas those reared on *Malus domesticus* laid only 180 (Kovačevič, 1956). Corn or cotton squares were inferior to an artificial medium for *Heliothis zea* (Lukefahr and Martin, 1964). Obviously, larvae reared on certain foods accumulated more reserves or utilized more food than those reared on others; the adults laid eggs accordingly.

Carnivorous and omnivorous species also lay varying numbers of eggs when fed on different foodstuffs. For *Culex pipiens*, canary blood seemed to have a higher nutrient value than human blood (Tate and Vincent, 1936) since more than twice as many eggs were laid by females fed on the former rather than on an equal amount of the latter (Woke, 1937b). The anautogenous females of *Culex pipiens berbericus* laid 187 eggs when fed on canary and 58 when fed on man. For *Aedes aegypti* females, rabbit, canary, turtle, frog, and guinea-pig blood were more nutritious than human blood: the females laid nearly twice as many eggs when fed on any species other than man (Woke, 1937a). It is interesting to note that blood of poikilotherms was as nutritious as blood from canaries or guinea pigs. The flea *Xenopsylla cheopis* laid more eggs and matured them faster if fed on adult rather than on baby mice

(Buxton, 1948). In all of these cases the specific factor or factors which favor egg production at a higher rate are not known.

Water is usually taken in with the normal foods in most insect species; therefore, its effect on egg maturation cannot be determined precisely. However, it has been shown that, for some autogenous species which do not need to ingest food of any kind and whose water requirement could therefore be studied, water consumption increased the total egg output. For example, in the beetle *Bruchus quadrimaculatus*, egg production increased by about 25 per cent if the animal was given access to water (Larson and Fisher, 1924). In the moth *Ephestia cautella*, egg production likewise increased after water was supplied; however, no effect was noted in *Ephestia kühniella* (Norris, 1934). The absolute lack of dependence on drinking water for maximal egg output in *E. kühniella* is probably related to its normal habitat of dry flour. Egg production was increased 3 to 4 times in *Pyrausta nubilalis* (Kozhantshikov, 1938) and tripled in *Agrotis orthogonia* (Jacobson, 1965) after drinking of water had been allowed. These few examples illustrate the fact that water is essential for maximal egg production in at least some species. Water probably is essential for egg maturation because eggs contain a considerable amount (30–80 per cent), but the females do not have large water reserves.

One of the specific requirements for reproduction in insects is an adequate supply of carbohydrates. Although carbohydrates *per se* do not seem to be directly used in yolk formation in the oocytes (eggs contain very little of it), they are possibly necessary for utilization of dietary protein reserves (energy supply) in both autogenous and anautogenous species. This is shown, for example, in the autogenous Ichneumonid *Diadromus pulchellus* which matures no eggs if not given some supply of carbohydrates (Labeyrie, 1960a). After only sugar or honey had been supplied, egg production increased in *Chrysopa californica* (Hagen, 1950), *Rhagoletis lingulata* (Kamal, 1954), *Lampetra equestris* (Doucette and Eide, 1955), *Anthrenus verbasci* (Blake, 1961), and *Hippelates pusia* (Schwartz and Turner, 1966). In the butterfly *Colias philodice eurytheme*, a threefold increase over the number of eggs laid by females that had been fed only on water was observed in animals which had been fed on sugars (Stern and Smith, 1960). *Trichoplusia ni* fed on various sucrose concentrations showed an increase in total egg output with increased sucrose concentration: on 1–4 per cent sucrose, 200–400 eggs were laid by a female, and on 8–16 per cent, approximately 500–600 eggs were laid (Shorey, 1963). Different sugars may be of variable quality with respect to egg production, as is exemplified by the Syrphid *Sphaerophoria scuttellaris* (Lal and Haque, 1955). This autogenous species laid, on the average, 37.2 eggs when fed on sucrose, 14.7 eggs on fructose, and 6 eggs on maltose.

Many species survive on a carbohydrate diet alone for many weeks but do not produce or lay any eggs. These same species, on the other hand, when fed on only proteins, do not produce any eggs either and die within a few days. Apparently, only a combination diet of sugars and proteins allows egg maturation in the ovaries in these cases. This applies to many species, but has been demonstrated particularly well in the flies *Musca domestica* (Roubaud, 1922; Glaser, 1923), *Lucilia sericata* (Dorman *et al.*, 1938), *Calliphora erythrocephala* (Thomsen, 1952), *Phormia regina* (Rasso and Fraenkel, 1954), and *Lucilia cuprina* (Webber, 1958). In *Culicoides obsoletiformis*, sugar supplementation of the blood meal increased egg production (Amosova, 1959). Furthermore, in *Aedes diantaeus* and *A. intrudens*, more blood per egg produced had to be ingested when sugar had not been given than when it had (Volozina, 1967). These findings on the effect of carbohydrates on egg maturation in various species lead to the hypothesis that primarily energy is supplied with sugar ingestion. No conclusive information is available from any species as to how sugars may actually facilitate the use of proteins in yolk formation.

Just as qualitatively different sugars may have, in some cases, different effectiveness in promoting egg maturation, different proteins may also be more or less nutritious. For example, *Drosophila melanogaster* females laid about 500 eggs if fed on bakers' yeast but only 400 if fed on brewer's yeast (Robertson and Sang, 1944). *Drosophila ampelophila* laid more eggs when fed on yeast than when fed on potato (Guyénot, 1913). Potatoes probably contain fewer proteins than yeast although yeast proteins may be of higher nutritive value for this species. Certain proteins facilitated egg laying in *Culex pipiens* while others did not (Huf, 1929). Greenberg (1951) reported for *Aedes aegypti* that bovine plasma albumin, casein, gelatin, or globulin facilitated egg production; all of these proteins were superior to erythrocytes. As early as 1922, Gordon mentioned that *Stegomyia calopus* laid eggs when fed on whole blood of bats and birds but not when fed on washed erythrocytes. Obviously, proteins from red blood cells are, in some way, less nutritious than other proteins. Liver proteins were superior to a variety of other proteins in promoting egg laying in *Phormia* (Rasso and Fraenkel, 1954), but eggs were still produced on several other proteins. The same applies to *Lucilia* (Webber, 1958). Varying numbers of eggs were also laid by *Protophormia terraenovae* (Harlow, 1956), *Musca autumnalis* (Wang, 1964), and *Musca domestica* (Monroe and Lamb, 1968) depending on the kind of protein fed.

Larval growth can be promoted by the ten so-called essential amino acids if a number of additional substances—e.g. carbohydrates, vitamins, minerals, salts, etc.—are given. A mixture of these ten amino acids in the proper proportions facilitated egg maturation in adults of *Aedes aegypti* (Dimond *et al.*, 1956; Singh and Brown, 1957), *Drosophila melanogaster* (Sang and King, 1961), *Anthonomus grandis* (Vanderzant, 1963), and *Exeristes comstockii* (Bracken, 1965). The elimination of any of these amino acids drastically reduced egg production and/or completely prevented egg laying. Yeast hydrolysates supported egg growth in *Dacus dorsalis*, as did amino acids given together with carbohydrates, minerals, and vitamins (Hagen, 1958). Even single amino acids added to a suboptimal diet caused an increase in egg production. For example, Greenberg (1951) found that the addition of DL-isoleucine (1 per cent) to a variety of dietary proteins markedly improved egg laying in *Aedes aegypti*. This finding was confirmed by Dimond *et al.* (1955). Tryptophan is essential in the food of *Protophormia terraenovae* (Harlow, 1956).

As is implicit in the evidence presented above, a supply of dietary minerals is essential for some insects. It has been the practice, when rearing insects on artificial diets, to add a variety of salts to the standard medium, and better results have generally been obtained when these salts were supplied. Under normal conditions the insects naturally take in minerals with their food. Some insects need specific ions in their diet, as has been shown for *Culex pipiens*. This species could produce eggs on milk alone if it contained 10 per cent $FeNO_4$ (Huff, 1929). *Culex* laid eggs and larval growth was favored even on plant proteins (lentils) as long as Fe ions had been added (De Boissezon, 1933). However, if the species was fed hemoglobin without any additional Fe, egg maturation could occur normally. Egg maturation was accelerated in *Phormia* by the addition of a salt mixture to the food (Rasso and Fraenkel, 1954). Potassium, magnesium, and phosphorous were essential for yolk deposition in *Drosophila melanogaster* (Sang and King, 1961) and oviposition in *Myzus persicae* was doubled if K and Mg ions were added to the diet (Dadd and Mittler, 1965). The mode of action of mineral salts in egg maturation of many species is not understood, but the following example elucidates one possible mechanism. A low K content of the soil results in a high soluble N content of the sap of the plants; aphids feeding on these plants thus take in more N, and, as has been shown, the fecundity of these animals is higher than that of controls (El-Tigani, 1962; Van Emden, 1966). In another aphid, *Neomyzus circumflexus*, which was fed on an artificial diet and reared over many generations, it has been shown that Na, Fe, Zn, Mn, Cu, and Ca are essential for continued high reproductive rates (Ehrhardt, 1968); Fe and Zn were particularly important. Omission of any one of these ions did not immediately cause the cessation of reproduction but did so after several generations. This indicates that they are required only in trace amounts and can be carried over from one generation to the next.

Studies on nutritional requirements in insects must involve the search for essential vitamins; however, relatively few papers have dealt with adult vitamin needs. The study of an animal's vitamin requirement is complicated by two factors. Firstly, vitamins can be stored during larval stages and carried over into the adult, so their absence from the adult food may not have any effect. Secondly, many species have microorganisms in their intestine or fat bodies which may produce essential vitamins. Since different insects have different symbionts, the vitamin requirements then vary from species to species.

In 1946 De Meillon and Golberg showed that *Cimex lectularius* fed on folic acid or thiamine-deficient hosts produced considerably fewer eggs than those fed on adequately nourished hosts. *Drosophila* also require folic acid in their diet for a normal rate of egg maturation (King and Sang, 1959; David, 1964). Aminopterin (an antagonist to folic acid) treatment of the adult female of *Drosophila* caused degeneration of follicle and nurse cells in the ovarioles. Presumably DNA and RNA synthesis was impaired (David, 1964, 1965), since it is known that aminopterin administration in immature *Musca domestica* blocks the action of folic acid in DNA and RNA synthesis (Perry and Miller, 1965). This suggests that folic acid is required by the adult insect for the rapid synthesis of proteinaceous yolk by the ovarioles during egg maturation periods.

Different insects have different vitamin requirements, however. In the case of *Tribolium castaneum*, folic acid and thiamine do not seem to be essential for egg maturation since their omission from the diet did not affect fecundity (Applebaum and Lubin, 1967). Similarly, treatment of females of *Aedes aegypti* with folic acid antagonists (aminopterin and methotrexate) in concentrations of up to 2 per cent in the food was found to be ineffective. The insensitivity of *Aedes* to aminopterin may be associated with the fact that the nurse cells in this species contribute relatively little to the oocyte yolk (Akov, 1967).

During the period of egg growth, *Anthonomus grandis* females need ascorbic acid, B vitamins (Vanderzant *et al.*, 1962), inositol, and choline (Vanderzant, 1963). Omission of any of these vitamins resulted in a low rate of egg laying. B vitamins are essential for accelerated egg growth in *Phormia* (Rasso and Fraenkel, 1954) and in *Drosophila* (Sang and King, 1961). Vitamin deficiencies in *Exeristes comstockii* resulted in rather low egg production (Bracken, 1965, 1966) (Fig. 43). Upon addition of House and Barlow's (1960) vitamin mixture to the normal food, a high rate of egg laying was resumed within a few days. In contrast to results for other species, the absence of RNA and cholesterol from the adult diet of *Exeristes* did not affect egg production. Omission of biotin or pteroylglutamin acid from the diet of the larval cockroach *Blattella germanica* did not prevent the animals from becoming adults, but these females did not produce eggs (Gordon, 1959). Trace amounts of biotin are essential for growth of the larvae of the fruit fly *Anastrepha ludens* (Benschoter and Paniagua, 1966). If, however, it was fed to adults in concentrations higher than 0.1 per cent, their reproductive capacity was reduced; with dosages exceeding 0.5 per cent, it was completely lost. Upon removal of the biotin from the diet, the females recovered completely and resumed egg laying. A 2 per cent biotin content in the food of *Musca* induced sterility (Benschoter, 1967). The beetle *Cryptolaemus montronzieri* can live on a diet composed of casein, corn oil, saccharose, and various vitamins but would not mature eggs unless vitamin E had been added to its food (Chumakova, 1962). Vitamin E was likewise essential for the production of offspring of *Agria affinis* (House, 1966).

Vitamin requirements may be different in larvae and adults as is perhaps exemplified by the following case. Larval growth could be maintained in *Drosophila* without any apparent defects if carnitine replaced choline (Fraenkel *et al.*, 1955). It was, however, later noticed that carnitine-raised females did not lay as many eggs as choline-raised animals did; also,

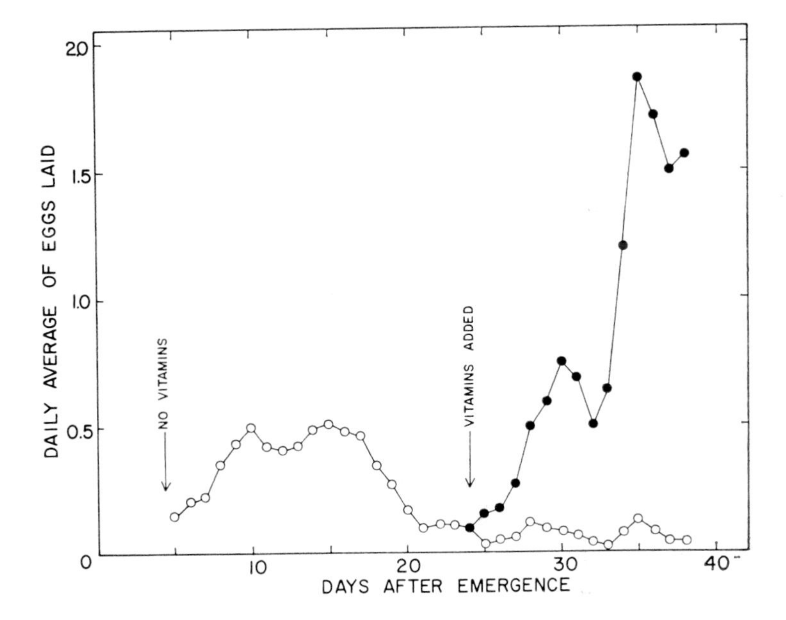

FIG. 43. Average number of eggs laid by females of *Exeristes comstockii* with and without vitamins in the food. After addition of vitamins, fecundity rose dramatically within a few days. (Adapted from Bracken, 1965.)

carnitine-raised males were sterile (Geer, 1966). If choline was replaced by carnitine in the adult diet, egg laying was also considerably impaired. These results may indicate that the requirements of adults for choline are higher than those of larvae; and, since carnitine may only partially substitute choline, the deficiency becomes more apparent in the adults. It might be worth mentioning that *Blattella germanica* and *Palorus ratzeburgi* could not be reared on carnitine (Fraenkel *et al.*, 1955).

Wheat germ oil and vitamin E promoted egg growth in *Acheta domesticus* in an indirect way (Richot and McFarlane, 1962). In this species, males fed a vitamin E-deficient diet did not mate. Since mating is essential for normal egg production, the effect of vitamin E deficiency became apparent, although the vitamin did not seem to be necessary for egg maturation as such (Meikle and McFarlane, 1965). This change in a male's behavior caused by the deficiency of a specific food component, as reported here, seems to be the only definite case. The fact that the males of some species do not mate if they are starved cannot be taken as evidence that specific deficiencies have caused the change in behavior; general inanition may be responsible for this (Johansson, 1964). However, males of most species are not impaired in their sexual behavior by food deprivation.

Studies on lipid and sterol requirements for reproduction in insects have become rather interesting in recent years. As was reported by Grison (1944), Colorado beetles laid only about one-tenth of their normal egg complement if fed on old leaves, but the addition of 2 per cent lecithin to the old leaves restored the females' capacity to lay eggs at a high rate: these animals laid even more eggs than females fed on young leaves (Grison, 1948, 1952). The addition of proteins to old leaves did not improve egg production in this species. It therefore appears that lecithin is the essential factor for normal fecundity in the Colorado beetle. Best egg laying was reporting for *Blattella germanica* when cholesterol had been added to the artificial food medium (Chauvin, 1949). Cholesterol was also essential for good egg production in *Drosophila* (Sang and King, 1961) and in *Anthonomus grandis* (Vanderzant, 1963; Earle *et al.*, 1967). Sterols are probably required in only minute quantities in many insect species, as is exemplified by the case of *Musca.* Lack of dietary cholesterol in the adult female did not cause a reduction in the actual number of eggs laid; however, only about 20 per cent of these eggs hatched (Monroe, 1959, 1960). When the larval diet was supplemented with relatively large quantities of cholesterol, the resulting adults laid a larger number of viable eggs than the animals which had received no sterols (Monroe *et al.*, 1961, 1967). Moreover, a number of these adults were able to lay eggs even when fed only water or carbohydrates: the females had become autogenous (Robbins and Shortino, 1962; Davies *et al.*, 1965). Clearly, sterols had been carried over from the larval stages. In this connection should be mentioned a rather unique type of reproduction which is found in the rabbit flea, *Spilopsyllus cuniculi.* This species produces eggs only during the last 10 days of the host's pregnancy (Mead-Briggs and Rudge, 1960). No flea that fed on male or post-partum female rabbits ever laid eggs. It appears that during late pregnancy of the host, the female flea takes in a factor which promotes egg maturation (Mead-Briggs, 1964).

The timing of the flea's reproduction seems to be an adaptation to the specific requirements of the immature fleas, which need the dried blood found so abundantly in the nest of a doe rabbit. The peculiar timing of reproduction in the rabbit flea suggests that steroid hormones of the host trigger egg maturation in the flea. It was shown that the doe's level of corticosteroids increases markedly during the last 10 days of pregnancy. Subsequently, it was also found that fleas that fed on male rabbits which were given daily injections of 10 μg of hydrocortisone or 1.3 mg of cortisone matured eggs like those fed on pregnant does (Rothschild and Ford, 1965a, b). Fleas sprayed with hydrocortisone also matured eggs. Whether steroids directly cause the maturation of eggs in the flea by acting as gonadotropic hormones or whether the flea's own endocrine glands become active after treatment remains to be shown. Hydrocortisone and cortisone do not appear to be the only vertebrate steroids affecting the fleas' reproductive physiology. According to a more recent report, the ovaries of the flea regress after parturition of the rabbit, presumably caused by ingestion of blood containing high levels of progesterone. Indeed, spraying with progesterone did induce regression of the ovaries (Rothschild and Ford, 1966).

Is there a possibility that other blood-sucking insects have evolved a similar sensitivity to vertebrate hormones and are able to make use of them? The tsetse fly *Glossina austeni* may be one such case. Females fed on pregnant goats produced 1.54 more offspring than those fed on non-pregnant hosts (Nash *et al.*, 1966). This finding suggests that *Glossina* females obtain a factor with the blood of the pregnant goats which promotes fecundity. This may be one or more of certain steroids which are available at higher levels in these hosts than in non-pregnant animals. On the other hand, *Glossina morsitans* that fed on male guinea pigs produced larger pupae than did those that fed on pregnant animals; this was presumably related to the higher protein content of the male blood as compared to the pregnant female blood (Langley, 1968).

A further unique case illustrating the role of sterols in insect reproduction is that of *Drosophila pachea* (Heed and Kircher, 1965), a species which lives exclusively on the Senita cactus (*Lophocerus schottii*) in the Mexican Sonoran desert. Rearing on artificial medium was possible only after pieces of the cactus has been added. The responsible factor was identified as $_{\Delta}7$-stigmasten-3β-ol and named Schottenol (Fig. 44). The synthetic sterol supports larval growth and egg maturation just as the addition of cactus stem to the medium does. If

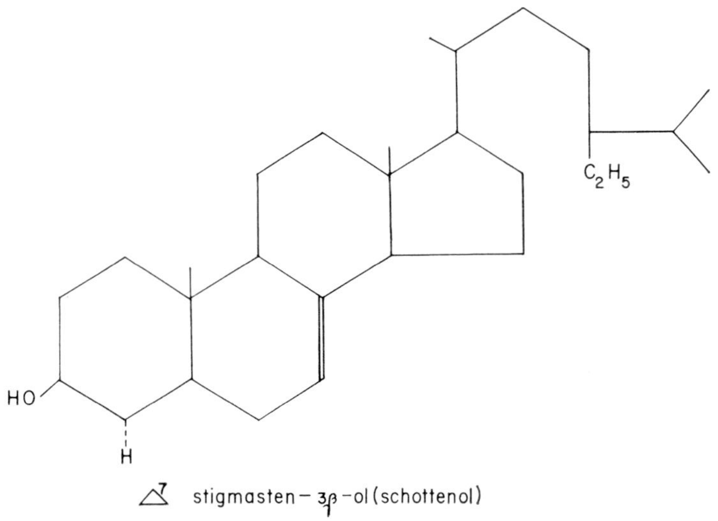

HO
H
C_2H_5
Δ^7 stigmasten-3β-ol (schottenol)

FIG. 44. Schottenol, a steroid required by *Drosophila pachea* for growth and reproduction.

the animals were reared on the specific sterol in their larval stages but were deprived of it as adults, no eggs were laid. Two additional sterols could support the species, namely $_{\Delta}7$-cholesten-3β-ol and $_{\Delta}5,7$-cholestadien-3β-ol. A variety of other sterols tested did not promote egg maturation. *Drosophila pachea* had found its ecological niche—the Senita cactus—a plant unsuitable for all other species of *Drosophila* since it contains two alkaloids which appear to be toxic to all but *D. pachea* (Kircher *et al.*, 1967).

Within the class Insecta, species have evolved very far apart and have become quite specialized with respect to food requirements. Insects have successfully occupied the most remote ecological niches and have adapted to unique environmental conditions, as is so beautifully demonstrated by the rabbit flea and *D. pachea*. Despite the multitudinous reports on nutritional requirements in adult insects, no generalization can therefore be made. The best we can do at this stage of research on food requirements is to compile the available information. This compilation of the data, however, merely shows how scanty our knowledge of the subject is, and at the same time leaves one with a feeling of frustration.

2. FOOD QUANTITY

The effects of quantity and quality of food eaten by an insect frequently cannot be separated. Those species which require specific nutrients for egg maturation will often mature more eggs if given a greater amount of the essential foodstuffs. For example, *Cimex lectularius* laid 216 eggs on 195 mg of blood and 541 eggs if allowed to take 335 mg (Titschak, 1930). This observation was basically confirmed by Khalifa (1952). *Rhodnius prolixus* (Buxton, 1930; Hase, 1934) and *Triatoma infestans* (Goodchild, 1955) laid more eggs if allowed to ingest large quantities of blood than if fed only small amounts; the minimum amount which allowed some eggs to mature was about 51 mg of blood in *Rhodnius* and 149 mg in *Triatoma* (Goodchild, 1955). For complete maturation of one batch of eggs, one

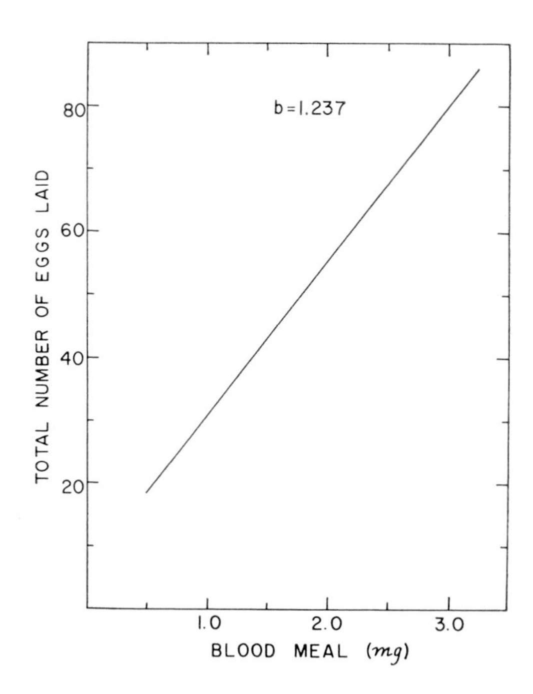

FIG. 45. Relationship of the number of eggs produced in the first batch and the weight of the blood meal in *Aedes aegypti*. (Based on data given by Roy, 1936.)

blood meal was sufficient in the mosquitoes *Stegomyia fasciata* (Macfie, 1915), *Anopheles maculipennis* (Nicholson, 1921; Hecht, 1933b, c), and *Culex pipiens pallens* (Hosoi, 1954). In some of these species it was also noted that the larger the meal, the more eggs per batch were produced. In *Aedes aegypti*, for example, the number of eggs per cluster was proportional to the amount of blood ingested at a single meal (Roy, 1936) (Fig. 45). The minimum quantity of blood necessary for any egg maturation in this species was 0.40 mg (Colless and Chellapah, 1960). However, the proportionality between the amount of blood ingested and the number of eggs laid no longer held if the animals took more than approximately 3.0 mg (Woke *et al.*, 1956). Since the total number of eggs possible per batch is determined by the number of oocytes present in the ovaries, suboptimal food quantities lower egg production by allowing maturation of only a fraction of the primary oocytes. The effect of different meal sizes on total egg production in blood-sucking insects is probably correlated with the amount of protein ingested. This was shown to be the case at least in *Aedes elutes*, where dilution of the blood resulted in a lower egg yield (Yoeli and Mer, 1938).

As in blood-sucking species, there is a correlation between quantity of food ingested and number of eggs produced in the phytophagous species *Leptinotarsa decemlineata* (Grison, 1958), *Sitona decipiens* (Markkula and Roivainen, 1961), and *Ctenicera destructor* (Doane, 1963).

Suboptimal qualitative and quantitative feeding of the larvae or nymphs may affect egg laying in the ensuing adults. Females from well-fed larvae of *Aedes*, for example, needed only one blood meal for the production of one raft of eggs, whereas those from poorly fed larvae usually had to feed twice before they laid eggs (MacDonald, 1956). Females obtained from well-fed larvae also laid more eggs than those from poorly fed ones (Mathis, 1938). This would indicate that the latter females carried over fewer reserves from the larval stages than did better fed animals. As in *Aedes*, *Drosophila* larvae reared on a nutrient-poor diet gave rise to females which laid fewer eggs than those obtained from larvae reared on a rich

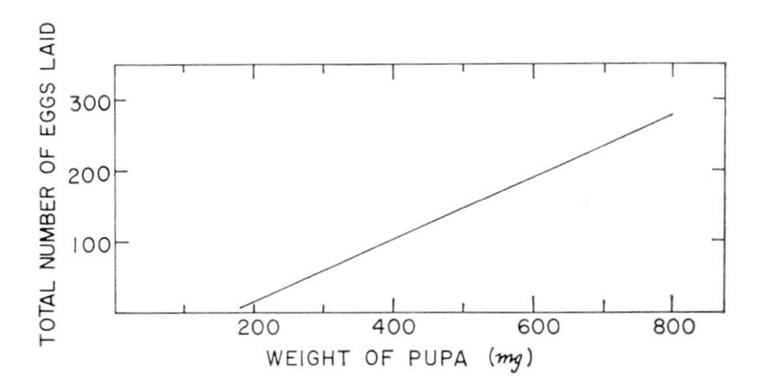

FIG. 46. Relationship of pupal weight and total number of eggs laid by a female of *Lymantria monacha*. (Modified from Zwölfer, 1933.)

diet (Alpatov, 1932). Several reports indicate that not only do females reared from poorly nourished larvae carry over smaller amounts of reserves—thus affecting the egg output—but also the number of ovarioles per ovary may be reduced when larval food supply is limited. Consequently, fewer eggs are produced by these animals. For example, pupae of *Lucilia* which weighed only 10 mg had 100 ovarioles whereas those weighing 30 mg had 200 (Webber, 1955). Similarly, *Aedes* females weighing 1.2 mg had 77 oocytes; in those weighing 2.4 mg, 118 oocytes were counted (Colless and Chellapah, 1960). Similar results have been obtained in a variety of autogenous and anautogenous species, such as *Lymantria monacha* (Zwölfer, 1933), *Ephestia kühniella* (Norris, 1933), *Porthetria* (= *Lymantria*) *dispar* (Schedl, 1936; Lozinsky, 1961), *Lucilia serricata* (Hobson, 1938), *Trichogramma evanescens* (Salt, 1940), several *Anophelidae* (Shannon and Hadjinicalao, 1941), and *Pristophora erichsonii* (Heron, 1955). Data collected from *Lymantria monacha* (Fig. 46) (Zwölfer, 1933) may serve to illustrate this point.

From these data it appears that total egg production of a female basically depends on two interdependent variables: the number of ovarioles present in the ovaries and the quantity of nutrients taken in by larvae and adults.

Parasitized females of the cicada *Chloriona* spec. (Lindberg, 1939), of the Delphacidae (Hassan, 1939), and of the bee *Andrena vaga* (Brandenburg, 1953) do not lay eggs. This is the result of "starvation" of the host caused by the enormous growth of the parasite leaving insufficient nutrients for the host's egg maturation. The female strepsipteran *Elenchus tenuicornis*, which lives in species of the Delphacidae, practically starves its host to death.

For many species it has been shown that virgins do not lay the eggs they have matured, but rather retain them in the genital ducts or ovaries where they are resorbed in many cases. The materials freed during oosorption can essentially be regarded as nutrients, i.e. they serve as reserves. As was pointed out in several hymenopteran parasitoids this phenomenon may best serve the species since the nutritional material may thus be saved for newly forming eggs which can later be fertilized (Flanders, 1942, 1957). Probably as a consequence of oosorption, the virgin females of many species live longer than the mated ones, which lay most of their eggs early in their lives. Among the Lepidoptera, both long- and short-lived species exhibit these characteristics. For example, *Fumea crassiorella*, when mated, live only 2 days on the average, but virgins live 5.4 days (Matthes, 1951). Further, *Pammena juliana* mated females live 10–12 days and lay about 180 eggs whereas the virgins live up to 30 days during which time they lay only about 80 eggs (Müller, 1957). Similar data could be quoted either for other Lepidoptera or other orders. *Schistocerca gregaria* females resorb a high proportion of their mature eggs if not mated (Highnam and Lusis, 1962; Highnam *et al.*, 1963b). Sometimes, as has been reported for several Acrididae, unfavorable conditions for egg laying (a dry season) apparently can cause oosorption (Phipps, 1966). Furthermore, the virgins of the coleopteran *Oryzaephilus surinamensis* reportedly live three times as long as the mated females but lay only about one-third as many eggs as do the mated females. These few examples may suffice to make clear that generally a low rate of egg laying, which in virgins is often accompanied by oosorption, can be associated with a long life span of the animal.

3. AUTOGENY

Many insect species are able to lay eggs (ranging in number from only a few to several hundred) without having ingested any proteinaceous foodstuffs. This phenomenon is particularly well known among those species with a short life span, such as many Diptera, Lepidoptera, and Coleoptera. The short-lived Ephemeroptera are a prime example (Needham *et al.*, 1935). The ability of insects to mature and lay eggs without protein ingestion was termed autogeny by Roubaud in 1929. Roubaud (1929) found a strain of *Culex pipiens* in a cave near Paris which could be reared without having to feed on birds, their normal food source. However, these autogenous females were able to lay a still greater number of eggs if allowed to feed on birds. This strain was subsequently named *Culex pipiens* var. *molestus*. The phenomenon of autogeny had already been discovered in *Aedes scutellaris* by Sen in 1917, but no analysis of it had been attempted at that time. Almost simultaneous with the findings of Roubaud, Huff (1929) observed a female of *Culex pipiens* which laid some eggs without a blood meal. It soon became apparent that larval feeding influenced the manifestation of autogeny in *Culex pipiens*, i.e. if the larvae were fed a nutrient-rich diet, a greater proportion of the adults laid eggs without feeding, and also more eggs were laid (De Boissezon, 1929; MacGregor, 1932; Weyer, 1934; de Buck, 1935; Déduit and Callot, 1955). Autogenous females of *C. pipiens* laid only 5 to 15 eggs if the larvae were reared on a poor diet, but 60 to 80 if reared on a rich diet (Möllring, 1956). As in *C. pipiens*, qualitative and quantitative larval feeding conditions determined the expression of autogeny in *C. fatigans* (Weyer, 1934), *Aedes scutellaris* (Sen, 1917; Laven, 1951), *Theobaldia subochrea* (Marshall and Staley, 1936), and *A. togoi* (Laurence, 1964). It is known that autogenous species and strains of mosquitoes accumulate considerably more reserves during their larval life than do anautogenous ones, a situation which may be related to the somewhat longer larval developmental periods in autogenous strains (Möllring, 1956; Twohy and Rozeboom, 1957). Even in *A. aegypti*, which was thought to be strictly anautogenous, Lea (1964a) succeeded in isolating an autogenous strain in the eighth generation after the larvae had been consistently fed on a high protein diet. Autogeny in mosquitoes is, however, controlled not solely by larval nutritional conditions, but also by the genetic background, as was demonstrated by certain experimental crosses. Crosses between autogenous and anautogenous strains of *C. pipiens* showed that a recessive gene controls autogeny, since autogeny disappeared in the F_1 generation but reappeared in F_2 (Roubaud, 1930, 1933; Weyer, 1935). Occasionally, however, autogenous females appear in the F_1 generation. Furthermore, in the F_2 generation, there was no 3:1 ratio of anautogenous to autogenous progeny (Kitzmiller, 1953), as would be expected if a single recessive gene were involved. It was therefore suggested that there are several genes controlling autogeny in *C. pipiens* which are located on two chromosomes (Spielman, 1957). Autogenous and anautogenous strains of both *A. scutellaris* and *Theobaldia subochrea* could also be crossed; here, as in *C. pipiens*, it was found that autogeny was genetically determined (Laven, 1951). It seemed to Laven that a pair of recessive homozygous genes for autogeny manifests itself in these species.

Nothing is known about the genetic control of autogeny in other Diptera or in many species of Lepidoptera and Coleoptera which exhibit this phenomenon. In these cases it appears that the presence of adequate nutritional reserves will allow the manifestation of autogeny. The extreme is seen in a species like *Musca domestica* which is thought to be strictly anautogenous, but which may become autogenous if the larvae are well fed (Robbins and Shortino, 1962; Larsen *et al.*, 1966). It is likely, however, that autogeny is basically under genetic control in species other than those of the genera *Culex*, *Aedes*, and *Theobaldia*.

Most autogenous species lay relatively few eggs when unfed and require proteinaceous foods for the production of additional batches of eggs. On the other hand, some species will not eat at all or require only water or carbohydrates (nectar) during their adult life even though they may produce up to several hundred eggs, as do many moths. A list of reported cases of autogeny has been assembled in order to illustrate the occurrence of the phenomenon among the various orders (Table 6). Females of those species which normally feed on proteins but which can lay a few eggs even when they are not given proteins are listed along with those which do not feed at all. In some species, e.g. the cockroaches *Diploptera punctata* (Engelmann, 1960b) and *Nauphoeta cinerea* (Roth and Stay, 1962b), autogeny may manifest itself if the females mate but not otherwise. The given list of autogenous species is certainly incomplete, especially for the Lepidoptera and Coleoptera, about which many casual observations on feeding are found in the extensive literature.

The significance of autogeny for the propagation of the species is ecologically rather interesting. Autogeny allows the species of mosquitoes to lay eggs even in the absence of a host for feeding. Among mosquitoes, the phenomenon is found more often in the temperate and cold regions of the earth, where hosts and food sources for adults are sometimes scarce, than in the warm and tropical regions. One could therefore postulate that, in some cases, autogeny evolved as a consequence of adverse climatic conditions. The following cases may serve to support this hypothesis. The mosquitoes *A. impiger* and *A. nigripes* are voracious biters if they have a chance to find a host, yet they live in the high Arctic where the mammals on which they would feed are very scarce. If they have no opportunity to feed on blood and are confined to feeding on nectar, they can none the less lay from one to a few

TABLE 6. Selected list of reported autogenous species
(authors are not necessarily the first for a given species)

Species	Authors
Dictyoptera	
Diploptera punctata	Engelmann, 1960b
Nauphoeta cinerea	Roth and Stay, 1962b
Pycnoscelus surinamensis	Roth and Stay, 1962b
Hemiptera	
Haematosiphon inodorus	Lee, 1954
Neuroptera	
Chrysopa californica	Hagen, 1950
,, *carnea*	Hagen and Tassan, 1966
Coleoptera	
Acanthoscelides obtectus	Carle, 1965
Anthrenus verbasci	Blake, 1961
Aphodius howittii	Crane, 1956
,, *tasmaniae*	Maelzer, 1960
Bruchus quadrimaculatus	Larson and Fisher, 1924
Ctenicera destructor	Doane, 1963
Ernobius mollis	Gardiner, 1953
Lampyris noctiluca	Naisse, 1966
Prionoplus reticularis	Edwards, 1961
Trogoderma anthrenoides	Burges and Cammell, 1964
,, *versicolor*	Norris, 1936
Ephemeroptera	
Majority of species known	
Diptera	
Aedes aegypti	Lea, 1964a
,, *detritus*	Vermeil, 1953
,, *impiger*	Corbet, 1964, 1965
,, *nigripes*	Corbet, 1964
,, *scutellaris*	Sen, 1917; Laven, 1951
,, *taeniorhynchus*	Lea and Lum, 1959
,, *togoi*	Laurence, 1964
Ceratitis capitata	Hanna, 1947
Chaoborus crystalinus	Möllring, 1956
Chortophila brassicae	Missonnier and Stengel, 1966
Clunio marinus	Caspers, 1951
,, *pacificus*	Oka, 1930
,, *tsushimensis*	Oka and Hashimoto, 1959
Culex fatigans	Weyer, 1934
,, *pipiens*	Roubaud, 1929
,, *tarsalis*	Moore, 1963
Culicoides bambusicola	Lee, 1968
,, *bermudensis*	Williams, 1961
,, *dendrophilus*	Amosova, 1959
Dasyneura leguminicola	Guppy, 1961
Dixa aestivalis	Möllring, 1956
Hippelates pusia	Schwartz and Turner, 1966
Lampetia equestris	Doucette and Eide, 1955
Megarhinus brevipalpis	Muspratt, 1951
Mochlomyx culiformis	Möllring, 1956
Musca domestica	Robbins and Shortino, 1962
Opifex fuscus	Haeger and Provost, 1965
Pegomyia betae	Missonnier, 1961
Prosimulium fuscum	Davies, 1961

Diptera (cont.)	
Rhagoletis cingulata	Kamal, 1954
Sphaerophoria scuttellaris	Lal and Haque, 1955
Theobaldia subochrea	Marshall and Staley, 1936
Wyeomyia smithii	Price, 1958
Lepidoptera	
Achroa grisella	Oldiges, 1959
Alabama argillacea	Lukefahr and Martin, 1964
Bupalus piniarius	Eidmann, 1931
Cacoecia murinana	Franz, 1940
Chorizagrotis auxiliaris	Jacobson, 1960
Colias philodice	Stern and Smith, 1960
Dasychira pudibunda	Eidmann, 1931
Dendrolinus pini	Eidmann, 1931
Ephestia cautella	Norris, 1934
,, *kühniella*	Norris, 1933
Galleria mellonella	Allegret, 1951
Heliothis virescens	Lukefahr and Martin, 1964
,, *zea*	Lukefahr and Martin, 1964
Lasiocampa pini	Eckstein, 1911
Lymantria dispar	Schedl, 1936
Lymantria monacha	Eidmann, 1931
Panolis flammea	Oldiges, 1959
,, *piniperda*	Eidmann, 1931
Prionoxystus robiniae	Solomon, 1967
Pyrausta nubilalis	Kozhantshikov, 1938
Serobipalpa ocellatella	Robert, 1965
Sphinx pinastri	Eidmann, 1931
Hymenoptera	
Apanteles ruficus	Hafez, 1947
Caraphractus cinctus	Jackson, 1966
Chrysocharis laricinellae	Quednau, 1967
Diadromus pulchellus	Labeyrie, 1960
Leptomastix dactylopii	Lloyd, 1966
Meteorus loxostegei	Simmonds, 1947
Mormoniella vitripennis	Moursi, 1946
Pristiphora erichsonii	Heron, 1955

eggs (Corbet, 1964, 1965). The propagation of the species is thus assured. Another example of the occurrence of autogeny in unfavorable locations is that of *A. detritus* (Vermeil, 1953). This halophilic species is found in ponds in the Sahara of Tunisia. These animals rarely have a chance to ingest blood since very few vertebrates are available, yet they lay as many eggs without proteinaceous foods as when they have had access to them. The evolution of autogeny under desert conditions has certainly assured the survival of this species.

The correlation between adverse climates and the occurrence of autogeny in some mosquito species may not apply for others, as is illustrated by the finding that *Culicoides bermudensis* (which lives in tropical regions) does not feed at all, not even on carbohydrates, and still produces a large number of eggs (Williams, 1961). It may not apply for other Diptera or for other orders of insects. No extensive systematic studies of this question have been made in groups other than the few mosquito species mentioned.

4. MECHANISMS INVOLVED

There are several pathways by which nutrition or nutritional factors may affect egg maturation. Mechanisms may differ in the various species. Necessary amounts of lipid and proteinaceous materials must be available for the formation of yolk and, in addition, specific compounds like vitamins, sterols, salts, etc., may be needed for specific metabolic processes associated with yolk formation and its incorporation into the oocytes. Furthermore, in species in which egg maturation is controlled by hormones, nutritional factors may affect the endocrine system rather than the process of the formation and deposition of yolk. Factors supplied with the food may possibly influence the endocrine system and yolk formation simultaneously. In short, the influence of feeding and nutritional factors on egg maturation is a complex phenomenon.

We have practically no detailed knowledge of the mode of action of any of the essential foodstuffs in association with egg maturation, with the possible exception of the role of folic acid in yolk deposition in *Drosophila melanogaster* (David, 1965). In this case it is reasonably certain that its absence from the food impairs DNA and RNA synthesis in the nurse and

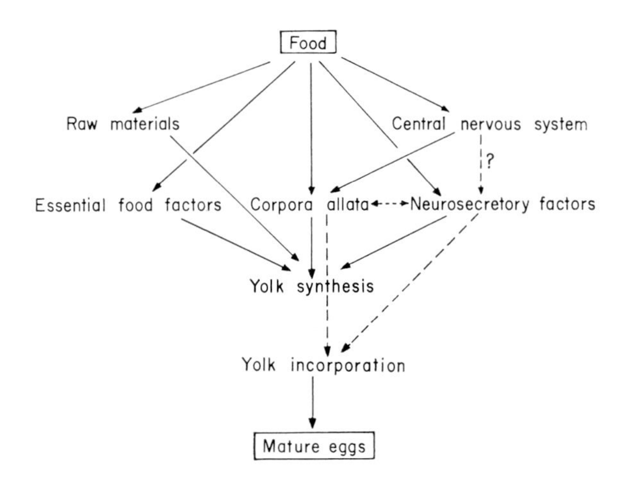

FIG. 47. Diagram illustrating the various pathways by which food intake can influence egg maturation. Stippled arrows indicate pathways for which the available evidence is somewhat circumstantial. The diagram does not claim to be precise in any of the details.

follicle cells, i.e. folic acid is essential for the synthesis of proteinaceous yolk. We can propose a crude scheme for the pathways by which nutritional supplies exert their control on egg maturation as such, as well as on the endocrine system. This scheme is not claimed to be accurate in any detail (Fig. 47), but it may serve a heuristic function. As to the pathway by which the endocrine glands (cf. p. 155)—notably the corpora allata—are influenced by food intake, Johansson (1964) proposed that first the central nervous system is affected, the corpora allata being then activated by the brain. The brain supposedly is the integrating organ. This is certainly true for the initial activation of the corpora allata by distention of the intestine in mosquitoes. However, in other species, and also for enhanced activation of this gland in mosquitoes, it has been suggested that nutritional factors influence the corpora allata directly via circulation (Engelmann, 1965a).

B. Mating Stimuli

It has been reported for many insect species that mating is one of the decisive factors which influence the total number of eggs a species lays. Mating, however, not only influences egg maturation as such, but also often stimulates oviposition. The majority of publications do not make clear whether mating merely stimulated laying of the matured eggs, whether the production of mature eggs in the ovaries was accelerated or the number of eggs increased, or whether all of these processes were influenced. In a few cases, even though no direct observations have been reported, we can infer which of these possibilities applies; in only a very few instances can we be certain of the nature of the mating stimuli. Although it has been shown that mating does not influence egg maturation in Lepidoptera—probably because egg maturation in many species is complete or nearly complete at emergence—virgins usually lay only a fraction of their mature eggs as compared with mated females. Virgins often retain their unlaid eggs in the ovaries or genitalia; some may lay these eggs shortly before death. The subject of oviposition stimuli will be dealt with in the pertinent chapter (p. 193ff.).

In many dipteran species, as in the Lepidoptera, the total number of eggs laid increases significantly after mating, as compared to that laid by virgins. Frequently, within hours

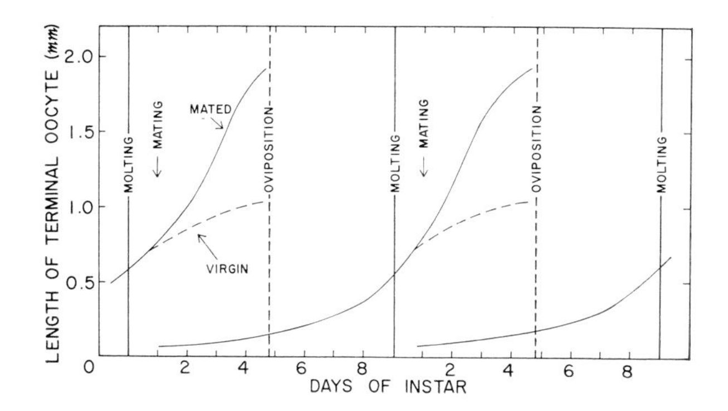

FIG. 48. Illustration of the dependence of molting and egg maturation cycles in *Thermobia domestica*. Mating is essential for complete egg maturation. In virgins, the partially matured eggs degenerate. (Modified from Watson, 1964.)

after mating, a sudden increase in egg laying is observed. In other species, both virgins and mated females lay a similar number of eggs, but virgins generally lay their eggs later in life; often a considerable delay in oviposition has been noted. In virgins which have not laid all their matured eggs, the ovaries or genital ducts are greatly distended by these eggs at the time of death. From these observations one may deduce that in Diptera and Lepidoptera mating stimulates not vitellogenesis but only oviposition. Mating may, however, in both of these orders, indirectly influence egg maturation. For example, virgins of *Anopheles punctulatus* retain most of their mature eggs in the oviducts and this appears to inhibit the maturation of subsequent oocytes in the ovarioles (Roberts and O'Sullivan, 1948). A similar observation has been made in *Heliothis zea* (Callahan, 1958). It is possible that this phenomenon is to be found in many other species, but precise observations are not available.

In contrast, a stimulation of the maturation of oocytes—i.e. yolk deposition—by mating has been shown in a number of hemimetabolous species. For example, mating is absolutely essential for full maturation of the oocytes in the thysanuran, *Thermobia domestica* (Watson, 1964). In this species, one set of eggs matures between moltings (Fig. 48); but, unless the females have had a chance to mate, all partially matured eggs degenerate and are completely resorbed by the time the animals molt again. Certainly, mating between molts in Thysanura would be essential since the spermathecae which contain the spermatozoa are shed with the rest of the cuticle during molting (Sahrage, 1953); unlike other species which store sperm for long periods, fertilization could thus not take place. Mated females of *Melanoplus bilituratus* laid nearly twice as many eggs as did virgins (Riegert, 1965). In this species, mating speeded up egg maturation processes; consequently, more eggs were produced in a lifetime. On the other hand, virgins of another grasshopper, *Gomphocerus rufus*, laid as many eggs as did mated females (Loher and Huber, 1964).

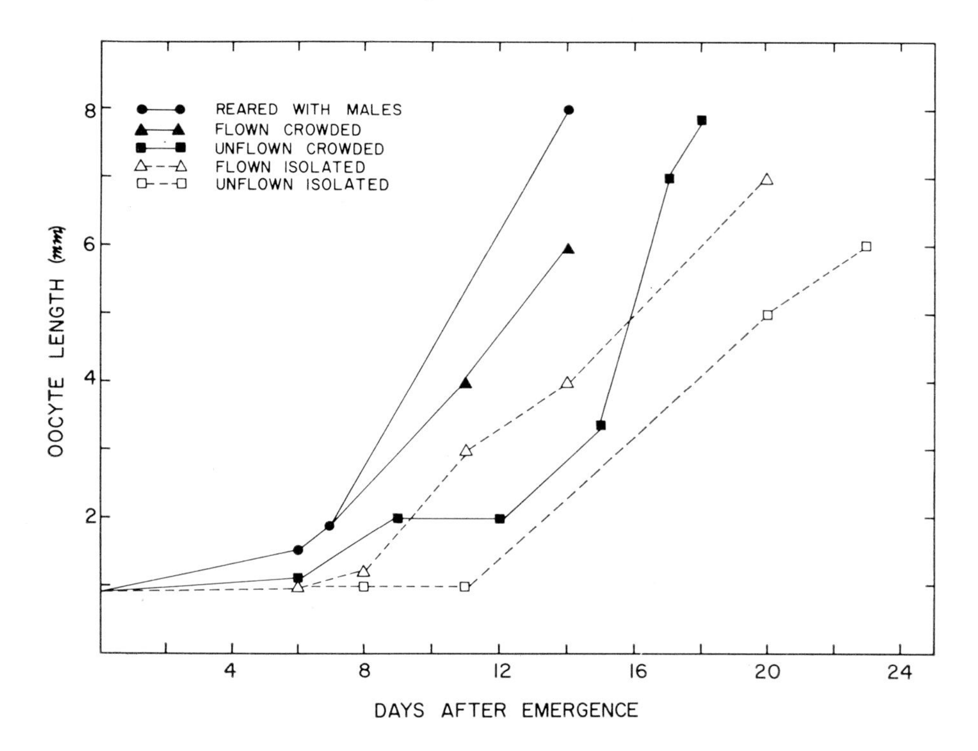

FIG. 49. Growth rates of the proximal oocytes in isolated or crowded, flown or unflown females of *Schistocerca gregaria* as compared with those of females reared with males. Note that mated females mature their eggs twice as fast as isolated unflown virgins. (Modified from Highnam and Haskell, 1964.)

Extensive research on the reproductive capacity of injurious locusts has been performed, and the effects of mating have been carefully recorded for many cases. In *Schistocerca gregaria*, mating accelerated egg maturation (Norris, 1954; Hamilton, 1955; Highnam and Haskell, 1964) (Fig. 49) although the fecundity of virgins was as high as or even higher than that of mated females (Norris, 1952; Hamilton, 1955); virgins of this species live longer than mated animals and consequently make up for the low rate of egg maturation. Even though more eggs are resorbed in the ovaries of virgins than in those of mated females (Highnam *et al.*, 1963b), the total egg output is not reduced. In *Locusta migratoria* mating like-wise speeds up the process of egg maturation and thus shortens the preoviposition periods (Mika, 1959; Quo, 1959).

Both total egg production and rate of egg maturation are stimulated by mating in *Peri-*

planeta americana, as was first shown by Griffiths and Tauber (1942). Females of this species kept with males laid 20 egg cases whereas virgins laid an average of only 9.7, even though the latter females lived almost twice as long as the mated females. This observation was basically confirmed by Pope (1953) and Roth and Willis (1956). According to Roth and Willis (1956), mated females of *Periplaneta* laid a total of approximately 940 eggs in a lifetime while virgins laid only 490 eggs. The same authors found that mated *Blatta orientalis* females laid 160 eggs and virgins 114, which indicates that in this species mating stimuli promote egg maturation less effectively than in *Periplaneta*. In *Blattella germanica*, as in *Periplaneta*, mating may double the egg output of the females (Willis *et al.*, 1958).

Among cockroach species the effect of mating on egg maturation rates appears to be species-specific (Table 7). While it has been shown that mating may have no effect on egg maturation in *Pycnoscelus surinamensis*, it is essential for the maturation of the first batch of eggs in *Diploptera punctata*. The effectiveness of mating on the rate of egg maturation in cockroaches may not necessarily be indicated by the total egg output of the female since virgins often retain some of their mature eggs in the ovarioles and resorb them during the following weeks and months. It was noticed in *Leucophaea* that during periods of egg resorption the rate of egg maturation was slowed down; consequently, fewer eggs were matured in a unit time (Engelmann, 1957a). Similarly, most virgins of *Byrsotria*, *Nauphoeta*, and *Blaberus* did not ovulate all the mature eggs, but rather resorbed some of them (Roth and Stay, 1962b). In all these cases the presence of sperm in the spermatheca seems to be involved in the control of oviposition (p. 197). To sum up, in cockroaches mating can stimulate egg production as well as deposition; the degree to which mating is effective varies according to the species.

As to the mechanisms by which mating causes an accelerated maturation of oocytes, we do have some information from the grasshoppers and the cockroaches. Females of *Locusta migratoria manilensis* which had mated with castrated males matured eggs as fast as if they had mated with normal males (Quo, 1959). This would indicate that the mechanical stimuli received by the female during mating are primarily responsible for egg growth

TABLE 7. The effect of mating on egg maturation in species of cockroaches

Species	Acceleration (averages)	Author
Pycnoscelus surinamensis (parthenogenetic strain)	none	Roth and Stay, 1962b
Pycnoscelus surinamensis (bisexual strain)	1–2 ds	,, ,, ,,
Byrsotria fumigata	approx. 1 d	,, ,, ,,
Blattella germanica	2 ds	Roth and Stay, 1962a
,, *vaga*	2 ds	,, ,, ,,
Blaberus craniifer	approx. 8 ds	Roth and Stay, 1962b
,, *giganteus*	approx. 8 ds	,, ,, ,,
Nauphoeta cinerea	12–20 ds*	Roth, 1964
Leucophaea maderae	30 ds plus†	Engelmann, 1960b; Roth and Stay, 1962b
Diploptera punctata	completely dependent	Roth and Willis, 1955 Engelmann, 1959a

* Many virgins only partially mature their eggs.
† Many virgins do not mature any eggs even after several months.

stimulation. However, it may also be that substances in the seminal fluid or from the spermatophore trigger an accelerated growth of eggs. Pure mechanical stimuli alone are responsible for the induction of egg maturation in the viviparous cockroach, *Diploptera*. In this species, in which mating is a prerequisite for egg maturation (Stay and Roth, 1958), the implantation of an artificial spermatophore into the bursa copulatrix simulated the natural mating act (Engelmann, 1958, 1959a). If, however, the nerve cord was cut prior to or shortly after mating, copulation was ineffective. This shows that information from mating is received in the genitalia and transmitted to the brain via the ventral nerve cord (Engelmann, 1959a; Roth and Stay, 1961). In *Leucophaea*, egg maturation was also accelerated either by implantation of an artificial spermatophore into the bursa copulatrix or by mating with castrated males (Engelmann, 1960b). When females of *Nauphoeta* were mated with castrated males or were separated from normal males 10–15 min after onset of copula, i.e. before a spermatophore or sperm mass had been transferred, egg maturation was stimulated just as after a normal mating (Roth, 1964a). In all the examples given, the actual transfer of sperm did not constitute the mating stimulus. Mechanical stimuli which are relayed via the brain to the corpora allata cause the liberation of gonadotropic hormone at a higher rate, consequently, egg maturation proceeds faster than in virgins (cf. p. 160).

Some of the blood-sucking Hemiptera lay few or no eggs if not mated. This is particularly striking in *Haematosiphon inodorus* (Lee, 1954), *Hesperocimex sonorensis* (Ryckman, 1958), and *Cimex lectularius* (Davis, 1964, 1965a, b). In these species egg maturation rather than oviposition itself is stimulated. The mechanism by which mating stimulates egg maturation, however, has been elucidated only for the true bed bug (Davis, 1965a, b). As has been discussed in the section on hormonal involvement in egg maturation, the mating stimulus in this species does not appear to be of a mechanical nature. Only after spermatozoa have reached the corpora seminales at the base of the ovarioles is egg maturation stimulated. It has also been shown that the injection of the sperm mass alone does not initiate the migration of the spermatozoa to the corpora seminales in *Cimex*. Environmental variables may play an important additional role as is exemplified in *Prostemma guttula* (Carayon, 1952c). This species mates in the fall but does not mature eggs until the spring; undetermined factors are essential for the stimulation of egg maturation here.

C. Environmental Factors

A variety of extrinsic variables, such as temperature, light, humidity, and food availability, as well as interactions with other animals, may influence egg production and fecundity of a female. The literature on these aspects is vast and deals largely with injurious insect species. The environment of any species consists of many different variables which may change independently, to a certain degree. The manipulation of one single parameter, both in nature and laboratory, entails changes in others; the proper evaluation of the obtained data is, therefore, often very difficult for the investigator.

1. TEMPERATURE

Optimal temperatures for egg output vary greatly among species, possibly reflecting the temperatures the species normally encounter during reproductive periods. As illustration of this point, data for some species have been compiled in Table 8. This shows also that the temperature range in which a given species may oviposit can be rather broad, as in *Bruchus obtectus* (Menusan, 1935), or relatively narrow, as in *Thermobia domestica* (Sweetman, 1938).

High temperatures may interfere with certain reproductive processes indirectly related to actual egg maturation. For example, females of *Panolis flammea* mated less frequently at high temperatures than at the optimal one; this consequently reduced the number of eggs laid since mating stimulates egg laying in the species (Zwölfer, 1931). In *Ephestia kühniella* no sperm are transferred at temperatures of 30°C or above (Norris, 1933); furthermore, the beetle *Bruchus obtectus* lays many more unfertilized eggs if kept at temperatures above 30°C rather than below (Menusan, 1935). On the other hand, migratory locusts generally reproduce best, both in the laboratory and in the field, at temperatures above 30°C (Hamilton, 1936; Popov, 1954); *Schistocerca gregaria* and *Locusta migratoria* rarely if ever lay eggs at temperatures below 20°C (Hamilton, 1936). The tropical or subtropical cockroach *Nauphoeta cinerea* matured no eggs below temperatures of 15°–17°C (Springhetti, 1962), and similarly *Musca domestica* laid no eggs below 14°C (Feldman-Muhsam, 1944). On the other hand, *Ptinus tectus* still laid a few eggs at 5°C (Ewer, 1942; Howe, 1951), and *Macrosiphum euphorbiae* (Barlow, 1962) produced viable larvae at this temperature. The temperatures recorded for these last two species are among the lowest at which an insect species was found to reproduce. Since temperatures fluctuate during the day–night rhythm, for short periods insects can probably sustain lower temperatures than those just mentioned and still lay eggs as soon as the temperature rises.

TABLE 8. Egg production or fecundity as dependent on environmental temperature (selected list)

Species	Optimal temperature (°C)	Range (if known)	Author
Panolis flammea (Lept.)	14–18	8–27	Zwölfer, 1931
Coccinella septempunctata (Col.)	16–20	—	Bodenheimer, 1943
Sitophilus granarius (Col.)	20	—	Surtees, 1964
Bruchus obtectus (Col.)	21–30	8–40	Menusan, 1935
Leptinotarsa decemlineata (Col.)	22	—	Grison, 1957
Spalangia drosophilae (Hym.)	24	—	Simmonds, 1953
Sitona lineata (Col.)	24.5	—	Anderson, 1934
Calandra oryzae (Col.)	25–29	—	Birch, 1945
Schistocerca gregaria (Orth.)	32–38	—	Hamilton, 1936
Locusta migratoria (Orth.)	32–38	—	Hamilton, 1936
Piesma quadrata (Hem.)	37–40	—	Schubert, 1927
Thermobia domestica (Thys.)	37	32–41	Sweetman, 1938

An interesting observation on fecundity has been reported by Oldiges (1959) for *Galleria mellonella*. Females obtained from larvae reared at 26°C laid more eggs than animals reared at either lower or higher temperatures. At 26°C the larvae consumed more food and consequently became heavier than those reared at other temperatures. The resulting heavier adults laid more eggs than did the lighter ones, a phenomenon which has been reported for a number of other species (cf. p. 122).

The few cases mentioned illustrate some of the principal findings concerning temperature effects. As with any of the environmental factors, the primary site of action of temperature with respect to egg production can only be surmized. Temperature affects a multiplicity of enzymatic reactions, each of which has its own characteristics. The final result observed is the overall balance among them. This poses an almost insurmountable problem, yet one must also ask whether we would actually gain a deeper insight into the primary processes of egg maturation even if we were to know all the details on temperature affected enzymatic reactions.

2. HUMIDITY

As is the case for reports of temperature effects on egg maturation, many rather incidental reports are to be found in the entomological literature on humidity effects. In many cases, the optimal relative humidity at which egg maturation and egg laying are observed is related to the species' natural environment. Again, most reports simply record oviposition, which may not actually reflect the degree of egg maturation. The effects of humidity are intimately related to those of temperature since any change in temperature is accompanied by a change in relative humidity. It appears that at low humidity the animals lose much water through evaporation or excretion and not enough remains for egg maturation. Rarely can the animal provide supplemental water rapidly enough through metabolic pathways; this is, however, an aspect which is open for further research.

A variety of difficulties are encountered when one tries to assess the humidity effects on egg maturation in insects. Firstly, blood-sucking species and those which feed on fresh plants obtain their necessary water supply with their food. Since food is essential for egg maturation, lack of water or the effects of a change in relative humidity cannot be observed. Secondly, some species drink water; consequently, variations in the environmental humidity have no effect. This has been shown to be the case in *Ephestia cautella and E. elutella*, in which no detrimental effects of low relative humidity were observed as long as the animals had access to water (Norris, 1934).

For a great many species—if not the majority—80–90 per cent relative humidity seems to be optimal for egg laying and rates of oviposition decrease below this value. This has been reported, for example, for the orthopterans *Schistocerca gregaria*, *Locusta migratoria* (Hamilton, 1936; Norris, 1957), and *Chortoicetes terminifera* (Andrewartha, 1945), the thysanuran *Thermobia domestica* (Sweetman, 1938), the hymenopteran *Apanteles angaleti* (Subba Rao and Gobinath, 1961), the lepidopteran *Panolis flammea* (Zwölfer, 1931), and the coleopterans *Cryptolestes ferruginea*, *C. minutus*, *C. turcicus* (Bishop, 1959), and *C. pusillus* (Ashby, 1961). Grain weevils like *Calandra granaria* (Eastham and McCully, 1943), *C. oryzae* (Howe, 1952), and *Cryptolestes* spec. (Bishop, 1959), which feed on dry grains and which rarely if ever have access to water, require high relative humidities for oviposition. *C. oryzae* and *Rhizopertha dominica* would not lay eggs when the moisture content of the grain was below 10 per cent and 8 per cent respectively (Birch, 1945). *Caryedon gonagra* laid equal numbers of eggs at 90 per cent or 50 per cent relative humidity (Cancella, 1965). *Sitona lineata* (Andersen, 1934), *Ephestia cautella*, and *E. elutella* (Norris, 1934) did lay eggs at 30–35 per cent relative humidity, but fewer than at higher relative humidities. Few species lay any eggs below a relative humidity of 50 per cent. However, one report states that *Trogoderma anthrenoides* still laid some eggs at 0 per cent relative humidity (Burges and Cammell, 1964).

3. PHOTOPERIODIC INFLUENCES

Some insect species are influenced by short photoperiods in the fall to cease egg maturation and oviposition, others by long photoperiods at the beginning of summer. Day length can, however, affect the reproductive behavior of a species in both direct and indirect ways. For example, for certain species food may cease to be available in the fall, for others in the summer.

Consequently, since food is essential for reproduction, the species does not reproduce during this period, i.e. reproduction may be indirectly controlled by the photoperiod via the food plant. Prevailing high and low temperatures, which may affect the reproductive capacity of a species, may coincide with long or short photoperiods. In other words, the species may react to a variety of environmental cues which may or may not be causally related to photoperiodic changes.

Among all climatic changes, the seasonally changing length of photoperiod is the most consistent variable in nature, and it is not surprising that through evolution some insects have made use of this cue to adjust to approaching adverse conditions. Arrest of reproductive activities may, in many cases, be a true diapause, a period which is preceded by preparatory physiological changes in the animal. This preparatory period makes the species partially independent of the environment and can be viewed as an adaptation to adverse conditions. Diapause is a phenomenon different from both arrest of growth and non-adaptive torpor induced by low temperatures or drought. The difficulties involved in the recognition of a true diapause have been pointed out by Lees (1955) and Danilevskii (1965), while eco-physiological aspects have been discussed by Andrewartha (1952).

The grasshoppers *Orthacanthacris aegyptia* (Grassé, 1922) and *Anacridium aegyptium* (Colombo and Mocellin, 1956) overwinter as adults and do not reproduce during this period. Even if relatively high temperatures are experienced, no eggs are laid for at least 4 months. However, if the females of *Anacridium* were exposed to long photoperiods in the fall, eggs began to mature within the following 25 days (Geldiay, 1966). This certainly suggests that in this species reproductive diapause is caused by the prevailing short days in the winter.

In contrast to these species, accelerated egg maturation was found after short day treatment (7–8 hr day length) in the locust *Schistocerca gregaria* while long days delayed egg maturation (Norris, 1957). This species experiences short days in nature during the rainy season, which is the one most favorable for reproduction. Availability of food (cf. p. 193) and moist sand for oviposition thus coincide with short photoperiods. For another locust, *Locusta migratoria*, it has been reported that the females lay more egg pods in winter than in summer, even if food is available in equal amounts throughout the year (Cassier, 1965a). In this case it is also thought that fecundity is affected by the different photoperiods.

Hibernation, which may be a reproductive winter diapause has been reported for several species, but often no further analyses have been made. In most of these cases, day length may be the decisive environmental cue. For example, *Sitona cylindricollis* does not mate or oviposit in winter even at high temperatures (Herron, 1953). *Eurygaster integriceps* (Zwölfer, 1930; Vodjdani, 1954), *Phytodecta olivacea* (Waloff and Richards, 1958), *Nysius huttoni* (Eyles, 1963), and others also do not reproduce in winter. Adults of *Chloriona smaragdula* and *Euidella speciosa* (Strübing, 1960), *Stenocranus minutes* (Müller, 1962), and *Pyrrhocoris apterus* (Sláma, 1964b) can be induced to go into diapause by short-day treatments. Cessation of egg laying in winter has been reported for *Anopheles maculipennis* even if the females were feeding (Swellengrebel, 1929; Hecht, 1933a, b), although egg laying resumed in spring when feeding was allowed. Since egg maturation after feeding was not observed in winter, the phenomenon was termed "gonotrophic dissociation". An influence of day length can be suspected. *Culex pipiens* can be reared throughout the year, and no hibernation is observed when they are treated with prolonged illumination (Tate and Vincent, 1936). An interesting observation was made on the mosquito *Culex tarsalis*, which lives in the climatically moderate regions of California where reproduction might therefore be possible throughout the year. As judged from animals caught in the field, however, the number of females feeding on blood declines in the fall and is very low in the winter. In the majority of females, the fat bodies become prominent in fall although no eggs mature (Bennington *et al.*, 1958); it is possible that the older parous females do not survive through the fall and only young nulliparous animals which had not engorged on blood enter diapause (Burdick and Kardos, 1963; Bellamy and Reeves, 1963). The implication is that once an animal feeds on blood, egg maturation follows. The females of this species resume feeding on blood in mid-winter. Following feeding, eggs are laid in February and March under prevailing light conditions which in the fall apparently had induced diapause (Bellamy and Reeves, 1963). The average temperature rises only little in January and February and it is questionable whether such a temperature rise could cause a change in the females' behavior. In the laboratory, *Culex tarsalis* could be induced to go into diapause during short-day treatment even at high temperatures (Harwood and Halfhill, 1964). Autogenous and anautogenous females of this species behave similarly with respect to diapause in fall and winter (Moore, 1963). Why diapause in *C. tarsalis* is broken in mid-winter during prevailing short-day conditions is still unknown. Is it the accumulated energy debt which causes them to resume feeding? Is it that induction and termination of diapause are controlled by different mechanisms? Diapause is induced as the day length becomes progressively shorter and, conversely, it is broken during increasing day lengths. This may be the factor that causes the behavior of the species to change. Other species of mosquitoes exhibit different reactions to changes in photoperiod and temperature. For example, *C. pipiens* does not lay eggs under short photoperiods and low temperatures but does lay when the temperature is raised even during short-day conditions (Eldridge, 1968). *C. quinquefasciatus* will reproduce under both short-day and low-temperature conditions; no gonotrophic dissociation is observed (Eldridge, 1968).

A phenomenon similar to that in *C. tarsalis* can be observed in the Colorado beetle *Leptinotarsa decemlineata*, i.e. winter diapause is induced by a short photoperiod but its termination seems to be brought about by different means. When the soil temperature rises in spring to about 9°C, these beetles begin to crawl towards the surface. Diapause is broken at a temperature of approximately 18°C (Le Berre, 1965). Again, is it the accumulated energy debt which makes the animals responsive to higher temperatures?

Reproductive summer diapause (aestivation), an ecologically interesting phenomenon, has been reported for the alfalfa weevil, *Hypera postica*, by Snow (1928) and Manglitz (1958) but no inducing mechanisms were known at the time. *Hypera* emerges in early summer, feeds, but does not oviposit; instead the female goes into a reproductive diapause until fall. Day length of more than 12 hr seems to induce summer diapause in this species (Huggans and Blickenstaff, 1962; Guerra and Bishop, 1962). Under laboratory conditions, long days induce diapause in the trichopteran *Limnophilus rhombicus* (Novak and Sehnal, 1965). This species emerges in May but does not lay eggs until fall of the same year. Long days can be suspected of inducing aestivation in the beetles *Otiorrhynchus cribricollis* (Andrewartha, 1933), *Sitona lineata* (Andersen, 1935), *Listroderes obliquus* (Dickson, 1949), and in some populations of the lepidopterans *Agrotis infusa* (Common, 1954) and *Euplagia quadripunctata* (Walker, 1966).

In the Chrysomelid *Galeruca tanaceti*, photoperiods of more than 15 hr daily induce reproductive diapause even at temperatures of 20°C (Siew, 1966). Shortening of the day length at the rate of 2 hr per 2 weeks successfully induced egg maturation. It has been observed that these animals eat little during diapause. In other words, this species aestivates even though food is abundantly available at this time of the year in England.

Successive generations of the bivoltine coccinellid, *Coccinella novemnotata*, may undergo either aestivation or hibernation (McMullen, 1967). This species, found in the San Joaquin valley of central California, feed on aphids which live on alfalfa. Alfalfa grows in this region in fall and spring, the rainy seasons, and therefore allows aphid propagation at these times only. McMullen shows that either long or short days induce reproductive diapause in this coccinellid. These are the conditions the species experiences in nature during the summer and winter. Even though irrigation now enables alfalfa to be grown all year round in central California making food available throughout the year, this coccinellid is still found only in the winter months and not during the summer. We have to assume that summer and winter diapause are caused by different mechanisms. The interaction of additional factors may play an important role (p. 139).

Aestivation in the alfalfa weevil, *Hypera postica*, has metabolic characteristics similar to winter diapause in other species. These consist primarily of feeding and the accumulation of reserves before entering diapause (Tombes, 1964a, b, 1966). The same characteristics have been observed in the beetle *Nebria brevicollis* (Ganagarajah, 1966).

A large body of literature deals with photoperiodic influences on the mode of reproduction in aphids. This is an exceedingly complex situation which will be dealt with elsewhere (p. 204ff.).

4. GROUP INTERACTIONS

The effect of population density on egg maturation and rate of oviposition is different in different species. It has been reported for *Drosophila melanogaster* that females living in groups lay fewer eggs than do solitary individuals, even though adequate food is available (Pearl, 1932). Collisions, interference with feeding, etc., seem to be the reasons for the low egg yield. Similarly, females of *D. simulans* lay fewer eggs if kept in groups than if kept in isolation (Sameoto and Miller, 1966). Egg production per female in *Trichogramma evanescens* was lowered even if only two animals were kept together, irrespective of whether the additional animal was another female or a male (Lund, 1938). In the beetles *Leptinotarsa decemlineata* (Grison and Ritter, 1961) and *Bupalus piniarius* (Klomp and Gruys, 1965), individuals interfere with one another in feeding, an essential factor in reproduction; consequently, grouped animals lay relatively fewer eggs.

Group effects on reproduction have been studied to some extent in several species of migratory locusts. It was reported for *Schistocerca gregaria* that females under crowded conditions matured their eggs faster than those in isolation (Husain and Mathur, 1945), particularly when older males were present in the colonies. Crowded conditions indeed facilitate mating; but, even when mating was prevented, egg maturation was still accelerated (Norris, 1954)—possibly because the crowded animals were more active than the isolated ones. It is known that activity facilitates the liberation of neurosecretory materials necessary for egg maturation processes (Highnam and Haskell, 1964) (Fig. 49). Even though crowded *Schistocerca* females matured eggs at a faster rate, their total egg output was below that of isolated females (Papillon, 1960). Isolation in *Locusta migratoria* (Norris, 1950; Albrecht *et al.*, 1958) and *Nomadacris septemfasciata* (Albrecht, 1959; Norris, 1959a) caused accelerated egg growth as compared to that in grouped females. This behavior is precisely opposite to that observed in *Schistocerca*. Isolated pairs of *Nomadacris* produced nearly twice as many eggs in a life span as did crowded females. When the latter species were kept in relative isolation for several generations, more ovarioles developed per ovary than when the females were reared in large groups.

As is apparent from these reports, animals in isolation or in pairs had a higher total egg output than did crowded ones. Under crowded conditions, the females' activities—like feeding, resting, pairing, and grooming—are interferred with; this in turn reduces egg production.

5. AVAILABILITY OF HOSTS

The availability of hosts for parasitic insects, or of plants for oviposition for many other species, often determines the number of eggs laid. Mature eggs are stored in the ovaries or genitalia and only a few eggs are laid if a suitable substrate is not available. In parasitic Hymenoptera mature eggs may be resorbed; and in other insects further egg maturation may be prevented by the presence of stored fully grown eggs. In other words, the total egg output is reduced because a substrate is temporarily unavailable.

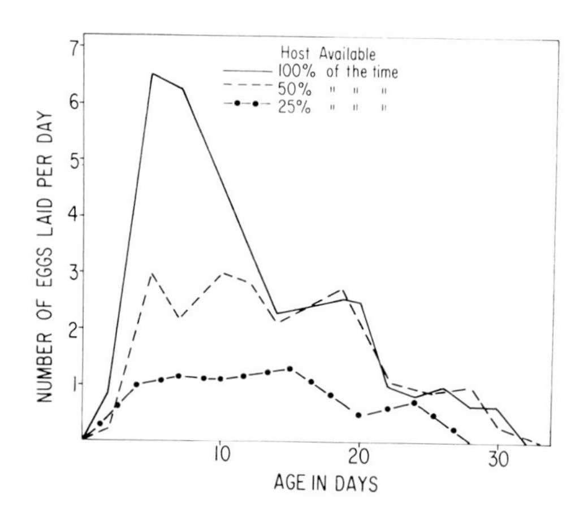

FIG. 50. Effect of host availability (percentage of time) on egg maturation and egg laying in the hymenopteran *Diadromus pulchellus*. The reduced fecundity when hosts are infrequently available is not caused by the unavailability of substrate for egg laying. (Modified from Labeyrie, 1960a.)

The hymenopteran *Diadromus pulchellus*, which does not need to feed as an adult, laid more eggs when the host was present all the time than when the host was present only a fraction of the time (Labeyrie, 1960a, 1962) (Fig. 50). Not only were fewer eggs laid, but also fewer eggs were matured in the ovaries of females which had only infrequent access to the host. In this case it cannot be argued that mature eggs or eggs being resorbed blocked the maturation process of subsequent eggs in the ovarioles since, as Labeyrie (1964) reported, if the hosts were perceived all the time but were inaccessible for oviposition, eggs matured in the ovaries at the same rate and in equal numbers as they did in those animals which had direct access to the hosts; if no hosts were available, these eggs were later resorbed. In this species, olfactory perception of the host seems to stimulate egg maturation. The precise mechanisms involved remain to be worked out.

Diadromus uses *Acanthoscelides obtectus*, a bean weevil, as host. This weevil produces more eggs if the bean seeds are perceived (Voukassovitch, 1949), and it has been shown that both differentiation of oocytes and vitellogenesis is favorably influenced by the perception of the seeds (Labeyrie, 1960b). Perception of the host plant may also stimulate egg production in the ovaries of *Acrolepia assectella*. An average of 233 eggs were laid when the host plant (onion) was continuously present, whereas only 78 eggs were produced when the plant was presented only every 3rd day (Cadeilhan, 1965a, b). In this case it is not clear, however, whether the actual perception of the host plant alone stimulated egg maturation.

These few reports on the sensory perception of a host and its influence on egg maturation are opening up a research field which has barely been touched. It should be particularly rewarding to work on possible endocrine links in the chain of events.

6. INTERACTIONS OF DIFFERENT ENVIRONMENTAL FACTORS

In nature, any species may encounter a multiplicity of environmental variables by which it can be affected. Relatively little, however, is known about how the individual animal integrates the information received by its various sensory receptors. For example, short days caused the beetle *Leptinotarsa decemlineata* to go into diapause in the fall, and conversely, artificial long days could prevent diapause (DeWilde, 1958). However, deprivation of food likewise induced diapause (Faber, 1949) even when the animals was experiencing long days (De Wilde, 1958). Obviously, one factor overrides another in this case. Under prolonged illumination or constant light, only a few females of the moth *Mamestra brassicae* mated; consequently, only these mated females laid the full complement of eggs (Bonnemaison, 1961). This species, as most nocturnal moths, requires dark for normal mating behavior and oviposition. Only prolonged illumination (more than 14 hr) will induce the female of the mosquito *Culex tritaeniorhynchus* both to feed enough to build up reserves for egg maturation and to display normal mating behavior (Newson and Blackeslee, 1957; Eldridge, 1962).

Both photoperiod and temperature affect reproduction in *Culex pipiens* (Eldridge, 1966). Under short-day conditions, eggs mature only at high temperatures, whereas with long days oviposition is observed even at relatively low temperatures. Length of illumination per day and temperature seem to have a certain synergistic effect in this species.

A photoperiod of 6–13 hr, i.e. short day, induced diapause in the bug *Lygus hesperus* even at temperatures of 27–32°C (Beards and Strong, 1966; Leigh, 1966). In California, this diapause is broken again in November or December during the rainy season when food becomes available. Is it the increased humidity during the rainy period, or the availability of food, or the accumulated energy debt which causes the termination of the reproductive diapause? Certainly, it is not the photoperiod (short day) or temperature (relatively high) which terminates this diapause in nature.

One of the fascinating cases of reproductive diapause is found among the coccinellid beetles. The species *Coccinella septempunctata* is found throughout much of the world in northern temperate and southern subtropical or tropical regions. Depending on the region, the species may be univoltine, bivoltine or multivoltine (see Hagen, 1962) (Fig. 51). Populations in the Euro-Siberian region have either one or two generations per year, and the adults hibernate (Bodenheimer, 1943; Sundry, 1966). Long days plus the availability of food prevent diapause (Hodek, 1962). In hot and dry regions, an aestivo-hibernation can be found, i.e. the female goes into reproductive diapause during early summer and stays in it until the next spring. In the coastal plains of the Mediterranean region, there may be a second generation in the fall after the females have come out of aestivation (Bodenheimer, 1943). Aestivation can be attributed to the rather dry season and lack of food. All the evidence available suggests that both photoperiod and food availability control diapause in this species, irrespective of where the populations are found. An extreme situation, where possibly up to twenty successive generations occur, is found in India. High temperatures and good food sources

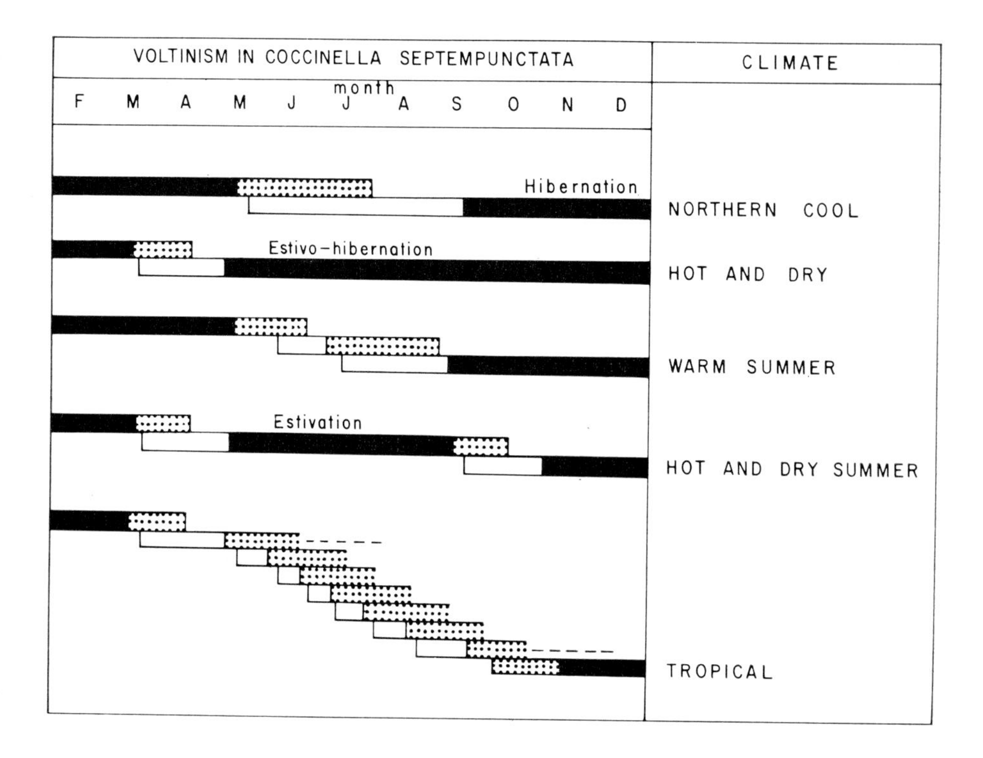

FIG. 51. Voltinism in *Coccinella septempunctata* as induced by various climatic conditions in different parts of the world. (Modified from Hagen, 1962.)

during most of the year apparently allow this high rate of propagation (Puttarudriah and Basavanna, 1953). One of the best examples of how the availability of food can influence reproductive behavior of coccinellids is illustrated by *Hippodamia convergens*. This species emerges in California in May. It then used to migrate to the mountains and adjacent valleys, where it stayed reproductively inactive until the next spring (Hagen, 1962). With the introduction of extensive irrigation systems and the spread of the alfalfa aphid, *Therioaphis maculata*, food became available throughout the summer. As a consequence, *H. convergens* has become multivoltine and no longer migrates to the aestivo-hibernation site. In another population of this species in Arkansas, aestivo-hibernation is observed when food becomes scarce, but not otherwise (Stewart *et al.*, 1967). Population of coccinellid beetles are presumably mixed in their genetic background with respect to voltinism. Depending on the environmental conditions, one or the other strain is favored. This is borne out by experiments reported by Bonnemaison (1964c) for *Coccinella septempunctata*. About 85–95 per cent of the females of a population taken from the Paris region were univoltine, i.e. adults underwent an aestivo-hibernation period. If the animals were then reared under long photoperiods for several generations, only up to 15 per cent went into aestivation. In other words, a strain had been selected for which was not responsive to long days.

The carabid beetle *Pterostichus nigrita* also has an aestivo-hibernation period. It is interesting here that following long days in the summer, short days induced oocyte growth to some extent. Only after a succeeding long-day exposure, however, was growth completed and eggs laid (Thiele, 1966). Long days following a period of short days induced egg maturation in 90 per cent of the females of the beetle *Agonum assimile*; but, if the population was continuously exposed to long-day conditions, only 50 per cent of the females laid eggs (Thiele, 1966). Apparently a population mixed with respect to expression of the aestivo-hibernation phenomenon is to be found in nature; 50 per cent of the animals may have an obligatory diapause induced by long day. In still another carabid, *Pterostichus oblongopunctatus*, successions of long and short photoperiods or vice versa did not cause egg maturation—rather the exposure to cold terminated the diapause (Thiele, 1968). It is interesting to note that all of these Carabidae lay their eggs in early summer; yet we observe that these species use different environmental cues in timing their reproductive season.

D. Hybrid Vigor (Heterosis)

Heterosis is defined as the superiority of a hybrid over its parents in any measurable attribute. In animals, hybrid vigor has generally been measured by the rate of growth, survival, fecundity, or hatchability of eggs. In the present context, a full discussion of the concept will not be given; for more details, the interested reader is referred to the pertinent literature (Dobzhansky, 1951; Gowen, 1952; Lerner, 1954).

The literature dealing with hybrid vigor in insects is almost entirely confined to a few *Drosophila* species, presumably because of the ease of rearing them and also because of the number of highly inbred lines which are cultured in many laboratories and which may serve as controls. A discussion of heterosis in the context of species propagation most likely has only theoretical value since, with rare exceptions, natural populations exhibit a balanced genetic polymorphism which has evolved to suit the species. It is thought that the superiority of heterotic animals over pure inbred lines is based on precisely such a balanced genetic polymorphism (Dobzhansky, 1951). A decline in the productivity of a population is a typical laboratory phenomenon which is due to inbreeding over many generations. In nature, this can occur only in small, isolated regions.

Nevertheless, a consideration of heterosis in insect populations is of some value. For example, Hadorn and Zeller (1943) mentioned that the total egg production of females from inbred *Drosophila melanogaster* was approximately 900, whereas 1gl/Cy hybrid females laid 1800 eggs in a lifetime. Similar observations were made by Gowen and Johnson in the same species (1946). The reproductive capacity of inbred lines may vary greatly, apparently the result of a variety of factors which become apparent to different degrees in different genetic combinations. In any reported case of hybrid vigor, it is difficult to pin down its actual causes. Sondhi (1967) reported that blood transfusions from hybrid to highly inbred flies improved fecundity and longevity in the recipient flies: apparently the physiological properties (no precise agent is known) of the hemolymph of hybrids are superior to those of inbred animals. Hybrids of inbred lines of *D. subobscura* lived longer than either of the parents and also laid nearly twice the number of eggs as did the females of the inbred lines laid (Clarke and Maynard Smith, 1955). Selection of lines in this species caused egg output to decline rapidly to about 20 per cent of the original number, apparently due to inbreeding and accumulation of detrimental factors (Hollingworth and Maynard Smith, 1955). This may have been caused in part by a reduced "athletic ability" during courtship resulting in less success in mating. It has been reported, for example, that inbred males could not follow the rapid side-to-side courtship movements of the female; consequently, the female did not accept the courting male (Maynard Smith, 1956).

Inbreeding of various strains of *Oncopeltus fasciatus* also resulted in reduced fecundity, i.e. fewer eggs were laid and many nymphs did not survive to become adults (Turner, 1960); crossbreeding restored the vitality of the colony.

Research on hybrid vigor in insects apparently was never very active and only sporadic reports appear in the literature. Interesting research subjects are posed by the phenomenon, particularly in biochemical terms, but very little has been done so far.

CHAPTER 8

HORMONAL CONTROL OF EGG MATURATION

THE very nature of reproductive processes in insects, i.e. the periodicity of egg maturation, suggests hormonal control. Egg production is coordinated with the animal's external and internal environment such as availability of food, long or short photoperiods which may affect the food supply, mating, the presence of embryos in viviparous species, possibly humidity, etc. The process of yolk deposition in the maturing oocytes has to be controlled "economically" according to the species' needs.

During the last 30 years detailed investigations of hormonal involvement in egg maturation have been made in comparatively few species. Since insects differ enormously in their biology and reproductive physiology, comparisons and analogies are often unsound. Various factors, such as food supply or mating, essential for the activation of endocrine glands and egg maturation, are of varying importance in different species studies. A common principle has not been found and probably should not be expected to exist. Clearly, detailed investigations at all levels of sophistication must be carried out in more species before we can be reasonably certain how hormones control reproductive processes in insects. This chapter attempts to synthesize present knowledge on the subject.

A. Anatomy of the Neuroendocrine System

The neuroendocrine system in insects consists of corpora allata, corpora cardiaca, prothoracic glands, and neurosecretory cells of both the brain and ganglia of the ventral nerve cord. In the present context, the neurosecretory system of the brain and suboesophageal ganglion, the corpora cardiaca functioning as the release organ of neurosecretory materials, and the corpora allata will be considered.

Corpora allata are found in insects of all orders (Hanström, 1942; Cazal, 1948; Bitsch, 1962; Cassagnau and Juberthie, 1966). In the majority of the studied species of both lower and higher orders, the glands are paired; however, some species among Hemiptera, Plecoptera, and Diptera possess an unpaired gland (Hanström, 1942). Embryologically they have arisen as ectodermal invaginations between the mandibular and first maxillar segments (cf. Hanström, 1942). In Machilidae the paired corpora allata stay close to the site of origin (Bitsch, 1962), whereas in species of other orders they move to a dorsal and posterior position during embryonic and larval development. Both the unpaired and paired glands are generally located within the head capsule posterior to the brain and corpora cardiaca (Figs. 52, 53). In larval cyclorraphan Diptera, the corpus allatum is incorporated into the ring gland and makes up its dorsal part; following metamorphosis and the degeneration of the lateral portions of the ring gland, it moves posteriorly (Fig. 54).

Histologically, the corpora allata appear as solid glandular tissues in most species

(Fig. 55); in Phasmida (Pflugfelder, 1937), Machilidae (Bitsch, 1962), and in the cockroach *Diploptera punctata*, however, the corpora allata have retained their original vesicular structure. The fine structure of the glands with respect to their activity stages has been studied in only a few species. Suffice it to mention here that electron micrographs of active corpora allata show peculiar ergastoplasmic whorls in females of *Leucophaea maderae* (Scharrer, 1964), and a highly developed smooth endoplasmic reticulum in females of *Drosophila melanogaster* (Aggarwal and King, 1966) and in males of *Schistocerca gregaria* (Odhiambo, 1966d). These structures may be related to hormone synthesis.

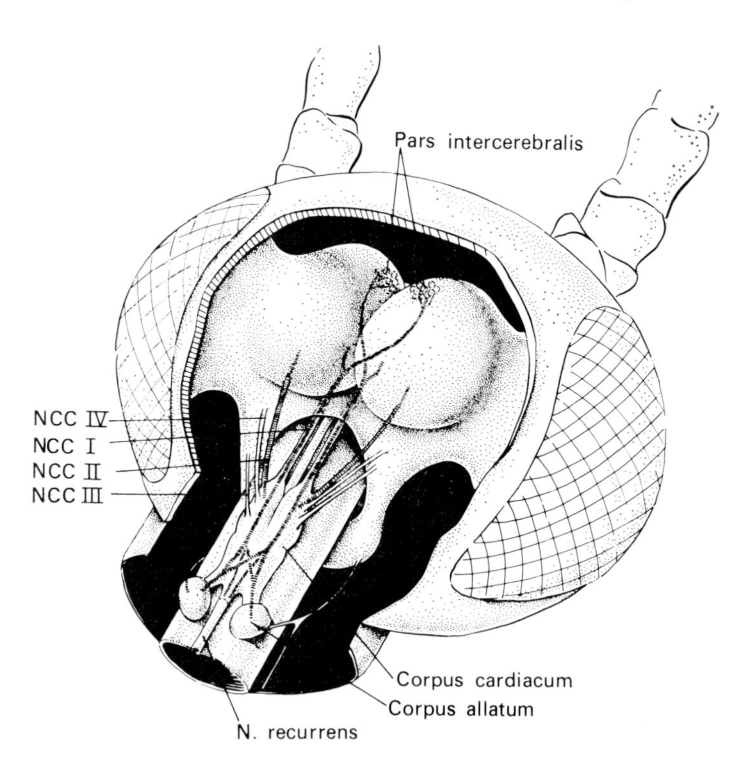

FIG. 52. Semidiagrammatic illustration of the brain and retrocerebral complex of a dictyopteran species.

The originally paired corpora cardiaca of the majority of species studied are fused to form a more or less single organ of glandular appearance. They generally are closely associated with the hypocerebral ganglion with which they originate as invaginations of the foregut during embryonic development (Hanström, 1942). They are thought to be of nervous nature. It is well documented that the corpora cardiaca function as storage and release organs for neurosecretory materials (see Scharrer and Scharrer, 1963), as it has been histologically demonstrated in many species that the neurosecretory materials found here are derived from the brain. Following the severance of the nerves leading to the corpora cardiaca in *Leucophaea*, the glands become depleted of neurosecretory materials (Scharrer, 1952). By the use of darkfield illumination, the flow of some material along the nervi corporis cardiaci could be directly observed in *Calliphora erythrocephala* (Thomsen, 1954). Other inclusions resembling primary neurosecretory granules seem to arise in the cardiacum cells themselves, as demonstrated in *Leucophaea* (Scharrer, 1963), *Calliphora*, (Normann, 1965), the aphid *Myzus persicae* (Bowers and Johnson, 1966), and *Hypera postica* (Tombes and Smith, 1966). Thus the corpora cardiaca may function as an independent neuroendocrine organ. Experimental evidence for this has recently been obtained from studies in the cockroach *Leucophaea* (Engelmann and Müller, 1966). The corpora cardiaca and the hypocerebral ganglion are closely associated and cannot be recognized anatomically as separate organs in many species. Because of the consequent difficulty of surgical removal of the corpora cardiaca without destruction of other nervous pathways, experimental results are open to several interpretations, and the function of the corpora cardiaca in reproduction is still unclear.

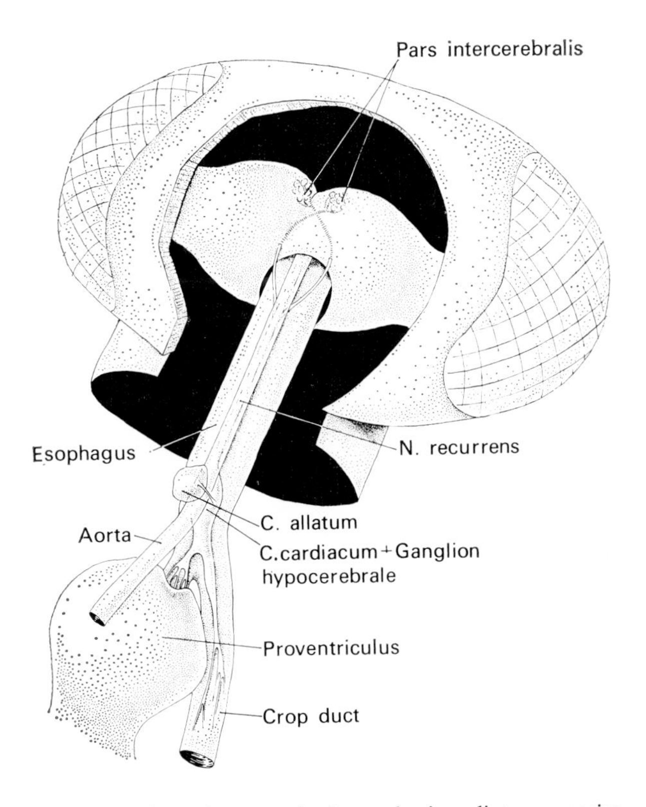

FIG. 53. Brain and retrocerebral complex in a dipteran species.

The detailed innervation of the corpora allata and corpora cardiaca has been worked out in only a few species. However, a great many species from most orders have been investigated to some extent by Cazal (1948). He finds that, in the vast majority of species, two pairs of nerves innervate the corpora cardiaca (Fig. 52). The median pair, the nervi corporis cardiaci interior (or NCC I), are thought to arise in the neurosecretory cells of the pars intercerebralis, whereas the lateral pair, the nervi corporis cardiaci exterior (or NCC II), originate in the lateral group of neurosecretory cells of the protocerebrum. A third pair of nerves (NCC III) leading from the brain to the corpora cardiaca have been found in some, but not all, species (Dupont-Raabe, 1956, 1964; Willey, 1961). They originate in the tritocerebrum close to or from a group of neurosecretory cells. A fourth pair of nerves (NCC IV) lead into the corpora cardiaca from the deutocerebrum and were found in Orthoptera, Coleoptera, Heteroptera, Isoptera, and Dictyoptera (Brousse-Gaury, 1967a, b). The nerves NCC III and NCC IV are about 15–20 μ in diameter; they both carry stainable neurosecretory materials.

In all species examined, the nerves coming from the pars intercerebralis are known to decussate within the frontal part of the protocerebrum. In a thorough study of the stomodeal nervous system in *Periplaneta americana* and other cockroach species, Willey (1961) found that apparently some of the fibers leaving the area of the pars intercerebralis do not decussate. He also states that non-neurosecretory fibers join the NCC I after the chiasma from other parts of the protocerebrum. This finding has been confirmed for *Periplaneta* and has been extended to *Blabera fusca* (Brousse-Gaury, 1968). From no other species have similar

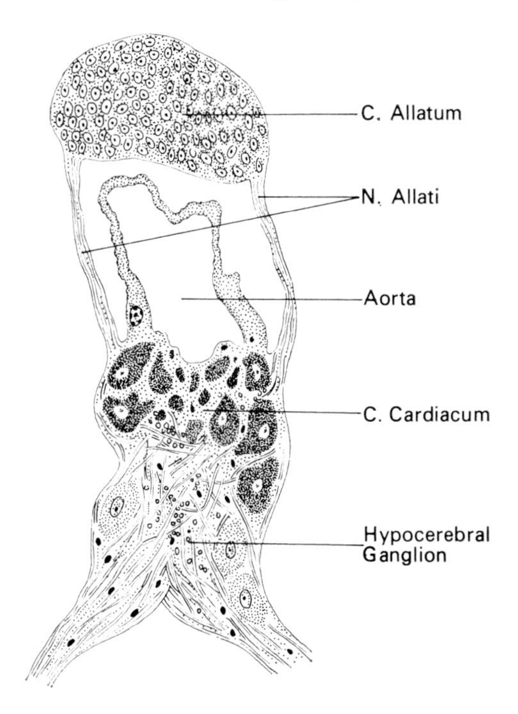

FIG. 54. Section through the corpus allatum and corpus cardiacum of the adult dipteran *Eristalis tenax*. (Adapted from Cazal, 1948.)

findings been reported. Thus, the presence of both neurosecretory and non-neurosecretory fibers in the NCC I makes any interpretation in terms of the function of this nerve rather difficult.

Upon entering the corpora cardiaca, the NCC I and II fuse into a single nerve; the further course of individual axons is consequently difficult to determine. In the cockroach *Leucophaea*, this single nerve splits into three smaller nerves in the posterior part of the corpora cardiaca (Fig. 52). One of these nerves reaches the corpus allatum of the ipsilateral side, another decussates and reaches the corpus allatum of the contralateral side, and the third fuses with the nervus recurrens (Engelmann, 1957a). In *Periplaneta* this is not found (Willey, 1961), but rather a nerve which might be analogous to the decussating branches in *Leucophaea* connects the corpora allata.

A small nerve leading from the suboesophageal ganglion to either of the corpora allata has been found in several species (Engelmann, 1957a; Willey, 1961; Raabe, 1964, Cassagnau and Juberthie, 1967). In some animals this nerve enters the corpus allatum via the corpora cardiaca. It is uncertain, however, whether this nerve is found in the majority of species of the various insect orders since detailed anatomical studies have not been made in many species. In *Hypera postica* many axons which contain neurosecretory materials literally ensheath the corpus allatum (Tombes and Smith, 1966). These axons are believed to originate in the protocerebrum. A similar situation may exist in the cricket *Gryllodes sigillatus* in which axons of the allatum nerves superficially pass over the corpus allatum in bundles or networks and emerge as the NCA 2 which leads down to the suboesophageal ganglion

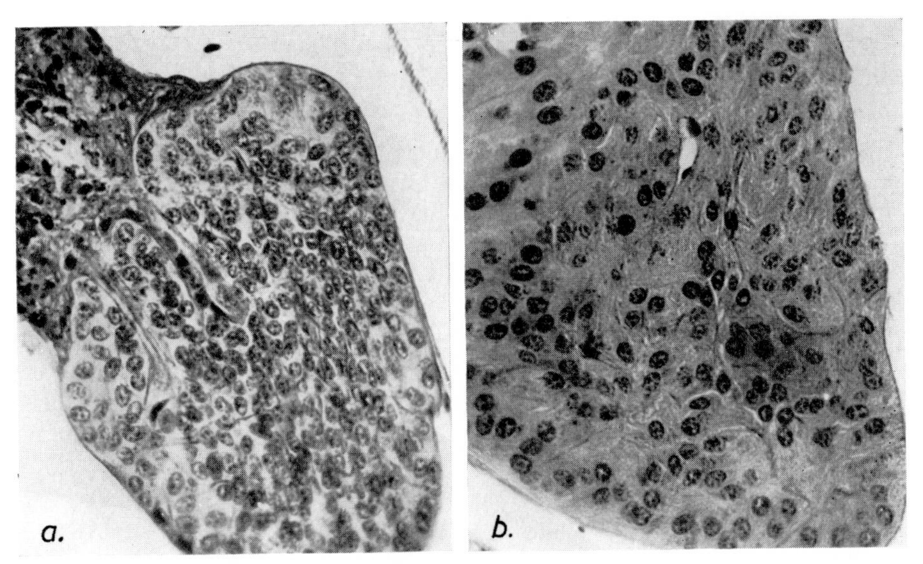

FIG. 55. Sections through corpora allata of *Leucophaea maderae*. (a) Inactive gland in which the nuclei are densely packed. (b) Active gland in which there is abundant cytoplasm. Note that the size of the nuclei in the active gland is hardly different from that of the inactive gland.

(Awasthi, 1968). As this short description of the major nerves associated with the retrocerebral endocrine system indicates, it may be difficult to ascribe definite functions to specific nerves.

Three groups of neurosecretory cells have been implicated in the control of egg maturation in various species. Most widely acknowledged are the neurosecretory cells of the pars intercerebralis of the protocerebrum (Figs. 52, 56). Neurosecretory materials from these cells enter the corpora cardiaca as has been shown in many species. These materials can also reach the corpora allata and may be deposited or released there. Even in the species in which neurosecretory granules rarely if ever could be found in the corpora allata with light microscopic techniques, such granules could be demonstrated with electronmicrographs. This was shown, for example, in the cockroach *Leucophaea* (Scharrer, 1964). Such anatomical findings suggest that neurosecretory materials have a functional significance in relation to corpus allatum activity cycles. The lateral group of neurosecretory cells from which the NCC II presumably originate have been implicated in the control of the corpora allata in

Schistocerca paranensis (Strong, 1965b). Lastly, several groups of neurosecretory cells in the suboesophageal and ventral ganglia deserve attention. As Scharrer (1955) has shown in the cockroach *Leucophaea*, stainable materials accumulate after ovariectomy in some of the cells of the suboesophageal ganglion. This has never been observed in normal females or in any male. Scharrer suggested that the function of these cells is altered under this experimental condition and that demonstrable materials thus accumulate. The cells are appropriately called "castration cells". The actual function of these cells is not known. Two groups of neurosecretory cells, which change their tinctorial properties during egg maturation cycles, are also found in the suboesophageal ganglion of *Locusta migratoria* (Fréon, 1964a, b). Furthermore, neurosecretory cells in the ganglia of the ventral nerve cord of *Leucophaea* show changing amounts of stainable materials during egg maturation cycles, which suggest a role in reproductive processes (Bessé, 1965, 1967). In *Locusta* active neurosecretory cells of the pars intercerebralis have both a highly developed endoplasmic reticulum and an active Golgi complex such structures indicate an involvement of the cells in hormone production (Girardie and Girardie, 1967).

B. *Egg Maturation Controlled by the Corpus Allatum*

1. EXPERIMENTAL PROOF

The corpus allatum as the source of the gonadotropic hormone in insects was first mentioned by Wigglesworth in 1936. The peculiar biology of the female of *Rhodnius prolixus* aided in the discovery of this basic principle which, as will be shown, operates in many other species. *Rhodnius* females produce a set of eggs only after a blood meal. However, decapitation (including the removal of the single corpus allatum) prevented egg maturation if the operation was performed shortly after feeding. If, on the other hand, the corpus allatum remained in the decerebrated animal or if the corpus allatum was removed a few days later, egg maturation proceeded normally (Fig. 57). After parabiosis of two fed females, only one of which retained her corpus allatum, egg maturation occurred in both animals. These operations employing methods of classical experimental biology firmly established for the first time that a blood-borne factor is involved in egg maturation in an insect. Together with earlier observations on insect molting (Wigglesworth, 1934), this report finally made insects a valid subject of research for endocrinologists.

Soon after the first report by Wigglesworth in 1936, Weed (1936) found the same principle to be at work in the grasshopper *Melanoplus differentialis*, i.e. allatectomy prevented egg maturation while reimplantation again allowed the maturation of a set of eggs.

During the subsequent 30 years, allatectomy and reimplantation of the glands have been performed in many species by many investigators. For the majority of the species studied, it has been established that the corpora allata are involved in promoting egg maturation (cf. Table 9). Several insect orders, e.g. Ephemeroptera, Odonata, Plecoptera, Isoptera, Mallophaga, Neuroptera, Mecoptera, Trichoptera and Hymenoptera, are not represented in Table 9 because experimental evidence for corpus-allatum-stimulated egg maturation has not yet been given for any of their members. However, in some cases histological observations of the corpora allata strongly suggest their function in promoting egg maturation. Obviously, many species of these orders do not easily lend themselves to experimentation because of size or difficulties in the rearing of even unoperated animals over the necessary time periods, but it will certainly be possible to find species within these orders which are suitable for experimentation.

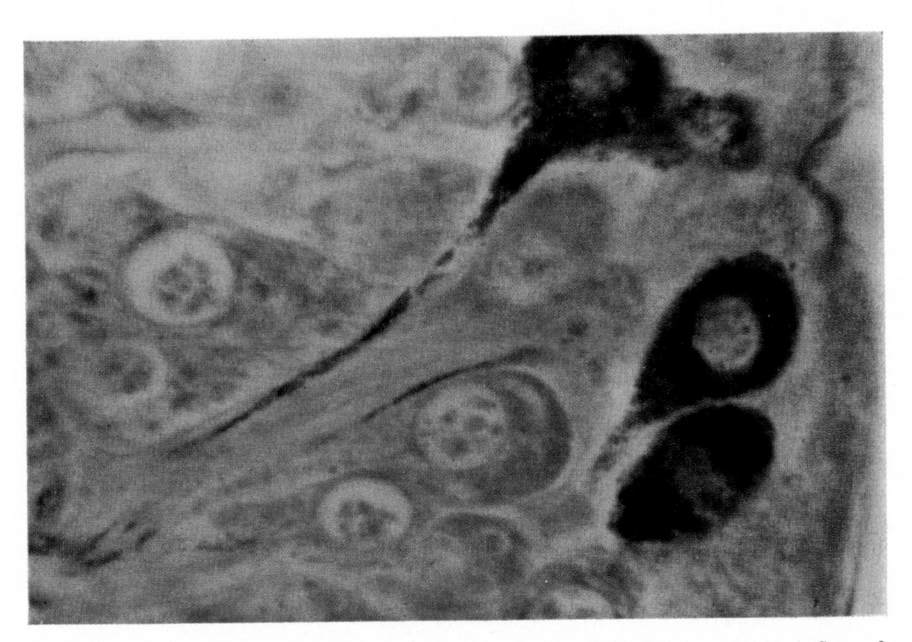

FIG. 56. Neurosecretory cells of the pars intercerebralis of *Gryllus campestris*. A flow of stainable material down the axon is visible.

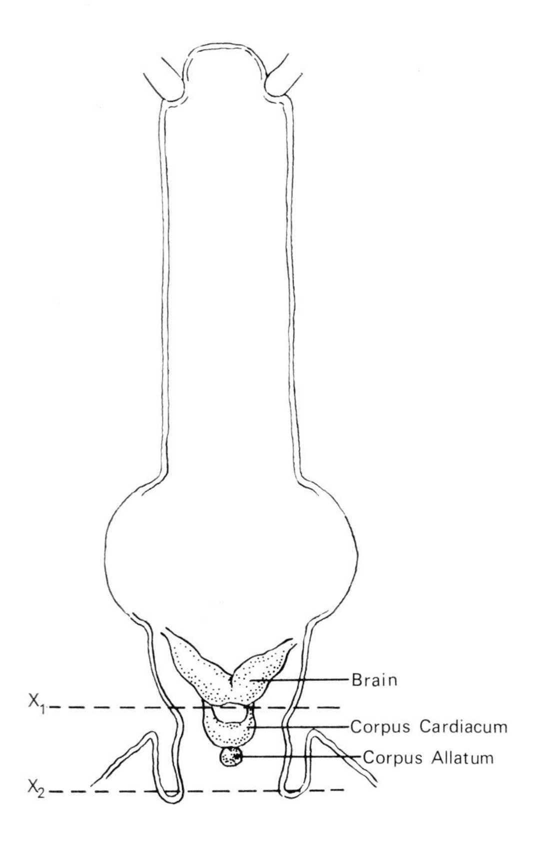

FIG. 57. Position of the brain, corpus cardiacum and corpus allatum within the head capsule and neck region in *Rhodnius prolixus*. X_1 = decapitation which removes only the brain. X_2 = decapitation which removes all endocrine glands. (Modified from Wigglesworth, 1936.)

In species of the Phasmida and some of the Lepidoptera, allatectomy did not prevent egg maturation and oviposition. Operations performed even during the larval periods in *Carausius morosus* (Pflugfelder, 1937) or *Bombyx mori* (Bounhiol, 1936), for example, had little effect on egg maturation in the pupae or the resulting adults. Even in isolated abdomina of *Phryganidia californica* (Bodenstein, 1938) and *Hyalophora cecropia* (Williams, 1952), egg maturation could occur. There are, however, reports for the Lepidoptera, *Bombyx mori* (Yamashita *et al.*, 1961), *Galleria mellonella* (Röller, 1962), and *Leucania separata* (Tsiu-ngun and Quo, 1963) that allatectomy performed in larvae reduced the total egg output. The earlier in larval life the animals were operated upon, the fewer eggs were produced (Röller, 1962). Egg output was similarly reduced after allatectomy in *Carausius* (Neugebauer, 1961). This may indicate that in these species the corpora allata do not control egg maturation directly but rather do so indirectly through their effect on certain phases of metabolism. Detailed biochemical investigations on normal and operated adults are necessary to gain a better understanding of the role of the corpora allata in egg maturation in insects of these two and possibly other orders. Control of egg maturation in certain lepidopteran species is,

TABLE 9

(a) *Species in which experimental evidence has been given for egg maturation controlled by the corpus allatum.**

Order	Species	Reference
Thysanura	*Thermobia domestica*	Rohdendorf, 1965
Orthoptera	*Anacridium aegypticum*	Geldiay, 1967
	Euthystira brachyptera	Müller, 1965
	Gomphocerus rufus	Loher, 1965
	Locusta migratoria	Joly, L., 1960
	Melanoplus differentialis	Weed, 1936; Pfeiffer, 1939
	Schistocerca gregaria	Highnam *et al.*, 1963; Pener, 1965
	,, *paranensis*	Strong, 1965a
Dermaptera	*Anisolabis maritima*	Ozeki, 1949
Dictyoptera	*Blattella germanica*	Roth *et al.*, 1962a
	,, *vaga*	,, ,, ,, ,,
	Diploptera punctata	Engelmann, 1959a; Roth *et al.*, 1961
	Leucophaea maderae	Scharrer, 1946; Engelmann, 1957a
	Nauphoeta cinerea	Lüscher, 1968
	Periplaneta americana	Girardie, 1962
Hemiptera	*Adelphocoris lineolatus*	Ewen, 1966
	Cimex lectularius	Davis, 1964
	Oncopeltus fasciatus	Johansson, 1954, 1958
	Pyrrhocoris apterus	Sláma, 1964
	Rhodnius prolixus	Wigglesworth, 1936
Diptera	*Aedes aegypti*	Gillett, 1956; Lea, 1963
	,, *sollicitans*	Lea, 1963
	,, *taeniorhynchus*	,,
	,, *triseriatus*	,,
	Anopheles maculipennis	Detinova, 1962
	Calliphora erythrocephala	Thomsen, 1940, 1943†
	Culex molestus	Clements, 1956; Larsen and Bodenstein, 1959
	,, *pipiens*	Larsen and Bodenstein, 1959
	Drosophila hydei	Vogt, 1943
	,, *melanogaster*	Vogt, 1940; Bodenstein, 1947
	Lucilia sericata	Day, 1943
	Musca nebulo	Deoras and Bhaskaran, 1967
	Phormia regina	Orr, 1964a
	Sarcophaga bullata	Wilkens, 1965, 1968†
	,, *securifera*	Day, 1943
Coleoptera	*Carabus catenulatus*	Joly, P., 1950
	Dytiscus marginalis	Joly, P., 1942, 1945
	Leptinotarsus decemlineata	De Wilde, and Stegwee, 1958
	Tenebrio molitor	Mordue, 1965b; Lender and Laverdure, 1965
Lepidoptera	*Pieris brassicae*	Karlinsky, 1963, 1967a

(b) *Insects in which the gonadotropic function of the corpora allata seems to be absent or is uncertain.*

Order	Species	Reference
Phasmida	*Carausius morosus*	Pflugfelder, 1937; Neugebauer, 1961
	Sipyloridea sipylus	Possompès, 1958
Diptera	*Calliphora erythrocephala*	Thomsen, 1943, 1952; Possompès, 1955
	Sarcophaga bullata	Wilkens, 1965, 1968.
Lepidoptera	*Bombyx mori*	Bounhiol, 1936
	Galleria mellonella	Röller, 1962
	Hyalophora cecropia	Williams, 1952
	Leucania separata	Tsiu-ngun and Quo, 1963
	Phryganidia californica	Bodenstein, 1938

* Only the first and/or second major publications are listed.

† Corpus allatum control of egg maturation in these species is incomplete.

on the other hand, completely under the influence of the corpora allata. Allatectomy prevented any deposition of yolk in *Pieris brassicae*, a species which requires food, or at least water, after emergence to assure yolk deposition in the oocytes (Karlinsky, 1963, 1967a, b, c). *Pieris* appears to be the only species of the Lepidoptera known to depend on the corpora allata for egg maturation.

Whereas allatectomy had no effect on egg maturation in some species or prevented it entirely in many others, its effect was rather peculiar in *Calliphora erythrocephala*. According to Thomsen (1940, 1943), allatectomy performed in adult females a few hours after emergence prevented egg maturation in nearly all animals, although some matured their eggs fully. These results could possibly be interpreted to the effect that, in those animals which matured eggs after this operation, the corpora allata had already liberated enough gonadotropic hormone to initiate egg maturation. Similar results have been obtained for the flesh fly *Sarcophaga bullata* (Wilkens, 1965, 1968). Strangely enough, however, if allatectomy was performed in late last instar larvae of *Calliphora*, the majority of the resulting imagines were able to mature eggs fully (Possempès, 1955). Allatectomy in these cases had been successfully performed, as Thomsen ascertained. Results of allatectomy in pupae of *Sarcophaga* were comparable to those obtained in *Calliphora* (Wilkens, 1968). Several interpretations of these findings are possible: the corpus allatum is not the primary source of a gonadotropic hormone; some other tissues may take over the role of the corpus allatum under experimental conditions; or possibly the neurosecretory cells of the pars intercerebralis produce the true gonadotropic hormone, but normally have to be stimulated by the corpus allatum hormone after metamorphosis (see p. 170). This latter possibility is inherent in the findings of Lea and Thomsen (1962) that nuclei of the neurosecretory cells become enlarged after a feeding of meat and the consequent activation of the corpora allata. As this brief summary suggests the situation in these two flies is not clear; the topic will be taken up again (p. 170).

2. MORPHOLOGICAL EVIDENCE

During periods of egg maturation, i.e. yolk deposition in the oocytes, the size of the corpora allata in *Rhodnius prolixus* increases markedly (Wigglesworth, 1936). The change in size is basically due to increased volume of cytoplasm and, to a lesser degree, to the increase in nuclear number and size. This holds true in practically all species examined; consequently, glands with widely spaced nuclei are considered active, and those with packed nuclei, inactive. A set of micrographs of sections from the cockroach *Leucophaea maderae* (Fig. 55) serves as an example. This correlation between egg maturation and increased cytoplasmic volume of the corpora allata has been found in members of the Orthoptera (Huignard, 1964; Highnam and Haskell, 1964; Cassier, 1965b; Strong, 1965a, b; Loher, 1965), Dictyoptera (Scharrer, 1952; Lüscher and Engelmann, 1955; Engelmann, 1957a, 1959a), Hemiptera (Wigglesworth, 1936; Nayar, 1953, 1956; Johansson, 1954; Dixon, 1963), Isoptera (Pflugfelder, 1938; Deligne and Pasteels, 1963), Diptera (Thomsen, 1940; Vogt, 1942; Mednikova, 1952; Doane, 1960a; Wilkens, 1968), Coleoptera (Palm, 1949; de Wilde, 1954; Lender and Laverdure, 1964a, b; Mordue, 1965a, b, c; Siew, 1965a, b), and Hymenoptera (Palm, 1948; Pflugfelder, 1948; Lukoschus, 1956). In Phasmida and Lepidoptera noticeable growth of the corpora allata also occurs after emergence and is correlated with egg maturation periods (Pflugfelder, 1937a, b; Kaiser, 1949; Tsiu-ngum and Quo 1963). Furthermore, an enormous increase in the size of the nuclei of the corpora allata has been reported for some Lepidoptera (Kaiser, 1949) during periods of egg maturation. This observation is remarkable because egg maturation proceeds in females of the Phasmida and many known Lepidoptera without the participation of the corpora allata. If one accepts the hypothesis that the gland size, or rather the nucleo-cytoplasmic ratio, can be equated with specific activities of this endocrine gland, it seems that the corpora allata are functional in some aspects of the animals' metabolism which are perhaps indirectly related to reproduction. The work of Röller (1962) and Yamashita *et al.* (1961) seems to substantiate this interpretation.

Further evidence for the view that the nucleo-cytoplasmic ratio of the corpora allata is indicative of their activity may be the frequent observation that ovariectomized animals have larger glands than do normal females, possibly because of lack of feedback inhibition. This has been reported, for example, in *Melanoplus differentialis* (Pfeiffer, 1940), *Drosophila* spec. (Vogt, 1942; Doane, 1960a), *Lucilia sericata* (Day, 1943), *Calliphora* (Thomsen, 1943), *Rhodnius* (Wigglesworth, 1948), *Leucophaea* (Scharrer and von Harnack, 1961), and *Gryllus domesticus* (Huignard, 1964). Many authors consider the enormously enlarged glands to be hypertrophied. No deviation from the norm was found in the ovariectomized grasshoppers *Schistocerca gregaria* (Highnam, 1962b), *Locusta migratoria* (Joly, 1964), and *Gomphocerus rufus* (Loher, 1966b). The time lapse after castration may be of importance for the expression of hypertrophy, as is exemplified by the work of Scharrer and von Harnack (1961) on *Leucophaea*. In this species, growth of the corpora allata beyond the volume found in normal, active animals was observed only infrequently and only several months after an operation; it was never found in castrated males.

In this context, it is interesting to note that sterile *fes/fes* mutants of *Drosophila melanogaster* have enormously enlarged corpora allata (King *et al.*, 1966). Not only the size but also the submicroscopic structure of the gland differs from those of wild type females. In the corpora allata of mutants, the mitochondria are swollen to a considerable degree. Upon implantation of wild type ovaries in which vitellogenesis was taking place, the corpus allatum of the mutant became normal in appearance.

The unusual growth of the corpora allata in all cases mentioned can be linked to ovariectomy. The question of whether the excess size of the glands indicates storage or increased hormone production remains to be answered; various authors have taken one or the other view. Is it actually necessary to assume that hormone storage is associated with hypertrophy of the gland cells? Theoretically, the hormone could be stored equally well in small cells if indeed it is stored in the glands at all. Moreover, one may postulate that an active cell in which hormone production is proceeding at a very high rate is larger than a relatively inactive one which only stores the hormone. Evidence for the view that corpora allata of ovariectomized animals are capable of high activity will be given in a later section. Suffice it to state here that corpora allata seem to be more active in animals with a high hemolymph protein level than in those with a low protein concentration. And indeed castrated females generally have a high blood protein level, indicative of the good nutritional status of the animals, since egg maturation, which would have drained off proteins, has not taken place (p. 182).

Cyclic changes in the corpus allatum volume which are correlated with egg maturation cycles have been reported for the cockroaches *Leucophaea* (Lüscher and Engelmann, 1955; Engelmann, 1957a; Scharrer and von Harnack, 1958) (Fig. 58) and *Diploptera* (Engelmann, 1959a). In these viviparous species, the activity cycles of the corpora allata are well defined because these glands are clearly activated by mating, inactive during the extended periods of pregnancy, and reactivated shortly thereafter. Similar cycles, though less pronounced since the corpora allata do not drop to complete inactivity, are found in *Culex molestus* (Larsen and Bodenstein, 1959), *Calliphora* (Strangways-Dixon, 1961b, 1962), *Schistocerca gregaria* (Highnam, 1962b), *Schistocerca paranensis* (Strong, 1965a), *Euthystira brachyptera* (Müller, 1965b), and *Sarcophaga* (Wilkens, 1965, 1968). Even totally isolated corpora allata of *Leucophaea* undergo cyclic volumetric changes in a neat correlation with induced cycles of egg growth in the ovaries (Engelmann, 1962). It is tempting to propose that the cyclic volumetric changes of the corpora allata represent cycles of hormone production.

Attempts have indeed been made to quantitate the hormone production of the corpora allata by a determination of the volume or nucleo-cytoplasmic ratio (Engelmann, 1957a). It was postulated that the larger the gland, the more hormone is produced; consequently, egg growth is faster. It was emphasized that the volume of the corpus allatum does not necessarily indicate the activity level of the gland; rather it is the cytoplasmic volume which denotes activity. The nucleo-cytoplasmic ratio was considered to be the decisive measurement because an increase in the size of the entire gland may simply reflect an increase in the number of nuclei (such an increase is unlikely to indicate hormone production of the gland cells). This idea was implicit in earlier published observations. Termite queens, particularly egg-laying ones, have enormously enlarged corpora allata in comparison to those of the workers (Pflugfelder, 1938; Deligne and Pasteels, 1963). It was recognized that in termites, as well as in other species examined, the increase of the corpora allata is primarily caused by an increase in cytoplasm. Likewise, the honeybee queen has larger corpora allata than do the workers, at least during her early life period (Pflugfelder, 1948; Lukoschus, 1956).

What physiological criteria do we have to measure actual hormone production of a gland of a given size? Pflugfelder (1938) believed that the relatively large corpora allata of termite queens are merely caused by the enriched diet provided by the workers; accordingly, the larger corpora allata of queens may not indicate greater hormone production but rather reflect an increased general metabolism. Further support for this was obtained by Pflugfelder through observations of honeybee queens and workers at various ages. He found that older queens have even smaller corpora allata than do workers of the same age, a finding confirmed by Lukoschus (1956). Pflugfelder concluded that the size of the corpora allata cannot be equated with hormone production since older queens lay as many or more eggs than do young ones. This argument may not be conclusive, however, since other hormones or pheromones important in maintaining the social structure in the colony may actually interfere with the expression of hormone action. Workers are prevented from maturing eggs even though the corpora allata might produce gonadotropic hormone. However, though non-egg-laying workers may have large corpora allata, egg-laying workers have still larger ones (Dreischer, 1956). This latter observation again supports the theory of a correlation between gland size and quantity of gonadotropic hormone produced. The issue does not yet seem to be settled.

Additional arguments against a proportionality between gland size and hormone production can be given as follows. In females of the cockroach *Leucophaea*, egg maturation proceeds at the same rate during the first and second cycles, i.e. in both cases yolk deposition lasts for about 3 weeks, even though the corpora allata of the animals in the second cycle are twice as large as are those during the first egg maturation period (Engelmann, 1957a) (Fig. 58). In the case of *Leucophaea* one may be misled by the fact that the time lapse between emergence and ovulation is 10 to 14 days longer than between parturition and the second ovulation. Close observation of egg growth in the ovaries shows that no egg maturation takes place during the 8 to 12 days following emergence, whereas it begins immediately after parturition (Fig. 58). These findings seem to indicate that the quantitative correlation

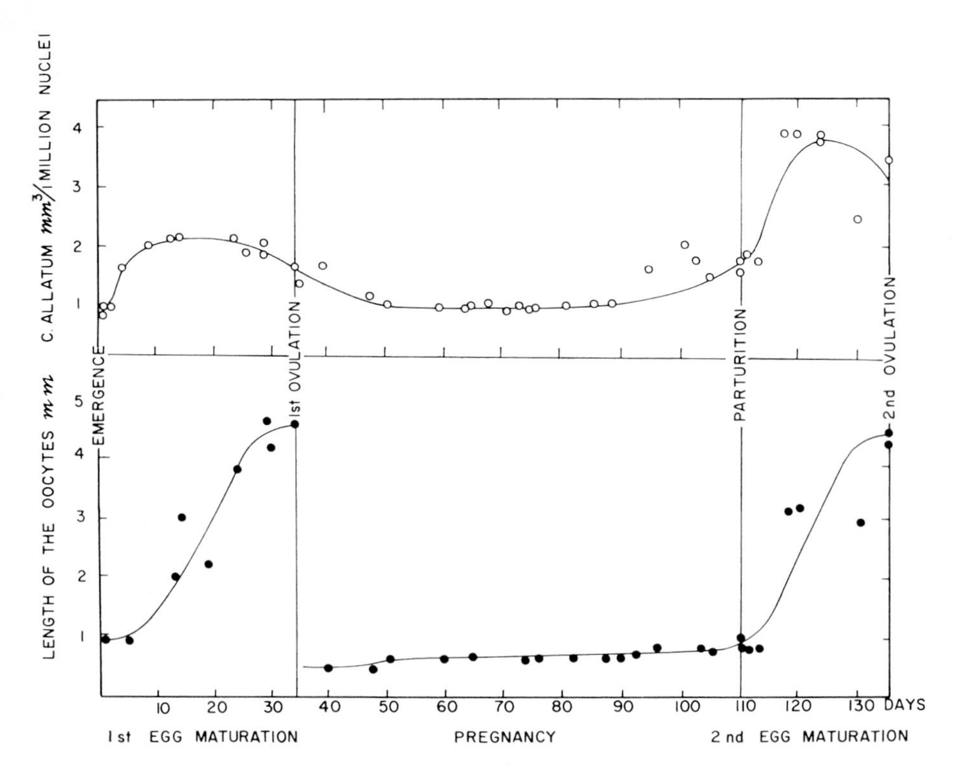

FIG. 58. Diagram indicating the morphological changes in the corpora allata of *Leucophaea maderae* (upper curve) which are correlated with the growth of the terminal oocytes of the ovarioles (lower curve) during the first preoviposition period, pregnancy, and second preoviposition period. (From Engelmann, 1960b.)

between gland size and hormone production does not hold in an absolute sense, not even in *Leucophaea* for which it was first postulated (Engelmann, 1957a). Likewise in *Melophagus ovinus* (Ulrich, 1963) and in *Tenebrio molitor* (Mordue, 1965a), no clear-cut correlation between gland size and egg maturation was found. In the flesh fly *Sarcophaga* there is an even less clear correlation between corpus allatum size and hormone production (Wilkens, 1965, 1968). It should be emphasized that, beyond a certain size, the corpus allatum in *Sarcophaga* and *Leucophaea* as well as other species, no longer seems to be correlated with accelerated egg maturation. Certainly, one can surmise that the growth rate of the oocyte is limited because of its structure and the limited rate of protein and lipid incorporation. In other words, regardless of the quantity of hormone present in the animal, it cannot be used because egg maturation can be stimulated no farther. In other words, we do not have a true bioassay to determine the quantity of gonadotropic hormone output by a gland of a given size.

To summarize the discussion on corpus allatum size and hormone production, there is evidence that active corpora allata have a certain characteristic appearance, i.e. the nucleo-cytoplasmic ratio decreases because of an increased volume of the glands due to the quantity of hormone produced and released. Yet no clear-cut quantitative relationship can be established by using only histological criteria.

C. *Activation of the Corpora Allata*

1. FEEDING STIMULI

As has been shown in the section on feeding and egg output (p. 113 ff.), the majority of species do not mature eggs or oviposit if they have not had the opportunity to feed (see also Johansson, 1964); a protein diet is required for egg maturation. The actual mechanism by which egg maturation is impaired by lack of food or stimulated by feeding is only vaguely understood. Basically, lack of food and reserves could affect reproduction in either or both of the following ways: firstly, there could simply not be enough reserves to synthesize proteinaceous and lipid yolk for egg maturation; secondly, the endocrine glands could be restrained in their hormone production because of a lack of reserves or a lack of mechanical feeding stimuli. We shall be concerned here primarily with the effect of feeding on the endocrine glands' activity.

What is the mechanism involved in the activation of the corpus allatum in anautogenous species? The nutritive value of the food is certainly important, as will be shown for several species. On the other hand, Detinova (1953), probably on the basis of the work of Wigglesworth (1934) on *Rhodnius*, proposed for *Anopheles maculipennis* that the mechanical stretching of the midgut or stomach by a blood meal triggers the activation of the corpus allatum. Detinova showed for this species that after the intestine had been filled with a solution of non-nutrient value, egg maturation was initiated; in this species yolk deposition normally starts only after the ingestion of a blood meal. In *Culex pipiens* (Larsen and Bodenstein, 1959) an unusual distention of the intestine was caused by fruit juice ingestion after anal occlusion. Some egg maturation, which normally would never have occurred on fruit juice, was observed; egg maturation did not, however, proceed beyond an initial stage, apparently because the juice lacked sufficient nutrient value. These few observations would indeed speak in favor of the concept that the corpus allatum activation occurs via nervous pathways. Similar experiments on *Phormia regina*, however, did not initiate any egg maturation (Orr, 1964a). Specifically, proteins had to be supplied to activate the corpus allatum in the latter case.

For the activation of the corpus allatum in *Culex*, the actual process of feeding apparently was not essential since a blood or milk enema initiated egg maturation. It was implied that the distention of the intestine supplied the necessary stimulus (Larsen and Bodenstein, 1959). However, since both milk and blood have nutritional value and thus could affect corpus allatum activity by raising the blood protein level, the experiment is not unequivocal. However, injection of whole blood into the hemocoel did not activate the corpus allatum; probably the injected blood could not be utilized. From these observations Larsen and Bodenstein (1959) postulate a mechanism triggered by the distention of the intestine; the information is then processed in the brain, which in turn activates the corpus allatum. Whether the corpus allatum is finally activated by neurosecretory materials released from the pars intercerebralis or via nervous pathways cannot be determined from the data available on this species. It is also quite likely that mechanical distention of the intestine as well as nutrition supplied with the ingested food are important in the activation of the corpus allatum in *Culex*.

Whatever the final step in the activation of the corpus allatum may be, in unfed *Culex* females and probably also in those from other species, the brain appears to restrain these glands. Implantation of active corpora allata (Larsen and Bodenstein, 1959) into unfed females allowed at least some degree of egg maturation, which indicates that sufficient reserves were available to assure a certain degree of yolk deposition. Similarly, in starved females of *Oncopeltus fasciatus* which normally do not mature eggs (Johansson, 1954, 1958), implantation of corpora allata from fed females caused egg maturation in some but not all animals. Also, in about 50 per cent of starved *Leucophaea* females, egg maturation could be induced by implantation of active corpora allata (Johansson, 1955; Engelmann, 1965a). Severance of the nervi allati in starved *Oncopeltus* females was followed by egg maturation in 50 per cent of the animals (Johansson, 1958). The same result was obtained if the brain and corpus allatum with severed nervi allati were implanted. However, after implantation of the brain together with the intact nerves of the retrocerebral complex, only 10 per cent of the females matured eggs. From these experiments Johansson (1958) concluded that in *Oncopeltus* the nutritional effect of food intake on the corpus allatum is mediated through the central nervous system via the nervi allati. Further, more indirect support for this may come from experiments on *Exeristes comstockii.* Normally, females of this species do not mature eggs when fed on sucrose alone; yet when farnesylmethyl ether was also supplied with the sugar, eggs did mature, just as after feeding on proteins (Bracken and Nair, 1967). This again shows that females of this species have enough reserves to mature eggs, but normally the corpora allata do not become active unless proteins have been ingested. The corpus allatum could be either controlled, i.e. restrained, by ordinary nerves or activated by neurosecretory materials passing down the neurosecretory axons into the corpus allatum. Whether food intake stimulates the corpus allatum or whether it causes the cessation of inhibition cannot be concluded with certainty at present.

2. THE INTERNAL NUTRITIONAL MILIEU

The examples given above seem to indicate that in some species the central nervous system mediates the mechanical stimuli necessary for activation of the corpora allata. It may also mediate the effect of nutrition as such. However, nutritional factors supplied with the food need not affect the brain first but could affect the corpora allata directly as some recent experiments have illustrated.

In *Leucophaea* (Engelmann, 1965a), implantation of active corpora allata into normally fed pregnant animals whose own corpora allata are inactive resulted in egg maturation in about 80 per cent of these females whereas only about 44 per cent of starved animals matured eggs. Comparable results were obtained by severance of the NCC I in fed and starved *Leucophaea* females (Engelmann, 1965a). These results are similar to those obtained by Johansson (1958) in *Oncopeltus* and by Larsen and Bodenstein (1959) in the starving anautogenous *Culex* as quoted above. Are these data evidence that nutritional factors possibly affect the corpora allata directly without the mediation of the central nervous system? This idea, originally drawn from experiments performed in *Leucophaea* (Engelmann, 1965a), can be further substantiated by data obtained in several other insect species.

What are the underlying events? Starving animals with limited protein reserves may not be able to sustain the high metabolic activity in the corpus allatum cells. It seems possible, then, that the corpora allata are directly influenced: inactive glands *in situ* or implanted isolated ones could not become active. However, implanted active isolated glands may remain active long enough to cause egg maturation in a certain percentage of the operated animals. From this we could further postulate that cyclic volumetric changes of the corpora allata in association with egg maturation cycles are caused by a periodic drain of reserve proteins by the growing oocytes. Indeed, this may be suggested for *Calliphora* (Strangways-

Dixon, 1961b, 1962), *Schistocerca gregaria* (Highnam, 1962b), *Phormia* (Orr, 1964b), *Pyrrhocoris apterus* (Sláma, 1964), *Musca* (Bodnaryk and Morrison, 1966), and *Periplaneta* (Mills *et al.*, 1966). In each of these cases, the blood protein level decreased with the approach of complete egg growth and was paralleled by a decreased corpus allatum volume. In completely isolated corpora allata of *Leucophaea* (Engelmann, 1962) the cyclic volumetric changes persisted through several egg maturation cycles (Fig. 59), probably because of fluctuations in the protein level of the hemolymph. This observation on isolated glands affected by the changing internal environment may be the strongest evidence for the theory of direct control of the corpora allata by the nutritional milieu. In *Pyrrhocoris* (Sláma, 1964a) cyclic fluctuations in O_2-consumption, normally paralleled by egg maturation periods, persisted after the nervi allata had been severed. This is presumably in accord with the activity rhythms of the corpora allata and with the blood protein levels which were maintained in those animals whose corpora allata had been denervated. However, no such fluctuation in corpus allatum size and blood protein level was found in *Gomphocerus rufus* (Loher, 1965, 1966b).

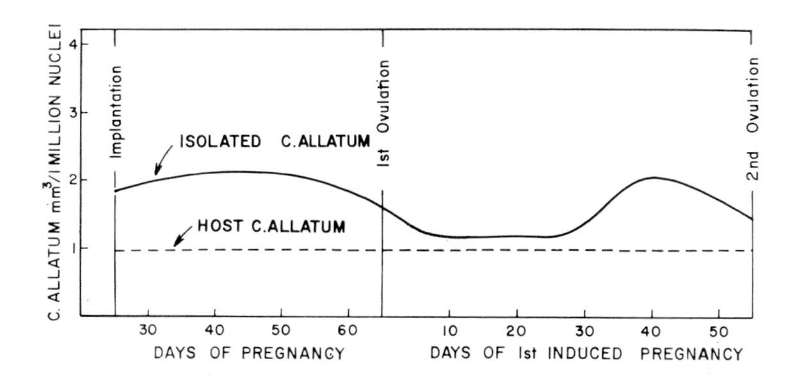

FIG. 59. Pattern of activity cycles of isolated corpora allata of *Leucophaea maderae* as compared with the activity level of glands *in situ*. (Modified from Engelmann, 1962.)

Studies on *Calliphora* (Strangways-Dixon, 1962) give further experimental evidence for the dependence of corpus allatum activity on blood protein concentration. If a female of this species was forced to ingest proteins during periods when she would not voluntarily eat them, the corpus allatum did not undergo a decrease in volume, but remained as large as during its very active periods. Additional indirect evidence for control of corpus allatum size by the hemolymph protein level came from studies on the honeybee by Müssbichler (1952). Here, the corpora allata became enlarged only after pollen or casein had been fed, presumably because the blood protein level was then raised. Conversely, the corpora allata of parasitized *Andrena vaga* (Brandenburg, 1956), are very small, and no eggs mature, probably because of the almost complete drainage of nutritional material by the parasite, *Stylops*. The former animals are in a state of severe starvation and parasitic castration is achieved.

As was reported in an earlier section, corpora allata of ovariectomized females can surpass in size those of normal egg-maturing animals. For example, the corpora allata of some *Leucophaea* females attain an enormous size several months after the operation (Scharrer and von Harnack, 1961). The question then arose whether this increase in size, identified as hypertrophy, denotes hyperactivity or storage of unused hormone. A negative feedback mechanism from the ovaries, implying that there is normally partial inhibition of the corpora allata by maturing oocytes, has been postulated by several authors for a variety of species. It is conceivable that increased activity of the gland may be reflected in the increased size of the corpora allata; a cell in which the mechanism of hormone production is fully operative may be larger than one which is merely moderately active. Why then should a corpus allatum of an ovariectomized female become more active than one of an egg-maturing animal? The answer may lie in the fact that females which do not mature eggs do not use nutritional reserves to the same degree that normal animals do. In other words, the glands are bathed in a nutrient milieu richer than that of unoperated controls. This idea was indeed supported by the observation that, in castrated females of *Schistocerca* (Highnam, 1962b), *Pyrrhocoris* (Sláma, 1964b), *Phormia* (Orr, 1964b), and *Leucophaea* (Engelmann, unpublished), the blood protein concentration is rather high for prolonged periods. On the other hand, the enormously high blood protein concentration in castrated *Gomphocerus* was not paralleled by hypertrophy of the corpora allata (Loher, 1965, 1966b).

Furthermore, in the sterile adp^{fs} mutant of *Drosophila melanogaster*—in which mutant sterility resides in the ovaries—the fat bodies store nearly twice as much ether-soluble lipids as do those of wild-type females (Doane, 1960, 1961, 1963). Hypertrophy of the corpus allatum is reported in this case. One could postulate that the corpus allatum of mutants is exposed to a milieu rich in nutrients which allows higher activity than that in the wild-type females. This interpretation would be in line with thoughts expressed earlier in this chapter.

Considerable work has been done on feeding-induced egg maturation in mosquitoes. Experiments on several autogenous and anautogenous species or strains, in all of which the corpus allatum is essential for egg maturation, have partly clarified the way in which the corpora allata are activated in these cases. Ligation of the neck at various time intervals after feeding, ligations in autogenous species at various intervals after emergence, and homo- and hetero-transplantation of corpora allata (Clements, 1956; Gillett, 1956; Larsen and Bodenstein, 1959; Detinova, 1962; Lea, 1963) were methods used. Corpora allata from anautogenous *Aedes aegypti* transplanted into neck-ligated autogenous *Culex molestus* promoted egg maturation (Larsen and Bodenstein, 1959), as did corpora allata from anautogenous *A. aegypti* and *A. sollicitans* transplanted into allatectomized autogenous *A. taeniorhynchus* (Lea, 1963). The transplanatation of active corpora allata from autogenous *C. molestus* into anautogenous *C. pipiens* also induced eggs to mature to some extent (Larsen, 1958; Larsen and Bodenstein, 1959), whereas active corpora allata from the autogenous *A. taeniorhynchus* did not cause egg growth beyond an initial stage in the anautogenous *A. aegypti* or *A. sollicitans* (Lea, 1963). In both anautogenous and autogenous mosquito species, ligation experiments clearly showed that egg maturation proceeds only if the corpus allatum is present. In anautogenous species, the animal's own corpus allatum is not activated unless the female has had a chance to obtain a blood meal. In autogenous species, the corpus allatum is apparently activated shortly after emergence without the stimulus of food intake. The internal nutritional milieu allows its full activation. The important factor in the manifestation of autogeny in autogenous species is indeed the amount of reserves carried over from the larval period, as was demonstrated in *Culex pipiens molestus* (Clements, 1956), *A. taeniorhynchus* (Lea, 1964b), and *A. togoi* (Laurence, 1964). When the larvae of these species were fed a protein-poor diet, only a small percentage of the resulting females matured eggs.

Obviously, two factors must be considered in the discussion of endocrine activity related to autogeny: nutritional status, i.e. available reserves for yolk formation, and activation of the corpus allatum. It seems possible that the corpus allatum in some unfed anautogenous species is at least mildly active and that it liberates some gonadotropic hormone, but yolk formation from the limited reserves is stimulated only by highly active corpora allata. Normally, as soon as more proteins, the building materials for yolk, become available, yolk formation is initiated. In other words, the activity of the corpora allata in unfed females does not become apparent. As is outlined below, this hypothesis explains the otherwise puzzling results obtained in some species.

Allatectomy shortly after emergence and before feeding did not eliminate egg maturation in all operated females of *Calliphora* provided the animals were fed meat thereafter (Thomsen, 1940, 1948a, b). Apparently the corpus allatum had secreted gonadotropic hormone in at least some animals by the time of the operation which was before proteins had been fed. Females receiving only sugars never matured eggs. Similar observations were made in *Sarcophaga* (Wilkens, 1965, 1968): here, about 70 per cent of the allatectomized females partially or fully matured one set of eggs if they were fed meat after allatectomy. Sugar-fed females of *Sarcophaga*, unoperated or allatectomized, did not mature eggs even within 2 or 3 weeks. If meat was given to sugar-fed females 14 days after allatectomy, again yolk was deposited in the oocytes in many of the operated flies. Yolk was deposited by both meat-fed and defined-medium-fed females. The defined-medium was made up of amino acids, vitamins, and salts only and did not contain any known gonadotropic hormone simulating substances. Such results seemed to show that some of the females had produced and released gonadotropic hormone by the time of emergence and that this hormone was not metabolized within the 2-week period. In this context it is noteworthy that, as already mentioned, the corpus allatum of a newly emerged *Sarcophaga* female appears "inactive", i.e. it is small, and the nuclei are densely packed; some females produced gonadotropic hormone none the less. In the light of more extensive work in *Sarcophaga* which involved removal of the neurosecretory cells of the pars intercerebralis in conjunction with observations on protein metabolism, some alternative explanations for these results are possible (p. 170) (Wilkens, 1965, 1968).

Additional support for the above hypothesis can be found in experiments involving several species of mosquitos. Females of the anautogenous *Culex pipiens*, *Aedes aegypti*, and *Anopheles labranchiae* matured eggs even though their neck was ligated only 2 to 3 min after feeding (Clements, 1956). It is difficult to conceive that the gonadotropic hormone would have been produced and released within the very short time interval between onset of feeding and ligation. More likely, the hormone was already present before feeding began, but its effect could be noticed only after food was supplied. In contrast to these cases, in *Phormia* the corpus allatum must be present for at least 2 days after feeding to assure full egg maturation (Orr, 1964a); no hormone seems to be liberated before feeding.

These outlines on the control of the corpora allata by food intake in the few species cited suggest that two synergistically functioning mechanisms operate. They are thought to function in the following way. In species which periodically ingest large amounts of food, e.g. mosquitoes and blood-sucking Hemiptera, the distention of the intestine may activate the corpora allata or at least initiate their activation. However, unless the animal has ingested proteinaceous food, egg maturation can proceed only to a limited extent. In other species, presumably nutrients *per se* permit activation of the corpora allata. In both cases nutritional factors supplied in the food may directly affect the endocrine glands without mediation of the central nervous system. The theory of direct control of corpus allatum by nutritional factors (Engelmann, 1965a) is admittedly based on results from relatively few species; additional observations are necessary before definite conclusions can be drawn.

One must also bear in mind that many operations involving endocrine glands affect food intake; consequently, the results are often causally related not so much to the operations as to the effect of partial starvation in the animals (see p. 174f.). These observations on feeding and the nutritional status of the animals (reserves) also lead one to reconsider the oft-postulated factors which are supposedly released from the maturing eggs in the ovaries and which supposedly partially or totally inhibit the corpora allata. The observed restraining of corpus allatum activity in those cases could well be only a reflection of the nutritional status of the animal. The postulated negative feed-back mechanism may not exist at all. At any rate, no factors from the ovaries or mature eggs have been found so far in any species.

3. MATING STIMULI

Mating affects the total egg output in many insect species, according to reports by many investigators. However, the observation of the total number of eggs laid by a female does not allow us to draw conclusions concerning the nature of the mating stimulus since mating may stimulate either egg production in the ovaries or the deposition of mature ova or both. It is known for a great number of species that virgins "reluctantly" lay their eggs (see p. 129). In this section, only the effects of mating on the neuroendocrine system and the consequent stimulation of oocyte growth will be considered. Presumably stimulation of the corpora allata is involved in many insect species; however, in only the few cases which will be discussed below do we actually know details of the mechanisms.

Most of this work was done on several species of cockroaches. For example, in *Periplaneta* mating accelerates egg maturation and thus increases the total number of eggs laid during the animal's lifetime (Griffiths and Tauber, 1942; Roth and Willis, 1955). Similar observations have been made in the viviparous cockroaches *Diploptera* (Stay and Roth, 1958; Engelmann, 1959a; Roth and Stay, 1961), *Leucophaea* (Engelmann, 1960b), and *Nauphoeta cinerea* (Roth, 1964a). Since it is known that egg maturation in several species of cockroaches is under the control of the corpora allata (Scharrer, 1946; Engelmann, 1957a, 1959a; Girardie, 1962; Roth and Stay, 1962a), one may deduce from these observations that mating stimuli normally activate these endocrine glands.

The effect of mating on the speed of egg maturation varies considerably in the various cockroach species studies. For example, mating is absolutely essential for the initiation of corpus allatum activity and oocyte growth in *Diploptera* (Engelmann, 1958, 1959a). However, the speed of egg maturation is not at all influenced if females of the parthenogenetic strain of *Pycnoscelus surinamensis* are mated (Roth and Willis, 1961). Other species fall between these two extremes; mating speeds up egg maturation by 2 days in *Blattella germanica* (Roth and Stay, 1962a), whereas in *Leucophaea* the rate of egg maturation in mated females is in the average twice as high as that in virgins (Engelmann, 1960b; Roth and Stay, 1962b). In the latter species, many virgins do not even mature eggs within 5 or more months.

The following questions must be asked in this context; where are the mating stimuli perceived and how do they reach the corpora allata? What is the mechanism involved in the activation of the corpora allata? Should the activation of the corpora allata be interpreted as a cessation of inhibition of the glands or as an actual stimulation? Attempts to obtain answers to such questions have been made in only a few species. For *Diploptera* it was established (Engelmann, 1959a; Roth and Stay, 1961) that purely mechanical stimuli can cause the maturation of a full batch of eggs. In this species, the implantation of an artificial "glass spermatophore" into the bursa copulatrix simulated the natural mating act while castrated males stimulated egg maturation just as normal males did in both *Leucophaea* (Engelmann, 1960b) and *Diploptera* (Roth and Stay, 1961). Primarily, the deposition of the spermatophore (and, to a lesser degree, the actual mating by either castrated or normal males) activated the corpora allata. The transfer of spermatozoa apparently was not essential to promote egg growth.

Severance of the ventral nerve cord prior to or shortly after mating in *Diploptera* (Engelmann, 1959a; Roth and Stay, 1961) prevented the transmission of mating information to the corpora allata. Transection of the ventral nerve cord in *Leucophaea* (Engelmann, 1960b) before or within 2 days after mating likewise prevented the mating stimuli from "coming through". It is peculiar that the actual mating lasts only 1 to 2 hr, yet in *Leucophaea* the corpora allata were maximally activated only if the nerve cord remained intact for about 2 days afterwards. It is obvious that the mating information travels craniad along the ventral nerve cord and reaches the corpora allata via the brain. As to the actual mechanism by which mating stimuli normally reach the corpora allata, no clear answer can yet be given. However, several explanations could account for the observations just mentioned. One may postulate involvement of humoral agents traveling along the ventral nerve cord since the nervous connection to the brain must be intact for periods of days, i.e. the release of some hypothetical neurohumor or neurosecretory factor from the last abdominal ganglion has to be continuously stimulated by the implanted spermatophore. However, since the removal of the spermatophore shortly after mating (Roth and Stay, 1961) did not abolish the mating effect, this assumption may not be entirely correct. One could likewise postulate that the corpora allata have to be stimulated nervously for several days in order to become sufficiently active to cause egg maturation. It is also possible that in these two species of cockroaches an inhibitory center in the brain has to be influenced for prolonged periods in order to cease completely the inhibition of the corpora allata. Whatever the precise nature of the stimuli may be, they do not reach the corpora allata directly but rather are relayed by the brain. Since severance of the nervi allati in virgins of *Leucophaea* and *Diploptera* caused the activation of the corpora allata in nearly 100 per cent of the cases, whereas normally only 30–40 per cent of *Leucophaea* virgins and no *Diploptera* virgins matured eggs (Engelmann, 1959a, 1960b, 1965a), the information obtained from these two species of cockroaches favors the viewpoint that the corpora allata are inhibited in virgin females. An activation of the corpora allata through mechanical stimulation during severance of the nerve cord or nervi allati probably can be ruled out since isolated corpora allata remained active for several months and did not require repeated stimulation (Engelmann, 1957a, 1962; Roth and Stay, 1961). There is, however, evidence for *Leucophaea* that the abdominal ganglia contain certain control centers for the corpora allata (Engelmann, 1964); these centers are related to inhibition rather than activation.

Mating and feeding, which operate synergistically, are essential for egg maturation in the bed bug *Cimex lectularius* and other Cimididae (Mellanby, 1939; Davis, 1964, 1965a, b). In these species the sperm are injected into the organ of Ribaga or spermalege during copulation (haemocoelic insemination), and they migrate through the haemocoel to the oviducts, conceptacula seminis, and corpora seminales of the ovarioles (see p. 102). In a series of experiments, which will be reported below, Davis (1964, 1965a, b) attempted to answer the question of how mating activates the corpora allata. Fed virgins or starved mated females matured only a few eggs (Fig. 60); however, fed virgins into which active corpora allata had been implanted matured eggs almost as well as did the unoperated mated fed females

(Davis, 1964). Fed females which were artificially inseminated, i.e. a mixture of sperm and seminal fluid was injected into the organ of Ribaga, matured as many eggs as the normally mated females. The injection of either sperm or seminal fluid alone did not promote egg maturation; in the first case, the spermatozoa remained in the spermalege. This indicates that, in *Cimex*, it is the seminal fluid which causes the sperm to migrate through the haemocoel to the oviducts within approximately 3 hours after mating or injection (Davis, 1965a). Only if the sperm reach the upper parts of the reproductive tract (corpora seminales) will the corpora allata be activated (Fig. 61); if for some reason they stay in the conceptacula, no egg maturation occurs. Spermatozoa could be activated *in vitro* by 0.07 M sodium citrate; after these activated spermatozoa had been injected, they migrated through the haemocoel to the oviducts in the same way spermatozoa from seminal fluid would. Semen did not have to be injected into the spermalege itself, i.e. even after injection directly into the haemocoel, the activated spermatozoa migrated to the oviducts and caused the activation of the corpora allata (Davis, 1965a).

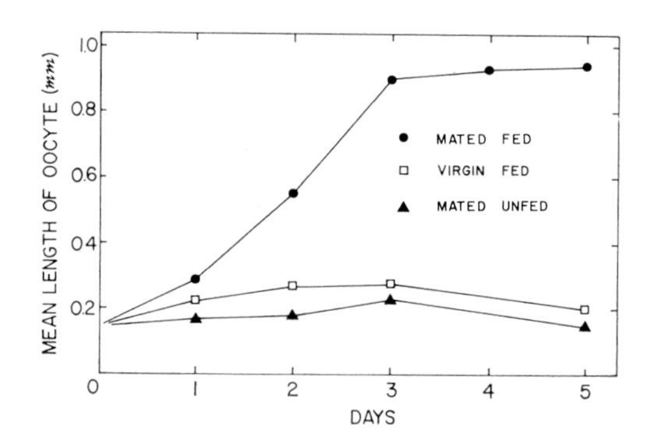

FIG. 60. Mean growth curves of the terminal oocytes in mated or virgin, fed, or unfed females of *Cimex lectularius*. The diagram illustrates the synergistic action of mating and feeding in this species. (Modified from Davis, 1964.)

It takes approximately 8 hr after mating for the stimulus to reach the corpora allata and for enough hormone to be liberated to assure the maturation of the eggs; ligation of the neck after this time no longer prevented egg maturation (Davis, 1965b). Removal of the conceptacula seminis or transsection of the oviducts prior to mating did not prevent the mating stimulus from reaching the corpora allata; the sperm still gathered around the oviducts and migrated to the corpora seminales. This definitely showed that the conceptacula seminis are not essential for the perception of the mating stimuli and also that the spermatozoa must reach the corpora seminales in order to be effective. A stimulus for the corpora allata apparently originates there and is initiated by the presence of the spermatozoa. This stimulus could be mediated either neurally or humorally. Cutting the thoracic nerve cord within approximately 3 hr after mating prevented activation of the corpora allata, which strongly suggests that a neural mechanism is involved. It should be mentioned here that the period within which the operation must be performed if activation of the corpora allata is to be prevented coincides with the time needed for the spermatozoa to reach the genital tract. How the mating stimuli reach the thoracic ganglionic mass is somewhat uncertain since egg growth was stimulated to some extent in both virgins and mated females by cutting

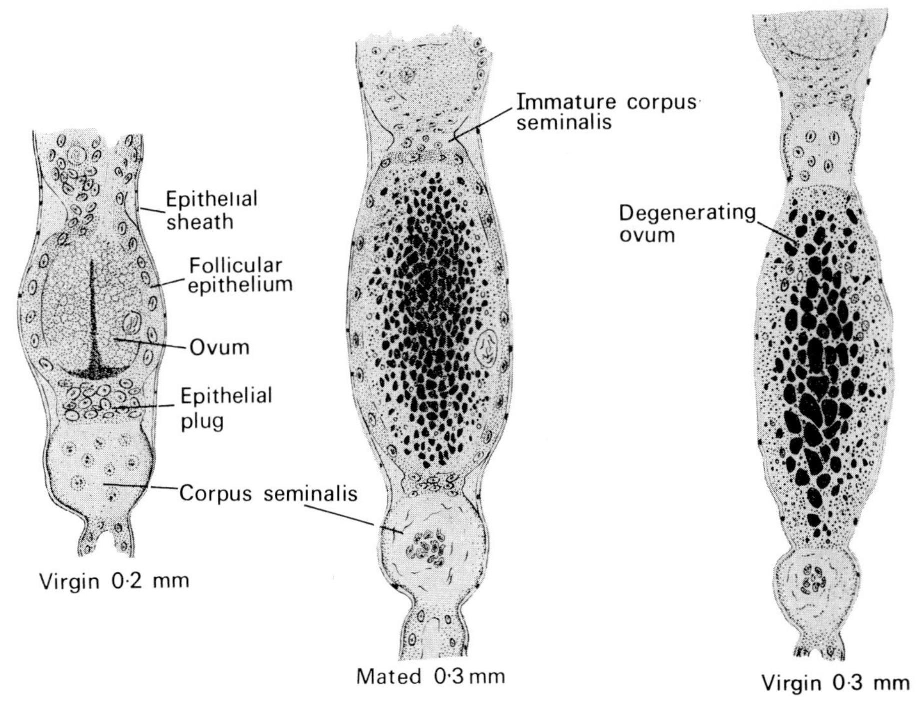

FIG. 61. Section through ovarioles illustrating stages of development of the corpus seminalis and oocyte in virgin and mated *Cimex lectularius*. In virgins the ovum degenerates and is resorbed, whereas in mated females it continues to grow to mature size. (From Davis, 1964.)

any one of several pairs of the abdominal nerves. It is highly probable, however, that the mating stimuli are carried to the thoracic ganglionic mass by abdominal nerves; only about two-thirds of the operated animals matured eggs after severance of these nerves had been followed by mating. One hundred per cent of the control animals matured eggs after mating (Davis, 1965b). Severance of the abdominal nerves in virgins of *Cimex* may provide a stimulus for the corpora allata just as cutting the ventral nerve cord does in *Leucophaea* (Engelmann, 1960b). In both cases, the operation may have had a specific effect on the corpora allata, or there may have been a non-specific wounding effect. The latter is the less probable explanation since other types of wounding did not initiate egg maturation in either *Cimex* or *Leucophaea*.

Very little specific information concerning the activation of corpora allata by mating stimuli is available in other insect species. One can, however, deduce from a number of reports that mating stimuli indeed activate the corpora allata. For example, Watson (1964, 1967) has shown in the firebrat *Thermobia domestica* (syn. *Lepismodes inquilinus*) that virgins do not mature eggs beyond a certain subliminal size. Since egg maturation seems to be under the control of the corpora allata in this species (Rohdendorf, 1965), the glands are obviously activated by the mating act. Mating sped up egg maturation to a certain extent in *Schistocerca gregaria* (Norris, 1954), as did a slight wounding of the animal. Similarly, enforced activity or repeated electrical shocks caused in both *Schistocerca* (Highnam, 1961, 1962a) and *Locusta migratoria* (Highnam and Haskell, 1964) (Fig. 49) more rapid egg growth than untreated virgins showed. Mating also accelerated egg maturation in *Locusta migratoria manilensis* (Quo, 1959). In the latter species, mating with a castrated male had the same effect on egg maturation speed as did mating with a normal male, which suggests that only mechanical stimuli received during mating are responsible for the effect. The transfer of spermatozoa seems unimportant for acceleration of egg maturation in this species.

Correlated with mating, a release of neurosecretory materials from the corpora cardiaca is observed in *Schistocerca* and *Locusta* (Highnam, 1961, 1962a; Highnam and Haskell, 1964). This observation implies that neurosecretion plays a role in egg maturation, be it a direct effect on the ovaries or an indirect one on metabolism and corpus allatum activity. Two fuchsinophilic cells of the suboesophageal ganglion in *Locusta* accumulate neurosecretory materials during the peak egg growth following mating (Fréon, 1964a, b); this again suggests that mating somehow affects the release or accumulation of neurosecretory materials. The significance of the observed changes in the neurosecretory system in correlation with certain phases of egg maturation must, however, be further elucidated. The precise physiological counterpart of any observed histological change has to be found since egg maturation certainly involves multiple and complex changes in the animal's metabolism. This word of caution on the interpretation of changes in the neuroendocrine system which are observed after mating also applies to *Tenebrio molitor*, where quite similar correlations have been found (Mordue, 1965a).

In the sterile mutant *fs* (2) *adp* of *Drosophila melanogaster*, it was shown that mating affected the size (activity) of the corpora allata. Virgins of this mutant had considerably smaller corpora allata than did mated females, in which the gland hypertrophied (Doane, 1960); hypertrophy itself seems to have been possible because of the exceptionally rich nutritional milieu in the sterile mutant. In *Drosophila virilis*, the corpus allatum was also larger in mated females than in virgins (Doane, 1961). Mating clearly increased the production of mature ova in protein starved females of *Drosophila melanogaster*, implying that the corpus allatum had been stimulated (Merle and David, 1967). Nothing is known about the pathway by which mating stimulates the corpus allatum to increased activity in *Drosophila*.

4. ENVIRONMENTAL STIMULI

A variety of environmental factors may influence corpus allatum activity; thus, the total egg output may be changed by the altered living conditions (Norris, 1964). It is known that extrinsic factors effectively control the timing of reproductive periods in many insect species; but, unfortunately, extremely little is known in any one species about the mechanisms involved in the effect of single environmental components; even less is known about the interaction of the various extrinsic factors. In view of the paucity of knowledge on this subject, the effect of environmental changes on the endocrine system can be little more than surmised.

(a) *Photoperiodic influences*

The Colorado beetle *Leptinotarsa decemlineata* shows positive geotactic behavior and hibernates in the soil in autumn, during which period the corpora allata are relatively small; no eggs mature in the ovaries (De Wilde, 1954). Short days seem to be the environmental cue to initiate digging, but availability of food exercises a decisive influence on diapause (De Wilde, 1958), which is shown by the fact that a large percentage of starving animals go into diapause even under long-day conditions (Faber, 1949). In normal animals, light intensities of as little as 0.1 lux, if sustained for 15 hr or more per day, are sufficient to prevent the animals from going into diapause (De Wilde and Bonga, 1958): 15 hr of light is close to the critical photoperiod. Animals treated with light in excess of 15 hr daily will feed and lay eggs even at a temperature of as low as 17°C (De Wilde *et al.*, 1959). Since egg maturation in beetles is under the control of the corpora allata (Joly, 1945), these observations lead one to surmise that the corpora allata are activated by long-day conditions. It is unclear, however, whether long days stimulate the corpora allata to activity or whether short days inhibit them. Do they need an activating stimulus at all or do other factors, such as changed nutritional conditions in association with the onset of long days, allow the corpora allata to become active?

The complete diapause syndrome in *Leptinotarsa*, which is induced by short days, consists of small (inactive) corpora allata and consequent cessation of egg maturation, cessation of feeding, digging behavior, and reduced O_2-consumption. In an attempt to elucidate further the phenomenon of the diapause syndrome, De Wilde and Stegwee (1958) allatectomized beetles of both sexes and found that even under long-day conditions oxygen consumption

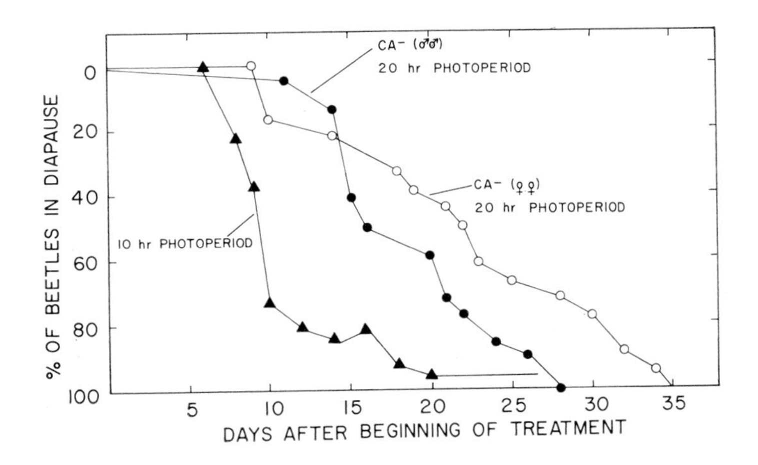

FIG. 62. The effect of allatectomy in male and female beetles of *Leptinotarsus decemlineata*. Diapause is induced even under long-day conditions. The treatment was started the first day after emergence. (Modified from De Wilde and De Boer, 1961.)

dropped, feeding stopped, and the animals began to dig in the soil. The effect of allatectomy closely resembled that of a natural diapause although, under these conditions, the animals did not become positively geotactic until slightly later than did the unoperated beetles (Fig. 62). It appeared then that diapause in *Leptinotarsa* is an endocrine deficiency syndrome. Reimplantation of active corpora allata into allatectomized females reversed the effect: the animals left the soil and began to feed and to lay eggs, some even laying quantities comparable to those of normal females. It is interesting to note that implantation of active corpora allata into naturally diapausing beetles (10 hr daylength) did not break their diapause. This was interpreted to mean that either the implanted corpora allata were inactivated by some humoral factor or that the corpora allata needed a substratum which was lacking in the short-day-treated animals. Further experimental analysis of the function of the neuroendocrine system is essential for a full understanding of diapause in *Leptinotarsa*. An interesting phenomenon which likewise is not clearly understood (De Wilde *et al.*, 1959) is the apparent insensitivity of the beetles to short day once the natural diapause has been broken. Obviously, the physiological conditions of postdiapausing females differ from those of prediapausing beetles. Apparently, as this brief report indicates, there is more to diapause in *Leptinotarsa* than the mere inactivation of the corpora allata. At any rate, it has clearly been shown that long photoperiods stimulate corpus allatum activity or at least are associated with active glands.

Apart from the relatively well-worked-out studies on the Colorado bettle, little is known about the effect of photoperiods on the endocrine system of other insects. A few reports make some suggestions, but the manner in which endocrines are involved remains to be shown in each case. The question of whether photoperiods affect the endocrine system directly or whether endocrines merely respond to changing conditions in the animal's metabolism must again be raised; it has not yet been answered for any species.

Egg output in *Anopheles maculipennis* gradually decreases in late summer (Detinova, 1962) and, since egg maturation depends on corpus allatum activity in this species, involvement of the endocrine system is suggested. Short-day treatment of adult *Culex tarsalis* (Harwood and Halfhill, 1964), even at high temperatures, was followed by a decrease in egg output in the majority of females; typically, as in naturally diapausing animals, the fat body size increased. The effect of short-day treatment was further enhanced by low temperatures. Short days induced reproductive diapause in *Pyrrhocoris apterus* (Sláma, 1964b) and in *Anacridium aegyptium* (Geldiay, 1965). Since the corpora allata are essential for egg maturation in these species, the corpora allata are apparently inactivated by short photoperiod; or, inversely, long days permit the activation of the corpora allata. In *Anacridium*, possibly the neuroendocrine system of the brain is also involved since it has been demonstrated that brain cautery prevents egg maturation just as does allatectomy (Geldiay, 1967).

The red locust, *Nomadacris septemfasciata*, which experiences only slight changes in photoperiods in equatorial Africa, reproduces in summer, which is the rainy season and thus favorable for egg laying (Norris, 1959a). Adults which emerged in late summer exhibited a reproductive diapause until the onset of the next rainy season. It was shown in laboratory colonies that the longer the photoperiod experienced by the adults, the shorter the diapause. Even a slight lengthening of the photoperiod from 12 to 13 hr of light, which is the natural shift in the equatorial region in spring and early summer, drastically shortened the period of diapause, and egg maturation followed (Norris, 1959b, 1962). Once again, there is no definite information whether the changing photoperiods affect the endocrine system, which in turn regulates egg maturation. Possibly the corpora allata are inactivated under short-day conditions, as these glands are then small and their nuclei are densely packed (Strong, 1966b).

Reproductive summer diapause (aestivation) is reported for several insect species. The possibility of endocrine involvement in this phenomenon has barely been investigated, yet it seems plausible that endocrines are affected by long-day conditions, possibly in a way corresponding to that of short-day conditions, but with a reversal of long- and short-day effects. The few available reports are mentioned below. In the beetle *Galeruca tanaceti*, large amounts of stainable neurosecretory materials are found in the pars intercerebralis during periods of aestivation and ovarian inactivity, while considerably less material is seen during oviposition (Siew, 1965b). High metabolic activity in these cells has been shown by a high turnover rate of radioactive cystine during periods of egg maturation and oviposition (Siew, 1965c). This implies a functioning of these cells in the reproductive processes, be it direct or indirect, and would corroborate the findings in *Schistocerca*, where autoradiographic studies showed a high metabolic activity in the neurosecretory cells during periods of egg maturation (Highnam, 1962b; Hill, 1962). In addition, a correlation was found between corpus allatum size and egg maturation, i.e. the glands were large during periods of yolk deposition in the fall and small during summer diapause (Siew, 1965a). Histological changes in the neurosecretory system have been found in *Nebria brevicollis* (Ganagarajah, 1965). Unfortunately these observations were based only on classical histological findings. The correlation with reproductive processes alone does not permit one to draw any definite conclusions about the function of neurosecretion in the egg maturation of these species.

The alfalfa weevil *Hypera postica* aestivates under long-day conditions, as can be shown in both field and laboratory populations. During this aestivation, the corpora allata are small, and the neurosecretory system looks relatively empty (Tombes and Bodenstein, 1967). The treatment of diapausing females with *trans-trans*-10,11-epoxyfarnesenic acid methyl ester, a compound which mimics the action of the corpus allatum hormone, breaks diapause, and egg maturation resumes (Bowers and Blickenstaff, 1966). A positive dose-response curve was obtained, thus indicating that the corpus allatum hormone is most likely involved in breaking the natural diapause. There is a further suggestion of a corpus allatum function in the metabolism of the prediapause animals, as activity of these glands can be associated with an accumulation of lipid reserves (Tombes and Dunipace, 1967).

(b) *Chemical and group stimuli*

In several insect species, particularly locusts, grouping or crowding may either increase or decrease the total egg production. Grouping interferes with eating behavior and facilitates mating, both of which affect egg production. However, it is apparent that an additional group factor, distinct from the above, influences egg output in some cases. The difficulty of its recognition lies in the fact that the effect of crowding or grouping *per se* is frequently inseparable from the effects of mating and feeding in a group. The nature of the stimuli is often unknown; they could be chemical, olfactory, or mechanoreceptory. Since egg maturation in locusts is under the control of the endocrine system, notably the corpora allata and neurosecretory system, one may assume that certain group stimuli somehow reach the endocrine system. The way in which this takes place is unknown.

Precopulatory and preoviposition periods are shortened in *Schistocerca* by the presence of old males in the same cages (Husain and Mathur, 1945; Norris, 1954). Even when copulation was prevented (Norris, 1962b), the virgins, while in the presence of the old males, still produced more eggs than they would have normally. For the same species it has been shown that enforced activity will enhance egg maturation under both crowded and isolated conditions (Highnam and Haskell, 1964). Now, because individuals reared in groups are generally more active than isolated animals, the effect of group rearing may simply be the consequence of enhanced locomotor activity. Maximum speed of egg maturation was achieved if the females were both crowded and periodically forced to fly (Fig. 49). Highnam showed that the neuroendocrine system appeared empty shortly after the animals were made to fly. Since Highnam (1964) observed that emptying of the neuroendocrine system was correlated with egg maturation, he favored the idea of neurosecretory involvement in the promotion of egg maturation. Crowding in *Locusta migratoria* had an inhibitory effect on egg maturation but enforced activity in both crowded and isolated females did accelerate egg maturation, just as in *Schistocerca* (Highnam and Haskell, 1964). The mechanism by which crowding affects egg maturation in these species is obscure, but again the neuroendocrine system is implicated.

It has been shown for *Leptinotarsa* (Grison and Ritter, 1961) that beetles in groups of five ate less than did isolated ones and, possibly as a consequence, produced only about one-quarter the egg mass of isolated females. It is believed that the beetles interfere with each other in their daily activities. Moreover, according to Grison and Ritter, certain sensory stimuli received during encounters with other beetles inhibit some ill-defined physiological functions. Similarly, in *Drosophila melanogaster*, the greater the population density, the lower the egg production per female, even though abundant food was available (Pearl, 1932). It was assumed that collisions among the flies interfered with activities such as food intake, oviposition and energy output. The sum of these components resulted in a lowered egg output. The endocrine system may be only secondarily affected.

Rather interestingly, terpenoid extracts from aromatic desert shrubs such as *Commiphora myrrhae* stimulated egg maturation in *Schistocerca* (Carlisle *et al.*, 1965). A single contact with the aromatic oils hastened egg maturation. Whether the animals perceived the chemicals olfactorily or whether they needed direct contact is uncertain. This seems to be the first report in an insect species of a chemical that has been perceived from the environment accelerating reproductive processes. The peculiar phenomenon may be of great ecological interest—as will be outlined below. In the Near East, *Schistocerca* females usually mature eggs at the onset of the rainy season. However, even if rainfall is delayed, or even in widely separated areas where rainfall commences at different times, the locusts mature their eggs more or less at the same time. Since the aromatic desert shrubs like *Commiphora* species come into leaf and flower, presumably controlled by photoperiod, shortly before the rain normally begins, it is conceivable that the bursting buds spray the oil and thus provide an environmental cue for the stimulation of egg maturation. No detailed information on endocrine involvement is available, but it can be surmised that the endocrines are "turned on" by external stimuli.

Definite group effects on reproductive processes are to be found in the social insects. Since only scanty experimental information is available on the endocrine control of reproduction in these insects and since the social phenomena are highly complex, the matter will be dealt with in a subsequent chapter (p. 232).

D. *Neurosecretion and Egg Maturation*

Although we know that in the several species studied food intake stimulates corpus allatum activity by either nervous or direct nutritional pathways, we cannot be entirely certain whether or not neurosecretory materials, particularly those from the pars intercerebralis of the protocerebrum, act as intermediary links in corpus allatum stimulation. Since the discovery of stainable materials in the neurons of the pars intercerebralis (Hanström, 1938), a vast literature has accumulated on the topic of neurosecretory neurons and their stainable inclusions. Various types of neurosecretory neurons have been classified by many investigators on the basis of their tinctorial properties. Undoubtedly, some of these classes represent different activity stages of the same neurons.

The area of the pars intercerebralis of the protocerebrum has frequently been implicated as the control center for reproductive processes since this is the apparent origin of the nervi corporis cardiaci interior which contain axons that make up the nervi allati (see Highnam, 1964, 1965). However, it is somewhat uncertain whether all neurons of the nervi allati originate in the pars intercerebralis since a detailed study on *Periplaneta* (Willey, 1961) showed that the nervi corporis cardiaci I contain not only the axons arising in the pars intercerebralis, but also non-neurosecretory fibers which apparently originate in other brain areas. This finding has been confirmed and extended to another cockroach, *Blabera fusca* (Brousse-Gaury, 1968). To date, no additional studies are available, but one suspects that similar anatomical details do exist in other species. Therefore, any interpretation of data concerning the direct control of the corpora allata by the neurosecretory cells of the pars intercerebralis should only be tentative. It will be difficult technically to determine the function of certain nerve fibers within the NCC I in the context of corpus allatum control.

Neurosecretory materials, which may or may not be stainable with the usual staining procedures, are synthesized in the perikaryon of the cell, transported the length of the axon, and presumably released at the axon terminal. The stainable material is deposited in the corpora cardiaca, which function as neurohemal organs. Neurosecretory materials have been demonstrated within the corpora allata of several species with both the light and the electron microscope (see Highnam, 1965). Where their presence can be correlated with reproductive processes, a functional significance in the control of the corpus allatum seems indicated. Signs of activity of the neurosecretory cells, such as changes in stainable materials in association with egg maturation, have been sought in many species, but often, even after very careful investigation, no indications of activity changes in these cells could be found (Normann, 1965). Recently, however, such attempts have been successful in *Schistocerca* (Highnams, 1962a, b; Highnam and Lusis, 1962), *Locusta* (Highnam and Haskell, 1964; Cassier, 1965), *Tenebrio* (Mordue, 1965a), *Calliphora* (Bloch *et al.*, 1966), and *Polistes* spec. (Strambi, 1967). In all these cases the neurosecretory system is reported to contain large amounts of stainable material when the ovaries are inactive, i.e. when no yolk is being deposited, and to have less material in periods of egg maturation. In animals in which egg maturation can be induced or enhanced by crowding, mating, or enforced activity (e.g. *Schistocerca, Locusta,* and *Tenebrio*), a release of neurosecretory material is observed: the neurosecretory system, i.e. neurosecretory cells and corpora cardiaca, looked "empty" (Highnam, 1961, 1962a, b; Highnam and Lusis, 1962; Highnam and Haskell, 1964). In contrast to these species, the neurosecretory cells in *Hypera postica* (Tombes and Bodenstein, 1967) are filled more during reproductive periods than during diapause.

However, static histological pictures of certain cells in the brain do not permit one to draw definite conclusions about dynamic events in hormone production by these same cells. The rate of production or release of biologically active materials could be changed in different ways by various treatments or "demands" (Highnam, 1965). Earlier interpretations of the biological role of materials demonstrated in certain neurons favored the idea that those cells which were completely filled with material were active while those which looked empty were synthetically inactive. This may still apply in some cases, such as the parasitized *Polistes* (Strambi, 1965), but the opposite interpretation is now favored by many investigators (see Highnam, 1965). Good support for the second view was found by the application of labeled amino acids in *Schistocerca.* A high rate of incorporation of S^{35}-labeled cystine into the neurosecretory cells was observed when the cells appeared histologically "empty" (Highnam, 1962b), and a low rate of incorporation when they appeared "full". Obviously, at least in this case, an "empty" system is metabolically active whereas a "full" system is inactive. Even though good evidence for the metabolic activity of neurosecretory cells is available, we still do not know whether a hormone or hormones are produced by these same cells during their metabolic "activity" phases, nor do we know at what rate the hormones might be produced. Nonetheless, these observations contribute to a better understanding of the matter.

Another demonstration of induced activity in the neurosecretory cells of an insect is that given by Lea and Thomsen (1962). Here, in meat-fed females of *Calliphora* which are just beginning to mature eggs, the nucleoli of the neurosecretory cells become enlarged, a finding which does not hold for sugar-fed flies. Interestingly enough, the nuclei and nucleoli of neurosecretory cells in allatectomized meat-fed flies are approximately as small as those of the sugar-fed control flies; however, after reimplantation of corpora allata, they become enlarged. These findings may indicate that the corpus allatum of meat-fed animals exerts a stimulatory influence on the neurosecretory cells of the brain; from this one may postulate reciprocal action of neurosecretory cells and corpora allata. In any case, active corpora allata seem to stimulate the neurosecretory cells to higher metabolic activity, either directly or indirectly.

What kind of experimental evidence for the actual functioning of neurosecretory cells in insect reproductive processes do we have? Have we really observed a correlation of events in only certain cells (shown by histological means) with egg maturation periods? The first experimental clue for neurosecretory involvement was given in studies on *Calliphora* by Thomsen in 1948 and 1952. In this species, removal of the neurosecretory cells eliminated egg maturation completely, whereas extirpation of other brain areas did not, although egg maturation was considerably retarded in most operated animals of the latter group. Reimplantation of neurosecretory cells into females previously deprived of them resulted in the resumption of egg maturation in some flies; most females still matured no eggs. Since Thomsen and later Possompès (1956) have shown that allatectomy in *Calliphora* does not prevent complete egg maturation in many females, one is led to assume that in this species it is neurosecretion from the pars intercerebralis which contains the true gonadotropic hormone. From these results arose the question of what the mechanism of control of egg maturation is. Control of protein metabolism by neurosecretory materials was suggested; it was then thought that activated metabolism may allow, in turn, an activation of the corpus allatum, but it remains questionable whether the corpus allatum is indeed essential. Similarly, as in *Calliphora*, the autogenous mosquito *Aedes taeniorhynchus* did not mature any eggs in the ovaries after destruction of the pars intercerebralis; after reimplantation of the pars intercerebralis, egg maturation resumed (Lea, 1964b, 1967). This observation was understood to mean that neurosecretion controls corpus allatum activity and can thus be quoted in support of Thomsen's hypothesis. Unlike the situation in *Calliphora*, in *A. taeniorhynchus* the corpus allatum is indeed essential for egg maturation. Allatectomy is followed by complete cessation of egg maturation (Lea, 1967). In this species, the implantation of active corpora allata into pars intercerebralis cauterized females did not compensate for the defect, nor did the implantation of pars intercerebralis into allatectomized females. Replacement therapy was possible only by implantation of the originally removed tissues (Lea, 1967). This finding would then indicate that in *A. taeniorhynchus* there are two different hormones, each furnished by a separate endocrine organ, which are essential for egg maturation. The exact role of the respective hormones is thus far uncertain, and Lea does not speculate on their modes of action. Support for this novel interpretation of egg maturation as a process controlled by two hormones was independently given by Wilkens (1967, 1968), who worked with the fleshfly *Sarcophaga bullata*. It was also found here that egg maturation was abolished after destruction of the neurosecretory cells of the pars intercerebralis and that even highly active corpora allata could not repair the defect. Nonspecific brain surgery still allowed egg maturation, i.e. did not duplicate the effect of neurosecretory cell removal. For this species, the hypothesis was advanced that the corpus allatum, which appears to be active even after removal of the neurosecretory cells, promotes the protein biosynthesis essential for egg maturation, whereas the neurosecretory hormone acts as a true gonadotropic hormone, facilitating the incorporation of yolk into the oocytes (see p. 183). This hypothesis can, at this time, be applied only to the species *Sarcophaga* and represents the reverse of the hypothesis suggested by data obtained in *Schistocerca* (Highnam, 1964).

The control of egg maturation by two hormones has been suggested in a third case, that of *Locusta migratoria* (Girardie, 1966). In this species also, both the corpus allatum and neurosecretion from the pars intercerebralis are essential for egg maturation. Neither endocrine organ can replace the other. The precise role of the two hormones is, however, uncertain.

In addition to the cases mentioned so far, cautery of the pars intercerebralis completely prevented egg maturation in adults of *Schistocerca gregaria* (Highnam, 1962a, b; Hill, 1962), *Anacridium* (Geldiay, 1965), *Gomphocerus* (Loher, 1965, 1966a, b), *Schistocerca paranensis* (Strong, 1965a, b), and *Aedes aegypti* (Lea, 1967); it did so in *Tenebrio* (Mordue, 1965b) when operated either as pupae or as young adults.

Replacement therapy, i.e. reimplantation of a brain, allowed partial egg maturation in some female *Schistocerca gregaria*; in no case did complete egg maturation occur (Highnam, 1962a). Cautery of the pars intercerebralis reduced fecundity in *Carausius morosus* and *Clitumnus extradentatus* (Dupont-Raabe, 1952), *Oncopeltus fasciatus* (Johansson, 1958), *Locusta migratoria* (Girardie, 1964), *Aedes triseriatus* and *A. sollicitans* (Lea, 1967), and *Musca nebulo* (Deoras and Bhaskaran, 1967). In all these cases, the eggs completed their maturation. After reimplantation of the neurosecretory cells in *Musca nebulo*, fecundity was not increased. This observation suggests that there may be other parameters associated with severe brain surgery, as will be discussed further at several points below.

The reports so far mentioned seem to indicate that neurosecretion from the pars intercerebralis controls certain phases of reproductive processes in those species where the presence of these cells is essential; in other species, however, neurosecretion may play only a minor role. As will be exemplified even further, there now appear to be various patterns of neuro-endocrine control of egg maturation among insects, which range from no involvement of neurosecretion to the complete dependence upon it. One of the extreme cases is that of the cockroach *Leucophaea*. Approximately 30 per cent of the females of this species mature eggs after cautery of the pars intercerebralis (Engelmann, 1965a; Engelmann and Penney, 1966). This low percentage of reproducing females is, however, not caused by the lack of a neurosecretory hormone. It was noted that *Leucophaea* females did not accept the courting males after destruction of the pars intercerebralis; consequently, the failure to mate, an act essential for corpus allatum activation in most animals, may explain the failure of a high proportion of operated females to mature eggs. Similarly, in unoperated *Leucophaea* virgins only 30 to 40 per cent of the animals mature eggs as compared to nearly 100 per cent of the mated females (Engelmann, 1960b). Thus, a correlation between the function of the pars intercerebralis and egg maturation has not been established in this species.

An interesting effect of cautery of the pars intercerebralis in *Tenebrio molitor* females has been reported by Mordue (1965c). In this species the corpora allata remained small and egg maturation did not take place if the operation was performed in pupae or young adults; however, the same operation performed in older females resulted in a slight hypertrophy of the corpora allata, judging from their size. Strangely enough, egg maturation was not observed in the latter cases. It was also reported that these animals had an elevated protein concentration in the hemolymph. One could assume that the increased size of the corpora allata represented highly active glands (in accordance with observations in other species); however, since Mordue noted that no eggs matured, one would have to assume that the release mechanism for the hormone was impaired. Is it possible then that in *Tenebrio* destruction of the pars intercerebralis did not prevent hormone production, but only its release? Or could this finding fit Wilkens' interpretation (1967, 1968) of the situation in *Sarcophaga*, in which he postulated that neurosecretion is essential for the incorporation of yolk?

Although these reports on several insect species suggested that normally the median neurosecretory cells of the pars intercerebralis liberate a substance which is involved in egg maturation, possibly by activating the corpora allata, a series of experiments on *Schistocerca paranensis* (Strong, 1965b) indicated instead that very likely the lateral neurosecretory cells activate the corpora allata. The function of the median group remains obscure in this species. This report is the first to implicate a function for the lateral neurosecretory cells in reproductive processes. Strong arrived at his conclusion in the following manner. Bilateral destruction of the pars intercerebralis, which always included cautery of the lateral neurosecretory cells, did not allow activation of the corpora allata. The corpora allata remained

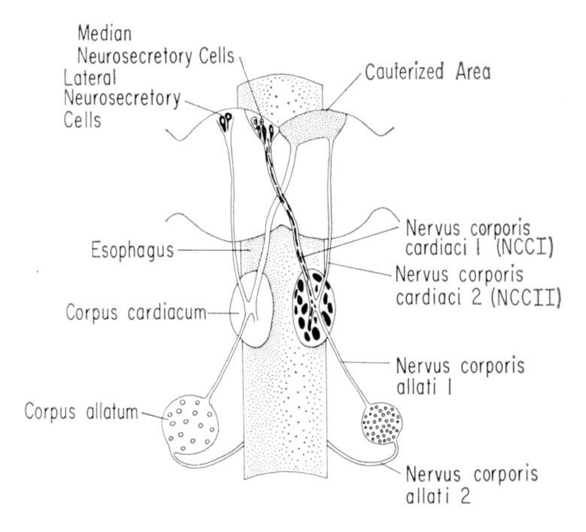

FIG. 63. Summary of the effects of brain cautery on corpus allatum activity in *Schistocerca paranensis*. Note that after unilateral destruction of the pars intercerebralis the corpus cardiacum of the contralateral side is depleted of neurosecretory materials and that the corpus allatum of that side becomes active. (Adapted from Strong, 1965c.)

small, and no eggs matured. However, after unilateral destruction the contralateral corpus allatum increased in volume and egg maturation followed; the ipsilateral gland stayed small (Fig. 63). If the median neurosecretory cells were the activation center for the corpora allata, unilateral brain cautery should actually have resulted in activation of the ipsilateral corpus allatum since the NCC I, which decussates from the other side, remained intact. That the fibers from the median neurosecretory cells decussate within the protocerebrum is also shown by the absence of neurosecretory material in the contralateral corpus cardiacum after unilateral brain surgery. Since fibers from the lateral neurosecretory cells do not decussate, these experiments strongly suggest that they transmit stimulatory information to the corpus allatum on the same side. However, one could still argue that some non-neurosecretory nerve fibers, which represent the pathway by which the corpus allatum is activated, join the NCC I after their decussation. This possibility has not been excluded for *Schistocerca paranensis*. Up to the present time, nothing is known about whether the lateral group of neurosecretory cells have a similar stimulatory function in other species; therefore, further detailed studies along the line of Strong's approach are necessary. Since reports for most species deal with bilateral destruction of the pars intercerebralis, which presumably included the lateral neurosecretory cells on both sides, one must be cautious in interpreting the effects of these operations.

The inhibitory effect on egg maturation following cautery of the pars intercerebralis is often interpreted as meaning that normally substances travel down the axons into the corpora allata; these glands are then stimulated upon the release of the substance. This seems plausible since, in several cases, neurosecretory materials have been shown within the corpora allata. Yet one could also postulate that non-neurosecretory nerve fibers, which have been demonstrated within the corpora allata, or within the nervi allati of some species (see Scharrer, 1964), transmit the information from the brain. Whether the corpora allata are stimulated by neurosecretory materials or solely by nervous action has not been unequivocally shown for any species. It appears, at any rate, that at least in some species, stimulatory information from the brain, whether neurosecretory or nervous, reaches the corpora allata via their innervation. This is shown by the fact that isolated corpora allata became or remained inactive or only mildly active; consequently, egg maturation proceeded at a limited rate, at best. As examples of this, the following cases can be quoted: *Locusta* (Staal, 1961), *Schistocerca paranensis* (Strong, 1965a, b), and *Schistocerca gregaria* (Pener, 1965, 1967). The same results seem to occur in *Tenebrio* when the NCC I are severed shortly after emergence, although not when done later in life (Mordue, 1965b, c, 1967).

Experiments on *Leucophaea* do not support the hypothesis of stimulatory action by the brain on the corpora allata, either through direct nervous control or through direct neurosecretory channels. Severance of the nervi allati or the NCC I plus NCC II resulted in activation rather than inhibition of the corpora allata (Engelmann, 1957a), and egg maturation always followed the operation. The severance of the NCC II alone had no effect. Identical operations produced the same results in the viviparous cockroach *Diploptera* (Engelmann, 1959a; Roth and Stay, 1961). In *Leucophaea*, even totally isolated corpora allata, which definitely were not reinnervated by the brain, induced growth of up to five batches of eggs within 8 or 9 months (Engelmann, 1962). These isolated glands caused the maturation of a greater egg mass than innervated corpora allata would do within the same time interval. The isolated glands underwent cyclic volumetric changes in accord with the induction of egg maturation (Fig. 59). If any stimulation from the brain is essential for the activation of the corpora allata in this species, however, it can reach these glands via the circulation. This may also hold for *Gomphocerus rufus* in which isolated inactive corpora allata became active and induced egg maturation in a normal manner (Loher, 1966b). These reports demonstrate that both the disruption of an intimate structural connection between the corpora allata and the brain and the destruction of the neurosecretory cells of the pars intercerebralis can have different effects in different species, probably indicating different operative mechanisms.

One important question must still be answered: namely, whether the effects which follow cautery of brain areas in some insect species is indicative of a direct or an indirect role of neurosecretion in reproductive processes. This is further linked to the problem of making a proper control operation, since any lesion in the brain may have multiple effects due to the fact that relatively large regions—at any rate larger than would be desired—are destroyed.

In pioneering studies on *Schistocerca*, Hill (1962, 1965) showed that the rate of protein synthesis in the fat bodies as well as the concentration of proteins in the haemolymph were reduced in pars intercerebralis cauterized females. It appeared that the failure to mature eggs in pars intercerebralis operated animals was the consequence of low protein metabolism, which did not allow activation of the corpora allata. Following these studies, Highnam *et al.* (1963a, b) and Highnam (1964) suggested that the corpora allata facilitate the deposition of available proteinaceous yolk into the growing oocytes but are not involved in protein synthesis. This suggestion, as should be pointed out again, is contrary to that by Wilkens (1967a, b) based on results obtained in the flesh fly *Sarcophaga.* The initial findings by Hill corroborated a suggestion made by Thomsen as early as 1952 that a principle liberated from the pars intercerebralis regulates protein metabolism. Supporting data for this may be found in observations of low protein concentrations in the hemolymph of brain-operated females of *Gomphocerus* (Loher, 1965), *Leucophaea* (Engelmann, 1965a, b, 1966), and young females of *Tenebrio* (Mordue, 1965c).

The above-mentioned observations pose still another question: namely, whether the operations on the pars intercerebralis of the brain indeed interfere with protein metabolism and, consequently, stop yolk deposition in the oocytes, or whether other activities of the operated females are impaired. Does the operation specifically change the protein metabolism, or is protein metabolism only secondarily affected? Since operational techniques are very crude in view of the size of the brain, any operation also destroys areas of non-neurosecretory nature; therefore, results must be interpreted cautiously. A likely side effect of any operation in the brain is reduced food intake which in turn reduces metabolic activities, particularly that of protein synthesis; therefore, protein concentration in the hemolymph will be low. In other words, a low protein concentration in the blood may simply be the result of reduced food intake and not specifically the consequence of deficient neurosecretory activity. Several reports may indeed be interpreted to this effect. For example, pars intercerebralis operated adult females of *Calliphora* (Thomsen, 1952) are found to have almost completely exhausted fat body cells. Yet these animals do not mature eggs; thus, one would not expect the reserves to be drained off. Even though the animals seemed to feed, however, it is highly possible that the operated females ate less than the normal ones. (Thomsen gave no quantitative data on feeding.) Later, Strangways-Dixon reported (1961b) that, after removal of the neurosecretory cells of the pars intercerebralis, *Calliphora* females ate only a little protein and that their carbohydrate intake was also rather low. This operation definitely caused a deviation from the typical feeding pattern found in normal females. However, if one assumes that egg maturation did not occur in operated females of *Calliphora* because they did not eat adequately, an explanation must still be found for the observed egg maturation, even though limited, after reimplantation of neurosecretory cells (Thomsen, 1952). Did those reimplanted females eat more than before, or was the food then propagated more rapidly down the alimentary canal? Recent observations on brain-operated females of *Sarcophaga* (Wilkens, 1967a, b, 1968) again showed that food intake in these animals was reduced; however, the reduction in food intake alone could not be responsible for the lack of egg maturation since the unoperated females, when fed an amount similar to that eaten by the operated ones, did mature eggs.

Strong arguments for neurosecretory control of protein synthesis in *Schistocerca* have been brought forth (Hill, 1962, 1965; Highnam, 1961, 1962a, b, 1964, 1965). Here, as in *Calliphora,* cautery of the pars intercerebralis, which involved the destruction of a large brain area, did not allow egg maturation. One could argue that this was due to the animals having fed very little. Objections to this interpretation have been raised, however, by the investigators who believed that brain cautery did not influence food intake since feces production was comparable to that of normal unoperated females (Hill *et al.*, 1966). Additional controls were also made in *Schistocerca* by removing the frontal ganglion (which puts the animals into a state of semi-starvation). Following removal of the frontal ganglion only a little yolk (more, however, than after cautery of the pars intercerebralis) was deposited into the oocytes (Highnam *et al.*, 1965). From this experiment the hypothesis that neurosecretory materials specifically stimulate protein metabolism was thought to have been strengthened.

In *Leucophaea* (Engelmann, 1965a) and *Periplaneta* (Mills *et al.*, 1966), brain operations could be definitely linked to a changed feeding pattern. For *Leucophaea*, it was shown that any lesion in the brain (whether in the region of the pars intercerebralis, in the region lateral to this area, or in the olfactory glomeruli, either bilaterally or unilaterally) definitely caused the animals to feed less than they normally would (Engelmann, 1965a; Engelmann and Penney, 1966). This was true even if the operations had been performed several weeks before the observations on feeding were made, i.e. before emergence. These animals presumably had overcome the operational shock. After such operations, many animals did not eat at all for several weeks but later did resume feeding at a moderate rate (Engelmann, 1965a). Also, many animals had erratic and increased locomotor activity and could easily be startled by tactile stimuli. The increased locomotor activity certainly depleted the protein reserves more rapidly than in normal animals. It should be noted here that in *Schistocerca*, *Calliphora*, and *Tenebrio* this operation was usually performed after emergence, which is only a few days before observations started (Thomsen, 1952; Hill, 1962; Mordue, 1965c).

Food intake definitely affected the protein concentration in the hemolymph, as shown by observations on *Leucophaea*: starvation drastically lowered the protein level in the hemolymph, while renewed feeding replenished the proteins within a few days. These observations show that in any discussion on endocrine control of metabolism, other aspects of the animal's behavior and biology as well as its general reactions to operations, must be considered.

From the data available for *Tenebrio* (Mordue, 1965c), one could construe that, in this species also, feeding behavior was affected by thermocautery of the pars intercerebralis. Animals operated on either during the pupal stage or during the first few days of imaginal life, i.e. before they had ingested large amounts of food, invariably had small corpora allata and matured no eggs, at least within the test period. On the other hand, females operated on the eighth day after emergence had corpora allata similar to or even larger than those of the sham operated controls. Obviously, the animals operated on 8 days after emergence had already fed and had built up some reserves by the time of operation; therefore, the operation did not have adverse effects on the corpus allatum activity. It is mentioned, however, that these animals did not mature eggs, which possibly indicates an impairment of the hormone release mechanism following destruction of the pars intercerebralis, as has been previously discussed.

In *Leucophaea* no evidence for the control of the corpora allata by neurosecretion from the pars intercerebralis, either directly, or indirectly via protein metabolism, was found (Engelmann and Penney, 1966). After destruction of the pars intercerebralis, about 30 per cent of the unmated females matured eggs, which is the same percentage as found for unoperated virgins (p. 171); i.e. their corpora allata must have been activated despite the absence of this neurosecretion. Furthermore, a certain percentage of the operated females which had not matured eggs after some time matured eggs after implantation of active corpora allata. In each of these experimental series, the protein concentration in the hemolymph of those animals which matured eggs rose above that of the controls; this was not observed in the females which failed to mature eggs. Proteins characteristic of egg maturing animals were detected by immunological techniques in the hemolymph of those operated females which matured eggs, but not in those which did not mature eggs (p. 183) (Fig. 66). This would indicate that in *Leucophaea* neurosecretion from the pars intercerebralis does not control either quantitatively or qualitatively the protein metabolism associated with egg maturation. It is strongly suggested that in this species the corpus allatum hormone controls this phase of protein metabolism. Similarly, it seems that in *Rhodnius prolixus* neurosecretion from the brain does not control either corpus allatum activity or egg maturation because protein synthesis associated with yolk deposition proceeds in decerebrated animals much the same as in unoperated females—provided the animals had been fed prior to decapitation (Wigglesworth, 1936, 1948; Coles, 1965a, b).

E. Inhibition of the Corpora Allata

Fluctuations in corpus allatum size, which presumably represent activity cycles of these glands, could theoretically be caused by periodic stimulations or inhibitions. In most cases it is not possible to determine whether any factor (e.g. feeding or mating) actively stimulates the corpora allata or causes the cessation of an inhibition. Inhibitory substances released from the maturing ova have been postulated. This was based on the observation that castration frequently resulted in hypertrophy of the corpora allata while, on the other hand, decreased corpus allatum volume was correlated with the presence of oocytes approaching maturity. However, in view of the more recent findings on blood protein concentrations which fluctuate in association with egg maturation cycles, there is no reason to postulate an inhibitory substance originating in the growing oocytes (p. 160). The low nutritional milieu at the end of any egg maturation period simply does not allow the endocrine glands a high metabolic activity. The gland then has an inactive or mildly active appearance, i.e. the nucleo-cytoplasmic ratio is high (Engelmann, 1957a, 1965a). Since in *Leucophaea* even totally isolated corpora allata responded by volumetric changes to the fluctuating protein level in the hemolymph, it seems that nutritional factors directly influence these glands: the brain does not exert direct nervous control. It appears as though the corpora allata rather passively reflect changes in the nutritional milieu, a principle which may apply to most of the species studied so far, although it may not apply in the certain specific cases outlined below.

Only in one group of insects, namely, the viviparous Blattaria, has active inhibition of the corpora allata by the brain been definitely demonstrated. In these species, embryos develop within the brood sac of the female for approximately 2½ months, during which time there is no space for additional eggs to mature. In *Leucophaea* severance of the nervi corporis cardiaci I when the corpora allata are totally inactive (i.e. during pregnancy) resulted in an increase in the corpus allatum volume, with egg maturation following (Scharrer, 1952; Lüscher and Engelmann, 1955; Engelmann and Lüscher, 1957; Engelmann, 1957a). If the NCC I was severed unilaterally, only the ipsilateral corpus allatum became activated (Fig. 64). Destruction of a brain area ventral to the neurosecretory cells of the pars intercerebralis but not that of the neurosecretory cells themselves also caused the activation of the corpus allatum of the same side (Engelmann and Lüscher, 1957). These results were interpreted to mean that during pregnancy the corpora allata are inhibited from a certain brain region. The disruption of the pathways to the corpora allata released the inhibition of the glands, and an untimely egg maturation followed.

The question then arose as to the mode of inhibition of the corpora allata. Based on the results obtained in *Leucophaea*, it was reasoned that the corpora allata are probably inhibited nervously. If, on the other hand, neurosecretory materials were assumed to be involved in this process, it would be necessary to postulate that these materials exercise an inhibitory influence on the corpora allata. However, either bilateral or unilateral cautery of the pars intercerebralis with its neurosecretory cells did not release the inhibition of the corpora allata. Moreover, after unilateral severance of the NCC I, inhibitory neurosecretory materials could still potentially reach the same side via decussating nerves from the contralateral corpus cardiacum or via the circulation (Fig. 64); but in the latter case, the corpus allatum was activated (Engelmann, 1957a). Severance of the NCC I or the nervi allati during pregnancy in the truly viviparous cockroach *Diploptera* yielded comparable results (Engelmann, 1958, 1959a). Therefore, the interpretation as to the control of the corpora allata during pregnancy which was given for *Leucophaea* seems also to hold for this species.

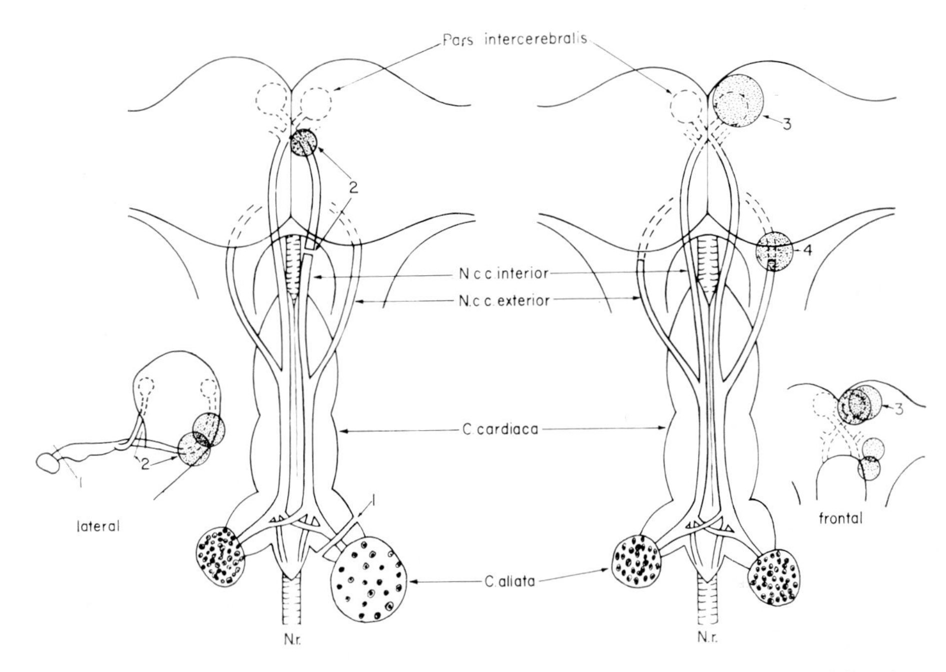

FIG. 64. The effect of various brain operations and transsections of nerves supplying the corpora cardiaca and corpora allata in pregnant females of *Leucophaea maderae*. Operations at locations 1 and 2 resulted in an activation of the corpora allata; those of 3 and 4 had no effect. (From Engelmann and Lüscher, 1957.)

The assumption of nervous control of the corpora allata from the brain is further strengthened by the discovery of non-neurosecretory nerve fibers within the corpora allata of *Leucophaea* (Scharrer, 1964). Furthermore, in recent studies on other cockroach species, non-neurosecretory fibers have been found in the NCC I, fibers which apparently originate in other brain areas (Willey, 1961; Brousse-Gaury, 1968). It is a possibility that these fibers are the ones which carry the message to the corpora allata. This finding corroborates the experimental results in *Leucophaea*, namely, that destruction of a certain brain area which does not contain neurosecretory cells lifted the inhibition of the corpora allata.

The above results show that the brain is obviously informed of the presence of developing embryos in the brood sac of these viviparous species. Basically, two theories have been advanced concerning the mechanism involved. The one claims a humoral pathway between the genitalia and the central nervous system (Engelmann, 1957a, 1964, 1965a), while the other proposes a purely nervous information channel (Roth and Stay, 1959, 1961, 1962a, b; Roth, 1964b). Evidence for both hypotheses is presented here.

Removal of the egg case from pregnant females of *Leucophaea* (Lüscher and Engelmann, 1955; Engelmann, 1957a) was followed by earlier egg maturation than would normally have occurred. The further advanced pregnancy was when the egg case was removed, the faster the next eggs matured; after ovulation, the ovaries are not immediately competent to respond to gonadotropic hormone (Scheurer and Lüscher, 1966). This basically applies for *Diploptera* (Engelmann, 1959a; Roth and Stay, 1961), *Blattella germanica* (Roth and Stay, 1959, 1962a), *Pyconescelus surinamensis*, and several additional viviparous species (Roth and Stay, 1962b; Roth, 1964b). If the natural egg case was replaced by a false ootheca or glass bead, no egg maturation took place in any of these species (Roth and Stay, 1959, 1962b). An interesting teratological specimen of the viviparous cockroach *Pycnoscelus* was discovered by Roth (1968). This female had two brood sacs each containing an egg case; the one egg case was extruded 10 days before the other, and it was noticed that egg maturation had begun shortly after the first ootheca had been released, i.e. egg maturation was initiated while the female was still "half" pregnant. These observations seem to be evidence for a purely nervous and non-humoral pathway of corpus allatum inhibition. One must, however, ask why the remaining egg case in the teratological specimen did not inhibit the corpora allata: the brood sac appeared to be properly innervated and the message should have been transmitted to the brain. This obviously calls for additional experimentation to search for a solution.

In *Leucophaea* it has been shown that distention of the brood sac would inhibit feeding in these females in a manner similar to that in females during a normal pregnancy (Engelmann and Rau, 1965). The continued inactivity of the corpora allata and consequent failure to mature eggs after replacement of the ootheca by a false egg case may thus be the effect of semi-starvation. Still, information on the filling of the brood sac by an egg case is transmitted via the ventral nerve cord; severance of the cord during pregnancy in all the viviparous species studied, including *Leucophaea* (Engelmann, 1960b, 1964), resulted in egg maturation (Roth and Stay, 1959, 1962a, b). After cord severance the formation could not "get through"; consequently, the brain released its inhibition of the corpora allata. It is peculiar, however, that in pregnant females of *Leucophaea* severance of any or all pairs of nerves from the last abdominal ganglion supplying parts of the genitalia did not result in an activation of the corpora allata (Engelmann, 1964). In *Leucophaea* it was also noted that the more posteriorly the nerve cord was severed, the higher was the percentage of females which remained pregnant and did not mature more eggs (Table 10). If all the abdominal ganglia were severed from the rest of the anterior central nervous system, only a slight delay in onset of egg maturation was noticed.

How can we interpret these results? If one assumes that inhibition of the corpora allata is brought about by a substance released by the eggs in the brood sac or by the distended brood sac itself, one must postulate that receptors, i.e. receptive neurons, for this agent are located in all abdominal ganglia, as well as in the brain itself. The action of all these receptive neurons is cumulative, i.e. the more of them are cut off from the inhibitory center in the brain, the better the inhibition of the corpora allata can be released (Engelmann, 1964). The fact that no or little inhibition of the corpora allata was observed after placing part or all of the egg case into the hemocoel in *Pycnoscelus* (Roth and Stay, 1959) or in *Leucophaea* (Engelmann, unpublished) may point to the possibility that the stretched brood sac liberates this agent. It is also possible that the implanted eggs with their embryos are "metabolized"

TABLE 10. Severance of the ventral nerve cord in females with inactive corpora allata (pregnant) in *Leucophaea maderae* (Engelmann, 1964)

Site of operation (abdominal ganglia)	No. of operated females	Results 4 to 5 weeks after operation	
		No egg maturation (numbers)	Initiated or complete egg maturation (numbers)
2/3	34	4	30 (88 per cent)
3/4	10	2	8 (80 per cent)
4/5	14	5	9 (64 per cent)
5/6	18	12	6 (33 per cent)
posterior 6	22	22	0 (0 per cent)

soon after implantation into the body cavity and that, therefore, within a short time after implantation, inhibitory substances are no longer released. In the case of the teratological specimen of *Pycnoscelus*, the second egg case carried beyond the normal gestation period does not liberate inhibitory agents any more.

It should again be pointed out in this connection that no agents from the egg case or brood sac which could be responsible for the inhibition of the corpora allata in viviparous cockroaches have been extracted so far. Recently, however, extracts have been prepared from house-fly eggs which, when injected, did inhibit the maturation of the next set of eggs (Adams *et al.*, 1968). Unfortunately, the mortality was rather high (on the over-all, between 30 and 50 per cent) and it is, therefore, not entirely clear whether those surviving females could be compared to normal or control injected animals. Certain toxic agents in sublethal dose presumably can inhibit the endocrine activity of the treated females, and we are uncertain whether or not we should term these agents "oostatic hormones". In *Leucophaea*, some egg extracts are rather toxic to the females (Engelmann, unpublished). In another dipteran, *Aedes aegypti*, no further eggs mature if the previously matured eggs are retained in the genital ducts; a so-called "gonotrophic dissociation" is observed (Judson, 1968). Can we postulate in this case that the mature eggs liberate an inhibitory substance for the corpus allatum or for the ovaries themselves? No information about the precise pathway of egg maturation in this species is available.

F. The Mode of Hormone Action in Egg Maturation

1. THE GONADOTROPIC HORMONE

From the many studies on the role of the corpora allata in reproduction in the female insect, it seems that the hormone or hormones liberated by these glands may have several functions. One of these, as was early recognized in insect endocrinology, is to promote the incorporation of proteinaceous and lipid yolk material into the growing oocytes (Wigglesworth, 1936; Weed, 1936; Pfeiffer, 1939; Thomsen, 1942); usually a rapid growth of the terminal oocytes in the ovarioles takes place. In connection with egg maturation, the accessory sex glands of females in many species (Wigglesworth, 1936; Scharrer, 1946; Thomsen, 1952; Engelmann, 1957b) and the glands of the oviducts in grasshoppers (Weed, 1936) are activated. The accessory sex glands of *Periplaneta* which secrete substances with which the females build the egg case are thought to be even more sensitive to the corpus allatum hormone than the ovarian tissues (Bodenstein and Sprague, 1959).

In several cockroach species one of the accessory sex glands furnishes a substance which apparently digests part of the secretion with which the spermatophore is fixed into the bursa copulatrix during mating (Engelmann, 1959a, 1960b; Roth and Stay, 1961); consequently, mated allatectomized females rarely, and then only several weeks after mating, lose the spermatophore, whereas unoperated females do so within a few days.

Oocytes of immature ovaries in various Orthoptera and Odonata are impermeable to vital dyes until yolk deposition begins, thus indicating that the permeability properties of the oocytes may also be under the control of the corpus allatum hormone (Iwanoff and Mestscherskaja, 1935). Dramatic changes in the appearance of the follicular epithelium of growing oocytes are known in many species (p. 52ff.). This may indicate that the corpus allatum hormone also acts directly on these tissues. From these reports, it appears that the corpora allata produce a true gonadotropic hormone which influences the activities of the ovaries and other tissues of the genitalia, as was originally proposed (see Highnam, 1964).

The corpora allata may not be the only source of a true gonadotropic hormone as data obtained by Wilkens (1967a, 1968) on *Sarcophaga* indicate. Here, the neurosecretory cells of the pars intercerebralis seem to furnish the gonadotropic hormone since the oocytes do not incorporate yolk unless these cells are present—irrespective of the availability of the corpus allatum hormone. Possibly the results on brain-cauterized *Tenebrio* (Mordue, 1965c) and the associated lack of egg maturation even in the presences of large corpora allata can be interpreted in the same manner. This interpretation of Mordue's results finds support from recent observations on the same species that both *in vivo* and *in vitro* yolk deposition does not proceed if either the brain or the corpora allata are absent (Laverdure, 1967a, b; Lender and Laverdure, 1967); in this species the fat bodies contain substances (stored neurosecretion or juvenile hormone?) that promote egg growth. It appears now that we are far from being able to generalize on the origin of a gonadotropic hormone in insect species. This word of caution is also based on the fact that relatively few species have been thoroughly worked through.

2. THE EFFECTS ON METABOLISM

Egg maturation certainly entails the activation of specific phases of the species' metabolism. The following questions must therefore be raised. Does the corpus allatum liberate a single hormone which activates the metabolism associated with egg maturation as well as promotes the deposition of yolk into the growing oocytes, or does the corpus allatum produce several hormones? Are the animals' changes in metabolism merely the result of increased "demands" by the growing eggs? To what extent does neurosecretion affect general or specific metabolism? These and several related questions have still not been satisfactorily answered for any species; in particular, the question of the existence of a metabolic hormone secreted by the corpus allatum is under considerable discussion at the present time.

As early as 1945, Pfeiffer reported an enormous increase in fatty acid content in the fat bodies of *Melanoplus differentialis* after allatectomy. Since this accumulation of fatty acids is not observed in ovariectomized females, it cannot be attributed to a non-utilization of reserves due to the lack of egg maturation. Rather, it suggests a direct influence of the corpus allatum hormone on lipid metabolism. Similarly, allatectomized females of *Periplaneta americana* stored 68 per cent more lipids than did unoperated females, the turnover rates of both triglycerides and phospholipids being slowed down in operated females (Vroman *et al.*,

1965). *In vitro* uptake of lipids by the ovaries in *Leucophaea* was stimulated by the corpus allatum hormone (Gilbert, 1967). This report definitely demonstrates that the corpora allata are involved in lipid metabolism and thus presumably facilitate egg maturation. In the earlier work of Pfeiffer (1945), it was suggested that the corpora allata affect general metabolism rather than specific phases.

Bodenstein's observations on *Periplaneta* (1953) and those of L'Helias (1953) on *Dixippus morosus* showed that the corpora allata are somehow involved in protein metabolism too. Allatectomy in *Galleria mellonella* (Röller, 1962) also seemed to indicate that the corpora allata play a role in protein metabolism since fewer eggs were laid after allatectomy but injection of protein hydrolysates into operated females increased the egg output, i.e. the defect could be remedied. In *Phormia regina* the fat bodies hypertrophied after allatectomy with considerable amounts of lipids, carbohydrates, and proteins being stored (Orr, 1964a, b). This was not observed after ovariectomy, which once again indicates that the corpora allata are involved in some phases of the fat bodies metabolism and that the effect of the corpora allata is not mediated by the ovaries. Orr (1964a, b), in his discussion, even goes so far as to suggest that the corpus allatum hormone only stimulates the animal's metabolism and has no gonadotropic function. However, one must then ask whether the specific structural changes observed in tissues of the genitalia are caused by increased general metabolism alone, or whether we must not postulate a true gonadotropic hormone.

The role of the corpora allata in metabolism was disputed by Highnam and his students (see Highnam, 1964), who claimed that neurosecretion from the pars intercerebralis controls protein metabolism. Their viewpoint was based on the observation that removal of the neurosecretory cells of the pars intercerebralis in *Calliphora* (Thomsen, 1952) allowed no egg maturation whatsoever, possibly because protein metabolism was impaired. Similarly, cautery of the pars intercerebralis prevented egg maturation in *Schistocerca* (Hill, 1962; Highnam, 1962a, b), *Tenebrio* (Mordue 1965c) and *Gomphocerus* (Loher, 1965, 1966). After cautery of the pars intercerebralis in *Schistocerca* and *Gomphocerus*, the hemolymph protein level was significantly reduced as compared to that of normal animals. A causal relationship was proposed between release of neurosecretory materials and protein metabolism since, contrarily, only a slight reduction in the protein level was noticed after allatectomy. Failure to mature eggs after brain surgery may be the consequence of a low protein level in the hemolymph which does not permit activation of the corpora allata. It was also shown that the incorporation rate of C^{14}-glycine into proteins of operated females of *Schistocerca* (Hill, 1965) was lower than that of normal animals. Following allatectomy of *Schistocerca* neurosecretory materials accumulated in the corpora cardiaca and neurosecretory cells; therefore, it appeared as if the corpora allata normally stimulate the release of this material. It was indeed suggested that the protein concentration in the hemolymph of allatectomized females was somewhat lower because little neurosecretory material had been released (Highnam *et al.*, 1963a). A certain reciprocity between the corpora allata and the neurosecretory system was claimed to exist (Highnam, 1964).

Interestingly, the implantation of corpora cardiaca into brain cauterized animals resulted in a short term (a few hours) rise in blood proteins in *Schistocerca*, implying that neurosecretory materials were released by the corpora cardiaca (Highnam *et al.*, 1963a) (Fig. 65). Whether this short-term rise is of any significance with respect to long-term reproductive processes will have to be investigated. Unfortunately, the implantation of active corpora allata into pars intercerebralis cauterized females and the subsequent effect on blood proteins have not been reported for *Schistocerca*. Furthermore, the results of brain operations in some of the species are difficult to interpret because of undesirable side effects (see p. 174). It is clear, however, that hormones derived from either the neurosecretory cells or the corpora allata do control certain phases of protein metabolism associated with egg maturation. Highnam and co-workers (1963a) and Highnam (1964) postulated (from work on *Schistocerca* primarily) that neurosecretory materials alone control protein synthesis and turnover rates, thus making proteins available; however, protein incorporation into the growing oocytes is controlled by the true gonadotropic hormone from the corpora allata.

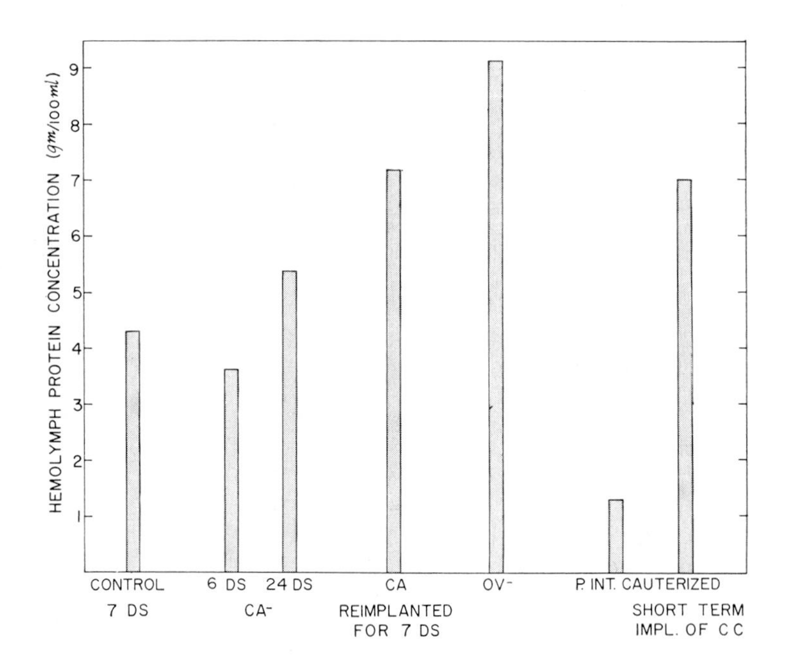

FIG. 65. The effects of various operations on blood protein concentrations in *Schistocerca gregaria*. (Based on data given by Highnam and co-workers.)

One could equally well interpret the results of the reported experiments to the effect that the corpora allata control certain phases of the protein metabolism associated with egg maturation. For instance, ovariectomy, which allowed high corpus allatum activity, or reimplantation of corpora allata into allatectomized females, caused a rise in the protein level of the hemolymph above that of the controls in *Schistocerca* (Highnam, 1962b) (Fig. 65). This argumentation is countered by the suggestion that the corpora allata cause the liberation of neurosecretory materials and that these in turn stimulate protein metabolism. This latter argument is based on histological observations only, and one must wait for experimental evidence on the question. Support for the view that the corpus allatum hormone controls protein metabolism comes from experiments performed in several other species. As in *Schistocerca*, the blood protein concentrations in ovariectomized females of *Phormia* (Orr, 1964b) and *Gomphocerus* (Loher, 1965, 1966) are rather high as compared to levels in the controls. One has to bear in mind that blood proteins are not used in castrated animals, therefore, the high protein level may only reflect the titer of unused proteins. On the other hand, allatectomy in *Periplaneta*, which likewise is followed by a non-utilization of blood proteins, resulted in a relatively low protein concentration in the hemolymph (Thomas and Nation, 1966a). This has also been observed in allatectomized *Leucophaea* females (Engelmann, 1965a). In *Leucophaea* the low protein level in the blood is not the consequence of a reduced food intake. In allatectomized females of *Periplaneta*, the rate of incorporation of H^3-labeled amino acids into the fat body was rather low (Thomas and Nation, 1966b), as was the case for C^{14}-labeled amino acids administered to operated females of *Nauphoeta* (Lüscher, 1968) and *Leucophaea* (Engelmann, unpublished). Following the implantation of active corpora allata into decapitated *Nauphoeta* females, Lüscher observed increased protein synthesis, an observation which could not be made after the implantation of brain or corpora cardiaca. The findings in these species may indeed serve as strong evidence for the view that the corpora allata control protein metabolism to a certain extent.

Quantitative assays of the total protein concentration reveal only the net result of protein biosynthesis and utilization; the results can be subject to several interpretations and are therefore not conclusive.

Qualitative assays may be more pertinent to an understanding of the hormonal effects on protein metabolism during egg maturation. The observation that in *Schistocerca* one particular protein is more predominant during egg maturation periods than at other times may point in this direction (Hill, 1962). In *Rhodnius prolixus*, two specific "yolk proteins" which were identified immunologically appeared in the hemolymph during periods of egg maturation. The biosynthesis of these "yolk proteins" is apparently under the control of the corpus allatum hormone (Coles, 1964, 1965a, b). These proteins could still be detected in the hemolymph after decapitation, which removed the brain but left the corpus allatum in the animal. Obviously, neurosecretion from the brain is not involved in protein metabolism and egg maturation in this species. Cellulose acetate and immunoelectrophoresis showed in *Leucophaea* that here also the biosynthesis of one specific female protein is controlled by a hormone liberated from the corpora allata (Engelmann, 1965b; Engelmann and Penney, 1966; Engelmann, 1966). In newly emerged females and males of this species, four or possibly five antigens are detectable in the hemolymph. Correlated with the onset of corpus allatum activity in the females is the appearance of a sixth antigen, which moves slightly cathodally in the electric field (Fig. 66). During the inactive phases of the corpora allata, i.e. during pregnancies, or in allactectomized females this specific female antigen is absent. It was never present in males of any age. It was, however, found in all pars intercerebralis-cauterized females which matured eggs, but not in any operated females which did not mature eggs. Therefore, neurosecretion from the pars intercerebralis could not have been responsible for the synthesis of this particular protein (Engelmann and Penney, 1966). In other words, in *Leucophaea* the biosynthesis of one specific female protein is apparently under the control of the corpus allatum hormone. This protein is always associated with egg maturation. Further support for this finding was obtained by the application of the synthetic juvenile hormone to allatectomized females. Under these conditions, the specific female protein was synthesized by the animals; egg maturation followed a few days thereafter. Immunological procedures in the fleshfly *Sarcophaga* also demonstrated that the corpora allata are associated with the production of one specific female protein (Wilkens, 1967); in its absence, no eggs mature in this species.

Immunologically, it has been demonstrated in *Rhodnius* (Coles, 1965a, b) and in *Leucophaea* (Fig. 66) (Engelmann and Penney, 1966), as well as in *Hyalophora* (Telfer, 1960) and *Sarcophaga* (Wilkens, 1967) that the oocytes selectively take up intact proteins from the

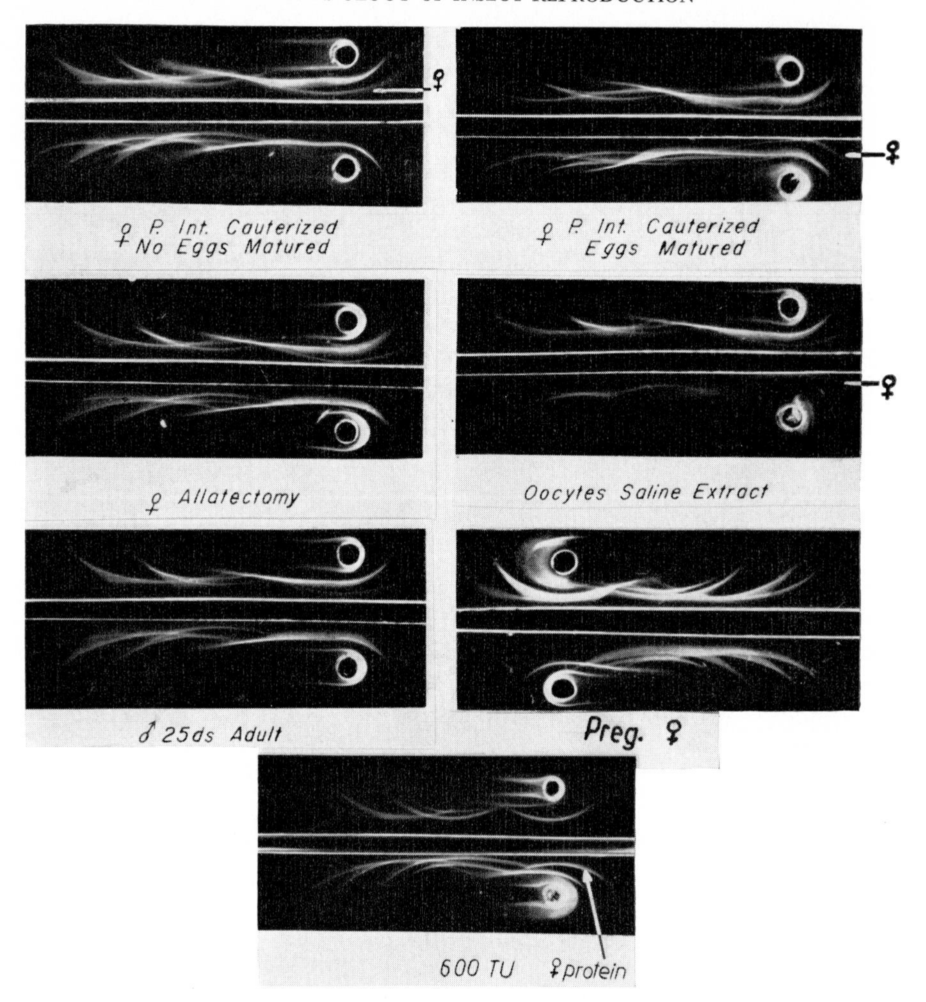

FIG. 66. Immunoelectrophoresis of hemolymph of *Leucophaea maderae*. In each of the upper traces, blood from egg maturing females has been electrophoresed as a control. The female protein is apparent in blood of all females that matured eggs, regardless of the operation. Allatectomized females were treated with the t,t,t-isomer of the juvenile hormone for 4 days.

hemolymph. Specifically, the female protein was taken up in *Hyalophora* against a concentration gradient. It appears that extraovarian proteins make up a large proportion of the egg proteins in certain species. In other species, the follicular epithelium and the nurse cells appear to contribute to the protein synthesis associated with egg maturation. Can we assume that here also hormones control the synthesis and incorporation of these proteins?

The unique appearance of certain proteins during the active phases of an endocrine gland and the association of these proteins with egg maturation lead one to assume that these specific proteins may have specific key functions in yolk formation. These studies are certainly an interesting start for further work on the primary processes of egg maturation. Conceivably, the turnover rates of all blood proteins are greatly increased during periods of egg maturation, but further experiments will have to prove the involvement of the hormones from either the pars intercerebralis or the corpora allata.

Qualitative changes in the blood proteins associated with egg maturation also occur in *Periplaneta*, as has been shown by conventional electrophoretic methods (Menon, 1966; Thomas and Nation, 1966a; Adiyodi and Nayar, 1967). One protein specific to egg-maturing females appears. This observation is similar to that made in *Leucophaea* using different electrophoretic methods (Engelmann and Penney, 1966). Specific changes in blood proteins are found during egg maturation in *Musca domestica* (Bodnaryk and Morrison, 1966) and *Locusta* (Minks, 1967). This has also been reported for *Phormia* (Chen and Levenbook, 1966), *Nauphoeta* (Adiyodi, 1967), and *Pieris brassicae* (Lamy, 1967). In none of these latter cases do we know anything specific about endocrine control mechanisms of changes in the blood proteins; we can only suspect that either the neurosecretory system or the corpora allata are involved since these changes occur in association with egg maturation. Also, we are uncertain about the specificity of the proteins because immunological techniques were not used for the demonstration of the protein changes. Nevertheless, these studies are of great interest. The occurrence of specific female proteins may be reflected by the sex differences of the fat body cells, the presumed sites of protein synthesis, as was shown histologically in *Locusta* (Lauverjat, 1967).

The many reports on the role of active corpora allata in egg maturation indicate that it may be at least twofold. Pfeiffer (1945) first suggested that the hormone from these glands simultaneously controls various phases of metabolism as well as the incorporation of yolk into the growing oocytes. This theory still seems best to explain the present data. A good deal of information is available on the control of protein metabolism, as was outlined above, but extremely little on the details of lipid or carbohydrate metabolism associated with egg maturation.

The hormone or hormones liberated by the corpora allata certainly control many more aspects of the animal's biology than those discussed in this chapter. Notably, the central nervous system seems to be a target organ since various behavioral patterns change under the influence of the corpus allatum hormone. This subject will be treated in the pertinent chapters on behavior and its control by hormones.

One interesting but scarcely explored aspect of hormone action in reproductive physiology is the interaction of various hormones. In the cockroach *Leucophaea* it has been found (Engelmann, 1959b) that when prothoracic glands were implanted into adult females, egg maturation was inhibited while the animals were stimulated to molt. Injections of ecdysone duplicated the action of the prothoracic glands. On the basis of histological observations of the corpora allata, it was concluded that ecdysone inhibits these glands from producing gonadotropic homone, i.e. ecdysone does not act on the target organ, the ovary. It appears that the animals cannot mature eggs at the same time as they prepare to molt. Analogous observations have been made on *Rhodnius* (Wigglesworth, 1952). Provided this line of thinking is correct, one is able to interpret the findings of Watson (1964) in the firebrat *Thermobia domestica*, in which egg maturation always alternates with molting but never occurs simultaneously. Is it the metabolic demands of molting that prevent egg maturation via the inhibition of the corpora allata? The answer is, however, not as simple as it may appear from the outlined cases, because in juvenile insects the corpora allata are active while the animals molt; yet egg maturation does not take place even in implanted competent adult ovaries (Engelmann, 1959b). This indicates that additional factors of the larval milieu must be responsible in this case for the lack of response to juvenile hormone. In any case, the effects of ecdysone in the inhibition of egg maturation cannot be interpreted as simple antagonistic action of the juvenile hormone, as has recently been suggested from findings in *Musca* and *Tribolium confusum* (Robbins *et al.*, 1968). The precise mode of hormone interactions is still to be worked out.

3. THE CHEMICAL NATURE OF THE CORPUS ALLATUM HORMONE

Highly purified extracts with juvenile hormone activity have been tested in several insects species but often only limited data are available with respect to their gonadotropic effects. Crude extracts of *Cecropia* silkmoths, which had high juvenile hormone activity in a variety of bioassays, promoted egg maturation in allatectomized females of *Periplaneta* (Chen *et al.*, 1962) after injection of 50 μl or after topical application. Neither peanut nor olive oils, usually used as carriers, nor extracts from other insects had any gonadotropic activity. When the true juvenile hormone, isolated from *Cecropia* extracts, became available another exciting phase of research could be opened up. The authentic juvenile hormone was identified as t,t,c-10-epoxy-7-ethyl-3,11-dimethyl-2,6-tridecadienoate (Fig. 67) (Röller *et al.*, 1967; Dahm *et al.*, 1968). The authentic compound has high juvenile hormone and gonadotropic hormone activity (Röller *et al.*, 1966). None of the synthesized isomers or related compounds have similar activity; the t,t,t-isomer is the most active synthetic hormone, but it has only approximately half the juvenile hormone activity of the authentic hormone (Dahm *et al.*,

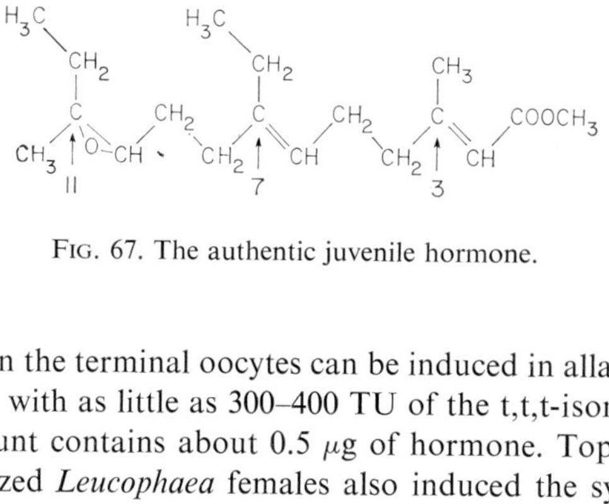

FIG. 67. The authentic juvenile hormone.

1968). Yolk deposition in the terminal oocytes can be induced in allatectomized *Periplaneta* and *Leucophaea* females with as little as 300–400 TU of the t,t,t-isomer (Engelmann, 1967, unpublished); this amount contains about 0.5 μg of hormone. Topical application of the hormone to allatectomized *Leucophaea* females also induced the synthesis of the specific female protein, an apparent prerequisite for egg maturation in this species (Fig. 66) (Engelmann, unpublished). Just as egg maturation was induced in treated animals, so was activity of the accessory sex glands in decapitated *Periplaneta* females (Bodenstein and Shaaya, 1968); the application of only 75 TU increased the level of protocatechuic acid glucoside in the left colleterial gland above that of the control. 450 TU mimic the increase in unoperated animals after metamorphosis. The long-standing discussion of whether the juvenile hormone is identical with the gonadotropic hormone has now been resolved. Just as the juvenile corpora allata liberate a gonadotropin after transplantation into the adult animals and vice versa, the authentic juvenile hormone has juvenilizing effects in the larvae and nymphs of several insect species and promotes egg maturation in the adults.

Synthetic compounds with juvenile hormone activity have been tested for their gonadotropic activity in a variety of insect species. For example, decapitated fed females of *Rhodnius* fully matured eggs after topical application of farnesol (Wigglesworth, 1961) or after injection of farnesyl methyl ether; the latter compound is the most potent compound among farnesol derivatives tested in this species. A minimum of 2 μl of this material caused a few eggs to mature; with increased amounts, more eggs matured; however, the operated females never matured as many eggs as unoperated females normally would. A definite dose response was obtained; but even after injection of 100 μl or more, still only two-thirds of the normal number of eggs were produced. Feeding farnesyl methyl ether in sucrose to adults of the parasitoid *Exeristes comstockii* promoted egg maturation at a rate even higher than that in animals fed only a protein diet (Bracken and Nair, 1967). Also egg maturation was induced in a certain percentage of allatectomized females of the cockroach *Byrsotria fumigata* after treatment with the same compound (Emmerich and Barth, 1968). In *Schistocerca* yolk deposition was induced in allatectomized virgin females after application of farnesol, which, in this case, also reduced the percentage of eggs which would normally have been resorbed in females kept without males (Highnam *et al.*, 1963a, b). As in *Rhodnius* and *Schistocerca*, when treated with farnesol or its derivatives, egg maturation could only partially be induced in *Calliphora* by treatments with farnesylphosphate and farnesylpyrophosphate (Weirich, 1963). Farnesol induced egg maturation in fed *Cimex* (Davis, 1964, 1965b) and accelerated it in the grasshopper *Euthystira brachyptera* (Müller, 1965b) to some extent. The highest gonadotropic activity of a variety of substances tested was obtained with 10,11-epoxyfarnesenic acid methyl ester in allatectomized females of *Periplaneta* (Bowers *et al.*, 1965). The same chemical was a potent stimulant for breaking reproductive diapause in the cereal leaf beetle *Oulema melanopus*; this also suggests that normally the corpus allatum hormone terminates diapause in this species (Connin *et al.*, 1967).

In contrast to these reports, no gonadotropic activity was obtained in allatectomized *Pyrrhocris* by the application of *Cecropia* extracts or most of several other substances which do mimic juvenile hormone activity in other species tested. Chemicals mimicking the juvenile hormone may indeed be active in certain species only and not in others. This is exemplified by the action of "juvabione" which almost exclusively affects *Pyrrhocoris* and no other species, even related Hemiptera (Bowers *et al.*, 1966). This may also, of course, apply to the gonadotropic activity of these chemicals. The true gonadotropic hormone or juvenile hormone is not identical to farnesol or its derivatives as now has been made abundantly clear (Röller *et al.*, 1966, 1967); possibly the juvenile hormone from *Cecropia* is active in all species of insects. At present nothing is known about whether species of different orders have the same authentic juvenile and gonadotropic hormone or whether in the course of evolution, several agents with juvenile hormone activity have evolved in the respective species.

G. Integration of Endocrine Control of Reproduction

As is apparent from the preceding sections, a variety of intrinsic and extrinsic factors influence egg output in the species studied. Total egg production reflects the hormone output of the corpora allata in those species in which egg maturation is stimulated by these glands. Food consumption is absolutely essential for the activation of the corpora allata in most insects, but some species mature at least one set of eggs without any food intake. Mating stimulates corpus allatum activity to a varying degree, as is known from detailed studies in a few species. Light or chemical perception is essential for the maturation of eggs in a few species, but these external stimuli have no apparent influence in many others. While it is clear that in most species the corpora allata are involved in the control of egg maturation, detailed information concerning the control mechanism is available for only a few species.

Any generalization on the principles involved is sure to be revised within a few years and may not apply in the literal sense for any particular species; it may be of some heuristic value to set down a general outline which would incorporate present knowledge (Fig. 68).

In species which need to feed in order to mature eggs, the protein level in the hemolymph rises shortly after the beginning of feeding but prior to the activation of the corpora allata. The level of proteins in the hemolymph probably reflects the general nutritional status of the animal. Autogenous species have more reserves than anautogenous ones, which apparently allows activation of the corpora allata without any food intake. The effect of nutrients on the corpora allata appears to be a direct one: the brain does not mediate the influence of nutrition on these glands. It is possible, however, that neurosecretory activity interacts in

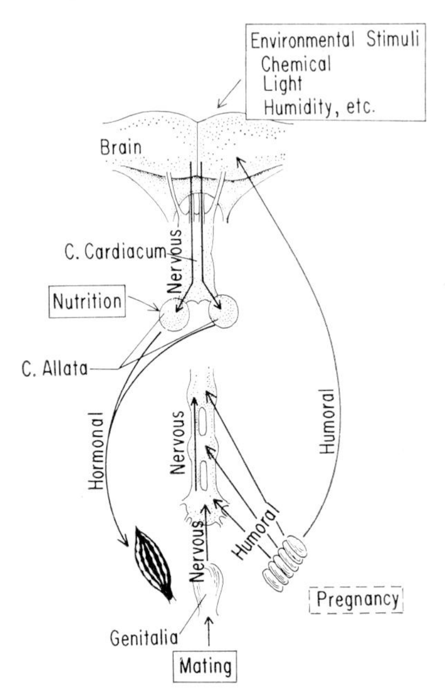

FIG. 68. Summary of the various extrinsic and intrinsic influences on the endocrine system and egg maturation in insect species. Only some of these stimuli are operative in any given species.

overall protein metabolism, but the available evidence for this is only indirect. It seems essential that the corpora allata be bathed in a nutrient-rich milieu in order to become active. In some species, the corpora allata become fully activated following feeding, probably as a consequence of both the mechanical stimulation during feeding and the elevated nutritional status following digestion of the food. In most species, however, additional stimuli are required for full activation of the corpora allata, one of which is certainly mating. Mating stimuli are received in the genitalia and apparently reach the brain via the ventral nerve cord or the abdominal nerves. These stimuli either cause the activation of the corpora allata or perhaps block the partial or total inhibition of these glands by the protocerebrum. It is difficult to determine which of these mechanisms applies. In either case, the corpora allata become active. Mating stimuli are rather ineffective unless the blood protein level has reached a certain height. This has been shown at least in *Leucophaea* (Engelmann, 1965) and *Cimex* (Davis, 1965). Similarly, light stimuli, which serve to activate the corpora allata at least in *Leptinotarsa* (De Wilde, 1958), are effective only if concomitant feeding is allowed.

An interesting and probably unique mechanism for control of the corpora allata in insects is found in several species of viviparous cockroaches. In these species, the corpora allata are inactive during the long periods of pregnancy, but they show a high degree of activity shortly thereafter. Their inhibition is apparently accomplished by the release of an agent from the brood sac or the egg case within the brood sac. This substance presumably affects neurons in the ventral nerve cord and brain which in turn influence a center in the brain to inhibit the corpora allata. According to the available data, the inhibition of the corpora allata by the brain during pregnancy is achieved nervously since severance of the nervi allati activated the glands, which then stayed active for prolonged periods. Excessive feeding did not activate the corpora allata, whereas in non-pregnant females it may do so, at least to some extent. Likewise, mechanical stimulation of the genitalia, which during other periods would definitely activate the corpora allata, had no influence on them. Apparently, if two stimuli of opposite nature (inhibitory and activating) are administered simultaneously, the central nervous system integrates "appropriately".

Nutritional factors, mating, light, and chemical stimuli activate the corpora allata either directly or indirectly, and the presence of eggs in the brood pouch of viviparous cockroaches inhibits these glands. A certain degree of interaction of these factors seems to occur, at least as far as is known. The end result is the integrated control of the corpora allata which apparently "serves the species best". Egg maturation, which can potentially take place if enough reserves are available, does not take place under unfavorable conditions.

For a few species of insects, it is now proposed that the endocrine control of egg maturation may be achieved according to either of several patterns. On the one hand, egg maturation is solely dependent on the corpus allatum hormone; species like *Rhodnius* and *Leucophaea* fall into this category. On the other hand, for species such as *Calliphora* and *Sarcophaga*, the neurosecretory cells of the pars intercerebralis furnish a hormone which promotes egg maturation. Then, in between these two extremes fall several species in which both the corpora allata and the neurosecretory cells are essential for egg maturation. Neither of the two sources of hormone can replace the other. This was demonstrated for *Locusta* and possibly applies to *Schistocerca* and some species of mosquitoes. Additional species appear to depend, to a certain degree, on either of the two hormones but are not completely prevented from maturing eggs if one of the sources is missing. As was pointed out at the beginning of this chapter, detailed and complete information on endocrine control of egg maturation is only available for a few species belonging to a few orders. Species of many orders have not been investigated in detail; therefore, we are far from a general understanding of egg maturation processes in insects. The development and accumulation of knowledge of these aspects in recent years probably justify an optimistic outlook for the not too distant future, however.

Addendum

The gonadotropic activity of the authenic juvenile hormone and its synthesized isomers, has been further established in *Periplaneta* (Röller and Dahm, 1968; Röller *et al.*, 1969) *Leucophaea* (Engelmann, 1969), and *Locusta migratoria* (Joly, 1969). Since then a small quantity of a second juvenile hormone has been identified in the *Cecropia* extracts. It is the lower homologue to the first JH: 10-epoxy-3,7,11-trimethyl-2,6-tridecadienoate (Meyer *et al.*, 1968). This hormone also has gonadotropic activity and causes the synthesis of the female specific protein in *Leucophaea* (Engelmann, unpublished).

CHAPTER 9

ENDOCRINE INFLUENCE ON REPRODUCTION IN THE MALE INSECT

CONTROL mechanisms of imaginal maturation processes in male insects are at best only partially understood in only a few species. Probably the best known case is that of the desert locust *Schistocerca gregaria*. After emergence, the male of this species shows a distinct external maturation which is correlated with the beginning of readiness to mate. The animals are, at emergence, greyish and light brown and have pinkish hind tibiae; upon the attainment of maturation, which is signalled by attempts to mate with mature females, the color of the abdomen changes to yellow, while the hind tibiae become bright yellow (Fig. 69). The maturation process just described is apparently accelerated by the presence of sexually mature older males in crowded conditions (Norris, 1952, 1954). Accelerated maturation is noticeable even when only two males are kept together, particularly if one of them is older (Norris, 1960).

Close observations of males revealed that mature males liberate a volatile substance which seems to be perceived by young males olfactorily as well as by physical contact (Norris, 1954; Loher, 1960). The perceiving male typically vibrates his antennae, palpi, and hind femora (Loher, 1960). Loher demonstrated then that the active principle, obviously given off only by mature males, is secreted by the hypodermis. It could be extracted and stored in diethyl ether, chloroform, benzene, and other fat solvents. It promoted maturation in young males even when it could only be perceived olfactorily.

The biological properties of the substance obtained from mature males fits the definition of a pheromone, a chemical releaser and promoter of behavioral and developmental processes which is transmitted from one to another animal of the same species (Karlson and Butenandt, 1959). Its biological function could best be described as a coordinator facilitating synchronized maturation of the males, which is of obvious advantage in breeding swarms of this locust.

From these observations arises the question of what controls secretion of the pheromone. Loher found in 1960 that sexual maturation did not occur after allatectomy. Such operated males practically never developed the characteristic coloration of sexually mature males, exhibited no sexual behavior, i.e. did not vibrate their antennae, palpi, or femora upon perceiving other males, nor did they mate. Allatectomy in mature males resulted in the loss of mature coloration and cessation of mating attempts. Reimplantation of active corpora allata restored the yellow mature coloration, and the males again mated readily upon meeting a receptive female. These series of experiments for the first time showed that an endocrine gland promoted maturation in an adult male insect species. Corpora allata from either a male or a female restored sexual activity in the recipient male (Pener, 1967). Even though the precise mode of control of maturation processes is not known in *Schistocerca*, one can

FIG. 69. The effect of allatectomy on coloration (indicating maturation) in males of *Schistocerca gregaria*. Control left side, operated right. (Courtesy of W. Loher.)

postulate that the sensory perception of the male pheromone by the young male is somehow processed by the central nervous system, which in turn activates the corpora allata. The corpus allatum hormone then causes, directly or indirectly, changes in the integumental coloration, promotes pheromone production, and makes the male receptive to mature females. This conclusion is further supported by the observation that in mature males the corpora allata show signs of activity, such as increased size and low nucleo-cytoplasmic ratio (Loher, 1960; Odhiambo, 1966b).

In males of *Locusta migratoria*, sexual behavior is abolished after allatectomy, as well as after cauterization of the neurosecretory cells in the pars intercerebralis. After reimplantation of corpora allata or brain tissues from the pars intercerebralis, sexual behavior resumed (Girardie and Vogel, 1966). These results were interpreted to the effect that normally neurosecretion from the pars intercerebralis stimulates the corpora allata which in turn affect the sexual behavior of the animal. Experiments involving brain surgery in combination with allatectomy and reimplantation of the corpora allata were not reported; therefore, the conclusions drawn may be subject to reinterpretation. As in *Schistocerca* and *Locusta*, sexual maturation of the male is not achieved after allatectomy in *Nomadacris septemfasciata*; the males do not mate and only resume sexual activity upon implantation of active corpora allata (Pener, 1968). It is peculiar that reports on endocrine control of sexual maturation processes in males are confined to migratory locusts. Is this the result of a selection of these species because of their availability, or do analogous mechanisms indeed not exist in species of other orders?

In insects, spermatogenesis does not seem to be under the endocrine control of either the pars intercerebralis or the corpora allata. This has been shown, for example, in *Rhodnius prolixus* (Wigglesworth, 1936), *Locusta* (Girardie and Vogel, 1966), *Schistocerca* (Cantacuzène, 1967), and *Scatophaga stercoraria* (Foster, 1967). Since in most species spermatogenesis does not occur in the adult, it is, however, possible that the hormone from the prothoracic glands may stimulate sperm production in the larvae and pupae. Results of experiments dealing with this possibility in several species are somewhat inconclusive. On the other hand, it has definitely been shown in *Drosophila* that complete spermatogenesis can occur in the absence of prothoracic glands (Garcia-Bellido, 1964a, b). Similarly, sperm differentiation was observed in the adults of the long-lived cockroach *Leucophaea* (Engelmann, unpublished).

The accessory sex glands of the male cockroach *Leucophaea* are active and secrete the materials used in spermatophore production throughout the life of both normal and allatectomized animals (Scharrer, 1946; Engelmann, unpublished). This has also been reported for many other species studied. However, in *Rhodnius* the accessory sex glands of allatectomized males are attenuated and show no signs of activity (Wigglesworth, 1936). The activity of the homologous glands in *Schistocerca* and *Locusta* appears to be likewise under the control of the corpora allata since they are small in operated males (Loher, 1960; Girardie and Vogel, 1966). The possibility remains, however, that there is no direct endocrine control of secretory activity even in these species and that the lack of activity is a secondary phenomenon following the ablation of the corpora allata. Possibly the metabolism of the animals is affected by the endocrine glands just as in the females, i.e. in their absence, protein metabolism may be lowered. This in turn may not allow the necessary high metabolic activity of the accessory sex glands simply because the starting material had not been made available. It should be pointed out here that the accessory sex glands in *Schistocerca* are not completely inactive in allatectomized males (Odhiambo, 1966b; Cantacuzène, 1967).

Electronmicrographs demonstrate that largely smooth surface endoplasmic reticulum is found in the cells of these glands after allatectomy, whereas in cells of normal animals there is abundant rough surface endoplasmic reticulum (Odhiambo, 1966a). This difference in ultrastructure possibly denotes a corpus-allatum-dependent RNA and protein synthesis in these glands.

As these brief outlines show, our knowledge of corpus allatum involvement in male maturation processes is limited to a few species. Besides the control of behavioral aspects (p. 190), the hormone appears to affect color changes (*Schistocerca*) and activity of the accessory sex glands (*Rhodnius*, *Schistocerca*, *Locusta*). One may perhaps postulate that the demonstrated effects of allatectomy are expressions of a generally reduced metabolism and thus are non-specific. This is certainly implicit in the results of the works of Odhiambo (1966a, b, c). The interpretation of the effects of brain cautery, particularly removal of the pars intercerebralis, must await further experimentation. It would be highly desirable to investigate a variety of long-lived species.

CHAPTER 10

OVIPOSITION

OVIPOSITION behavior in insects is nearly as diverse as the species themselves can be. The behavior is adjusted mostly to specific requirements and living conditions of the larvae. Ovulation and oviposition may not take place simultaneously, i.e. some species store a considerable number of mature eggs in the outer genital ducts until suitable conditions for oviposition arise. In most cases the literature does not, however, distinguish between these two processes. In the following account, emphasis has been placed on the physiology of egg laying and on the factors that influence it.

For example, members of the Phasmida apparently do not exhibit any kind of adaptive oviposition behavior. The females simply drop their eggs, which then fall among the dead leaves below the food plant. Eggs of the phasmids have a hard shell and do not require additional protection for survival. Similar situations have been observed in many species of several orders and are usually related to the species' particular habitat. The other extreme in oviposition behavior is to be found in the digger wasps. One of the earlier and more careful investigations of a digger wasp is that conducted by Baerends on *Ammophila campestris*. This species digs a burrow, provisions it with a paralysed caterpillar onto which it lays one egg, and then closes the burrow (Baerends, 1941). Later, the burrow is inspected and, after the larva has hatched, more prey is provided as needed; several inspections take place. After the larva has begun to pupate, no more prey is provided and the burrow is finally closed. A female may work on several burrows simultaneously and always provisions them as required. A closely related species, *A. adriaansei*, takes care of only one burrow at any one time and provisions this fully in one phase (Adriaanse, 1948). In both these species we thus observe a peculiar and highly efficient oviposition behavior which has been coupled with a kind of parental care.

Several groups of insect species, particularly those which are of economic or medical importance, have been studied in great detail with regard to their oviposition requirements. Grasshoppers and locusts, for example, generally deposit their egg pods into loose and moist soil; some such species are *Orthacanthacris aegyptia* (Grassé, 1922), *Schistocerca gregaria* (Bodenheimer, 1929; Popov, 1954), and *Locusta migratoria* (Kennedy, 1949). *Locusta* and *Nomadacris septemfasciata* (Woodrow, 1965b) may probe in dry or sodden soil but then will abandon the site without depositing any eggs. Relatively high salinity (0.05 M NaCl or higher) causes the females of *Locusta* not to lay their eggs (Woodrow, 1965a). Chemosensors seem to be located on the tip of the abdomen. In various species there may be some differences in the preference for certain soils. The two species of *Chorthippus parallelus* and *C. brunneus* were given a choice between moist-dry, loose-compact, and coarse-fine sand (Choudhuri, 1958); under these situations, *C. parallelus* preferred moist, loose, and coarse sand, whereas *C. brunneus* oviposited into relatively dry, compact, and fine sand. Similar variations with respect to oviposition preferences probably exist in nearly all species. Not only in grasshoppers and crickets do we find a preference for certain soil conditions but also in other insect species that oviposit into the soil as, for example, the June beetle, *Phyllophaga lanceolata* (Travis, 1939).

In another group of injurious insects, the mosquitoes, considerable details are known about the oviposition sites (see Clements, 1963). Several factors, such as salinity, pollution, reflectance, and texture of the surface, influence a species. Generally, waters of above 0.5–1.0 per cent NaCl are avoided by *Culex fatigans* and *Aedes aegypti* (Woodhill, 1941; Kennedy, 1942). Salt receptors have been found on the tarsi in several of the species studied, i.e. the animals sense the salinity when they settle on the water surface (Wallis, 1954; Hudson, 1956). *C. pipiens* and *A. aegypti* can discriminate between concentration differences of 0.02 M NaCl (Hudson, 1956). Certain populations of *A. aegypti* apparently do not settle on the water surface for oviposition but rather hover over the water; hygroreceptors do not seem to be located on the antennae (Bar-Zeev, 1967). *C. fatigans* prefers waters which contain organic materials to those which are clean (Zelueta, 1950), as does *Stegomyia fasciata* (Fielding, 1919). *C. tarsalis* oviposits preferably into waters which contain exuviae or pupae of its own species (Hudson and McLintock, 1967). The latter species lays less frequently into those waters in which other species live. Many more details on oviposition sites of mosquitoes are known, but in the present context, the few examples given may suffice to illustrate the situation.

Manduca sexta, the tobacco hornworm, oviposits preferably onto tomato rather than other solanaceous plants (Yamamoto and Fraenkel, 1960; Waldbauer, 1962). The female is attracted to tomato olfactorily, since, after amputation of her antennae, she oviposits indiscriminately. The leaves of the tomato plant contain an ethanol and water-extractable compound which is the attractant (Yamomoto and Fraenkel, 1960). *Leptinotarsa decemlineata*, a phytophagous beetle, normally oviposits on potato, which is the normal food plant of the larvae. However, if another solanaceous plant, *Solanum nigrum*, is available, the females will select this plant over the potato, even though *S. nigrum* is not a food plant (Hsiao and Fraenkel, 1968). The chemical which attracts the beetles has not been identified. Both olfaction and contact chemoreception stimulate *Phormia regina* to oviposit on the proper substrate (Barton Browne, 1960) and, in *Syrphus corollae*, odor from or contact with the aphid host elicits oviposition (Bombosch, 1962; Volk, 1964). Furthermore, depending on the species of the genus *Lucilia*, ammonium carbonate, which decomposes into the attractants CO_2 and NH_3, and also indole and sulphydryl mixtures are attractive and will stimulate oviposition (Cragg, 1956; Barton Browne, 1965). In the species of *Lucilia*, the texture of the sheep wool and the warmth of the sheep may act as additional stimuli.

As these few examples demonstrate, oviposition behavior may depend on a complex array of information that the species obtains from its environment. This may apply even more so in certain species of Odonata. Species of this order may simply drop their eggs into the water during flight, or deposit them into submerged plants, even going under water for up to 30 min for oviposition (see Corbet, 1963a). Species of the genus *Lestes* and *Sympecma* (Loible, 1958), as well as others, oviposit while in tandem with the males. The male of the species apparently guides the female to the oviposition site. This behavior is of some interest to evolutionary biologists since, within the order of Odonata, one can, in effect, observe the evolution of oviposition behavior from the more primitive to the complex. One may even find the evolutionary trend exhibited within one species, namely *Platycnemis pennipes*. In southern France, a population of this species was found in which the females oviposit singly, whereas females of other locations always oviposit while in tandem with the male (Heymer, 1967). The single oviposition behavior may be exhibited by the more advanced species. An interesting situation is found in species of the genus *Calopteryx* which is regarded as more highly evolved among the Odonata. Generally these species oviposit singly, but the females usually stay within the male's territory for oviposition; the male may chase other males while the female is ovipositing, as if he were guarding her (Buchholtz, 1951; Johnson, 1962a). Another interesting aspect of the oviposition behavior of Odonata should be mentioned. Among this group of insects, one finds species which oviposit either by simply dropping their eggs into the water or thrusting their ovipositor into sand under water (exophytic) or by depositing their eggs into plants (endophytic). Exophytic and endophytic oviposition are associated with structural differences of the ovipositor in the various species. Many of the exophytic species have only rudimentary ovipositors.

The oviposition behavior of insects can be peculiarly suited to the conditions of the species habitat or the substrate that they oviposit in. For example, species of the tettigoniid *Scudderia* insert their flat eggs into leaves of various plants in an interesting manner. The female first chews away the edge of the leaf and then parts the upper and lower surface of the leaf with her ovipositor (Fig. 70). This peculiar mode of oviposition certainly protects the eggs from desiccation, since the leaves stay fresh throughout the embryonic development of the species. Another curiosity found in the dipteran parasite *Villa quinquefasciata* may be worth mentioning, since it is a case where a parasite oviposits independently of the presence of the host (Biliotti *et al.*, 1965). The female of this species possesses a large ventral abdominal pouch which she fills with fine sand prior to oviposition. Then, while in flight, the female shoots her eggs—together with the sand particles—underneath stones, leaves, or debri; this behavior continues until there is no more sand in the pouch. The animal then refills her pouch and oviposition continues. What the biological significance of this rather odd behavior is can only be speculated upon. Certainly, any desired number of oddities could be assembled in this context since insects are so very diverse in their oviposition behavior.

A considerable number of reports are concerned with the effects of mating on egg deposition. As has been most often observed, the rate of oviposition is stimulated by mating; whether this is the result of a stimulated rate of egg maturation or whether it is oviposition *per se* which has been enhanced is often not possible to decide (p. 129). Apparently in the majority of the short-lived Lepidoptera, notably noctuid moths, mature eggs are produced during the later stage of the pupal life; thus, often a full complement of eggs is available on the day of emergence. Virgins of these species generally retain most of their eggs until shortly before death and then, if not mated, "reluctantly" lay them, as has been reported for *Cynthia*, *Cecropia*, *Lunas*, and *Promethea* silkmoths (Rau and Rau, 1914), *Porthetria dispar* (Klatt, 1913), *Trochilium apiforme* (Schulze, 1926), several other noctuids (Eidmann, 1931), *Bombyx mori* (Mokia, 1941), *Mamestra brassicae* (Bonnemaison, 1961), and many others. In all of these cases mating is followed by a burst in egg laying, which indicates that the eggs had simply been stored. Similarly, just as in Lepidoptera, one finds in Diptera that virgins often retain most of their mature eggs. As examples, the following can be cited: *Drosophila ampelophila* (Guyenot, 1913), *Musca domestica* (Glaser, 1923; Hampton, 1952), *Lucilia cuprina* (Mackerras, 1933), *Chrysomyia rufifacies* (Roy and Siddons, 1939), *D. melanogaster* (Chiang and Hodson, 1950), *Strumeta tryani* (Barton Browne, 1956a), *Culex pipiens* (Déduit, 1957a), and *Aedes aegypti* (Lang, 1956; MacDonald, 1956). Interestingly, in *A. aegypti* Gillett (1956) found a strain (Lagos) in which the virgin females lay most of their eggs, whereas in another strain (Nevala) the virgins do not lay. Gillett postulates the presence

FIG. 70. Oviposition of the tettigoniid *Scudderia* spec. The animal first chews away the edge of the leaf, then parts the leaf, and inserts its ovipositor to deposit one egg.

or the liberation of an ovulation hormone even in virgins in the first case, which does not occur in the second unless the females have mated. Also, virgin females of *Cochliomyia hominivorax* laid the same number of eggs as mated females did, except that oviposition was somewhat delayed (Crystal and Meyners, 1965).

The phenomenon of egg retention, as found in virgin females of Lepidoptera and Diptera, can also be observed in other insect species, such as the Lampyridae *Lampyris noctiluca* and *Phausis splendidula* (Schwalb, 1961). However, in many other species mating stimulates both egg maturation and egg laying. For example, the grasshopper *Gomphocerus rufus* (Loher and Huber, 1964) and the locust *Locusta migratoria* (Mika, 1959) lay nearly as many egg pods whether mated or not, although virgins lay in somewhat delayed and irregular cycles.

Mating is important in the stimulation of oviposition but it is also important in the inducement of the regular oviposition behavior. A few examples will be given to illustrate this. Virgins of *Porthetria dispar* lay irregularly scattered masses of eggs which are not covered with the glandular secretion and scales (so-called "Afterwolle") that are observed in the mated females (Schedle, 1936). Virgins of *Coccinella undecimpunctata aegyptica* scatter their eggs irregularly, whereas mated females lay theirs in an orderly fashion (Ibrahim, 1955). Or, *Anthonomus grandis* females, which normally deposit their eggs into a cotton square, if not mated, lay outside the square which causes the eggs to dry out in a short while (Mayer and Brazzel, 1963). Some of the virgin ovoviviparous cockroaches build irregular and incomplete egg cases which they may carry for variable amounts of time (Roth and Willis, 1952b, 1961; Sengel and Bullière, 1966). Often, virgin cockroaches drop single eggs, scattering them around, and do not build a protective ootheca around them. These eggs dry up in a few days. To sum up this brief account, mating in many species affects both oviposition rate and behavior. Eggs deposited by a virgin are often not viable because they have not been protected from environmental influences.

The question naturally arises as to what mechanisms are involved in the mating influenced change in oviposition behavior. In the bisexual strain of the cockroach *Pycnoscelus surinamensis*, Stay and Gelperin (1966) showed that the filling of the spermatheca with sperm caused the female to adopt the mated behavior. After removal of the spermatheca, severance of the spermathecal innervation, or mating with castrated males, the females behaved like virgins, i.e. they did not produce normal egg cases and single eggs were scattered around. These results could be confirmed in another cockroach, *Leucophaea maderae* (Engelmann, unpublished). Analogously, in the gypsy moth *Porthetria* extirpation of the spermatheca or its ligation caused a female to behave like a virgin (Behrenz, 1952). These results would indicate that the filling of the spermatheca is recorded by the nervous system, thus assuring the coordinated movements of the muscles involved in oviposition. The influence of mating may, however, be separated into two components, one of which causes release of the stored eggs and the other of which influences the orderly deposition of them. This is suggested by observations on *Porthetria* in which females that had been mated to castrated males (Klatt, 1920) deposited all of their eggs but in the disorderly fashion that virgins would.

In *Locusta migratoria manilensis* the entire process of oviposition appears to be controlled by the last abdominal ganglion since no oviposition took place after its removal whereas cutting the ventral nerve cord anteriorly did not affect oviposition behavior (Quo, 1959). However, the control of oviposition may be different in different species, as may be illustrated by *Polistes gallicus* which generally does not oviposit after ventral nerve cord severance (Gervet, 1964). Also, other nervous centers may be superimposed on the last abdominal ganglion, as is suggested by the observation that decapitation often induces oviposition. This has been reported, for instance, in *Mantis religiosa* (Chopard, 1914), *Aedes aegypti* (De Coursey and Webster, 1952), *Culex tarsalis* (Curtin and Jones, 1961), several crane fly species (Chiang and Kim, 1962), and *Plodia interpunctella* (Morère and LeBerre, 1967), as well as other species. Decapitation did not induce oviposition in *Anopheles quadrimaculata* (Curtin and Jones, 1961). In a variety of species the effect of decapitation on oviposition may be analogous to that of decapitation of a male *Mantis*. Here, decapitation is followed by an increased rate of copulatory movements (Roeder, 1935); in this case, an inhibitory center in the suboesophageal ganglion for certain nervous activity was removed by decapitation.

Nervous control of oviposition most likely is only part of the entire control mechanism which governs egg laying in any given species. Possibly the actions of humoral or endocrine agents are superimposed. This is suggested by the following reports. In *Drosophila melanogaster*, implantation of paragonia (male accessory sex glands) into virgins stimulated egg laying (Kummer, 1960; Garcia-Bellido, 1964c; Leahy, 1966), as did mating with sterile males which had functional paragonia (David, 1963). It appears that the paragonia of *Drosophila* and the accessory glands of *Aedes aegypti* (Leahy and Craig, 1965) contain a principle that causes oviposition. Extracts of males of *Drosophila* injected into virgins stimulated egg laying (Leahy and Lowe, 1967), as did reciprocal implantation of the glands in *Drosophila* and *Aedes* (Leahy, 1967). However, implantation of vesiculae seminales or fat body in *Drosophila* (Garcia-Bellido, 1964c) or testes in *Musca domestica* (Hampton, 1952) did not stimulate egg laying. The agent from the male accessory glands appears to affect mainly the release of the mature eggs and not egg maturation.

Other hormonal principles are possibly involved in other species. For example, in *Bombyx mori*, injections of blood from females or males or of a 2 per cent saline solution stimulated oviposition (Mokia, 1941). Does perhaps the hypertonic saline solution cause the liberation of some ovulation hormone? *In vitro*, brain extracts increased the motility of the oviducts in *Carausius morosus*, an observation which may be related to normal oviposition behavior (Enders, 1956). In the plant bug, *Iphita limbata*, blood from ovulating females, implantation of the neurosecretory cells, and extracts from fully grown but unlaid eggs induced ovulation and oviposition in females which had not been quite ready to ovulate (Nayar, 1958a, b). The results of a combination of these experiments in conjunction with histological observations led Nayar to postulate the following sequence of hormonal actions during ovulation. First, the fully grown eggs liberate an ovarian hormone which then causes the neurosecretory cells to release a principle; this, in turn, stimulates the motility of the oviduct, thus inducing ovulation. According to Nayar, females which were not yet ready to ovulate and which still had submature-sized eggs could be stimulated to oviposit. Since the time of this report, no analogous mechanism in other species of insects has been found.

The study of the endogenous control mechanisms leading to oviposition behavior poses many fascinating problems which invite analysis. In this context are mentioned a few examples which show that females with mature ova change their behavior. Females of *Gomphocerus rufus* that are ready to oviposit will not accept a courting male (Renner, 1952). Or, *Pieris brassicae* females become attracted to green, blue-green, or blue colors when they carry mature eggs; when they find such colors, they exhibit the typical preoviposition behavior, drumming the leaf with their legs (Ilse, 1937). *Pimpla ruficollis*, a parasite of *Rhyacionia buoliana*, becomes attracted to the food plant of the host (*Pinus silvestris*) only when ready to oviposit (Thorpe and Caudle, 1938); the young *Pimpla* female is repelled by the pine oil. The question in all of these cases is whether the endocrine activity of the animals causes the changed behavior or whether this change is brought about by the presence of mature ova. Furthermore, it is not clear whether the mature ova furnish a hormone causing these behavioral changes or whether it is merely the filling of the abdomen which induces this change.

Oviposition behavior often consists of a sequence of stimulus-response interactions. This is found particularly often in the more "sophisticated" cases, such as grasshoppers, mosquitoes, or parasites. In all of these instances, the animal receives chemical and tactile stimuli via antennae, tarsi, and ovipositor which make her "decide" to lay or not to lay the eggs. The parasite *Mormoniella vitripennis* illustrates this particularly well (Edwards, 1954). First the female is attracted by the odor of the host or the area in which the host lives, then

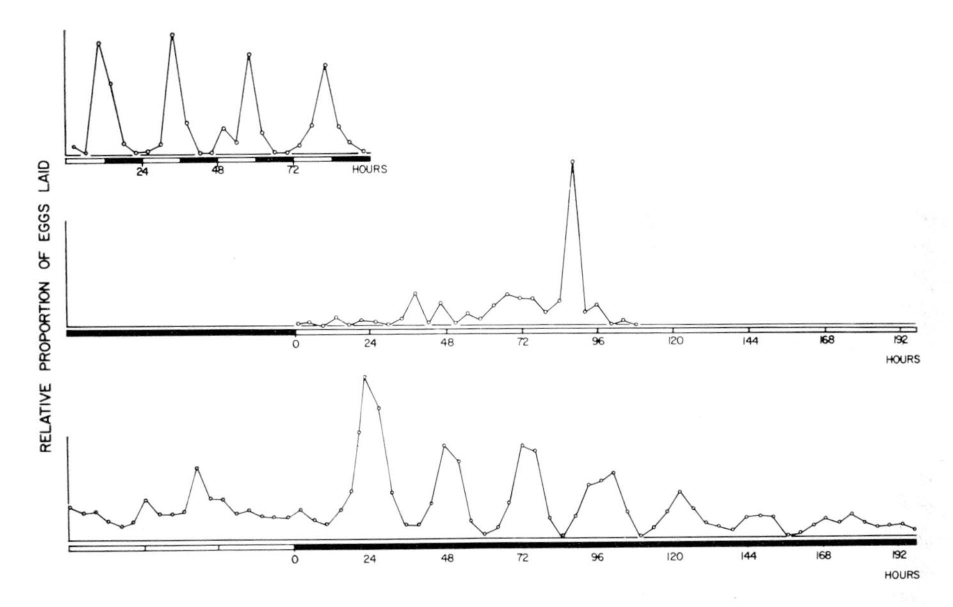

FIG. 71. Oviposition cycles of *Aedes aegypti* under a light–dark cycle of 12:12 hr (top graph). At sunset the majority of animals laid each day. In constant light no cyclicity is observed. An abrupt change from light to costant dark induced a cyclic egg laying (bottom graph). A change from dark to light did not induce cyclicity (middle graph). (Modified from Gillett *et al.*, 1959 and Haddow *et al.*, 1961.)

she recognizes the shape of the host; together these stimuli cause her to drill. If the host is alive—which is realized after insertion of the ovipositor—the number of the eggs deposited can vary according to the size of the host. Quite similar sequences of events can be established for many species; this short description may suffice here.

Entomologists and physiologists are attracted by yet another aspect of insect oviposition behavior—the often rhythmic circadian peak of egg laying. For example, many noctuids, such as *Porthetria*, oviposit at night (Klatt, 1920). Circadian oviposition behavior also appears in two moths which are injurious to European vineyards, *Polychrosis botrana* and *Clysia ambiguella* (Götz, 1941b). Götz obtained a continuous record of oviposition for these two species by simply positioning the females above a rotating sticky board onto which the eggs fell and then stuck to the glue. As seen in Fig. 37, *Polychrosis* oviposits during late afternoon and early night, as well as during early morning; *Clysia*, on the other hand, oviposits throughout the afternoon into the early night. Oviposition time as well as mating time (Fig. 37) is species specific. This also applies to the several species of mosquitoes for which records are available, i.e. *Aedes aegypti* oviposits in late afternoon (Haddow and Gillett, 1957), *Taeniorhynchus fuscopennatus* during early night (Haddow and Gillett, 1958), and *A. apicoargentus* around midday (Haddow *et al.*, 1960). Under constant light or constant dark conditions, *A. aegypti* did not exhibit any cyclic oviposition behavior; the animals must, therefore, receive some clue from the environment as to when to oviposit (Gillett *et al.*, 1959). In an attempt to find the environmental cue that sets the rhythm, Haddow *et al.* (1961) showed that a light stimulus of only 5 min during constant dark could establish an oviposition rhythm. Continuous light following a period of constant dark, however, did not induce rhythmic oviposition. Only if the continuous light was followed by constant dark was the circadian egg laying established; this periodic oviposition behavior lasted for about 4 days. From this, it appears that it is the shutting off of light which is the cue used by the animals (Fig. 71).

Parasitic Hymenoptera may not oviposit if only already parasitized hosts are available, as, for example, has been shown in *Nemeritis evanescens* (Williams, 1951). In this case the mature eggs may be resorbed. If no hosts were offered to females of several species of pteromalid Hymenoptera, the eggs were not laid but rather were resorbed (Flanders, 1939, 1950). In these cases the females could certainly make use of the materials freed during resorption of the eggs. Additional eggs could be matured which could then be laid when favorable conditions arose. One might, of course, also argue that simple storage of the eggs is more economical than the activation of the metabolic machinery involved in oosorption. In this connection, one can again counter with the hypothesis that mature eggs may have a limited viability period and, therefore, their resorption indeed "serves" the animal. Parasitic Hymenoptera may serve as an example of the situation in which availability of a suitable substrate for oviposition determines whether or not eggs are laid. Females of many other species are known to drop their eggs even under unfavourable conditions, but they usually have retained them for some time.

CHAPTER 11

HETEROGONY

THE phenomenon of heterogony, i.e. the alternation of parthenogenetic and bisexual reproduction in a given species, has been reported in species of the Aphidoidea, Cynipidae, and Cecidomyidae (see Lampel, 1962). An additional case, the beetle *Micromalthus debilis*, probably does not fall under this heading since only parthenogenetic reproduction is known for certain in this species. Although sexuals of both sexes have been reported, these sexuals may be nonfunctional. In heterogonic species, from one to many parthenogenetic generations in succession may occur, but, as far as is known, there is only one bisexual generation in any cycle. Heterogony may be obligatory, i.e. intrinsically controlled, or it may be facultative; in the latter case environmental factors, such as day length, temperature, or crowding, influence the appearance of either sexual or parthenogenetic individuals. Mechanisms for control of heterogony are incompletely understood in most of the species. In the following account, emphasis will be placed on those few cases for which we have some knowledge of the controls.

A. Aphidoidea

1. THE PHENOMENON

It has been known for many years that certain aphid species may occur in several morphs which are physiologically and morphologically distinct. Alate or apterous parthenogenetic viviparae and alate or apterous sexual females and males are known. The interest in the different reproductive forms, which often exhibit polymorphism, is reflected by the great number of publications on this topic. Many of the life cycles of the various species are described in reviews by Weber (1930), Marchal (1933), and Mordvilko (1935). The extensive information was then again reviewed by Bonnemaison (1951), Appel *et al.* (1957), Bodenheimer and Swirski (1957), Kennedy and Stroyan (1959), Lees (1966), Hille Ris Lambers (1966), and Lampel (1968). Through the many years of research on the different species, the terminology has become increasingly complex. There is a certain amount of confusion, since some of the same terms have been applied to different morphs, and some of these terms describe morphological characteristics, whereas others physiological or developmental stages within the life cycle. In the present context, the terms used are those of Lampel (1968), who tries to simplify and to bring order into the confusion. For a detailed description of the known species, the reader is referred to Lampel's recent book. Lampel (1965a, b, 1968) proposed a simplified terminology which does accommodate all forms on both primary and secondary hosts (Fig. 72). According to him all animals on the primary host have the prefix civis, those on the secondary host exsulis. One can thus describe a civis-fundatrix, civis-virgo, etc., as well as the exsulis-fundatrix, exsulis-virgo, etc. Whether this terminology will be successfully used must still be shown, but it certainly is an attempt to eliminate the often complicated terms. In the following account a few examples of heterogonic life cycles will be given, proceeding from the simple to the rather complex types. The various terms will be used as they become necessary to describe a particular morph.

A very simple case of heterogony is exhibited by *Acanthochermes quercus*, which lives on *Quercus petraea* or *Q. robur* in Italy. The fundatrix (arising from a fertilized egg) appears in April or early May; she parthenogenetically lays eggs which become either males or females (the sexuales). These females lay the fertilized eggs in May from which the fundatrix will hatch the following April (Fig. 73a). In this species, parthenogenesis strictly alternates with bisexual reproduction. Here, the fundatrix is at the same time the sexupara, known as a separate generation in most aphid species. The reproductive period in this species of aphid is

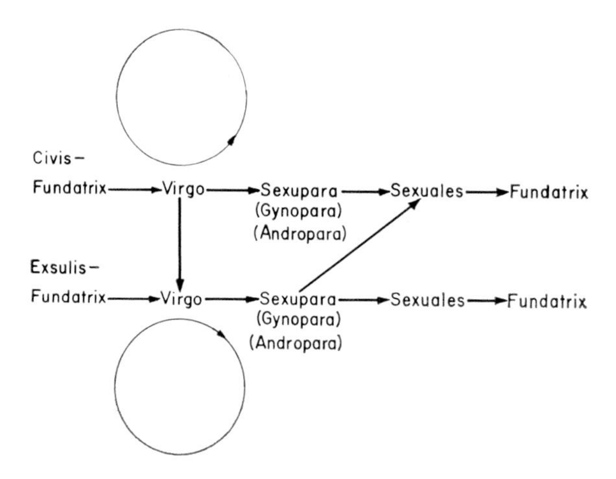

FIG. 72. Diagrammatic representation of the life cycles and succession of generations in aphid species. Individuals of the civis generations live on the primary host, those of the exsulis on the secondary host. (Adapted from Lampel, 1968.)

restricted to only 2 months, but the egg lasts for the rest of the annual cycle. A primitive heterogonic cycle has also been found in *Mindarus abietinus*, in which species a separate sexupara generation gives rise to the sexuales (Fig. 73b). As in *Acanthochermes*, the reproductive period is condensed into a short portion of the year, from April to June. Further complications in the life cycle may be seen in the European monoecious species *Dysaphis devecta* which has four generations: fundatrix, virgo, sexupara, and sexualis in succession (Fig. 73c). It appears that the sequence of generations is as obligatory in this species as it is in those mentioned above: there is no duplication of any of the generations. Eggs which overwinter are laid in June or July by the females of the sexualis generation. The virgo (fundatrigenia of some authors) can be either apterous or alate, whereas the male sexualis is alate, the female sexualis is apterous. It ought to be mentioned here that the sexupara of this species is either an andropara or gynopara: any given animal gives rise either to the male or to the female sexualis but never to both. This is in contrast to *Acanthochermes* and many other species in which the sexupara produces both males and females.

The heterogonic life cycle of many species, particularly those which migrate to secondary hosts (heteroecy) and then remigrate to the primary host either in the fall or the following

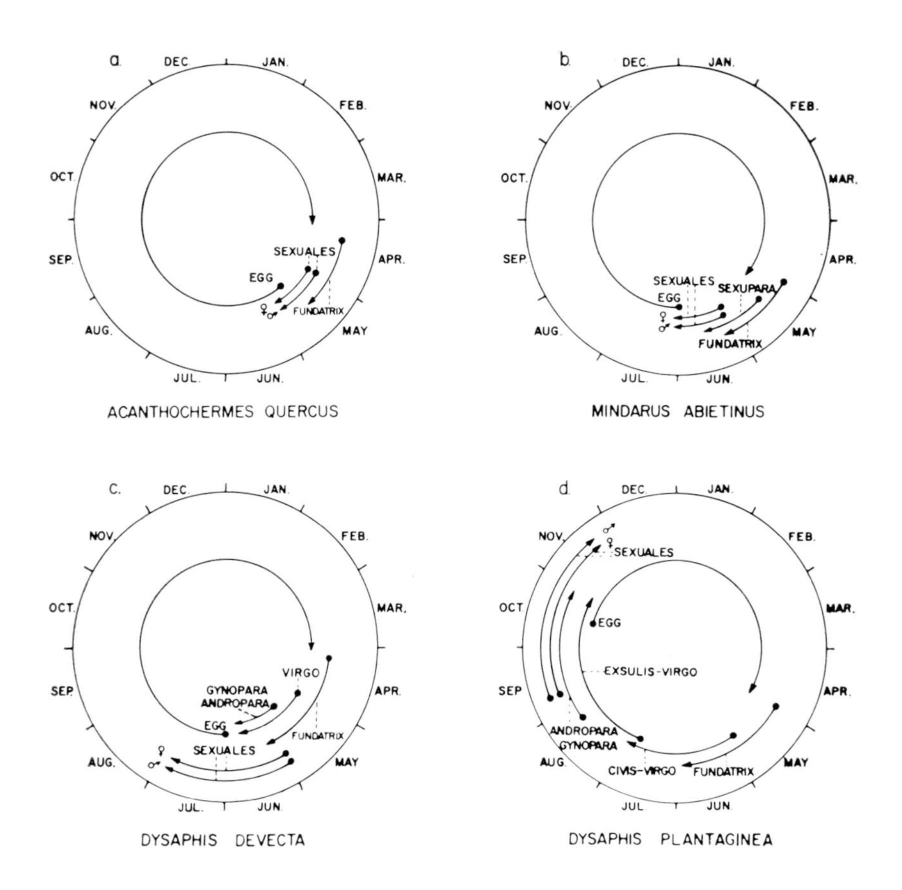

FIG. 73. Illustration of succession of generations in four species of aphids. Examples of increasing complexity are given. (Adapted from Lampel, 1968.)

year, can become very complex. *Dysaphis plantaginea* may serve as an example of a heteroecious species in which the entire life cycle is completed in 1 year (Fig. 73d). Several virgo generations occur on both the primary and the secondary hosts. *Sacchiphantes viridis* migrate to and from the primary host in two successive years (Fig. 74). Individuals of some species, in which the different generations live on different hosts, living habits can be rather diverse. This is shown, for example, by *Pemphigus betae*, the one virgo lives on beet roots, while the sexuales, fundatrix, and virgo of the following year (fundatrigenia) live on poplar trees (Parker, 1914; Harper, 1959), causing galls on the leaf petioles. Species of the oviparous Phylloxeridae and Chermesidae have the most complex life cycles among the Aphidoidea. Detailed descriptions have been given by Börner (1909), Annand (1928), and Steffan (1961, 1962a) among others. For example, in *Sacchiphantes viridis* (Steffan, 1961) the fundatrix (living on spruce) produces the virgo (allata migrans) which migrates to larch and produces the exsulis-virgo (hiemosistens) which overwinters. Several generations of exsulis-virgo (progedientes) may occur until the exsulis-sexupara arise; these migrate back to spruce giving rise to sexuales which lay the eggs yielding the civis-fundatrix. Quite similar are the heterogonic cycles of other Chermesidae such as *Adelges cooleyi* (Cumming, 1959) or *A. lariciatus* (Cumming, 1968).

In all the species so far mentioned, whether monoecious or heteroecious, a definite succession of one to many parthenogenetic and sexual generations occurs. These species are the so-called holocyclic. Among the Aphidoidea, several species exist which do not have a sexual generation; they reproduce exclusively by parthenogenesis and thus have only virgo generations. This is, for example, known in *Dysaphis tulipae*, *Sacchiphantes abietis*, and *S. segregis* (Steffan, 1961). These species are considered anholocyclic species which presumably have secondarily lost their sexual generation. In the strict sense, a discussion concerning the reproductive physiology of these species does not belong here; yet in some cases their existence points to important evolutionary aspects which will be dealt with under the pertinent heading below.

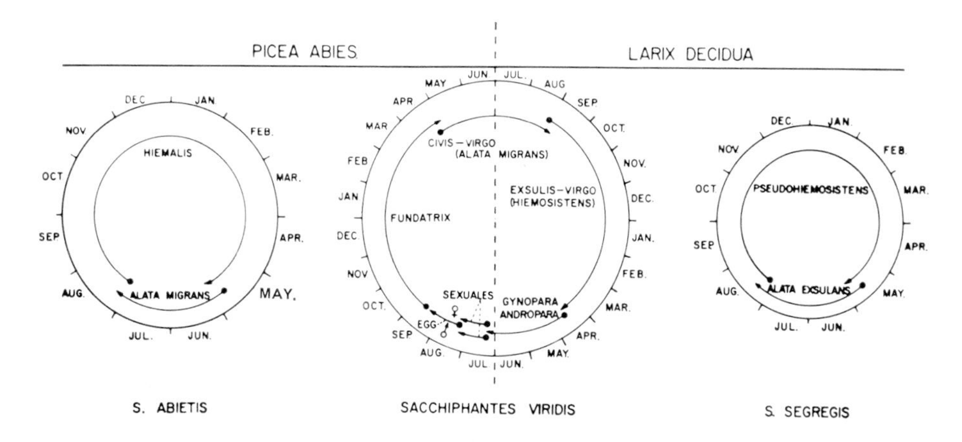

FIG. 74. Successions of generations in the biannual heteroecious *Sacchiphantes viridis* and the two anholocyclic derived species *S. abietis* and *S. segregis*. (Based on information given by Steffan.)

2. CONTROL OF HETEROGONY BY PHOTOPERIOD AND TEMPERATURE

The very observation that the number of virgo generations of a given species may be variable leads one to suspect that some epigenetic factors control the appearance of the sexuparae and sexuales. Photoperiodic changes throughout the year are the most consistent environmental changes and thus are used by many animals as a cue to adjust to "anticipated" bad climatic conditions. In many aphid species, the animals overwinter in the egg stage, particularly in the temperate and northern zones where, also, are found the most species with heterogonic life cycles.

Normally, males and oviparous females (sexuales) of *Aphis forbesi* appear in November in the climate of Tennessee (U.S.A.). When animals of this species were exposed to long days, no sexuales appeared even by December, but when short days ($7\frac{1}{2}$ hr of light) were given in February, ovipara and males were noticed by May already, instead of November (Marcovitch, 1923, 1924). Marcovitch found that basically the same results could be obtained in *Dysaphis plantaginea* (*Aphis Sorbi*), *Aphis fabae*, and *Capitophorus hippophaes*. These findings were confirmed for *Aphis fabae* (Davidson, 1929; Fluiter, 1950) and *Dysaphis plantaginea* (Bonnemaison, 1958), and the same principle could be extended to several other species, such as *Macrosiphum euphorbiae* (*M. solanifolii*) (Shull, 1928, 1929), *Aphis chloris* (Wilson, 1938), *Brevicoryne brassicae* and *Myzus persicae* (Bonnemaison, 1951), *Acyrthosiphum pisum* (Kenten, 1955), and *Megoura viciae* (Lees, 1959, 1960a, b). Under favorable conditions some of these species can be reared solely parthenogenetically. For instance, *Aphis fabae* were cultured under long days (16–18 hr of light) and 18–25°C for 248 generations without the appearance of sexual individuals (Müller, 1954).

During the investigations on some of these species, a second parameter appeared to be important in morph determination: temperature. It was found that high temperature may cancel the effect of short photoperiod, i.e. no sexuales appear. *Aphis fabae*, even when kept under only 8 hr of light on a 24-hr rhythm, did not produce males and oviparae if the temperature was above 19°C (Fluiter, 1950). Similarly in *Dysaphis plantaginea* (=*Sappaphis plantaginea*) was the short-day effect canceled when this species was reared at elevated temperatures (Bonnemaison, 1958, 1966a, b). In *Megoura viciae* a day length of less than 14 hr 55 min at 15°C results in the production of sexuales. Lees (1963) found that this critical photoperiod can be shortened by 15 min per 5°C rise in temperature. However, above 23°C no ovipara are produced and all offspring are virgines (virginoparae) or males.

The mechanisms for the determination of sex and of the female morphs, virgo or ovipara, are presumably different. Based on cytological studies in *Byrsocrypta ulmi* (=*Tetraneura ulmi*) (Schwartz, 1932) and in other aphid species (White, 1954), it is thought that sex determination in aphids occurs via the XO–XX-system, the male being XO. Recent studies on several Adelgidae have shown that such a system also applied in this group of Aphidoidea. In Adelgidae andropara and gynopara are cytologically different; in andropara, one of the large X-chromosomes is fragmented whereas in the gynopara the X-chromosome is not fragmented (Steffan, 1968a, b). Rearing of the individuals revealed that all animals of one lineage in any generation have either fragmented or non-fragmented X-chromosomes, i.e. a female lays eggs of only one type. Two lines of animals exist. Interestingly, the fragments of the X-chromosome are not lost during meiosis because Adelgidae have a diffuse kinetochore. This important finding indicates that sex determination takes place at the fertilization of the egg laid by the female sexualis. Essentially nothing is known about analogous systems in other species, but it is conceivable that male and female lines occur in other Aphidoidea, too. If this were indeed found to be so, it could explain why various authors have been uncertain about or have denied any influence of day length on male production. In *Megoura*, Lees (1959) has found that males are produced under nearly any photoperiod, provided that the temperature is suitable; on the other hand, von Dehn (1967) has found that for the same species more males are produced under short-day or constant dark conditions than under long-day conditions. Possibly strain differences can account for the difference in results. In some species it has indeed been observed that males appear more frequently under short-day conditions than under long-day photoperiods. This could, however, be due to the fact that more animals of the male line were determined to produce sexupara (andropara) under these conditions, and may not indicate any effect of illumination of sex determination. The matter is, however, not as clear cut as it might seem. For example, the older females of *Acyrthosiphum pisum* produce more males than do the younger ones (Kenten, 1955). Or, short-day conditions result in fewer males at slightly elevated temperatures than at lower temperatures in *Brevicoryne* and *Myzus* (Bonnemaison, 1951). What the detailed mechanisms in these cases are remains obscure. Furthermore, a 1:8 sex ratio in favor of the female observed in *Megoura* even under the most favorable conditions for male production. This sex ratio has to be explained if we are to assume that male and female lines exist as in the

Adelgidae. The major questions still to be answered are: how does day length or temperature cause the animal to change from parthenogenetic reproduction to gonochorism? How does elevated temperature interfere with effect of short illumination? How is the occurrence of a fragmented X-chromosome related to loss of one X in the male? It is conceivable that elevated temperature affects the chromosomal behavior, i.e. either a reductional or a non-reductional division would occur in the oocyte, but how does length of illumination affect chromosomal behavior? Another question is whether illumination affects the gonads directly or whether it alters the neuroendocrine system which in turn controls the production of the different female morphs. For *Megoura*, Lees (1963, 1966) favors this latter possibility.

This brings us to another consideration, namely, the pathway by which day length affects the animals. Obviously, one has to look at the effect of light on the food plant which may thus secondarily influence the species. This aspect has been investigated in a number of species which will be dealt with below. For *Macrosiphum* it has been shown that sexuparae always appear when the animals have been kept under short-day conditions, even though the food plants were under long-day illumination (Shull, 1928, 1929; MacGillivray and Anderson, 1964). No effect of nutrition, i.e. from the senescence of the plants, was noted on the appearance of sexupara in *Brevicoryne* (Bonnemaison, 1951), and Lees (1966, 1967b) argues against any influence of light via the food plant on the virgo of *Megoura*. As will be shown, however, this may not apply to other species. The effect of direct illumination on the virgo has certainly been shown for *Megoura* (Lees, 1961, 1964, 1966). Lees succeeded in exposing rather limited head regions to light by the use of fiber optics. Sixteen hr of light exposure of the head, particularly the midbrain, but not of other body parts, resulted in the production of virgines exclusively. After surgical blinding, the animals still responded to photoperiods; yet, after destruction of the brain, no response to long days was obtained; only ovipara and males were produced. The precise light sensor in *Megoura* is not known, but it could well be located in certain limited brain areas such as the neurosecretory cells of the pars intercerebralis. After the mother's brain had been treated short- or long-day length actually affects those embryos destined to give rise to either ovipara or virgines, it therefore seems likely that some hormone from the brain region determines the type of embryo. Exposure of the embryos themselves is ineffective. No specific hormone so far can be implicated in this process.

The time in the life cycle when the embryos either indirectly or directly perceive the light stimuli is critical for the species. Fluiter (1950) found that exposure of *Aphis fabae* to 8 hr of light daily at temperatures of 12–19°C yielded only ovipara, but at 22°C the same light exposure caused the production of only virgines. This knowledge was then used in the determination of the light sensitive period of this species. Animals which were transferred from 17°C to 22°C shortly after the last molt gave rise to oviparae for about 10 days, but thereafter to virgines. Clearly the first embryos had been determined to become oviparae by the time the mother became an adult, and only later did the higher temperature have an effect on morph determination. This would mean that the first embryos are determined during the mother's larval stages. The time of determination of virgo and ovipara in *Megoura* may be even earlier than that. Exposing the grandparent to long photoperiods before the birth of the mother and then switching to short-day lengths after her birth resulted in the mother first giving birth to virgines and then to oviparae (Lees, 1959, 1963). By the time the mother is born, she already contains a few embryos, which fact can explain the effect of light on the grandmother. If, however, the grandmother was exposed to short days and then the mother to long days, only virgines were produced. This would suggest that there is a mechanism for virgo production which is turned on by long-day length and that it is a system which cannot be turned off suddenly and switched to ovipara production. This same principle does not appear to be operative in *Dysaphis plantaginea*. Virgines of this species, when exposed to one 12-hr scotoperiod, will respond by the production of sexuparae (Bonnemaison, 1966c, 1967). If long periods of continuous light, under which condition only virgines are produced, are interrupted by 12 hr of scotoperiods, sexuparae will appear a definite time interval thereafter. It thus appears that the details of the mode of control of heterogony by day length in various species of aphids may be rather different. At this time, no generalization can be made on the basis of the few known cases.

3. CONTROL THROUGH CROWDING AND NUTRITION

The crowding of a species has frequently been discussed as a factor that can cause the production of sexuparae and sexuales. In some of these cases it is uncertain whether the effect is directly or indirectly caused by the shortage of food under crowded conditions or whether crowding as such is responsible. A further difficulty is encountered in the interpretation of the reported results because it is often unclear whether we are dealing with winged virgines (virginoparae) or winged sexuparae. Crowding probably results in the production of both.

For *Brevicoryne brassicae* it has been shown that crowding facilitates the production of sexuparae (Bonnemaison, 1951, 1958), and the application of short photoperiods to crowded populations of *Dysaphis plantaginea* accelerates even further the appearance of sexuparae (Bonnemaison, 1964a, b). In the latter species when the animals were crowded, even under long-day conditions, 50 to 60 per cent sexuparae were produced. An interesting case has been reported in *Pemphigus bursarius*, which is dioecious, lives on lettuce roots, and produces virgines during the early part of the year. Any virgo can yield either apterous virgines or alate sexuparae; the latter fly to the secondary host. It has been shown that the virgines can stay on the first host plant and even overwinter (Dunn, 1959), provided that the population density is kept low (Judge, 1968). The species is thus monoecious under uncrowded conditions. Development of large colonies on the plant roots usually coincides with the summer periods during which food quality becomes poorer; therefore the production of alate sexuparae which migrate appears to be advantageous for the species. Food plant conditions are, however, not primarily inducive to alate production in this species.

In the induction of sexuparae production, independence from the prevailing photoperiods may more often be found in subterranean species than in others. To exemplify this, another case may be mentioned. In *Eriosoma pyricola*, a species living on the roots of the pear tree, sexuparae production is controlled via the food plant; this can be shown as follows. Normally sexuparae appearance coincides with short photoperiods, conditions which cause the cessation of shoot growth (Sethi and Swenson, 1967). This observation led to the assumption that shoot growth may be causally related to sexupara production. However, regardless whether the animals were fed cut root pieces from quiescent plants or from plants grown under long photoperiod, no sexuparae were produced. This result is difficult to explain unless one assumes that under short-day treatment of the tree some substance, which disappears from the cut shoots after a short time interval, is produced which causes sexupara appearance. One could also argue that an agent produced by long-day-treated plants prevents sexupara production. Be that as it may, the nutrition controls heterogony in this subterranean species. Food factors may be operative in other species, too, yet it is often difficult to make a distinction between a direct effect of light on a species and an indirect effect via the plant, particularly in those species which live above ground. Subjecting the latter species to total darkness introduces, in addition, a factor which the species normally does not encounter, thus the results obtained are perhaps irrelevant.

4. INTRINSIC FACTORS AND "INTERVAL TIMER"

Control mechanisms for heterogony in Aphidoidea, which are based on the species' perception of environmental cues, such as day length, temperature, crowding, or nutritional factors, are not operative in a variety of species. As was pointed out earlier (p. 202), both *Dysaphis devecta* and *Mindarus abietinus* produce sexuparae and sexuales under long-day conditions while food is plentiful; the number of generations is rigidly fixed. Similarly, *A. farinosa*, *Lachniella costata*, and *Mindarus obliquus* complete their life cycles under long-day conditions in Europe (Hille Ris Lambers, 1960, 1966). This same situation applies to *Thuleaphis sedi* (Jacob, 1964). Arctic species complete their life cycles under practically 24 hr of illumination in midsummer, while equatorial species produce sexuales under nearly any prevailing photoperiod. Unfortunately, no experiments have been reported which deal with sexupara production in these cases, thus, because of lack of any contrary evidence, one has to postulate a genetic or intrinsic control of heterogony. It appears fairly safe to state that photoperiodic control of heterogony is more widespread among the species of the temperate zones where photoperiodic changes are indeed perceptible within relatively short time intervals. To be sure, the ability of a species to respond or not to respond to changing day length is genetically controlled in the last analysis. This is perhaps exemplified by the finding that, in *Myzus persicae* and *Brevicoryne brassicae*, clones were established which no longer responded to short photoperiods by producing sexuparae (Bonnemaison, 1958). In this connection, it is of interest that some animals of *Brevicoryne* overwinter as virgines in the Paris region; these may be the individuals from which the above-mentioned clones were derived.

Marcovitch (1924) made the important discovery that fundatrices or individuals of the first generation of *A. forbesi* did not respond to a short photoperiod and that only the summer generations could produce sexuales. This was found to apply also to *A. chloris* (Wilson, 1938), *Brevicoryne* (Bonnemaison, 1951), *Acyrthosiphum* (Kenten, 1955), *Dysaphis plantaginea* (Bonnemaison, 1958), and *Megoura* (Lees, 1960), and can probably be established for many additional species. Wilson (1938) developed the concept of a "time factor" which has to pass before the animals can respond. The "facteur fondatrice" of Bonnemaison (1951) and the "interval timer" of Lees (1960, 1961, 1966) describe the same concept. It appeared to Bonnemaison that this factor is a quantitative entity and is derived from the fundatrix; this factor gradually diminishes or is diluted during successive generations. It may counteract or make the animals unresponsive to any influences which cause the production of sexuparae. One can assume that this factor has evolved as a mechanism to prevent the production of sexuales in spring or early summer while short photoperiods still prevail. A more abundant production of offspring is thus guaranteed. This also means that the capacity to produce sexuales is latent in the virgo; as soon as a number of generations have passed, the animals become competent. In this context, it may be possible to answer an interesting question, namely, is it that a certain time must pass after the birth of the first generation, or is it that a concrete factor must be diluted? For *Megoura*, evidence for the first alternative has been accumulated (Lees, 1966). Lees isolated the first and last born larvae of a given female and did so repeatedly in the two lineages thus established. All animals were kept under 12 hr light and 15°C, which conditions normally induce the production of sexuparae. It was then shown that 66 to 76 days after the first larvae had been born, the offspring in both lineages became responsive to short-day treatment, i.e. the last born lineage underwent one generation less than the first born. This observation supports the concept of a "time interval" rather than a time factor. Whether this concept can be applied to other species still has to be shown.

5. POLYMORPHISM

Individuals of all generations, but particularly those of the virgo generation, in many aphid species may appear in either an alate or an apterous form. Alate individuals, whether of the parthenogenetic or the sexual generation, are of significant importance in the spreading of a species. New areas and food sources can be invaded which otherwise could have been reached only slowly or not at all. It is, furthermore, of considerable significance that winged forms most often arise from among crowded colonies, when food has theoretically become scarce. Thus we have to ask whether it is the crowding as such (tactile stimuli) or the scarcity of food which causes alate production. Besides these factors, it is conceivable that other environmental factors, such as photoperiod or temperature, may be influential in the control of wing formation. The literature pertaining to these aspects has recently been reviewed by Hille Ris Lambers (1966) and Lees (1966).

To make the story short, no conclusive evidence to date is available for any species that would unequivocally show that any factor other than crowding influences wing formation. In *Brevicoryne*, regardless of whether the animals were well or poorly fed, crowding caused the appearance of alatae (Bonnemaison, 1951), while isolation prevented it. This has also been observed in *Dysaphis plantaginea* (Bonnemaison, 1959), *Therioaphis maculata* (Paschke, 1959), *Megoura* (Lees, 1961, 1967a), and *A. fabae* (Way and Banks, 1967), to mention a few species. The effect of crowding in the production of alate virgines is on the mother in *Brevicoryne* (Bonnemaison, 1951), *A. craccivora* (Johnson, 1965), and *Megoura* (Lees, 1967a).

The question then arises: how can one define crowding? What is it that causes the production of winged offspring? Possibly, just the walking of the animals over one another, i.e. tactile stimuli, is enough. That this may be so is suggested by the observation that the simple brushing of isolated aphids of *A. craccivora* (Johnson, 1965) caused the production of alatae; however, brushing was not as effective as keeping several individuals together. Perhaps there was not enough tactile stimulation. For the virgines of *Megoura*, two animals already constitute a crowd (Lees, 1967a). Antennectomy in this species did not abolish the group effect. Further observation then revealed that there is probably more to the group stimulus than just the tactile stimulation by other individuals. Lee (1966) showed that the contact of an animal with other animals of its own species was 80 to 90 per cent effective in causing the production of alatae. Contact with other aphid species had varying results; for example, contact with *Acyrthosiphum* was 73 per cent effective, and that with *Aphis fabae* was 50 per cent. Contact with species of other orders had little or no effect, even though there had been considerable locomotor activity in these cases. This result could imply that only certain aphid species are peculiar in that they can exert the group effect which results in the determination of winged virgines. Perhaps not only tactile but also chemical stimuli are involved.

Apterous parthenogenetic individuals probably can be considered neotenic or paedogenetic forms of a species. It should, therefore, not be surprising to find the hypothesis advanced that the juvenile hormone from the corpora allata might be involved in the suppression of wing formation (Kennedy and Stroyan, 1959; Lees, 1961). Indeed, treatment of

Dorsalis fabae with farnesol did suppress wing formation (Dehn, 1963), supporting this idea. Likewise, when juvenile hormone from the *Cecropia* silkmoth was applied to apterous third instars of *Megoura*, wing formation was suppressed to some extent (Lees, 1966). Moreover, it has been shown in *Brevicoyne* that the winged individuals have smaller corpora allata than do the apterous females. The small size of the corpora allata is a morphological criterion considered to be representative of only moderately active glands (White, 1965). Conversely, aphids which had been isolated from the 1st instar on, and which remained apterous, had larger corpora allata than did the crowded ones. When the animals were treated with compounds of juvenile hormone activity from the first instar one, apterous animals occurred more frequently than they did in the controls (White, 1968). These few examples make it likely that the determination of winged or apterous forms occurs via the endocrine system, presumably the corpora allata.

6. SOME EVOLUTIONARY ASPECTS

The observation that a species lives on two different host plants in succession—heteroecy—lends itself to some interesting evolutionary thoughts concerning speciation. Heteroecious species appear to be more abundant in temperate than in tropical zones. Apparently, this does not merely result from the fact that more work has been done on aphid species in Europe and North America than elsewhere. Heteroecious species are also holocyclic and generally perform the different parts of their life cycle on the two hosts within either 1 or 2 years. Certainly, migration to a secondary host gives the species an advantage over those which do not migrate since a new food source can be exploited. Mordvilko (1928) saw in the evolution of heteroecy the adaptation of a species to changing food availability, an idea which is consistent with the fact that heteroecy is primarily found in temperate zones. The use of herbaceous plants as the usual secondary host seems to be advantageous for a continued high reproductive rate during the warm summer months. According to Mordvilko, some winged individuals may have fortuitously encountered another food plant, but if the sexual forms could not have lived on this secondary host, selective pressure would have favored their return to the primary host. He saw heteroecy as an endpoint in evolution. However, this does not appear to be the case, as will be outlined below (see also Kennedy and Stroyan, 1959).

Among the heteroecious species, there are those which obligatorily migrate to the secondary host and those which may or may not migrate, depending on the availability of the host (Müller, 1962). The virgines of *Dysaphis plantaginea* will no longer accept the primary host and will instead migrate to the secondary host where the sexuales are produced. Never were sexuales observed on the primary host. Also, individuals of *Myzus cerasi* which were transplanted back onto the primary host did not feed and died. It seems that, in these cases, a physiological change must take place during the imaginal molt of the virgo (fundatrigenia) which causes it to reject the food plant on which it developed as a larva. In other words, for the completion of the heterogonic life cycle, the species has to migrate back and forth between different food plants. *Myzus persicae* and *Aphis fabae* may be cited as species which can complete their full life cycle on either the primary or the secondary host. This can be shown experimentally by transferring the fundatrices to either of the hosts: they successfully complete their life cycle wherever they are put. This capacity to develop on either of the hosts is perhaps a relic of the original polyphagy of the species (Müller, 1962).

From this ability to develop completely on either of several host plants, it is probably not far to further evolution towards elimination of the sexual generation. The species, staying on one food plant, may become anholocyclic. It has indeed been observed that *Myzus persicae* is holocyclic in the greater part of Europe, but some populations around Paris (Bonnemaison, 1958) and most of the animals living in Israel (Bodenheimer and Swirski, 1957) are anholocyclic. Individuals of these latter populations cannot be induced to produce a sexual generation. Morphologically, these individuals are indistinguishable from the holocyclic animals. In this case, we are probably witnessing the evolution of a new species. This also appears to be taking place in several other species of aphids.

Some species of the Adelgidae, which have a most complex life cycle, presented interesting material for the study of speciation. Members of the Adelgidae, for example *Sacchiphantes viridis*, are dioecious and holocyclic; they have five generations, consisting of fundatrix, virgo (alata migrans) which migrates to larch, exsulis-virgo (hiemosistens), sexupara (remigrans) which migrates back to spruce, and sexualis (Fig. 74). Two anholocyclic species, *S. abietis* on spruce and *S. segregis* on larch, were described. A discussion then centered around the question of whether these are really separate species or whether they belong to *S. viridis*, since hardly any morphological differences were found. By the extensive rearing of all questionable species, Steffan (1961, 1962a, 1964) resolved the problem and brought evidence that they are indeed distinct species. Two species are found on spruce and larch respectively and do not migrate: they are truly anholocyclic. It is believed that *S. viridis* is the phylogenetically older species which even today still migrates between spruce and larch in a 2-year cycle. *S. viridis* is found only in regions where mixed populations of spruce and larch exist. On each of these hosts, then, a monoecious anholocyclic species evolved. Anholocycly seems to have evolved simultaneously with the adaptation to one of the hosts. From this it seems probably that heteroecy is not an endpoint in evolution. Rather, a species may revert to monoecy due either to favorable conditions or to selective pressure. These theoretical considerations, which are based on laboratory observations, could be further substantiated by field studies. In Montenegro was found a population consisting exclusively of *S. abietis* which corresponded to those in middle Europe (Steffan, 1962b). This isolated population had probably evolved together with an isolated population of the host plant, *Picea abies*; *Larix decidua* is not found in Montenegro.

B. Cecidomyidae

Heterogony has been described in several species of Cecidomyidae (order Diptera), all of which live on fungus hyphae. As long as larvae of these species can find suitable conditions, they reproduce paedogenetically, i.e. the mother-larvae again produce live larvae via parthenogenesis. The appearance of imagines has been described in *Miastor metraloas* (Springer, 1917), *Heteropeza pygmaea* (= *Oligarces paradoxus*) (Harris, 1924; Ulrich, 1934), *Mycophila nikoleii* (Nikolei, 1958), and *Tekomyia populi* (Nikolei, 1961). They probably occur in other species as well, yet no precise information is available. Imagines invariably appear during the summer when the substrate begins to dry out or food conditions begin to deteriorate. The description of the complete life cycle of *Heteropeza* may be taken as an example for similar cycles in those other species for which some of the details are not as well worked out (Fig. 75) (Nikolei, 1961; Camenzind, 1962; Ulrich, 1963). Mother-larvae again produce mother-larvae almost indefinitely; they have been reared paedogenetically for 262 generations (Ulrich, 1936). However, when food gets old or the animals are crowded, imagines may appear. Females arise via a so-called female imago-larva and pupa. Males can arise via a male-larva, a mixed male-female-larva, or a male imago-larva and pupa. Female imago-larva can still revert and become a parthenogenetic mother-larva (Fig. 75). Imagines

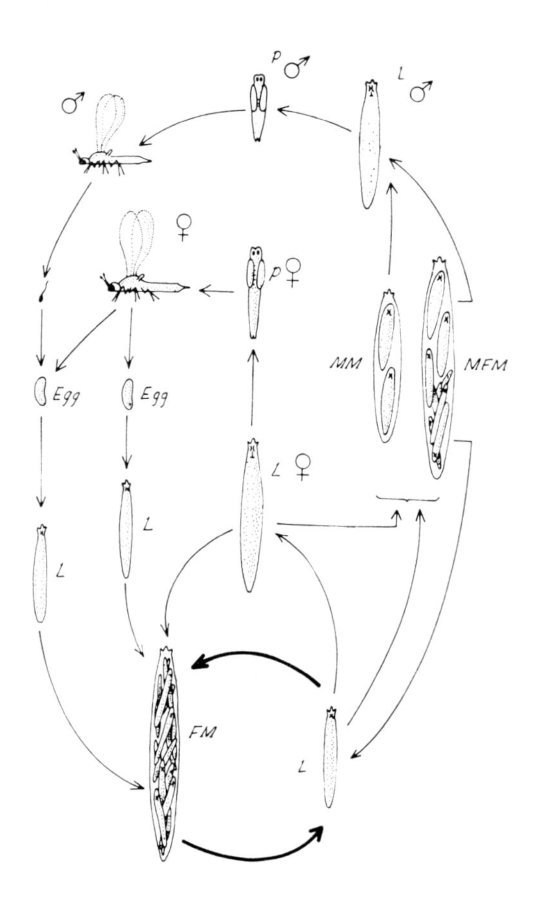

FIG. 75. Life cycle of the paedogenetic *Heteropeza pygmaea*. FM = female mother-larva producing only female-larvae. MFM = mixed male female mother-larva. MM = male mother-larva producing male-larvae which develop into pupae and male adults. (From Camenzind, 1962.)

lay a few eggs whether mated or not. In each case, mother-larvae hatch from these eggs and continue to give rise to mother-larvae as long as conditions are good. A direct development of imagines from the eggs to the larvae and pupae, i.e. the elimination of the paedogenetic generations, has not been reported.

It appears that the young paedogenetic larvae are pluripotent. Interestingly, certain qualities of food seem to favor the production of males, while others favor the production of mother-larvae (Camenzind, 1962). Unquestionably, the quality and the quantity of the food control heterogony in these species, a fact recognized by Springer (1917) and Gabritschewsky (1928, 1930) for *Miastor*. Sex determination also appears to be controlled via the quality of the food as discussed earlier (p. 15).

The production of winged sexual animals in these soft-bodied dipterans can be considered of adaptive value. Winged females have a chance to spread and thus preserve the species since flight gives them an extended access to new substrates for their paedogenetic larvae.

C. Cynipidae

Among the gall-producing Cynipidae (order Hymenoptera), several species are known to alternate strictly between a parthenogenetic spring generation and a bisexual summer generation. The first description of the phenomenon was given by Adler (1877, 1881). For *Neuroterus lenticularis* Doncaster (1910, 1911, 1916) reported that the spring females give rise either to male or female bisexual individuals, but not to both. In other words, a parthenogenetic female reproduces either via arrhenotoky or via thelytoky but not via deuterotoky. Determination of the sex of the heterosexual animal thus takes place at the fertilization of an egg from which hatches the parthenogenetic spring female. In several species, the heterosexual and parthenogenetic individuals are known as separate species. Only rearing, as has already been pointed out by Adler, can determine the true nature of a given species. Folliot (1964) presented a review of the life cycles of many species which have been reared; among these are *Neuroterus aprilinus*, *Andrius uruaeformis*, and *Cynips agama*. *Besbicus mirabilis* also has recently been described as a heterogonic species (Evans, 1967).

Unfortunately, most authors classify the parthenogenetic females as agamic, and write about agamic and heterosexual generations. The term agamic is misleading and therefore should not be used in this context. All individuals, whether parthenogenetic or bisexual, arise from gametes; they are not agamic. Parthenogenetic is the proper term, and the term agamic ought to be used in cases of species propagation, for example, by budding.

Nothing is known about the control of the succession of generations in these heterogonic species. In the absence of any contrary evidence, we can postulate that the strict alternation, one parthenogenetic and one bisexual generation, is genetically determined.

CHAPTER 12

VIVIPARITY

VIVIPARITY in species of insects appears to an outsider as a rare oddity with little significance. One is generally somewhat surprised when one learns of its widespread occurrence in many insect orders. One is furthermore astonished that relatively so little is known about the reproductive physiology and adaptive features of these species. The vast majority of the papers dealing with viviparous insects are restricted to descriptions and simple reports from which lengthy speculations concerning function and evolution have emerged. A fairly complete compilation of viviparous species is found in Hagan's (1951) book. Both the scarcity of information on the reproductive physiology in viviparous species and the peculiarities to be found among these species warrant a treatment of the subject in the present context.

A. *Classification and Occurrence*

Classifications of viviparity were attempted by Holmgren (1904) and Comstock (1925) on the basis of mode of reproduction: parthenogenetic or bisexual. Keilin (1916) subdivided viviparous insect species into those which incubate their offspring only during the embryonic life and those which incubate them in addition during part of or the complete larval life. These classifications appeared to be somewhat unsatisfactory as they did not adequately encompass the physiological aspects of reproduction and gave no indication of the extent to which viviparity had been perfected in these species. Following those attempts, Hagan (1931, 1951) reclassified the viviparous insect species. As a basis, he used the nutritional dependence of the embryo or larva on the mother and the type of auxiliary structures which serve the offspring in obtaining nutrients. The following types of viviparous reproduction were established.

1. *Ovoviviparity.* The embryonic development occurs within the mother, yet no nutrients are obtained from her; the embryos are deposited upon the completion of development. In this group belong, for example, *Blabera* and similar cockroaches, several Thysanoptera, the ephemeropteran *Chloëon*, several Lepidoptera, Diptera of the genera *Sarcophaga*, *Mesembrina*, etc., the coleopterans *Chrysomela*, *Phytodecta*, and others.

2. *Adenotrophic viviparity.* Special maternal structures serve to nourish the larvae in the uterus. Glossinidae and Hippoboscidae belong to this type.

3. *Haemocoelic viviparity.* Embryonic development is completed in the haemocoel of the mother; the embryos obtain nutrients through their egg membranes via simple osmosis. Several species of the Cecidomyidae and the Strepsiptera are grouped into this type.

4. *Pseudoplacental viviparity.* The embryos in the genital tract obtain nutrients via a placenta-like structure developed either by the mother or the embryo or both. The dermap-

teran *Hemimerus*, the dictyopteran *Diploptera*, and the hemipteran *Hesperoctenes* are examples of this type.

As has been pointed out by Hagan (1951), the placement of various species into one category or another, may be only tentative since, in many cases, no details are known concerning embryonic development. In the following account, some of the difficulties encountered with this classification system and the attempts to modify it will be discussed.

Many viviparous species are classified as ovoviviparous simply because there is no information on whether the embryos obtain nutrients from the mother, and also because no structural peculiarities seem to suggest that such is the case. An observation that embryos were present within the genital apparatus was often taken as evidence for ovoviviparity. This is, for instance, the case in several species of Thysanoptera (John, 1923; see Hagan, 1951), the ephemeropteran *Chloëon dipterum* (Calori, 1848; Bernhard, 1907; Grandi, 1941), the lepidopterans *Colias edusa* and *Parnassius apollo* (Pierce, 1911), the psocids *Hyperetes guestphalicus* (Jentsch, 1936), and *Pseudocaecilius hirsutus* (Wong and Thornton, 1968), and many others. Ovoviviparity has been reported in several Diptera, such as *Mesembrina meridiana* and *M. mystacea* (Cholodkovsky, 1908), *Musca larvipara* and *Dasyphora pratorum* (Portchinski, 1910), *Compsilura concinnata* (Hegner, 1914), *Calliphora icela* (Miller, 1939), several Sarcophagidae, and many others. In none of these cases, apparently, do the embryos obtain any nutrients from the mother. Yet no embryonic studies have been made on those species in which the embryos develop in an extension of the vagina, the uterus. Embryos of the fleshfly *Sarcophaga bullata* can develop completely and hatch after they have been removed from the uterus and kept on moist filter paper; they thus do not appear to need nutrition from the mother (Wilkens, 1966, unpublished). Among the Coleoptera, several termitophilic and myrmecophilic species appear to be ovoviviparous (see Hagen, 1951). *Micromalthus debilis* is a paedogenetic species, i.e. embryos develop within the ovaries of the mother-larva; viviparity and neoteny are thus combined (Barber, 1913; Scott, 1936, 1938; Pringle, 1938). This species is thought to be ovoviviparous. In several chrysomid species, as for example, the species *Chrysomela varians*, embryos develop within the ovary and hatch a few minutes after deposition (Rethfeldt, 1924).

This brings us to a consideration of what should, in the strict sense, be called ovoviviparous. In Hagan's definition, ovoviviparity applies only to those species in which the embryo obtain no nutrients but do hatch before deposition. In this sense, *Chrysomela varians* should not be considered as ovoviviparous but rather as oviparous, even though the embryos are fully developed by the time the eggs are deposited. In other words, we get entangled in semantics since this species essentially does not differ from others in which the embryos hatch within the genitalia just a few minutes earlier. This semantic difficulty, which entails the need for a modification of the term, was also felt by Roth and Willis (1958a); they therefore introduced the term false ovoviviparity for several species of cockroaches, such as *Nauphoeta* and *Leucophaea*. These authors meant to overcome the difficulty in terminology since, in these species, the eggs are first extruded during ovulation (as in oviparity) and then retracted. In this case there is, in my opinion, technically no major physiological difference between a cockroach which forms an egg case during ovulation and then retracts it for the completion of embryogenesis and, for example, an animal such as *Glossina* which also ovulates but whose egg is not visible from the outside. It is therefore suggested here that the term false ovoviviparity should not be used. Further, all species that complete embryonic development within the mother but whose embryos do not obtain nutrients, whether they are deposited before or after hatching should be classified as ovoviviparous. This would simplify the terminology and use only physiological criteria for classification. Within the cockroaches we find species, such as *Periplaneta americana*, which deposit egg cases as soon as they are formed. Then we find species which carry half of the egg case within the genital atrium practically throughout embryonic development and which should, therefore, be called ovoviviparous, e.g. *Blattella germanica*. Species such as *Panchlora viridis* (Riley, 1890), *Gromphadorhina laevigata* (Chopard, 1950), *Nauphoeta cinerea*, and *Leucophaea maderae* (Roth and Willis, 1955a, b), as well as several others, carry their egg cases completely concealed within a brood sac. In none of these species do the embryos obtain any nutrients from the mother except water, and they are thus classified as ovoviviparous; this is done even though all of these species extrude the egg case first and then, minutes to days (*Blattella*) thereafter, the embryos hatch. The embryos are fully developed by the time of the extrusion of the egg case. Parenthetically it could be added here that the truly viviparous cockroach *Diploptera* likewise extrudes its eggs case with the majority of the embryos only hatching thereafter.

Several species of Hemiptera could probably be classified as ovoviviparous; however, this classification is debatable since the embryos do not complete their development within the female. *Cimex lectularius* undergoes its initial stages of development within the ovariole and is deposited during blastokinesis (Abraham, 1934; Davis, 1956). This has also been reported for several other blood-sucking Hemiptera (Carayon, 1962). *Stricticimex brevispinosus* embryos are not deposited until the extremities have been formed (Carayon, 1959). Among the Dermaptera, one ovoviviparous species is known, *Prolabia arachidis* (Herter, 1943, 1965). Here also, as in the cockroaches, the eggs are laid before the fully developed embryos hatch. Hatching is facilitated by the female's eating away the chorion. This behavior seems to be part of the well-developed maternal care found among species of the earwigs.

True viviparity is observed when the embryos obtain from the mother nutrients without which they presumably would be unable to complete development. In rarely any case does this latter assumption seem to have been tested. Also, whether or not the embryos obtain nutrients is often only deduced from the presence of special structures. As mentioned earlier, Hagan (1951) distinguishes between adenotrophic, haemocoelic, and pseudoplacental viviparity. In the adenotrophic and pseudoplacental types, special structures serve the embryo; yet it has not been proven that the structures are the only pathways by which the embryos can obtain nutrients. Besides, as will be discussed below, a so-called pseudoplacenta may have other functions than nourishing the embryos. Haemocoelic viviparity is found in all Strepsiptera and some species of the Cecidomyidae. Strepsipteran species parasitize other insects; for example, *Elenchinus chlorionae* uses the Homoptera *Chloriona unicolor* or *C. smaragdula* as its host (Lindberg, 1939), *Acroschismus wheeleri* uses the Hymenoptera *Polistes* or *Andrena* (Hughes-Schrader, 1924), *Halictophagus tettigometrae* the hemipteran *Tettigometra impressifrons* (Silvestri, 1941), and several species of *Mengenilla* parasitize the thysanuran *Ctenolepisma* (Silvestri, 1942). In all of these species, as well as in others, the female lives within the host throughout her life, whereas the male undergoes complete metamorphosis and is winged. Upon maturation, the female penetrates the host's cuticle with her cephalothorax while her abdomen remains inside the host. The female then becomes attractive to the short-lived (2 to 4 hr) males. The male copulates with the cephalothorax by penetrating the cuticle of the brood canal with his aedeagus. The eggs are fertilized within the coelom where the embryos develop. Later the embryos are found in the brood pouches; from there, they enter the brood canal (Fig. 76) and are later released (Lauterbach, 1954). The larvae of the first stage, the triungulins, are the only free-living forms of the species; they

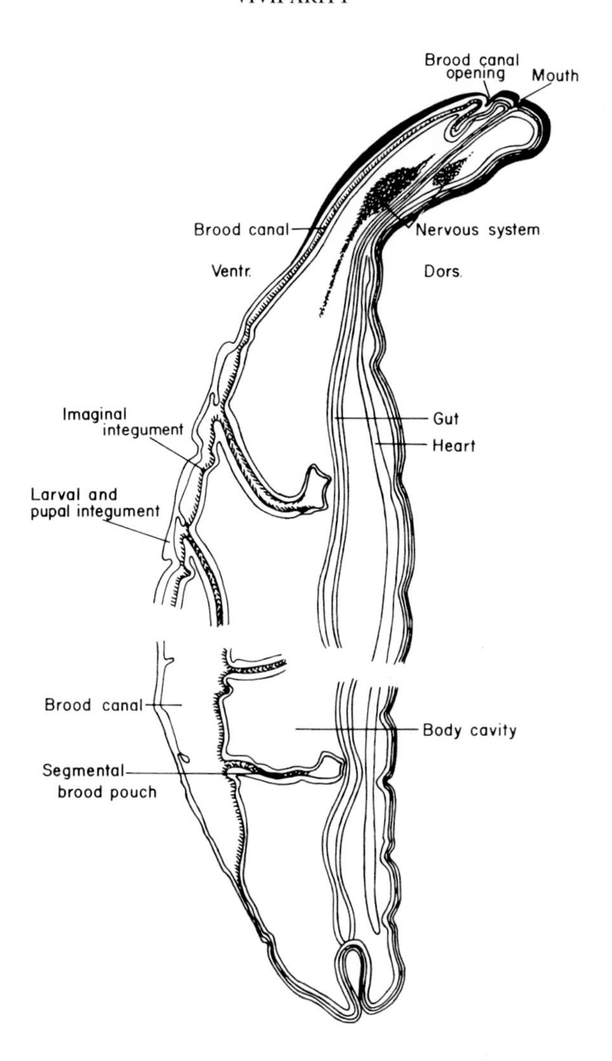

FIG. 76. Sagittal section through the strepsipteran *Stylops* spec. The embryos develop in the body cavity, then enter the brood pouch and brood canal; the larvae leave the female through an opening in the cephalothorax which penetrates the host cuticle. (From Lauterbach, 1954.)

seek new hosts to which to attach themselves and whose cuticle they penetrate. Generally between 1500 and 2000 triungulins are produced by a female, although it has been observed that *Corioxenos antestiae* produced up to 3700 larvae (Kirkpatrick, 1937).

Wagner (1863, 1865) was the first to observe a dipteran larva which contained developing larvae in its haemocoel. The species was probably *Miastor metraloas*. The phenomenon has been repeatedly observed in this same species (Meinert, 1864; Kahle, 1908; Springer, 1917), as well as in others such as *Heteropeza pygmaea* (Harris, 1924, 1925; Ulrich, 1934), *Mycophila nikoleii* (Nikolei, 1958), *M. fungicola* (Foote and Thomas, 1959), *M. speyeri* (Nicklas, 1960), *Tekomyia populi* (Nikolei, 1961), and *Henria psalliotae* (Wyatt, 1961). One may safely assume that the embryos obtain nutrients simply by osmosis through the egg membranes. Nutrition is indeed essential for normal development and normal growth since undernourished mother-larvae produce only a few small larvae. In both groups with haemocoelic viviparity, presumably no special structures serve the embryos in obtaining nutrients; the embryos are literally bathed in a nutrient-rich milieu.

Adenotrophic viviparity is confined to species of the Diptera. These are species which mature full-sized eggs, ovulate, and incubate them through larval development in the uterus. The larvae, which mature one at a time, are extruded; pupation usually follows within a few hours. This behavior has been observed in the *Hippobosca variegata* (Schuurmans, 1923), *Melophagus ovinus* (Pratt, 1899), and others (Hardenberg, 1929; Bequaert, 1952; Ulrich, 1963). In all of these species, the accessory sex glands function as "milk glands" from which the growing larvae obtain their nutrients (Pratt, 1899; Hardenberg, 1929; Ulrich, 1963). Larvipary, as above, has been reported in species of the Nycteribiidae, Streblidae, and Braulidae (see Hagan, 1951). Species of the Glossinidae likewise deposit only fully developed larvae which pupate soon after without having fed outside the mother. Only one larva is

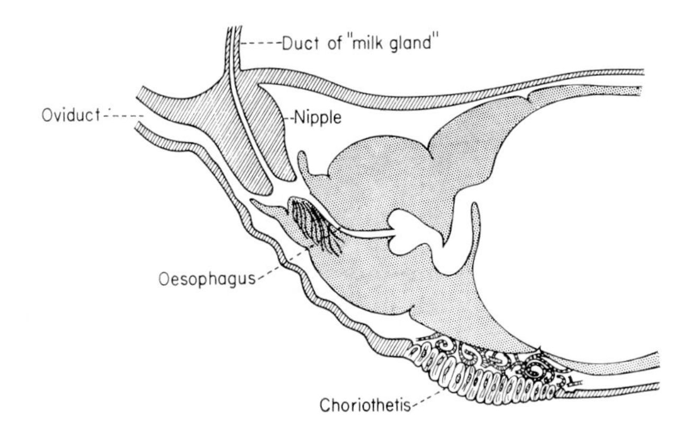

FIG. 77. Diagram illustrating the position of the larva in the uterus of *Glossina palpalis* during the entire larval period. The mouth of the larva is applied to the nipple into which the duct of the accessory sex glands opens. (From Hoffmann, 1954.)

found in the uterus at any one time. Just as in the species mentioned above, this larva is nourished from the "milk gland" (Minchin, 1905; Roubaud, 1909). The larva of *Glossina palpalis* grows within the uterus from 1.5 to 7.2 mm at the time that it is extruded (Hoffmann, 1954). During this entire period, the mouth of the larva is closely applied to a uterine nipple into which the duct of the "milk gland" opens (Fig. 77). It appears that the larva in the uterus continuously sucks at the nipple. In this species, during the period while a larva is developing in the uterus, another egg matures in the other ovary (Fig. 78) and is ovulated as soon as the larva is extruded. Jackson (1948) has described a peculiar structure on the floor of the uterus in *Glossina*. This structure, which was termed the "choriothetis", appears to liberate substances that aid in removing the egg chorion and the larval cuticles after the larva has molted. However, only some histological evidence and the fact that little remains of the chorion and the larval skins suggest this function for the choriothetis.

The third type of true viviparity, pseudoplacental, has been studied in detail in only a few species. In all of these cases, the embryos develop in the female genital tracts—ovariole or brood sac—and are not directly exposed to the food supply as they are in the haemocoelic type. Pseudo-placental viviparity has been found in the dermapteran *Hemimerus talpoides*,

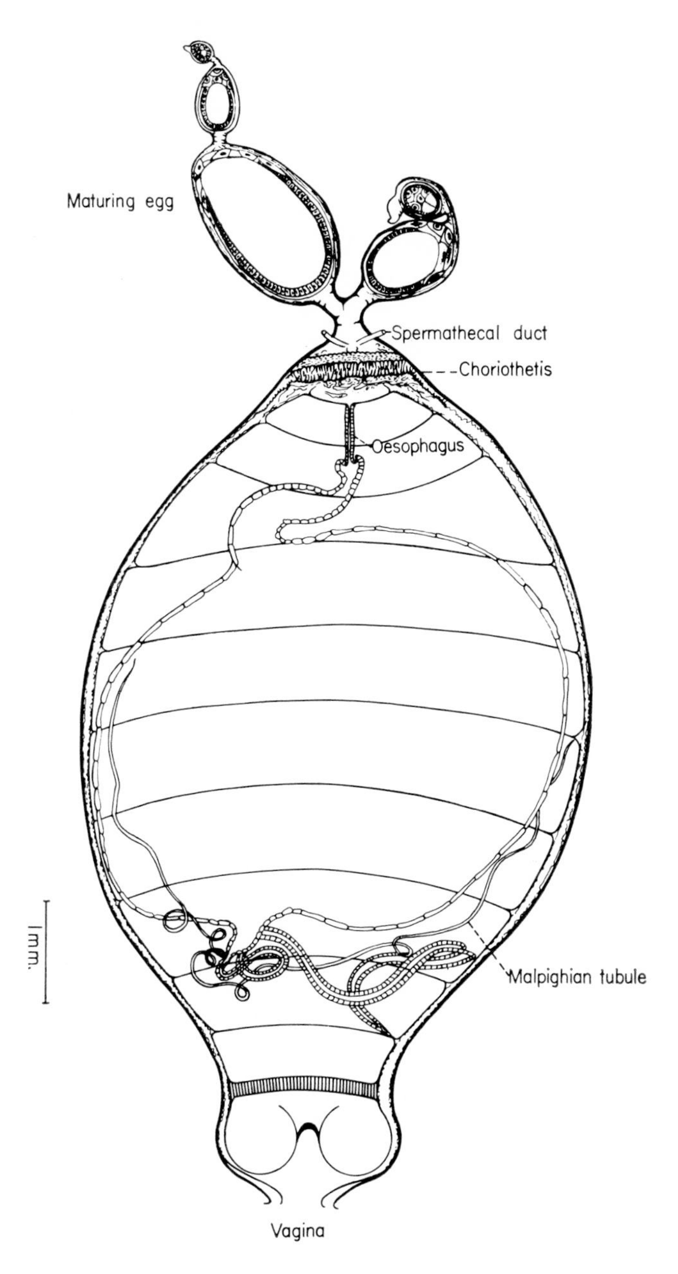

FIG. 78. Position of the larva in the uterus of *Glossina palpalis*. During the period of larval development, another egg matures in one of the ovaries. (Adapted from Hoffmann, 1954.)

an ectoparasitic species on rats in equatorial Africa; it was discovered by Hansen in 1894. A detailed description was then given by Heymons (1909, 1912). The egg of this species has no chorion and contains little yolk. During embryonic development the follicular epithelium thickens, particularly at the anterior and posterior poles. It is believed that this thickened epithelium (the pseudoplacenta of Hagan) provides the nutrients for the growing embryo. No recent study has appeared on the biology of this interesting species.

Hagan (1951) places the Aphididae among the species with pseudoplacental viviparity, even though, in the strict sense, no special placental structures are found; the embryos presumably obtain nutrients from a normal-looking follicular epithelium. The embryo develops within an egg that has no chorion.

An ectoparasite on bats, the polyctenid *Hesperoctenes fumarius*, represents unique material for the study of pseudoplacental viviparity (Hagan, 1931, 1951). Egg development begins in the ovariole; and, as development advances, the egg moves down the reproductive tract. The egg has little yolk and no chorion. Interestingly, during development, the pleuropodia grow out to form a sheath which eventually enwraps the embryo entirely. Presumably,

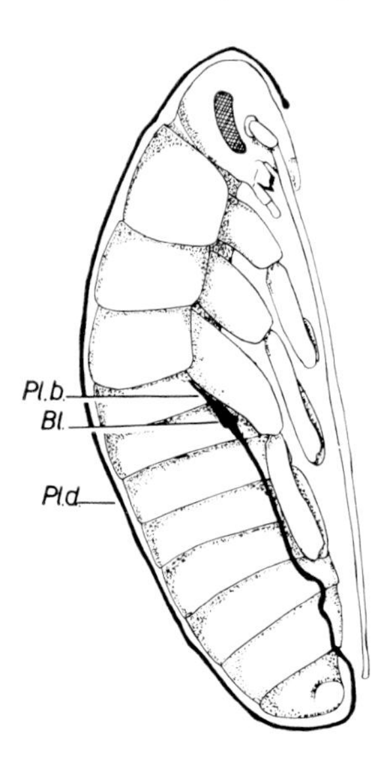

FIG. 79. Position of the pleuropodium in a late embryo of *Leucophaea maderae*. The tip lies near the anterior pole of the egg. It is also observed that the chorion breaks open at the anterior pole towards the end of the embryonic period. This suggests that the pleuropodia liberate a substance which serves to weaken the chorion.

the embryo obtains all its essential nutrients through this pseudoplacenta. Embryos at various stages in development can be found in a female at any time. A second species of the Hemiptera, *Physopleurella pessoni*, also appears to have a pleuropodial pseudoplacenta. It is the only species of the family Anthocoridae which exhibits this type of viviparity.

Finally, the only truly viviparous species of Blattaria ought to be considered, *Diploptera punctata* (= *D. dytiscoides*). The approximately twelve embryos per egg case do obtain other nutrients from the mother besides water. Their dry weight increases during development (Roth and Willis, 1955). The chorionated eggs ovulate and are then incubated in a brood sac. Hagan placed this species among the pseudoplacental type since he believes that the large thread-like pleuropodia function in the uptake of nutrients. As has been pointed out elsewhere, the ovoviviparous cockroaches *Leucophaea* (Engelmann, 1957b), *Nauphoeta* (Roth and Willis, 1958a), and others have similarly large pleuropodia (Fig. 79); yet in these latter species, the embryos do not take up nutrients from the mother. For *Leucophaea*, it has been suggested that the pleuropodia furnish enzymes which are functional in disrupting the chorion towards the end of embryonic development (Engelmann, 1957b). Perhaps one can also negate the nutritional role of the pleuropodia in other Blattaria; future research will have to show what the precise function of the pleuropodia is. Why should the embryos not obtain nutrients by osmosis through the egg membranes? In the absence of any conclusive evidence for or against the presence of a functional pseudoplacenta in *Diploptera*, this species remains listed as a pseudoplacental type. Possibly one may have to create a new category of viviparity for this species.

B. Control of Egg Maturation

The scarcity of information on the biology of viviparous species becomes particularly apparent when one tries to find experimental evidence for the control of egg maturation or of embryonic development. It is quite evident that some control exists since, in viviparous cockroaches or in some of the flies for instance, no eggs mature in the ovaries during gestation. Save for the Blattaria, nothing is known about these control mechanisms in any of the viviparous species.

At temperatures of 25–28°C, the period of gestation is about 2 to 2½ months in several of the viviparous cockroaches (Roth and Willis, 1955, 1964a; Engelmann, 1957a, 1959a). No egg maturation takes place in any of these species during this period. This is an essential adaptation for the species, since egg maturation itself requires only 2 to 3 weeks and no space would be available in the brood sac for incubation of a next batch. Indeed, if egg maturation was experimentally induced in *Leucophaea* during pregnancy, the first egg case was precociously extruded shortly before the next batch of eggs ovulated (Engelmann, 1962); the embryos were not ready to hatch and died within a few days. Viable offspring could not be produced if egg maturations followed one another in rapid succession. As outlined elsewhere (p. 177), the corpora allata of these cockroaches are inactivated during gestational periods; as a consequence, no eggs mature. Some controversy exists concerning the mode of inactivation of the corpora allata, however. One hypothesis favors the notion of the release of a substance by either the egg case in the brood sac or the brood sac itself; this substance causes certain centers in the ventral nerve cord and brain to inhibit the corpora allata (Engelmann, 1957a, 1960b, 1964). Up to now, no substance has been isolated. The other hypothesis favors an exclusively nervous control mechanism, originating in the distended brood sac in the genitalia of the female (Roth and Stay, 1959, 1962a; Roth, 1964b). Control via this latter mechanism is thought to apply also in *Blattella*. If indeed the mechanism is purely nervous, one has to explain the fact that denervation of the genitalia, i.e. severance of the nerves between the last abdominal ganglion and the target organ, does not lift the inhibition in a pregnant female. Replacement of the egg case by an artificial one (Roth and Stay, 1962a) with its resulting effectiveness in the inhibition of the corpora allata is not conclusive evidence for nervous inhibition since the females do not feed enough to support egg growth; this has been shown to be the case at least in *Leucophaea* (Engelmann and Rau, 1963).

In other viviparous insect species, we can only infer from the available observations that certain control measures for egg maturation do exist. As an example, the fleshfly *Sarcophaga bullata* does not mature eggs during pregnancy and, even if pregnancy is experimentally prolonged, eggs still do not mature (Wilkens, 1967). No mechanisms are known in this case. The question also arises: what controls the maturation of one egg in alternate succession in the ovaries of *Glossina* spec. or the Hippoboscidae?

In contrast to these cases, in which we find some kind of restraining control of egg maturation during certain periods of the animal's life, no such control of egg maturation is found in *Hemimerus* or *Hesperoctenes*, as eggs at various stages in development can be found in each ovariole. Also, in the cases of the Strepsiptera or the ephemeropteran *Chloëon*, no control mechanisms ought to be invoked because of the peculiar biology of these species: they have but one reproductive period.

C. Ecological Aspects

In several viviparous species, it is apparent that the number of offspring per female is rather low. For example, the viviparous Hippoboscidae and Glossinidae produce just one larva at any given time. *Melophagus ovinus*, which lives about 100 days, produces only about twelve larvae in this period (Bequaert, 1952), and *Hippobosca variegata* gives birth, on the average, to only 4.5 larvae (Schuurmans, 1923). *Glossina palpalis* produces six to twelve larvae in a lifetime (see Hoffmann, 1954). In all of these species, the propagation of the species is still assured, since the mortality rate of the hardened pupae is probably low. Viviparity, because it gives the growing larvae protection, allows a low reproductive rate. Similar considerations may apply to the viviparous cockroach *Diploptera*, as compared to the oviparous species. On the average this species has a total of thirty-five offspring (Stay and Roth, 1958), whereas *Periplaneta* or *Blatta orientalis*, for instance, produces several hundred eggs under favorable conditions. The latter species deposit their eggs outside; these eggs are thus exposed to a variety of adverse environmental influences. Low or high reproductive potential is, however, not necessarily linked with either viviparity or oviparity. This can be seen in species of Strepsiptera, which produce several thousand larvae, and in *Cloëon* (Bernhard, 1907), which deposits 600 to 700 larvae. In both these cases, however, the deposited larvae are not protected and are subject to predation, desiccation, or starvation.

Several interesting cases of viviparity have been found among the Diptera. These may lend themselve to an experimental approach to an understanding the evolution of viviparity. Portchinski (1885) reported that on the Crimean *Musca corvina* is viviparous during the summer months, but that in the early spring it is oviparous, just as the northern populations in Russia are normally. *Chloeopsis diptera* is reportedly oviparous in the north of Italy, but viviparous in the south (Giard, 1905). These rather odd findings suggest a faculative viviparity in warm southern climates. As to actual causes for this shift from viviparity to oviparity, no concrete suggestions can be made. More recently, a similar phenomenon has been reported in New Zealand for *Calliphora laemica* (Miller, 1939). *Calliphora stygia* can occasionally be found to produce live larvae in Australia (Murray, 1956; Norris, 1959). Normally, this latter species is said to be oviparous. May one assume that under certain climatic conditions fertilized eggs are simply retained and incubated by the female? This explanation may not, however, apply to *Musca corvina* which, when viviparous, produces one larva at a time, but when oviparous, lays twenty-four eggs in one batch. Additional observations could easily clarify the situation.

A kind of adaptation to viviparity has been seen in the viviparous species of Blattaria. These species generally have a thin and incomplete egg case which, when exposed to dry air, does not provide adequate protection from water loss. Internal incubation has become essential. Also, these species, with the exception of *Diploptera*, have a low initial water content in the eggs, whereas the majority of the oviparous species have a high initial water content. Eggs of the viviparous cockroaches gain additional water during early embryonic development. This may be advantageous for the female, since she thereby does not have to provide the large amount of water during the relatively brief egg maturation periods, but can instead provide it over a longer period of time. On the basis of the pattern of water uptake and the initial water content in the various species, Roth (1967c) proposed two alternative pathways for the evolution of ovoviviparity and viviparity in cockroaches. (1) Ovoviviparity arose directly from oviparous species which normally drop their egg cases in moist environments where these eggs can take up more water. (2) It may also have evolved via a species like *Blattella* which carries its egg cases half-way retracted; eggs of *Blattella* take up water from the mother during gestation. Viviparity as it is observed in *Diploptera* could then have evolved either from an ovoviviparous species like *Nauphoeta* or from a species like *Blattella*. Any of these pathways are feasible.

Hagan (1951) considers viviparity more highly evolved than oviparity. Viviparity seems to have evolved independently in several orders of insects. It occurs in both lower and higher orders of insects and, showing remarkable complexities, has been found in Dermaptera (*Hemimerus*), Dictyoptera (*Diploptera*), Hemiptera (*Hesperoctenes*), and Diptera (*Glossina* and *Melophagus*). Certainly these viviparous species have been successful in propagation, even though their reproductive capacity is relatively low. This in itself may illustrate the adaptive value of viviparity. The question is, however, whether viviparity can be termed an evolutionary advance or just another offshoot from the main evolutionary pathway. In any case, these considerations will always remain speculative and should be left at that.

CHAPTER 13

FUNCTIONAL HERMAPHRODITISM

In a discussion of reproductive phenomena, we certainly must include some of the unique modes of reproduction to be found among insect species. Hermaphroditism is one of these rare modes. Hermaphroditism ought not to be confused with either gynandromorphism or intersexuality (p. 20f.). In contrast to the latter two terms, hermaphroditism describes the normal mode of reproduction in a given species: hermaphroditic species may be both male and female simultaneously or in succession. Offspring having the same mode of reproduction are produced, often by self-fertilization.

Hermaphroditism has been described in *Perla marginata*. In the male of this species, the testes are fused anteriorly where they seem to form 70 to 100 ovarioles. However, as soon as the testes follicle have differentiated, oogenesis proceeds no further (Schoenemund, 1912; Junker, 1923). Oogenesis is not completed and the partially grown eggs degenerate; apparently no vitellogenesis takes place beyond an initial stage. No report is available whether any egg has ever matured in the functional male. The female is normal and has no identifiably testicular tissues.

Among species of the *Termitoxenia* (order Diptera), which live in termite colonies, apparently no males have ever been seen. Since mature sperm are found in the genitalia of all of the females, it was thought that this species is truly hermaphroditic (Wasmann, 1901; Assmuth, 1913, 1923). On the other hand, no spermatogenesis has ever been demonstrated in the unpaired organ which contains the sperm. Bugnion (1913) therefore concluded that the testis described by Assmuth is probably a spermatheca. Hermaphroditism presumably does not occur in this species.

The only certain case of hermaphroditism among species of insects, which has been known for a long time, is that of *Icerya purchasi* (Coccoidea). Pierantoni (1911) was the first to describe the occurrence of both male and female gonads in the same animal. Both gonads differentiate in the larval stages in this species, as in other insects. Definitely, spermatogenesis, as well as oogenesis, does occur in the hermaphroditic gonad (Pierantoni, 1914). The finding of hermaphroditism in *Icerya purchasi* was later confirmed by Hughes-Schrader (1925). All females have ovo-testes in which spermatogenesis usually precedes oogenesis (proterandry). No pure females were found so far. Self-fertilization is the common pattern of reproduction. Among 2500 individuals, only nine males were found and it is assumed that they arose from occasionally unfertilized eggs. That the males are haploid can be verified cytologically. Males can be produced by either mated or unmated hermaphrodites (Hughes-Schrader, 1926, 1928). Cytologically, no distinction between the spermatogenesis of males and that of hermaphrodites is possible. More recently it has been found that hermaphroditism also occurs in *I. bimaculatus*, a species from Kenya (Hughes-Schrader, 1963) and *I. zeteki* from Panama (Hughes-Schrader and Monahan, 1966). The biology of

I. zeteki appears to be similar to that of *I. purchasi*, i.e. occasionally males are produced via haploid parthenogenesis. It is speculated that hermaphroditism has arisen independently three times in the genus *Icerya*: *I. purchasi* in Australia, *I. bimaculatus* in Kenya and *I. zeteki* in Central America. In all these species the ovo-testes normally undergo only one phase of gamete production, first spermatogenesis and then oogenesis. Recently it has been shown, however, that there occasionally may be a second phase of spermatogenesis after the eggs have been produced (Delavault and Royer, 1966).

One is surprised by the rare occurrence of functional hermaphroditism among insects and by its definite appearance in only one genus. Why has it independently evolved several times just in this genus and in no other genus or order? This question has to remain unanswered.

CHAPTER 14

INSECT SOCIETIES

INSECT societies are found peculiarly among many species of the Hymenoptera and in all species of Isoptera. It is puzzling and has evoked speculations by many entomologists why social behavior is so common just in these orders of insects and not at all in others. As will be seen later, different characteristics of the species may have facilitated the evolution of societies several times; yet this in itself does not explain the evolution of social insect species. Since man himself is a social species, he has been interested from early times in those species which exhibit—at least superficially—certain similarities to human societies. As a consequence, probably no aspect of insect physiology has been treated so extensively and from so many different angles as those related to insect societies. The overwhelming wealth of research papers and reviews dealing with social insects gives ample proof of the interest of behaviorists, entomologists, and physiologists in the functioning of insect societies. While the works of earlier authors were largely descriptive (cf. Lubbock, 1884; Wheeler, 1910; Forel, 1921), reviews of more recent years reflect the interest in the experimental approach to questions of caste determination and control of the social structure. Caste determination in social insects has been reviewed by Wilson (1953, 1965, 1966), Goetsch (1953), Brian (1957, 1958, 1965a, b), Bier (1958), Allen (1965), and Weaver (1966). The production of supplementary reproductives in termites, ants, and wasps has been discussed in many of the reviews, but was stressed particularly by Plateaux-Quénu (1961). The biology of solitary wasps—presumably representing the evolutionary ancestors of social insects—has been reported in great detail by Olberg (1959), while the evolution of societies in wasps has been discussed by Evans (1958) and that in bees by Michener (1958).

In the present context, phenomena which do not have a direct bearing on reproduction will be either omitted or dealt with only marginally. This applies particularly to aspects of nest building, colonial behavior other than reproductive behavior, and details of swarming behavior. The significant feature of societies composed of different castes appears to be "obvious" with respect to propagation and preservation of the species. This apparent advantage has to be analyzed critically in the light of alternative possibilities, however. The advantage of a social system, frequently associated with polymorphism, probably lies in the apportionment of specific duties in the colony to specialists. This is achieved by various means, such as genetic, trophogenic, phermonal, and behavioral. The mode of caste determination is incompletely understood in the majority of the species. In some cases, we do have circumstantial information which allows a tentative conclusion as to the control of caste determination and structure.

A. Caste Determination and Control of Social Structure

1. GENETIC DETERMINATION

It is generally acknowledged that the determination of the male caste in hymenopterous species is genetic only, as Dzierzon (1845) had first postulated. Eggs of workers of the honeybee will generally be drones, since workers apparently do not mate. Exceptions to this are known. In the Cape honeybee, *Apis mellifera capensis*, workers give rise to diploid workers (Gough, 1928); the cytological basis for this appears uncertain. As was pointed out earlier (p. 14), development of haploid eggs is commonly found among Hymenoptera. Males can thus arise both from unfertilized eggs of the queen and from those of the worker. Just as in *Apis mellifera*, the unfertilized eggs of *Vespa germanica* develop into males (Marchal, 1896); these males often appear late in the season, probably when the sperm reservoirs are close to depletion. It is possible, however, that in this case also eggs laid by the workers contribute to the appearance of males since workers give rise only to males (Stadler, 1926). Similarly, workers of *Vespa orientalis* produce only males (Ishay, 1964), as do those of *Trigona gribodoi* (Bassindale, 1955).

According to Goetsch and Käthner (1937), virgins of *Lasius niger* produce only males, whereas Reichenbach (1902) and Crawley (1911) observed that unmated workers of this species can give rise to diploid workers. These divergent reports make it likely that, depending on certain conditions, workers of *Lasius niger* can produce either haploid males or diploid females. Indeed directly observed workers of the species *Oecophylla longinoda* laid parthenogenetically either male or female eggs (Ledouc, 1950). Similarly, virgin workers of *Formica polyctena* can lay either male or female eggs (Otto, 1960; Ehrhardt, 1962). In contrast to these species of ants, in *Odontomachus assiniensis* (Ledoux, 1954) and *Plagiolepis pygmaea* (Passera, 1966) the eggs laid by the workers develop only into males. From these few reports concerning the production of males, it is clear that they arise via arrhenotoky. Unfertilized eggs do not, however, necessarily give rise to males since, in some species, thelytoky has been reported.

There exists a long-standing discussion on the determination of the female castes in species of ants: blastogenic (genetic) *vs.* trophogenetic (cf. Wilson, 1953). The first hypothesis favors the idea that queen and worker caste are strictly genetically determined, whereas the latter claims that any larva can become either a queen or a worker, depending on the quality and quantity of food obtained. Blastogenic determination was favored by Forel (1921) and Wheeler (1937) primarily. Female larvae of species of *Camponotus* and *Myrmica* were subjected to semi-starvation (Ezhikov, 1934); small but fertile females were obtained but never any intermediates. This was taken as strong evidence for the genetic determination of the female caste. In another species, *Acromyrmex octospinosus*, Wheeler (1937) found so-called gynergates, mixtures of queen and worker as judged by the mixed head morphology. Wheeler had only preserved material available and from this he postulated the origin of the gynergates to be analogous to that of the gynandromorphs which he also found in the same colony. Only the head morphology was studied. Therefore, against Wheeler's hypothesis, the argument can be put forth that the gynergates represented not mixtures of queen and worker but rather true intercastes analogous to intersexes (Whiting, 1938). The discussion concerning *Acromyrmex* has to be left open at this point. The possibility of a blastogenic caste determination in certain species of ants remains to be verified experimentally. On the other hand, evidence for trophogenic determination has become rather compelling in a few species, as will be shown below.

Blastogenic determination of female castes appears to occur in the stingless bee *Melipona marginata* (Kerr, 1950a, b). In this species the worker:queen ratio is relatively stable at 3:1, and, since all larvae apparently receive the same kind of food, genetic caste determination is probable. The exact basis for caste determination in this species, which seems to be an exception to the commonly found pattern, has to be worked out and deserves particular attention.

2. TROPHOGENIC DETERMINATION

It has been known for some time that workers of a queenright colony of the honeybee *Apis mellifera* rear only workers, whereas those of a queenless colony rear queens, provided eggs or young larvae were present when the queen was removed. One of the distinctive features of a queen is that her ovaries are composed of 150–80 ovarioles; ovaries of the workers usually have only four to eight maximally about twenty (Becker, 1925; Hess, 1942; Snodgrass, 1956). The question naturally arises concerning the mechanism by which larvae with identical genetic backgrounds develop into either a queen or workers (cf. Chauvin, 1952; Ribbands, 1953). The crucial experiments for an answer to this question were performed by Becker (1925) and Zander (1925). Worker larvae of up to $3\frac{1}{2}$ days of age, when transferred to a queen cell, developed into queens, but, when transferred at the fourth or fifth day after hatching, they became workers even though they had received queen food. These observations lead to the likely conclusions that workers feed a larvae in the queen cells a special food and that determination of the female castes takes place during the first $3\frac{1}{2}$ days after hatching. Nutrition of the young larvae was indeed found to be the decisive factor for queen determination in the honeybee (Haydak, 1943; Gontarski, 1949), and much research centered around the search for the specific food constituents. The queen food, royal jelly, contains more lipids than does the worker food. It is also found to have a relatively high vitamin E and B content as well as biopterin (Butenandt and Rembold, 1958). Feeding worker larvae with vitamin E did not increase the ovariole number and thus was proven not to be the essential factor in queen rearing (Hess, 1942). Furthermore, it has been found that biopterin, neopterin, or pantothenic acid have no effect on queen production whereas feeding of royal jelly has (Rembold and Hanser, 1964). The active principle was found to be contained in a dialysate of the royal jelly; this principle could be isolated chromatographically, but no further characterization was made (Rembold and Hanser, 1964). This substance is unstable and cannot be stored for long periods, as had been established earlier (Shuel and Dixon, 1960). To sum up, the female castes in the honeybee are epigenetically determined.

Epigenetic factors appear both to determine the castes as well as to control the colonial structure in several species of ants. Emery (1896) believed that any egg can yield a larva which can develop into any caste, depending on the kind of food supplies. The young larvae which will develop into either workers or sexuals are indistinguishable in many species. Support for the trophogenic hypothesis was obtained by Wesson (1940) for *Leptothorax curvispinosus* in the following way. If larvae of this species were reared with an abundance of food, predominantly queens were obtained; but those larvae reared on a minimum diet became mostly workers. Queens could be reared in *Oecophylla smaragdina* by the feeding of flower buds and aphids to the larvae; without these foods, only workers were obtained (Bhattacharya, 1943). Just as has been seen with these species, overwintering third instar larvae of *Myrmica rubra* can become either queens or workers, depending on the amount of food obtained (Brian, 1951, 1954, 1956). Normally, these overwintering larvae develop from large eggs laid in summer and are destined to become queens the spring after they have undergone a diapause. Colonies of this species which contain an abundant number of larvae produce predominantly workers, presumably because those larvae are fed relatively poorly. These observations show that even the last instar larva is still undetermined. Later it was found in *Myrmica* that those eggs laid in early summer were usually small and developed into larvae which metamorphosed right after completion of development; they become workers. Temperature of 25°C favored metamorphosis (Brian, 1963). Young queens of this species laid more of the small eggs (worker biased) than did the older ones (Brian and Hibble, 1964). Thus, in this species several factors—food, size of the eggs, age of the queen, temperature—will determine whether the larva develops into either a queen or a worker. Certainly the females castes are not genetically determined. Rather similar observations were made for the small meadow ant *Formica rufa rufo pratensis* (Bier, 1954a). Eggs laid in the summer are small and are predispositioned to become workers; those laid in the fall are large and yield queens. The queen eggs have a more prominent pole plasm than do the worker eggs; this presumably indicates environmental influences on the queen during formation of the eggs. Significantly, abundant food supply to summer larvae (worker biased) causes them to develop into queens.

Changes in the environmental conditions do affect brood rearing and caste determination in several species of social insects, as is implicit in the above-mentioned cases. This has been more drastically shown in some species of army ants. *Eciton hamatum* and *E. burchelli* rear a sexual brood (few females and several thousand males) only during the dry season (Schneirla, 1949, 1956; Schneirla and Brown, 1952); the queens are reared by overfeeding. The exact causes for the sudden switch from a sequence of worker brood rearings to a sexual brood rearing are not understood, yet the correlation to environmental changes can be made. Similar observations could be made in a species found in Arizona, *Neivamyrmex nigrescens* (Schneirla, 1961).

Questions concerning caste determination in social insects are closely associated with those of maintenance of the social structure during the colonies life. Among species of ants we find those which have but one queen, others with several or even numerous queens. Species of the first category, such as *Atta sexdens* (Eidmann, 1935), *Eciton hamatum* and *E. burchelli* (Schneirla, 1938, 1956), *Anomma wilverthi* (Raignier and Van Boven, 1955), *Formica rufa rufa* (Gösswald, 1951), *Sphaerocrema striatala* (Soulié, 1964), and others, obviously control the sustained presence of only one queen. Workers of *Sphaerocrema* (Soulié and Dan Dicko, 1966) and those of *Leptothorax exilis* (Baroni Urbani, 1966) eliminate the surplus queens. If one places several queens of *Solenopsis saevissima* together into a worker colony, all but one of the queens are eventually killed by the workers (Wilson, 1966). Mechanisms by which the workers recognize the queens and modes of elimination of all queens but one are obscure. Perhaps pheromonal, nutritional, or behavioral means are operative. This is perhaps borne out by the observation that in *Camponotus ligniperda* and *C. herculeanus* monogyny is normally found, yet oligyny can occur in large colonies where encounters of queens are rare (Hölldobler, 1962). Even in oligynous and polygynous species, such as *Formica rufa rufo pratensis minor* and *major* (Gösswald, 1951), a set ratio of queen:workers seems to exist. This may be controlled by the availability of nutrients, since workers feed their queens.

Finally, worker egg laying and its control has to be considered in the present context, since it contributes to the fecundity of the colonies of some species. The contribution of workers to the overall fecundity of queenright colonies may range from none to egg laying by all the workers. The former pattern is found in the army ants (Schneirla, 1949), and *Formica rufa rufa* (Gösswald, 1951), to mention just two species. In *Oecophylla smaragdina*, the queen plays merely an auxiliary role in the propagation of the species, since practically all egg laying is done by the workers (Bhattacharya, 1943). Workers of *Formica rufa rufo pratensis minor* (Gösswald, 1951) lay a few eggs, those of *Formica fusca* on the average 0.1 per year, and those of *F. gagates* 5.02 per year (Bier, 1953), whereas egg-laying workers in *Myrmica rubra* are observed rather commonly (Brian, 1953). Worker eggs of the latter species are frequently fed to the larvae in queenright colonies. In the presence of a queen, workers of *M. rubra* and *M. scabrinodis* lay only small (worker biased) eggs but queenless workers can lay queen eggs (Carr, 1962). Since young queens, as well as old queens which had been dead for more than 1 day, had no effect on the type of eggs laid by the workers, it was assumed that normally mature queens do influence worker egg laying, be that influence pheromonal or nutritional. No potent volatile substances could be extracted, however (Carr, 1962; Brian, 1963). Queenless workers of *Formica rufa rufo pratensis* lay eggs and produce both new queens and workers (Bier, 1954b, 1956). Some kind of inhibition of worker egg laying is lifted by the removal of the queens. Similar findings are reported for *Plagiolepis pygmaea* (Passera, 1966). The question naturally arises as to the mechanisms of inhibition. Colonies of *Leptothorax tuberum unifasciatus* were divided by a wire screen; on the one side the colony was queenright, on the other queenless (Bier, 1954b). Queenless workers of this species laid eggs even though they were exposed to any volatile substances which might have been emitted by the queen. Pheromonal inhibition of the workers analogous to that found in the honeybees seems therefore unlikely. However, since food exchange was not possible through the screen (normally food passage between workers can be observed), oral distribution of the inhibitory pheromone could not occur in the queenless colony. From this observation, one cannot exclude pheromonal control. On the other hand, an alternative must be considered, namely, the action of a "profertile substance" (Bier, 1954b), which would normally be passed from the worker to the queen, but which is used by the workers themselves when queenless. Evidence which supports this alternative interpretation and which repudiates the hypothesis of pheromonal inhibition was obtained for the species *Myrmica ruginodis* (Mamsch, 1965, 1967). Workers of this species lay eggs only after the queen has been removed, and then a queenless colony of fifty workers may lay up to sixty eggs in a 5-day period (Fig. 80). The presence of a queen for at least 12 hr per day appears to suppress worker fertility completely. Just as does the presence of a queen, the presence of larvae which are being fed by the workers suppresses worker egg laying. In a normal colony with both queen and larvae present, worker fertility is inhibited by the queen and larvae. This hypothesis is furthermore substantiated because, in a large queenright colony without larvae, worker fertility is only incompletely inhibited. A dead queen does not inhibit egg laying in the workers. The conclusion from this work is that, normally, workers feed a profertile substance to both queen and larvae which, in a queenless colony where there are no larvae to be fed, promotes egg growth in the workers themselves. No evidence for pheromonal control of worker fertility is available. More likely, worker egg laying is trophogenically controlled.

3. PHEROMONAL CONTROL

In honeybee colonies, the social structure, i.e. one queen and many non-laying workers, appears to be rigidly controlled throughout the life of the colony. This control is obviously exerted by the queen, as her removal results in worker egg laying. Even though the capacity of workers to lay eggs may be different in various races, it is obviously checked by the queen (Ribbands, 1953). Even in a queenright colony (Anderson, 1963), workers of the Cape

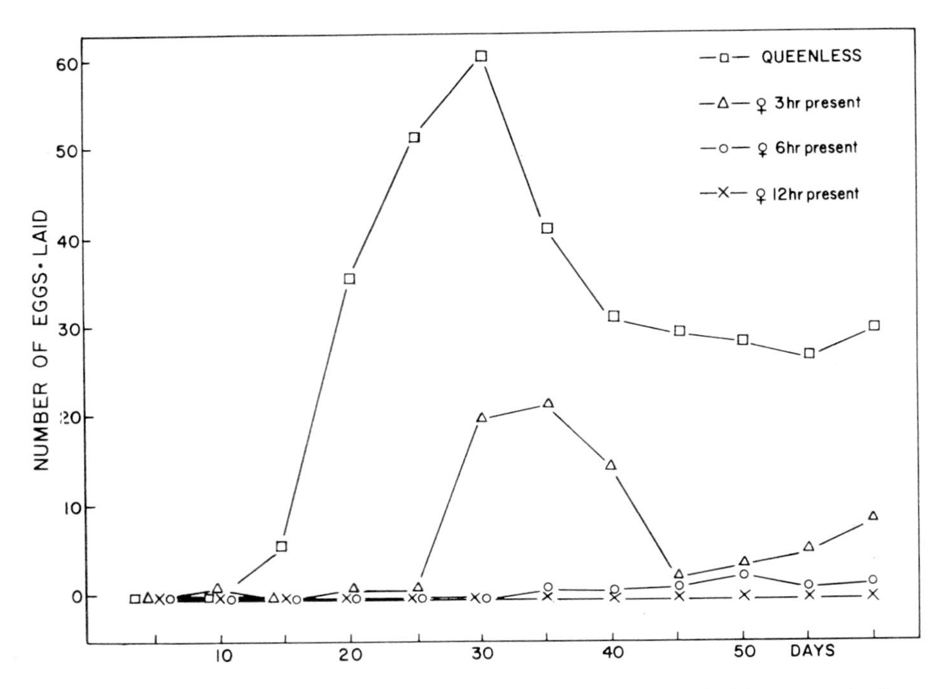

FIG. 80. Influence of the worker fertility by the presence of the queen in *Myrmica ruginodis*. Each point represents the collection of the eggs of 5 days. (Adapted from Mamsch, 1967.)

honeybee, *Apis mellifera capensis*, may lay occasionally; after removal of the queen, however, egg laying is considerably increased. One might also note that oviposition behavior of the workers differs from that of the queens; workers distribute their eggs irregularly on the comb, in contrast to the orderly fashion in which queens place their eggs (Seibert, 1922). If a normally fed colony becomes queenless within a few days close to 90 per cent of the workers begin to mature eggs as they simultaneously start to build emergency queen cells and raise queens if young larvae are present. How then does a queen prevent both queen cell construction and worker egg laying? In order to find the mode of worker control, Hess (1942) separated the queen from the workers by a single wire screen. The workers could lick the queen; they behaved queenright. If separated by a double grid, the workers behaved queenless. In contrast to these results, Müssbichler (1952), Pain (1954), and Groot and Voogd (1954) observed that workers behaved queenless even after separation from the queen with a single grid. From this it was concluded that workers have to come in actual contact with the queen (licking) to take up an ectohormone (pheromone). This inhibitory substance was thought to be distributed throughout the colony by the workers (Butler, 1956).

Following these preliminary experiments on the mode of worker inhibition, it was found that enthanol extracted queens did not inhibit worker egg laying, whereas reimpregnated dead queens did (Groot and Voogd, 1954; Voogd, 1955). The inhibition by impregnated queens was, however, only 39 per cent effective (Butler, 1957a) and fed ethanol extracts were either not potent at all (Vogel, 1956; Van Erp, 1960) or only slightly so (Butler, 1957a). In the search for the source of the inhibitory substance, it was found that smears from the mandibular glands of the queens but not from those of the workers inhibited queen rearing

(Butler and Simpson, 1958). One of the components of the extract from queens was identified as 9-oxodec-2-enoic acid (Butler *et al.*, 1959, 1961) and found to inhibit queen rearing and partial worker ovary development in quantities of 0.13 μg per bee. Simultaneously, Barbier and Lederer (1960) isolated and synthesized the same acid which was as potent as that of Butler *et al.* 9-oxodecenoic acid is considered to be the "queen substance". This substance is present in quantities of 130 μg per queen throughout her life and can already be extracted 5 days after emergence (Butler and Paton, 1962). Old superseded queens and swarm bees apparently contain only about one-half the amount of the queen substance found in egg-laying queens (Butler, 1957b).

Butler and Pain agree that the queen substance inhibits queen rearing (queen cell formation), but only incompletely inhibits ovary development in the workers. In the mandibular gland extract, Pain (1961b, c) and Pain and Barbier (1963) identify two pheromones, the queen substance (pheromone I) and another volatile compound attractive to workers (pheromone II) which, in synergistic action, inhibit both queen rearing and worker egg laying. Also, Butler (1961) observes the presence of a queen scent of the honeybee. The queen scent alone does not inhibit worker egg laying. In recent experiments it could then be shown that the volatile substance is 9-hydroxydecenoic acid (Butler *et al.*, 1964; Butler and Simpson, 1967) and, in combination with 9-oxodecenoic acid, almost completely inhibits queen rearing and worker egg laying olfactorily (Butler and Callow, 1968); no contact with these acids is necessary. This observation on the olfactory mode of action confirms those of Van Erp (1960) who found that extract impregnated pieces of wood, which could not be touched by the workers, inhibited queen rearing and about 50 per cent of the ovary development in workers. On the other hand, one has to reconcile these observations with those findings which failed to demonstrate olfactory control of worker physiology by the queens (Hess, 1942; Pain, 1954; Butler, 1954; and others).

While the function of the queen appears to be clear, one has to take into consideration other aspects of colony physiology. Thus it could be shown, for instance, that single workers of the honeybee rarely if ever mature eggs (Pain, 1960). Presumably, food exchange (proteins) between workers stimulates egg maturation; workers of larger groups mature eggs faster than those of smaller ones (Pain, 1961a). Another aspect of interest is that once workers have begun to lay eggs, they will inhibit egg laying in other workers in a fashion similar to that of queens (Velthuis *et al.*, 1965; Jay, 1968). The presence of a brood does not inhibit worker ovary development (Jay, 1968).

Finally, the endocrine aspects of the control of worker egg laying ought to be considered. If one assumes that the size of the corpora allata can give an indication of hormone production, one has to conclude that worker corpora allata are highly active. These glands are even larger in older workers than in queens (Fig. 81), yet in queenright colonies no eggs are matured by the workers (Pflugfelder, 1948, 1956; Lukoschus, 1956). Even though workers have large corpora allata, egg-laying ones have still larger glands (Müssbichler, 1952; Gast, 1967). This makes it likely that normally queens or rather the queen pheromones exert an inhibitory influence on the corpora allata (Lüscher and Walker, 1963). The question still has to be answered why, in queenright colonies, does the gonadotrophic hormone from the corpora allata not promote egg maturation. Gast (1967) proposes that the corpora allata of workers are indeed active, but no eggs mature since the neurosecretory cells of the pars interecerebralis are inactive. Presumably then, these cells are active in queenless workers. This interesting suggestion, based on histological evidence, ought to be tested experimentally. Alternatively, it is also possible to postulate that the corpus allatum hormone cannot express its action on the target tissues because of interference by the pheromones of the queen. Also, experimental evidence has to be obtained for the postulate that the corpora allata of queenright workers are actually active.

In a discussion on caste determination in termites, hypotheses basically similar to those which have been applied to bees, wasps, and ants have to be considered. To start with, in termites there are no female castes; workers and soldiers are generally of both sexes, which fact had already been recognized quite early by Lespés (1856) in *Termes lucifugus* and by Müller (1873a) in several species of *Kalotermes*. Termite colonies characteristically have only one or at the most a few sexuals. Primary reproductives undergo a regular metamorphosis, but break off their wings after the mating flight. Many species can, however, form so-called supplementary reproductives; these normally do not develop wings and a molt is necessary for their formation. It was Müller (1873b) who first recognized the nature of these supplementary reproductives in *Termes lucifugus*. New colonies are founded most often by isolated

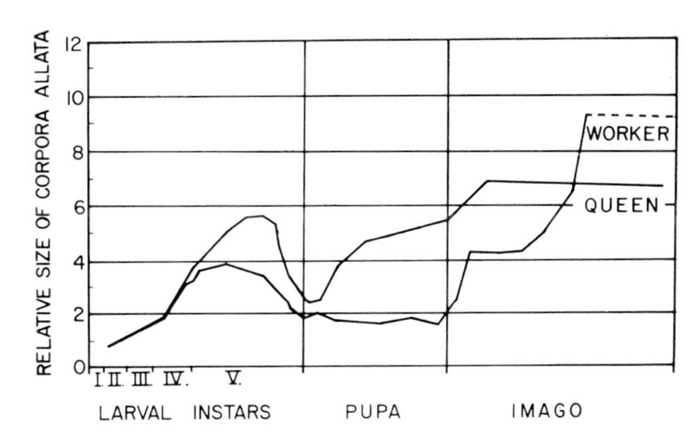

FIG. 81. Relative size of the corpora allata of queens and workers of the honeybee during larval, pupal, and early imaginal life. Size may represent activity. (Adapted from Lukoschus, 1956.)

pairs of primary reproductives which have been released from the colonies during swarming. Swarming in *Kalotermes flavicollis* generally occurs in spring or early summer, as it does in *Reticulitermes lucifugus* (cf. Grassé, 1949; Buchli, 1956a, b). Obviously, certain environmental cues, the nature of which is incompletely understood, determine the annual swarming period. In addition, the nutritional status of the colony and its size appear to be decisive factors for the annual appearance of sexuals (Buchli, 1958), since only well-fed colonies of at least 200 workers release winged imagines. Foundation of new colonies can take place by simple splitting of a large colony, as in *Anoplotermes* spec. and *Trinervitermes bettonianus*, where part of the colony walks off with the sexuals and founds a new colony (Grassé and Noirot, 1951). The phenomenon was termed sociotomy. Many aspects of termite biology have been discussed by Grassé (1949), Harris (1958), and Weesner (1960), and the reader is referred to these reviews for details not treated here.

Practically nothing is known about the mode of caste determination in termites except in the case of the supplementary reproductives; here, principally one species only, *Kalotermes flavicollis*, has been studied extensively. As in most species, shortly after the removal of both the primary sexuals, several supplementary reproductives arise in the colony, yet finally only one pair remains (Grassi and Sandias, 1896; Grassé and Noirot, 1946). Similar observations have been made for *Termopsis nevadensis* (Heath, 1927), *Reticulertermes angusticollis* (Castle, 1934), and others. From these early observations, one is lead to ask how the sexuals inhibit formation or stimulate elimination of additional sexuals. Related to this question is, of course, the problem of the competence of the individuals to respond to changed conditions in the colony. All larvae or nymphs are morphologically alike in several species of termites (Thompson, 1919; Bathellier, 1927), which leads one to conclude that sexuals, workers, and soldiers ontogenetically have a common type. In *Kalotermes flavicollis*, after removal of the sexuals, most of those larvae molting within a short time thereafter will be supplementary reproductives (Grassé and Noirot, 1946). A molt is essential for the formation of these sexuals. In *Kalotermes*, no true adult worker caste exists; Grassé and Noirot (1947) therefore introduced the term pseudergates for these non-differentiated individuals. Lüscher (1952a) then marked individuals with color spots shortly after hatching

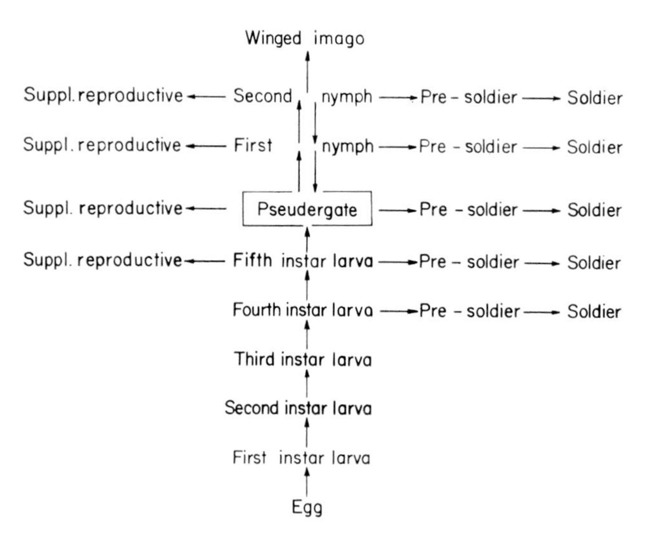

FIG. 82. Scheme of the possible developmental pathways of a larva of *Kalotermes flavicollis*. The pseudergate is the pluripotent stage. Soldiers develop only via two molts from either larvae or nymphs. (Adapted from Lüscher, 1961.)

and followed their fate throughout their lives (cf. Lüscher, 1961). He found that a fifth instar larva can molt either into a sixth instar larva, presoldier (then to soldier), or supplementary reproductives. A sixth instar larva, in addition, can molt into the first instar nymph (larva with wing buds) and further to the second instar nymph and imago. The sixth instar is, in other words, the pluripotent pseudergate which, in the majority of cases, will undergo stationary molts (Fig. 82). As was already known (Grassé and Noirot, 1946), pseudergates molting shortly after removal of the sexuals have a high competence to become supplementary reproductives. Lüscher (1952b) found that the competence of the pseudergates to become supplementary reproductives gradually decreases as time elapses since their last molt.

How in a colony is the formation of sexuals actually inhibited by the royal pair? This question centers around the problem of the mode (genetic, trophogenic, pheromonal) of control. Trophogenic control appeared to be one of the likely modes because one can easily observe that termites not only feed each other but also solicit the extrusion of feces droplets which are then readily taken up. Yet as will be shown below, current evidence favors an exclusively pheromonal control of caste determination. Alcohol or ether extracts of queens of *Zootermopsis angusticollis* fed to an orphaned colony on filter paper partially inhibited sexual formation (Castle, 1934, 1946; Light, 1942, 1944b). Observations in *Z. angusticollis* seemed to show that males and females probably liberate ectohormones (pheromones) suppressing the formation of their own sex (Light and Weesner, 1951); colonies headed by only females produced an excess of males, those headed by only a male an excess of females. In a series of observations on *Kalotermes*, Lüscher (1952b) then found that separation of a normal and orphaned colony with a wire grid did not suppress production of supplementary reproductives in the orphaned colony; however, all newly produced sexuals were eliminated within a short time. Elimination did not take place when the colonies were separated by a double grid, so that the pseudergates of the two colonies could not touch each other or the sexuals. The interpretation of these results is that pseudergates normally obtain with their proctodeal fluid (which cannot be obtained after separation with one grid) substances that inhibit differentiation towards the imago. Upon recognition of more than one pair of sexuals (touch with antennae), they are caused to eliminate the supernummerary ones. Apparently no olfactory pheromones exist in this species which might be functional in colonial control. This is, furthermore, amply shown by the following experiments. A queen can be fixed into a position between two colonies such that one colony has access to her head and the other to her abdomen. If a male was added on the side with the front part, no inhibition of sexual formation took place on this side. In the colony which contained only the female abdomen, however, production of female reproductives was considerably inhibited; male sexuals were produced (Lüscher, 1956a). Varnishing all of the female abdomen except the anus did not affect the inhibition, but closure of the anus did. Thus it seems that a substance is given off by the queen with the feces droplets which inhibits female production. Complete inhibition of sexual formation of both sexes was achieved if a male was kept with the female abdomen. In conclusion one can postulate the production of two sex specific pheromones in *Kalotermes*. This conclusion has, principally, also been reached by Grassé and Noirot (1960) for the same species and is analogous to that postulated for *Z. angustocollis* (Light and Weesner, 1951). Further observations in *Kalotermes* also show that the sexual male apparently is stimulated to release his pheromone either by another male or by the female, since, when kept alone with a worker colony he was rather ineffective in the prevention of the production of male supplementary reproductives. In this connection it must be pointed out, however, that in *Kalotermes* none of the inhibitory pheromones have been extracted so far (Lüscher, 1956a, 1961). An extract of the male head does, however, stimulate the production of female supplementary reproductives.

How is it possible that the inhibitory pheromones emitted by a single pair of sexuals can effectively inhibit sexual production in a large colony of 10,000 individuals? It is suspected that pseudergates participate in the distribution of the pheromones. Passage of these substances through the pseudergates seems possible. In order to test this possibility, Lüscher (1956b) positioned male or female pseudergates into a partition between an orphaned (abdomen side) and a normal (front part) colony. He found that, when males were used, these absorbed much of the male inhibitory pheromone; consequently male production was not inhibited but rather that of female supplementary reproductives, since the female inhibitory pheromone was passed through. If a female larva was used, the opposite was observed. Since olfactory inhibition did not occur, such a mode of inhibition of the production of supplementary reproductives would thus be more efficient than if each member of the colony had to come into contact with the royal pair. In summary then, in *Kalotermes* at least three pheromones are involved in the control of reproductives formation (Lüscher, 1961) (Fig. 83): a female and a male inhibitory, and a female stimulatory pheromone given off by the male. Furthermore, additional male and female pheromones function in signalling the presence of the sexuals; this may result in the elimination of the supernummary reproductives. If several sexuals are in the colony, they fight with each other. The injured individuals are then eliminated by the workers (Ruppli and Lüscher, 1964).

Mechanisms of caste control are not understood in species other than those mentioned. Supplementary reproductives have been found in most species examined. Generally it is thought that they arise from pseudergates via a molt. Yet in higher termites, the Termitidae,

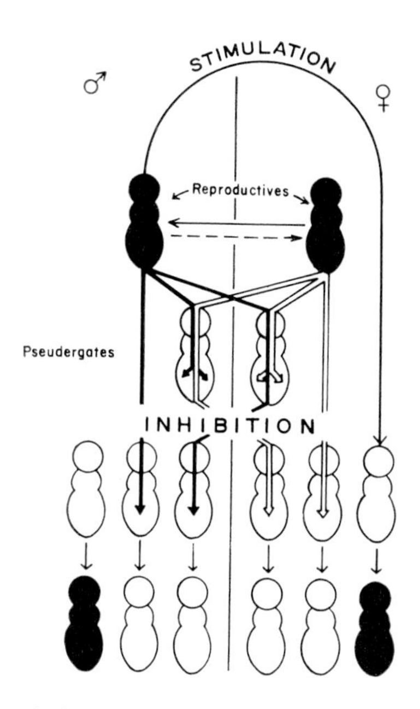

FIG. 83. Pheromonal control of castes by the royal pair in a colony of *Kalotermes flavicollis*, particularly the control of the production of supplementary reproductives. (From Lüscher, 1961.)

pseudergates may have lost the ability to become reproductives. Here, imagines which are found in the colonies at the time of the loss of the primary reproductives will become the supplementary reproductives (Noirot, 1954, 1956).

Another intriguing problem is how the perception of pheromones is translated into suppression or stimulation of the production of specific castes. Kaiser (1956) observed in *Microcerotermes ambionensis* that the corpora allata of pseudergates about to molt into supplementary reproductives were larger than those of other pseudergates. The same situation can be observed in *Kalotermes* (Lüscher, 1960). Interestingly, in normal colonies implantation of highly active corpora allata from reproductives into pseudergates did not induce the production of reproductives. Instead soldiers were invariably produced (Lüscher and Springhetti, 1960; Lebrun, 1967a, b). This can be interpreted to mean either that normally the corpora allata are not involved in the production of supplementary reproductives or that the interaction of certain pheromones with endocrine gland activity yields different results depending on the combination of the various agents. No definite conclusion can be drawn from the available data. The situation may be somewhat analogous to that found in the honeybee; the highly active corpora allata in workers do not cause egg maturation in the queenright colony.

4. DOMINANCE HIERARCHY

Social species with small colonies frequently exhibit a distinct division of labor; egg laying is often performed by only one or a few individuals. Thus, in *Polistes dubia* the overwintering foundress performs all duties at first (building, cleaning, foraging, fanning) until auxiliary females arrive (first brood); then the foundress concentrates more on egg laying, while the other females take over the worker duties (Steiner, 1932). Even when foundation of the nest in *Polistes gallicus* is performed by the joint efforts of several overwintering females, one of them lays more eggs than the others, thus exhibiting a kind of

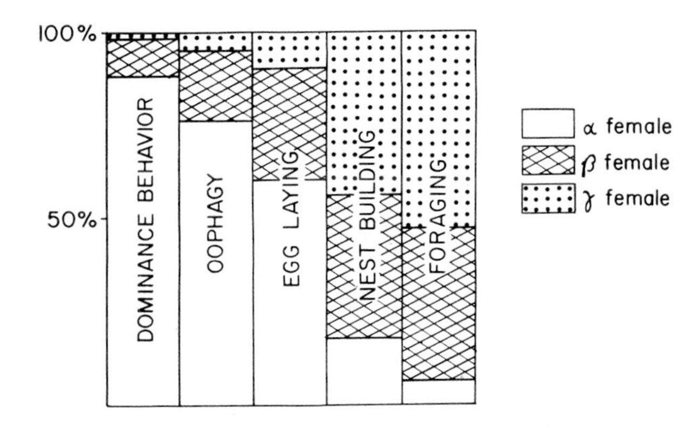

FIG. 84. Dominance hierarchy in *Polistes gallicus* and division of labor in a colony of three founding females. Each of the activities was observed 100 times and is expressed here as percentage spent by an individual at a certain activity. (Modified from Pardi, 1950.)

dominance hierarchy (Heldmann, 1936). Following these preliminary observations, an analysis of the situation in this species was begun by Pardi (1942, 1944, 1948a, b, 1950). In a colony founded by several females, the α animal lays the most eggs, eats more of the eggs than do the β and γ animals, and receives food from the other females (Fig. 84). The α female also participates least in nest building and foraging. Clearly, within a short time a social hierarchy is established which involves fighting and a differential food supply among the founding females. Functionally, a monogynic colony is established; no morphological distinction can be made between the females. Then, as soon as the first brood appears, these workers also establish a dominance hierarchy among themselves (Gervet, 1962). Generally workers build on the nest and forage. The dominant worker has a trophic advantage and may even lay some eggs. Egg laying occurs particularly when the dominant female has been lost, which again indicates that this worker is subordinant to the founding female. It is also subordinant to the auxiliary females when those are present. Workers do not mate and do not overwinter (Gervet and Strambi, 1965). In summer, probably associated with high temperature, the colony produces foundresses; in late summer the males appear (Pardi, 1950).

Naturally, the question arises as to what makes a female a dominant individual over others? Is it the reproductive capacity? Ovariectomy of the dominant female of *Polistes bimaculatus* did not, however, abolish her dominance (Deleurance, 1948, 1955). It appears, thus, that some internal factors which are non-related to ovarian development allow certain females to become dominant (Gervet, 1964b). Is it then the fact that the dominant female can guard her eggs better than the other females can theirs, a fact which allows her to remain dominant and which may be related to her fighting ability? Deleurance (1950) removed the eggs of the α female of *P. gallicus* and protected those of the β female from being eaten. Promptly the β female increased her egg-laying activity and could, apparently as a consequence, become the dominant female. From these outlines, it is evident that there are several characteristics which not only distinguish a dominant female from the others but which also aid in the maintenance of the hierarchy: dominance behavior, differential oophagy, i.e. eating eggs of other females (Gervet, 1962, 1964c), and the capacity to lay more eggs than others. The latter is probably only the consequence of the other features, primarily the ability to obtain nutritious foods. This may be further accentuated by the partialling out of nest duties, i.e. the subordinant females build on the nest and forage; thus their available energy is diverted into other channels.

Functional monogyny or oligyny is observed in several other small societies of bees and wasps. It is found, for example, in bumble-bees (cf. Cumber, 1949), some Polybiinae (Richards and Richards, 1951) and species of the Halictidae such as *Halictus marginatus* (Quénu, 1958a, b) and *Evylaeus nigripes* (Knerer and Plateaux-Quénu, 1966). It is tempting to postulate a control of egg laying such as is found in *Polstes*. However, no detailed analyses have been made in these cases and reports are generally descriptive only.

Obviously control of social structure and reproductive activity through behavioral means can only function in relatively small colonies. Establishment of dominance involves a considerable expenditure of energy during fighting, this energy loss would limit the effectiveness of the dominant animal in a large colony. In such a sizeable colony any one animal could only meet a limited number of individuals and therefore, in these cases, pheromonal or trophogenic control of the castes is probably a more economic means.

B. Fecundity

Propagation of a species depends largely on the egg-laying capacity of the females and the survival of offspring until they reach a reproductive age. One is generally amazed by the great number of animals in colonies of certain social insect species, and one gains the impression that social insects are more fecund than others. Some of the more spectacular cases may be listed here. For example, *Vespula vulgaris* can lay about 22,000 eggs per year (Spradbury, 1956), a number which is not reached by non-social species. Or, colonies of the small meadow ant, *Formica rufa rufo pratensis minor*, contain 25,000 queens and 1.5 million workers (Gösswald, 1951). In Europe and North America the honeybee queen, *Apis mellifera*, may lay about 120,000 eggs annually (Brünnich, 1922; Nolan, 1925) with a peak egg-laying activity in early summer (Fig. 85). During this peak period she may lay 1000 to 2000 eggs per day. In subtropical regions of the Mediterranean, *Apis* queens may lay up to 200,000 eggs per year (Bodenheimer and Ben Nerya, 1937). Just as in the northern regions, the queen here has an annual peak laying period in the spring and early summer and only lays few eggs during fall and winter. If, as has been estimated, a queen lives on the average for 3 to 4 years, she may lay a total of up to half a million eggs in her lifetime. The egg laying capacity of army ant queens such as *Eciton burchelli* is even greater. This queen lays

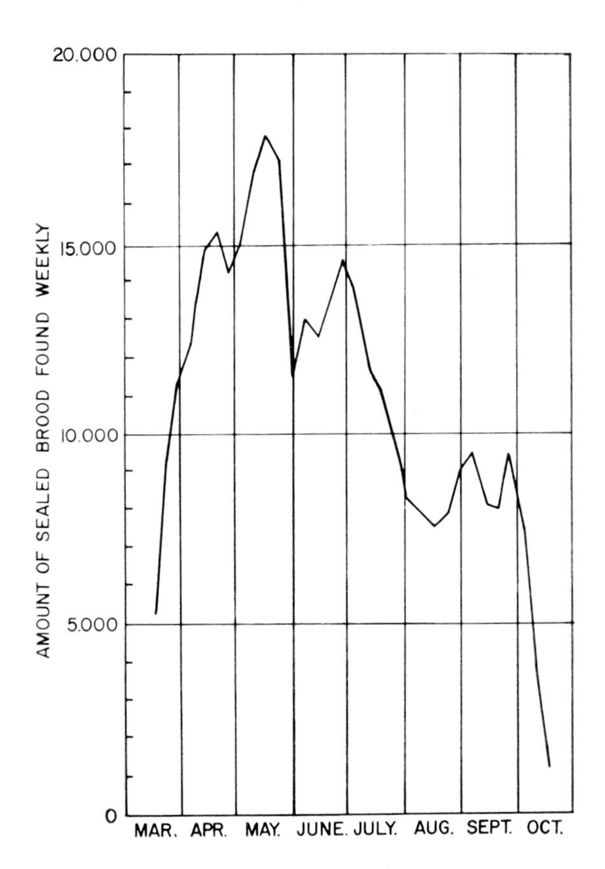

FIG. 85. Annual egg laying activity of a honeybee queen in North America. The pattern is typical for several colonies observed. In certain colonies in late summer, an increased oviposition rate is often observed, which here is only weakly noticeable. (From Nolan, 1926.)

about 120,000 eggs per brood, i.e. every 36 days (Schneirla, 1957a); this would mean that a queen lays over 1 million eggs per year. The egg-laying capacity of the queen of the African driver ant *Anomma wilverthi* is estimated to be 3 to 4 million per brood every 25 days, which amounts to about 40 million per year (Raigner and Van Boven, 1955). In this list of "truly fecund" females one should not omit the queen of the termite species *Bellicositermes natalensis*; during her peak period she lays about 36,000 eggs daily, i.e. about 13 million annually (Grassé, 1949). The life span of a queen of these species is not known for certain, but it is thought that she lives for up to 6 or 10 years.

The question naturally arises whether social species with this fantastic egg-laying capacity of the queen are indeed more fecund than other insect species. There is no doubt that the queen is highly fecund, but considering the great number of infertile females, the workers, the average fecundity calculated per female is less spectacular. The high fecundity of a queen is apparently only possible because she is fed a highly nutritious diet by the workers which themselves are infertile. Trophalaxis and food exchange has been known in social species for some time (cf. Janet, 1903) and has been repeatedly shown to occur in several species. Another prerequisite for a high egg-laying rate is the relatively long life span of a queen, which is associated with a gradual build up of a colony fit to perform the various duties essential for survival of the colony. It has indeed been observed that a queen of a large colony is more fecund than one of a small one of the same species. This is illustrated by the halictine species *Augochloropsis sparsalis*, as well as by *Lasioglossum imitatum* and *L. rhyticlophorum* (Michener, 1964), and *Myrmica rubra* (Brian, 1953b). However, if egg numbers are related to the total number of females in a colony then the ratio of eggs per female decreases as the colony size increases. Exceptions to this have been found in polygynic species like *Bombus americanorum*. Here, in colonies of 30 to 50 workers, 1.3 queens on the average are present, whereas in those of 110 to 130 workers, up to 60 queens or more are found (Michener, 1964); thus fecundity per female can be increased in larger colonies of this species.

Even though it is true that colony fecundity (queen egg laying) increases in larger colonies of most species, it can be still larger in queenless colonies in which many workers begin to lay eggs. The total number of eggs laid by queenless workers of the ant *Leptothorax tuberum* was in excess of that which a single queen could lay (Bier, 1954b); similar observations have been made for *Myrmica ruginodis* (Mamsch, 1967).

Occasionally one observes among social species what appears to be a considerable waste and loss of reproductive potential. The town-building ants of the genus *Atta* release many thousands of individuals during swarming periods, yet only a few are actually successful in founding new colonies; the majority of the swarming sexuals die soon after swarming. Also, foundresses will eat up to 90 per cent of their own eggs during the period between egg laying and emergence of the first brood (Goetsch, 1953). What the biological significance of this peculiar behavior could be remains obscure. One can visualize that it severely curtails the rapid rise of an efficient worker colony.

Interesting reproductive behavior has been observed in species of the Doryline ants of Africa and the New World. Both the driver ants of Africa and the army ants of America are monogynous species. As first analyzed in *Eciton hamatum* (Schneirla, 1938) and later found to apply in *E. burchelli* (Schneirla and Brown, 1952; Schneirla, 1956, 1957a, b), nomadic phases alternate with so-called statary ones. The hatching of the brood is correlated with the onset of the nomadic phase and, when the larvae begin to spin their cocoons, the statary phase begins (Fig. 86). During the statary phase, only small daily raids are conducted. It appears that brood rearing or rather the periodic egg laying of the queen determines the activity phases of the colony. The cycles are maintained by internal factors and not by extrinsic influences. Schneirla maintained that it is the hatching of the larvae—an event which necessitates a great deal of activity by the workers (feeding and foraging)—which stimulates the colony to become nomadic. Nomadic phases are not induced by the shortage of food, however. Essentially similar cycles are found in *E. vagens*, *E. dulcius*, *E. mexicanum* (Rettenmeyer, 1963), and the Arizonan species *Neivamyrmex nigrescens* (Schneirla, 1958, 1961). In the African driver ant *Anomma wilverthi*, Raigner and Van Boven (1955) recorded migrating rhythms of approximately 11, 20–25, and 50–56 days. Each of these migratory periods is correlated with specific events in the colonies of up to 20 million workers: egg laying, emergence of workers, and emergence of males. Again, as in *Eciton*, it is not the shortage of food which causes migration. Internal events determine the behavior of the colony.

C. *Evolutionary Aspects of Social Behavior*

The question of how insect societies may have evolved is of interest in the present context since it bears on aspects of reproduction. The subject has been reviewed repeatedly in recent years for various groups of social insects (Evans, 1958; Michener, 1958; Hamilton, 1964; Brian, 1965a, b; Wilson, 1966). It is clear that discussions of the evolution of insect societies and of the driving forces which enable their continued presence will always be theoretical considerations. Presumably, insect societies have arisen repeatedly and we will briefly point out some of the possible pathways.

All termites are truly social. It can be envisioned that here social contact is essential for survival, since these insects depend on proctodeal feeding during which intraintestinal symbionts are passed on; without these symbionts the animals apparently cannot utilize their food, cellulose. Interestingly, in the cases known, Isoptera seem to use the same pathway (proctodeal feeding) to control the social structure, i.e. to pass on inhibitory or stimulatory pheromones. Individuals of the termite colonies have an added advantage in that they can control their microclimate, construct protective nests, and can thus withstand unfavorable conditions. How these features could help to increase the reproductive capacity of

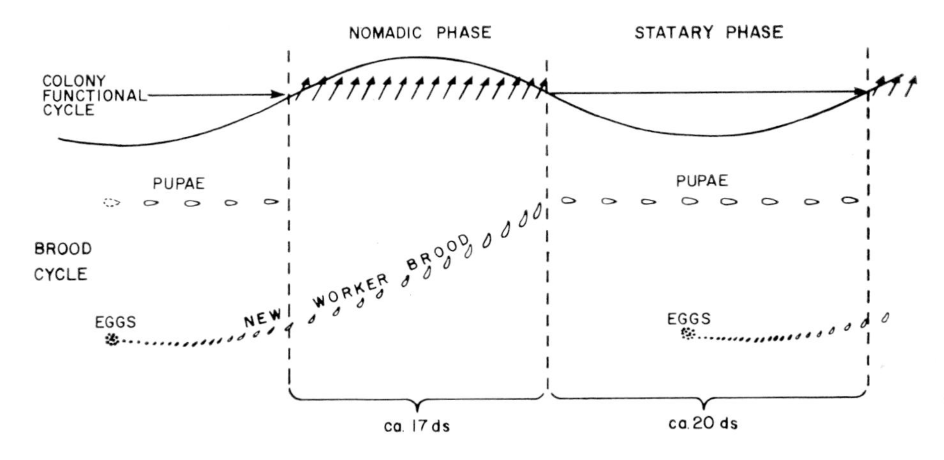

FIG. 86. Periodic activity changes in the army ants of the genus *Eciton*. The nomadic phase is induced by mass emergence of the new brood. Since egg laying of the queen is periodic, the phasic activity of the colony is basically determined by the queen's oviposition rhythm. (From Schneirla, 1957.)

the sexual individuals remains obscure. Many termites are slow in starting a colony and it often takes several years to build up a sizeable colony. Apparently the long life span of the termites is an added factor in the build up of a colony. Certainly the high fecundity of a queen of *Bellicositermes natalensis* is only possible because she is fed and attended by numerous workers. The social structure favors a large society.

Evolution of societies among some groups of Hymenoptera, particularly wasps, may take place via a loose association of the adults. One can envision that species having a subsocial structure, like that of the Sphecoidea which often nest closely together, may have been the ancestors of our present day social species. For example, several females of *Chloralictus opacus* and *C. phytidophorus* found small colonies together, yet a social structure (division of labor) is barely observable (Michener and Lange, 1958). Mated females of the species *Lasioglossum inconspicuum* and *L. zephyrum* may join in founding a colony (Michener and Wille, 1961; Batra, 1964, 1966). In these species an additional element enters, namely, the relatively long life span of the queen which allows a definite relationship between mother and daughter to be built up. Even though females of the first brood do lay eggs, they are less fecund and mainly perform the worker duties like nest building and foraging. Several other species of the Halictidae exhibit similar patterns. One of the salient features in these species is the fact that there is no definite mother–larva relationship; mass provisioning is common among Halictidae. Therefore, as has been pointed out by Michener (1958), the colony of foundress and daughters is essentially an association of adult females.

Foundation of a colony by an association of several females may be advantageous for the species since it may facilitate the rapid build up of a colony. This has indeed been observed among species of ants, for example *Lasius flavus* (Waloff, 1957). However, as soon as the first brood pupates, the founding females apparently become intolerant of each other, begin to fight, and eventually split, thus founding several daughter colonies. A detailed analysis of the biology of the species should be very rewarding. One can only speculate on what causes the change in behavior of the founding queens. Is it the appearance of pupae which relieves the females of larval care, is it the effectiveness of certain pheromones which were not apparent before, or is it the availability of more food (profertile substances of Bier) which facilitates aggressiveness? Colony foundation by several queens is also observed in species of *Polistes* as well as in other Hymenoptera. Here we witness the development of a dominance hierarchy which seems to allow the females to stay together.

Evans (1958) suggests a different pathway taken in the evolution of societies in the wasps. Important in this is the mother-larva relationship through progressive feeding and parental care. The first important step towards sociality is the provisioning of the larva with prey and, more so, the progressive provisioning as the larva grows. This is found in several subsocial species of the Sphecoidea, such as *Ammophila aberti* and *Bembix brullei*. In some cases a female takes care of several nests simultaneously. Among the Vespidae, for instance, females of *Synagris cornuta* and *Stenogaster micans* prepare the food to be given to the larvae by maceration. From here, one can envision a further step towards a true society in which the offspring stay in the nest and the life of the female is prolonged. The foundress then becomes the individual which lays most of the eggs, as is seen in *Polistes*. Further, true castes are established which are first physiologically and then morphologically distinguishable. In other words, true sociality involves division of labor which may or may not be associated with morphologically distinguishable castes. It is a possibility that division of labor among small insect societies (as seen in *Polistes*) evolved as a consequence of a dominance hierarchy (West, 1967). Under these circumstances it seems that the fittest individuals—the dominant ones—contribute most to the genetic pool of the species.

In a survey of social insects, one gains the impression that colonies of the higher evolved species are generally headed by only one egg-laying queen which can be extremely fecund. Certainly the high fecundity of the queen is only possible because she receives the highly nutritious food from the workers. Since the presence of a single egg-laying female in a colony—even though highly fecund—apparently does not proportionately improve the fecundity of the whole colony when calculated as eggs laid per female, other factors would seem to favor the establishment of monogynous colonies. The term haplometrosis has been introduced by Wasmann (1910) to describe a colony headed by one mated egg-laying queen; pleometrosis is the term used for colonies headed by several queens. The terms replace those of monogyny and polygyny respectively since the latter do not apply in the strict sense of the term to insect societies (cf. Richards and Richards, 1951).

In order for haplometrosis to be advantageous and increase the fitness of the species, the trait must outweigh the disadvantage that much of the genotype is lost due to the infertility of the workers (Hamilton, 1964). One can find some evidence that, during the evolution of social hymenopterous species, there was a trend from permanent pleometrosis as in *Belonogaster* and others to temporary pleometrosis as exhibited in many *Polistes* species toward haplometrosis as in the Vespidae (Evans, 1958; Yoshikawa, 1962). The means to achieve haplometrosis may be different in the various species. Exceptions to this evolutionary trend are found in the Polybiinae which are pleometrotic yet are considered to be highly evolved (Richards and Richards, 1951; Evans, 1958). Haplometrosis and pleometrosis can occur in species of the Halictidae, even in closely related species. For example, *Halictus marginatus* (Quénu, 1958a, b) and *Evylaeus nigripes* (Knerer and Plateaux-Quénu, 1966) are haplometrotic, whereas in colonies of *Chloralictus opacus* (Michener and Lange, 1958) and *Halictus maculatus* (Knerer and Plateaux-Quénu, 1966) pleometrosis is observed. In the latter species some inhibition of "workers" exists but it is incomplete. Colonies of the species of the first type generally live for several years and, during the final year, the workers begin to lay eggs from which males arise. For some reason, control of worker inhibition breaks down towards the end of the life of the colony. The same is found in the bumble-bees, such as *Bombus lucorum*, *B. terrestris*, *B. hortorum*, where workers often begin to lay eggs late in the season (Cumber, 1949); the colony becomes pleometrotic.

One can reason that an oligynous colony would make maximal use of the reproductive capacity of the egg-laying females and the availability of workers for nest duties; less of the gene pool would be lost than when only one female lays eggs. Such a colony would thus be fittest, but again, if several females are to be fed and cared for by the workers, fewer workers would be free to build and clean the nest. There seems to be a certain correlation of the occurrence of haplometrosis with the construction of enormous nests. Perhaps females (workers) which are engaged in construction and other nest activities are unable to participate in reproductive activities. The advantage of a large haplometrotic colony may thus not lie in the reproductivity of a single queen or whole colony but rather in the availability of a protective nest. This does, apparently, outweigh the overall reduced fecundity as calculated per female member of the society.

CHAPTER 15

GLOSSARY

Acroblast. Body within the sperm head which later gives rise to the acrosome.
Acrosome. Body within the head of a spermatozoon, usually located anterior to the nucleus. Its function is presumably to aid in the penetration of the egg membranes.
Aedeagus. The more or less extended portion of the phallus, functioning as a penis. Typically a sclerotized tube.
Aestivation. Summer diapause found in some insect species. It is apparently induced by drought or long-day conditions.
Alienicola. Individual of an aphid generation living on the secondary host. Syn. virginogenia. New term: exsulis-virgo (Lampel, 1968).
Allopatric. Referring to populations of species occupying mutually exclusive (but usually adjacent) geographic areas.
Ameiosis. Occurrence of only one division during meiosis instead of two.
Amphimixis. Union of egg and sperm pronuclei to form a zygote.
Anautogeny. Pertaining to female insects which are unable to reproduce without having ingested proteinaceous foods.
Androgenesis. Development of an individual from the fusion of two sperm pronuclei.
Andropara. Aphid of the sexupara generation which produces exclusively males (sexualis) via parthenogenesis.
Anemotaxis. Reaction and orientation to air currents. Insects generally face upwind.
Anholocyclic. Referring to aphid species which do not reproduce sexually. No fundatrix, sexupara, nor sexualis generations are known.
Apomixis. Development of an egg without fertilization.
Arrhenogenic. Referring to female which produces only males.
Arrhenotoky. Males arise from unfertilized eggs. Haploid pathenogenesis.
Autogeny. Pertaining to female insects which are able to lay at least some eggs without having ingested any proteinaceous foods.
Automixis. Union of two nuclei which arose from a division of a parent nucleus.
Bursa copulatrix. A more or less extended outpocketing of the outer female genital duct. Ectodermal origin.
Chemotaxis. Orientation towards and approach to a source of odoriferous chemicals.
Civis. New term in aphid terminology. It is used as a prefix in description of individuals of various generations living on the primary host plant (Lampel, 1968). Civis-fundatrix, civis-virgo, etc.
Common oviduct. Anterior ectodermal portion of vagina; gonoduct, uterus.
Conceptaculum seminis. In Cimicidae, the mesodermal pouch of the oviduct in which the spermatozoa assemble.
Deutertoky. Parthenogenetic production of both males and females by the same female.
Diapause. A period of arrest of growth which enables the species to survive unfavourable climatic conditions. Diapause is generally preceded by a preparatory period (accumulation of reserves). Diapause is not to be confused with torpor or quiescence.
Dioecious. Referring to aphid species in which certain generations live on either of two hosts (primary or secondary host). Heteroecious may be the preferable term.
Endophallus. The inner chamber of the phallus into which the ejaculatory duct opens; often an eversible sac or tube.
Epigenetic. Outside the genes. Mechanisms which affect gene expression.
Epistasis. The masking of the expression of non-allelic genes.
Ergate. The worker in species of ants (see Wheeler, 1910).
Exsulis. New term in aphid terminology. It is used as a prefix for individuals of the various generations living on the secondary host. Thus one distinguished exsulis-fundatrix, exsulis-virgo, etc. Cf. **Civis.**
Fundatrigenia. Parthenogenetic individuals which are the offspring of the fundatrix. The term is also used for

the subsequent parthenogenetic generations; it is then synonymous with virginogenia. The new term civis-virgo replaces both of these terms.

Fundatrix. In aphids, the viviparous parthenogenetic female which arises from the fertilized egg laid by the sexualis.

Germarium. Anterior portion of the ovariole where differentiation of the oocytes, nurse cells, and follicular epithelium takes place.

Gonapophyses. Pairs of appendages on the eighth and ninth abdominal segments of the female insect; they generally form the ovipositor.

Gonochorism. Determination and development of separate sexes, male and female, in a given species.

Gonopore. Opening through which the gametes are discharged.

Gynandromorph. In a bisexual species an individual which is composed of both male and female body parts. The animal tissues are genetically either male or female.

Gynogenesis. Parthenogenetic development of the egg after penetration of the egg membranes by spermatozoa the sperm pronucleus does not participate in the development.

Gynopara. Aphid of the sexupara generation which produces exclusively females (sexualis) via parthenogenesis.

Haplometrosis. The presence of only one fertilized egg-laying female per colony of social insects. Several males may be present (Wasmann, 1910). Monogyny refers to the presence of 1 pair of sexuals per colony.

Harpagone. Movable periphallic process having a clasping function. Syn. claspers.

Hermaphroditism. Condition in which an animal produces both male and female gametes. Self-fertilization may occur.

Heteroecy. Change of host plants. Observed among species of the Aphidoidea.

Heterogamety. Pertaining to the process of producing two kinds of gametes with respect to sex determiners, i.e. one type is female determining, the other male determining.

Heterogony. One or more parthenogenetic generations alternating with a bisexual one, e.g. in aphids and certain gall wasps.

Heterosis. Hybrid vigor resulting from an increased heterozygosity. For any measurable quality, the offspring are superior to either of their parents.

Hibernation. A form of diapause during the winter. It is distinguished from torpor in that a preparatory period precedes hibernation.

Hiemalis. Viviparous aphid generation that overwinters.

Holocyclic. Referring to aphid species which exhibit a succession of parthenogenetic and bisexual modes of reproduction. Heterogony.

Intersex. An individual which is a mixture of male and female body parts. Each cell, however, has an identical genetic background; environmental factors, such as temperature, hormones, or parasites influence the expression of the one or the other sex.

Intima. Innermost cuticular lining of organs of ectodermal origin; continuous with the cuticle of the body wall.

Mesomeres. The inner pairs of phallomeres.

Monoecious. Species of aphids living on only one food plant.

Monogynic. Referring to a female which gives rise to offspring of only one sex, male or female.

Nulliparous. Females which have not yet oviposited; the term is used primarily for mosquitoes.

Oolemma. Membrane surrounding the oocyte.

Organ of Ribaga. Organ of ecto-mesodermal origin on the abdomen in Cimicoidea into which the spermatozoa are injected. Cf. **Spermalege.**

Paedogenesis. Referring to viviparous larvae which produce offspring.

Paragonium. Accessory sex gland of the male Diptera; the secretion is mixed with the sperm mass.

Parameres. Pairs of phallomers which form the claspers in a certain species. Harpagone, gonostyle.

Parous. Females which have laid eggs at least once.

Parthenogenesis. Mode of reproduction. Development from an unfertilized egg.

Phallomers. Genital lobes or processes formed at the sides of the gonopore. They usually unite during ontogeny to form the phallus.

Phallus. The unpaired intromittent organ, including the phallobase and aedeagus.

Pleometrosis. The presence of more than one egg-laying fertilized queen in a colony of social insects (Wasmann, 1910). The term polygyny is misleading since it refers to one male having several females, which is not the case among social insects.

Pseudergate. Worker of the termite colony (Grassé and Noirot, 1947). No true worker caste exists in termites. The undifferentiated larvae perform the worker duties.

Receptaculum seminis. Dorsal outpocketing of the common oviduct or bursa copulatrix; it stores the sperm mass. Structure and occurrence are very variable according to species. Cf. **Spermatheca, Conceptaculum seminis** (Cimicidae).

Sexualis. The bisexual form of aphids, usually occurring in the fall or late summer.

Sexupara. Viviparous parthenogenetic female aphid which produces the bisexual forms, male and female.

Spermalege. Syn. to organ of Ribaga or organ of Berlese.

Spermatheca. Outpocketing of the female genital duct; it stores the spermatozoa. Cf. **Receptaculum seminis.**

Spermatogenesis. Process of formation of the spermatozoon from the spermatogonium through mature spermatozoon.

Spermatophore. Sperm-containing body built by the male either before or during copula and used in the transfer of spermatozoa to the female.

Spermiogenesis. Differentiation of spermatid to spermatozoon.

Sphragis. A structure found in some Lepidoptera adhering to the female after copula; it is formed from material secreted by the male. It ought not to be confused with a spermatophore.

Thelygenic. Referring to a female which, after mating, produces exclusively female offspring.

Thelytoky. Parthenogenetic production of female offspring. Diploid parthenogenesis.

Vesicula seminalis. A dilation or outpocketing of the vas deferens in which the spermatozoa are stored before copula.

Virginopara. Viviparous parthenogenetic female aphid producing only virginopara. Syn. virginogenia, fundatrigenia.

Virgo. Individual of the parthenogenetic generations of aphids which does not produce sexual forms (cf. Lampel, 1968).

Vitellarium. Sections of the ovariole where yolk is deposited in the growing oocyte.

Vitelline membrane. Membrane formed during vitellogenesis by the follicle cells. This membrane is thrown into innumerable folds, presumably in conjunction with proteinaceous yolk formation.

REFERENCES

Aboim, A. N. (1945) Développement embryonnaire et post-embryonnaire des gonades normales et agamétiques de *Drosophila melangaster*. *Rev. Suisse Zool.* **52,** 53–154.

Abraham, R. (1934) Das Verhalten der Spermien in der weiblichen Bettwanze (*Cimex lectularius* L.) und der Verbleib der überschüssigen Spermamasse. *Z. Parasitenk.* **6,** 559–91.

Adams, T. S., Hintz, A. M., and Pomonis, J. G. (1968) Oostatic hormone production in houseflies, *Musca domestica*, with developing ovaries. *J. Insect Physiol.* **14,** 983–93.

Adams, T. S. and Mulla, M. S. (1968) Ovarian development, pheromone production, and mating in the eye gnat. *Hippelates collusor*. *J. Insect Physiol.* **14,** 627–35.

Adams, T. S. and Nelson, D. R. (1968) Bioassay of crude extracts from the factor that prevents second matings in female *Musca domestica*. *Ann. Entomol. Soc. Am.* **61,** 112–16.

Adiyodi, K. G. (1967) The nature of haemolymph proteins in relation to the ovarian cycle in the viviparous cockroach, *Nauphoeta cinerea*. *J. Insect Physiol.* **13,** 1189–95.

Adiyodi, K. G. and Nayar, K. K. (1967) Hemolymph proteins and reproduction in *Periplaneta americana*: the nature of conjugated proteins and the effect of cardia-allatectomy on protein metabolism. *Biol. Bull.* **133,** 271–86.

Adler, H. (1877) Beiträge zur Naturgeschichte der Cynipiden. *Deut. Entomol. Z.* **21,** 209–47.

Adler, H. (1881) Ueber den Generationswechsel der Eichen-Gallwespen. *Z. Wiss. Zool.* **35,** 151–246.

Adriaanse, A. (1948) *Ammophila campestris* Latr. und *Ammophila adriaansei* Wilcke. Ein Beitrag zur vergleichenden Verhaltensforschung. *Behaviour* **1,** 1–34.

Aggarwal, S. K. and King, R. C. (1966) The ultrastructure of the thoracic endocrine system of female *Drosophila melanogaster*. *Am. Zoologist* **6,** 529.

Akov, S. (1967) Effect of folic acid antagonists on larval development and egg production in *Aedes aegypti*. *J. Insect Physiol.* **13,** 913–23.

Alam, S. M. (1952) A contribution to the biology of *Stenobracon deesae* Cameron (Braconidae, Hymenoptera), and the anatomy of its pre-imaginal stages. *Z. Parasitenk.* **15,** 159–82.

Albrecht, F. O. (1959) Facteurs internes et fluctuations des effectifs chez *Nomadacris septemfasciata* (Serv.). *Bull. Biol. Fr. Belg.* **93,** 414–61.

Albrecht, F. O., Verdier, M., and Blackith, R. E. (1958) Détermination de la fertilité par l'effet de groupe chez le criquet migrateur (*Locusta migratoria migratorioides* R. et F.). *Bull. Biol. Fr. Belg.* **92,** 349–427.

Aldrich, J. M. and Turley, L. A. (1899) A balloon-making fly. *Am. Naturalist* **33,** 809–12.

Alexander, R. D. (1957) Sound production and associated behavior in insects. *Ohio J. Sci.* **57,** 101–13.

Alexander, R. D. (1960) Sound communication in Orthroptera and Cicadidae. In *Animal Sounds and Communication*, W. E. Lanyon and E. N. Tavolga, Eds., pp. 38–91.

Alexander, R. D. (1961) Aggressiveness, territoriality, and sexual behavior in field crickets (Orthoptera: Gryllidae). *Behavior* **17,** 130–223.

Alexander, R. D. (1962) Evolutionary change in cricket acoustical communication. *Evolution* **16,** 443–67.

Alexander, R. D. and Brown, W. L. (1963) Mating behavior and the origin of insect wings. *Occasional Pap. Mus. Zool. Univ. Michigan* No. 628.

Alexander, R. D. and Moore, T. E. (1958) Studies on the acoustical behavior of seventeen-year cicadas (Homoptera: Cicadidae: Magicicada). *Ohio J. Sci.* **58,** 107–27.

Allegret, P. (1951) Retard de la nymphose chez *Galleria mellonella* L. après la sécrétion du cocon. Influence de l'alimentation. *C.R. Acad. Sci. Paris* **233,** 441–3.

Allen, M. D. (1965) The role of the queen and males in the social organization of insect communities. *Symp. Zool. Soc. London* **14,** 133–57.

Alpatov, W. W. (1932) Egg production in *Drosophilia melangaster* and some factors which influence it. *J. Exp. Zool.* **63,** 85–111.

Amosova, I. S. (1959) On the gonotrophic relations within the genus *Culicoides* (Diptera, Heleidae). *Rev. Entomol. USSR* **38,** 774–89.

Andersen, K. T. (1934) Kurze Mitteilungen über weitere Versuche zur Biologie und Ökologie von *Sitona lineata* L. *Verhandl. Deut. Ges. Angew. Entomol.* **9,** 42–49.

ANDERSEN, K. T. (1935) Experimentelle Untersuchungen über den Einfluss der Temperature auf die Eierzeugung von Insekten. II. Einfluss inkonstanter Temperaturen auf die Eierzeugung von *Sitona lineata* L. und *Calandra granaria* L. *Biol. Zentr.* **55,** 571–90.

ANDERSON, E. (1964) Oocyte differentiation and vitellogenesis in the roach *Periplaneta americana*. *J. Cell Biol.* **20,** 131–55.

ANDERSON, J. F. and HORSFALL, W. R. (1963) Thermal stress and anomalous development of mosquitoes (Diptera: Culcicidae). I. Effect of constant temperature on dimorphism of adults of *Aedes stimulans*. *J. Exp. Zool.* **154,** 67–108.

ANDERSON, J. F. and HORSFALL, W. R. (1965) Dimorphic development of transplanted juvenile gonads of mosquitoes. *Science* **147,** 624–5.

ANDERSON, J. R. (1964) Methods for distinguishing nulliparous from parous flies and for estimating the ages of *Fannia canicularis* and some other cyclorraphous Diptera. *Ann. Entomol. Soc. Am.* **57,** 226–36.

ANDERSON, N. H. (1962) Growth and fecundity of *Anthocoris* spp. reared on various prey (Heteroptera: Anthocoridae). *Entomol. Exp. Appl.* **5,** 40–52.

ANDERSON, R. C. and PASCHKE, J. D. (1968) The biology and ecology of *Anaphes flavipes* (Hymenoptera: Mymaridae), an exotic egg parasite of the cereal leaf beetle. *Ann. Entomol. Soc. Am.* **61,** 1–5.

ANDERSON, R. H. (1963) The laying worker in the Cape honeybee, *Apis mellifera capensis*. *J. Apic. Res.* **2,** 85–92.

ANDREW, C. G. (1966) Sexual recognition in adult *Erythemis simplicicollis* (Odonata: Anisoptera). *Ohio J. Sci.* **66,** 613–17.

ANDREWARTHA, H. G. (1933) The bionomics of *Otiorrhynchus cribricollis*. *Bull. Entomol. Res.* **24,** 373–84.

ANDREWARTHA, H. G. (1945) Some differences in the physiology and ecology of locusts and grasshoppers. *Bull. Entomol. Res.* **35,** 379–89.

ANDREWARTHA, H. G. (1952) Diapause in relation to the ecology of insects. *Biol. Rev.* **27,** 50–107.

ANNAND, P. N. (1928) A contribution toward a monograph of the Adelginae (Phylloxeridae) of North America. *Stanford Univ. Publ. Biol. Sci.* **6,** 1–146.

APLIN, R. T. and BIRCH, M. C. (1968) Pheromones from the abdominal brushes of male noctuid Lepidoptera. *Nature* **217,** 1167–8.

APPEL, O., BLUNCK, H. and RICHTER, H. (1957) *Handbuch der Pflanzenkrankheiten*, 5th ed., **5,** pp. 577.

APPLEBAUM, S. W. and LUBIN, Y. (1967) The comparative effects of vitamin deficiences on development and on adult fecundity of *Tribolium castaneum*. *Entomol. Exp. Appl.* **10,** 23–30.

ARTHUR, A. P. and WYLIE, H. G. (1959) Effects of host size on sex ratio, development time and size of *Pimpla turionellae* (L.) (Hymenoptera: Ichneumonidae). *Entomophaga* **4,** 297–301.

ASHBY, K. R. (1961) The life-history and reproductive potential of *Cryptolestes pusillus* (Schönherr) (Col., Cucujidae) at high temperatures and humidities. *Bull. Entomol. Res.* **52,** 353–61.

ASSMUTH, J. (1913) *Termitoxenia assmuthi* Wasm. Anatomisch-histologische Untersuchung. *Nova Acta Kais. Leop.—Carol. Deut. Acad. Naturforsch.* **98,** 187–316.

ASSMUTH, J. (1923) Ametabolie und Hermaphroditismus bei den Termitoxeniiden (Dipt.). *Biol. Zentr.* **43,** 268–81.

ASTAUROV, B. L. (1967) Experimental alterations of the developmental cytogenetic mechanisms in mulberry silkworms: artificial parthenogenesis, polyploidy, gynogenesis, and androgenesis. *Advan. Morphogenesis* **6,** 199–257.

AUBERT, J. F. (1959) Biologie de quelques Ichneumonidae Pimplinae et examen critique de la theorie de Dzierzon. *Entomophaga* **4,** 75–188.

AUBERT, J. F. (1961) L'expérience de la bourre de coton démontre que le volume de l'hôte intervient en tant que facteur essentiel dans la détermination de sexe chez les Ichneumonides Pimplines (Hym.). *Bull. Soc. Entomol. Fr.* **66,** 89–93.

AUBERT, J. F. and SHAUMAR, N. (1962) Nouvelle expérience sur le déterminisme du sexe chez les Ichneumonidae Pimplinae (Insectes Hyménoptères); ponte dans des cocons pleins et des cocons creux. *C.R. Acad. Paris* **255,** 2194–5.

AUBERT, J. F. and SHAUMAR, N. (1964) Première analyse des facteurs externes susceptibles de provoquer la fécondation des œufs diploides femelles chez les Ichneumonidae Pimplinae (Insectes Hyménoptères). *C.R. Acad. Sci. Paris* **258,** 3773–5.

AUGUST, C. J. (1967) The pheromone mediation of the mating behavior of *Tenebrio molitor*. M.A. Thesis, Univ. Calif. Los Angeles.

AWASTHI, V. B. (1968) The functional significance of the nervi corporis allati 1 and nervi corporis allati 2 in *Gryllodes sigillatus*. *J. Insect Physiol.* **14,** 301–4.

BAERENDS, G. P. (1941) Fortpflanzungsverhalten und Orientierung der Grabwespe *Ammophila campestris* Jur. *Tijdschrift Entomol.* **84,** 71–275.

BAIER, L. J. (1930) Contribution to the physiology of the stridulation and hearing in insects. *Zool. Jahrb. Abt. Physiol.* **47,** 151–248.

BALFOUR-BROWNE, F. (1922) On the life-history of *Melittobia acasta*, Walker; a chalcid parasite of bees and wasps. *Parasitol.* **14,** 349–70.

BALLARD, E. and EVANS, M. G. (1928) *Dysdercus sidae*, Montr. in Queensland. *Bull. Entomol. Res.* **18,** 405–32.

BARBER, H. S. (1913) Observations on the life history of *Micromalthus debilis* Lec. (Coleoptera). *Proc. Entomol. Soc. Washington* **15,** 31–38.

BARBIER, M. and LEDERER, E. (1960) Structure chimique de la "substance royale" de la reine d'abeille (*Apis mellifica* L.). *C.R. Acad. Sci. Paris* **250,** 4467–9.

BARLOW, C. A. (1962) Development, survival, and fecundity of the potato aphid, *Macrosiphum euphorbiae* (Thomas), at constant temperatures. *Can. Entomologist* **94,** 667–71.

BARNES, O. L. (1963) Observations on the life history of the desert grasshopper (*Trimerotropis pallidipennis pallidipennis*) in laboratory and insectary cages. *J. Econ. Entomol.* **56,** 525–8.

BARONI URBANI, C. (1966) Ulteriori osservazioni ed esperienze sulla monoginia dei Formicidi: saggio di un' analisi del comportamento in *Leptothorax exilis* Em. *Insectes Sociaux* **13,** 173–84.

BARTH, R. (1937a) Bau and Funktion der Flügeldrüsen einiger Mikrolepidopteren. Untersuchungen an den Pyraliden: *Aphomia gularis*, *Galleria mellonella*, *Plodia interpunctella*, *Ephestia elutella* und *E. kühniella*. *Z. Wiss. Zool.* **150,** 1–37.

BARTH, R. (1937b) Herkunft, Wirkung und Eigenschaften des weiblichen Sexualduftstoffes einiger Pyraliden. *Zool. Jahrb. Abt. Physiol.* **58,** 297–329.

BARTH, R. (1952) Die Hautdrüsen des Männchens von *Opsiphanes invirae isagoras* Fruhst. (Lepidoptera, Brassolidae). *Zool. Jahrb. Abt. Anat.* **72,** 216–30.

BARTH, R. H., JR. (1961) Hormonal control of sex attractant production in the Cuban cockroach. *Science* **133,** 1598–9.

BARTH, R. H., JR. (1962) The endocrine control of mating behavior in the cockroach *Byrsotria fumigata* (Guérin). *Gen. Comp. Endocrinol.* **2,** 53–69.

BARTH, R. H., JR. (1964) The mating behavior of *Byrsotria fumigata* (Guérin) (Blattidae, Blaberinae). *Behaviour* **23,** 1–30.

BARTH, R. H., JR. (1965) Insect mating behavior: endocrine control of a chemical communication system. *Science* **149,** 882–3.

BARTH, R. H., JR. (1968) The comparative physiology of reproductive processes in cockroaches. Part I. Mating behaviour and its endocrine control. *Advan. Reproductive Physiol.* **3,** 167–207.

BARTON BROWNE, L. (1958) The relation between development and mating in *Lucilia cuprina*. *Australian J. Sci.* **20,** 239–40.

BARTON BROWNE, L. (1960) The role of olfaction in the stimulation of oviposition in the blowfly, *Phormia regina*. *J. Insect Physiol.* **5,** 16–22.

BARTON BROWNE, L. (1965) An analysis of the ovipositional responses of the blowfly *Lucilia cuprina* to ammonium carbonate and indole. *J. Insect Physiol.* **11,** 1131–43.

BAR-ZEEV, M. (1967) Oviposition of *Aedes aegypti* L. on a dry surface and hygroreceptors. *Nature* **213,** 737–8.

BASSINDALE, R. (1955) The biology of the stingless bee *Trigona* (*Hypotrigona*) *gribodoi* Magretti (Meliponidae). *Proc. Zool. Soc. London* **125,** 49–62.

BASTOCK, M. (1956) A gene mutation which changes a behavior pattern. *Evolution* **10,** 421–39.

BATHELLIER, J. (1927) Contribution à l'étude systématique et biologique des termites de l'Indochine. *Faune Colonies Françaises* **1,** 125–365.

BATRA, S. W. T. (1964) Behavior of the social bee, *Lasioglossum zephyrum*, within the nest (Hymenoptera: Halictidae). *Insectes Sociaux* **11,** 159–86.

BATRA, S. W. T. (1966) The life cycle and behavior of the primitively social bee, *Lasioglossum zephyrum* (Halictidae). *Univ. Kansas Sci. Bull.* **46,** 359–423.

BAUMERT, D. (1959) Mehrjährige Zuchten einheimischer Strepsipteren an Homopteren. 2. Hälfte. Imagines. Lebenszyklus und Artbestimmung von *Elenchus tenuicornis* Kirby. *Zool. Beitr.* N.F. **4,** 343–409.

BAWA, S. R. (1964) Electron microscope study of spermiogenesis in a fire-brat insect, *Thermobia domestica* Pack. I. Mature spermatozoon. *J. Cell Biol.* **23,** 431–46.

BEAMENT, J. W. L. (1946) The formation and structure of the chorion of the egg in an Hemipteran, *Rhodnius prolixus*. *Quart. J. Microscop. Sci.* **87,** 393–439.

BEARDS, G. W. and STRONG, F. E. (1966) Photoperiod in relation to diapause in *Lygus hesperus* Knight. *Hilgardia* **37,** 345–62.

BECKER, F. (1925) Die Ausbildung des Geschlechtes bei der Honigbiene. *Erlanger Jahrb. Bienenk.* **3,** 163–223.

BEEBE, W. (1944) The function of secondary sexual characters in two species of Dynastidae (Coleoptera). *Zoologica* **29,** 53–58.

BEERMANN, W. (1955) Geschlechtsbestimmung und Evolution der genetischen Y-Chromosomen bei *Chironomus*. *Biol. Zentr.* **74,** 525–44.

BEHRENZ, W. (1952) Experimentelle und histologische Untersuchungen am weiblichen Genitalapparat von *Lymantria dispar* L. *Zool. Jahrb. Abt. Anat.* **72,** 147–215.

BELLAMY, R. E. and REEVES, W. C. (1963) The winter biology of *Culex tarsalis* (Diptera: Culicidae) in Kern County, California. *Ann. Entomol. Soc. Am.* **56,** 314–23.

BENAZZI, M. (1961) Nuove acquisizione nel domino della pseudogamia. *Monit. Zool. Ital.* **69,** 9–21.

BENNET-CLARK, H. C. and EWING, A. W. (1967) Stimuli provided by courtship of male *Drosophila melanogaster. Nature* **215,** 669–71.

BENNINGTON, E. E. SOOTER, C. A., and BAER, H. (1958) The diapause in adult female *Culex tarsalis* Coquillett (Diptera: Culicidae). *Mosquito News* **18,** 299–304.

BENSCHOTER, C. A. (1967) Effect of dietary biotin on reproduction of the house fly. *J. Econ. Entomol.* **60,** 1326–8.

BENSCHOTER, C. A. and PANIAGUA, G. R. (1966) Reproduction and longevity of Mexican fruit flies, *Anastrepha ludens* (Diptera: Tephritidae), fed biotin in the diet. *Ann. Entomol. Soc. Am.* **59,** 298–300.

BEQUAERT, J. C. (1952) The Hippoboscidae or louse-flies (Diptera) of mammals and birds. Part I. Structure, physiology and natural history. *Entomologica Am.* N.S. **32,** 1–209.

BERGER, R. S. (1966) Isolation, identification, and synthesis of the sex attractant of the cabbage looper, *Trichoplusia ni. Ann. Entomol. Soc. Am.* **59,** 767–71.

BERGER, R. S. and CANERDAY, T. D. (1968) Specificity of the cabbage looper sex attractant. *J. Econ. Entomol.* **61,** 452–4.

BERGERARD, J. (1958) Étude de la parthénogénèse facultative de *Clitumnus extradentatus* Br. (Phasmidae). *Bull. Biol. Fr. Belg.* **92,** 87–182.

BERGERARD, J. (1961) Intersexualité expérimentale chez *Carausius morosus* Br. (Phasmidae). *Bull. Biol. Fr. Belg.* **95,** 273–300.

BERGERARD, J. and SEUGÉ, J. (1959) La parthénogénèse accidentelle chez *Locusta migratoria* L. *Bull. Biol. Fr. Belg.* **93,** 16–37.

BERLAND, L. (1934) Un cas probable de parthénogénèse géographique chez *Leucospis gigas* (Hyménoptère). *Bull. Soc. Zool. Fr.* **59,** 172–5.

BERLESE, A. (1898) Fenomeni che accompagno la fecondazione in taluni insetti. Memoria I. *Riv. Pat. Veg.* **6,** 353–68.

BERNHARD, C. (1907) Über die vivipare Ephemeride *Chloëon dipterum. Biol. Zentr.* **27,** 467–79.

BESSÉ, N. DE (1965) Recherches histophysiologiques sur la neurosécrétion dans la chaine nerveuse ventrale d'une blatte, *Leucophaea maderae* (F.). *C.R. Acad. Sci. Paris* **260,** 7014–17.

BESSÉ, N. DE (1967) Neurosécrétion dans la chaine nerveuse ventrale de deux Blattes, *Leucophaea maderae* (F.) et *Periplaneta americana* (L.). *Bull. Soc. Zool. Fr.* **92,** 73–86.

BHATTACHARYA, G. C. (1943) Reproduction and caste determination in aggressive red-ants, *Oecophylla smaragdina*, Fabr. *Trans. Bose Res. Inst. Calcutta* **15,** 137–56.

BIER, K. (1953) Vergleichende Untersuchungen zur Fertilität der Ameisenarbeiterinnen. *Zool. Anz.* **150,** 282–8.

BIER, K. (1954a) Über den Saisondimorphismus der Oogenese von *Formica rufa rufo-pratensis minor* Gössw. und dessen Bedeutung für die Kastendetermination. *Biol. Zentr.* **73,** 170–90.

BIER, K. (1954b) Über den Einfluss der Königin auf die Arbeiterinnenfertilität im Ameisenstaat. *Insectes Sociaux* **1,** 7–19.

BIER, K. (1956) Arbeiterinnenfertilität und Aufzucht von Geschlechtstieren als Regulationsleistung des Ameisenstaates. *Insectes Sociaux* **3,** 177–84.

BIER, K. (1958) Die Regulation der Sexualität in den Insektenstaaten. *Ergeb. Biol.* **20,** 97–126.

BIER, K. (1962) Autoradiographische Untersuchungen zur Dotterbildung. *Naturwiss.* **49,** 332–3.

BIER, K. (1963a) Autoradiographische Untersuchungen über die Leistungen des Follikelepithels und der Nährzellen bei der Dotterbildung und Eiweissynthese im Fliegenovar. *Arch. Entwicklungsmech. Organ.* **154,** 552–75.

BIER, K. (1963b) Synthese, interzellulärer Transport, und Abbau von Ribonukleinsäure im Ovar der Stubenfliege *Musca domestica. J. Cell. Biol.* **16,** 436–40.

BIER, K. (1964) Gerichteter Ribonukleinsäuretransport durch das Cytoplasma. *Naturwiss.* **51,** 418.

BIER, K. (1965) Zur Funktion der Nährzellen im meroistischen Insektenovar unter besonderer Berücksichtigung der Oogenese adephager Coleopteren. *Zool. Jahrb. Abt. Physiol.* **71,** 371–84.

BIER, K., KUNZ, W. and RIBBERT, D. (1967) Struktur und Funktion der Oocytenchromosomen und Nukleolen sowie der Extra-DNS während der Oogenese panoistischer und meroistischer Insekten. *Chromosoma* **23,** 214–54.

BILIOTTI, E., DEMOLIN, D. and MERLE, P. DU (1965) Parasitisme de la processionnaire du pin par *Villa quinquefasciata* Wied. and Meig. (Dipt., Bombyliidae): importance du comportement de ponte du parasite. *Ann. Epiphyties* **16,** 279–88.

BINET, L. (1931) *La vie de la Mante religieuse*. Vigot Frères, Paris.

BIRCH, L. C. (1945) The influence of temperature, humidity and density on the oviposition of the small strain of *Calandra oryzae* L. and *Rhizopertha dominica* Fab. (Coleoptera). *Australian J. Exp. Biol. Med. Sci.* **23,** 197–203.

BISHOP, G. W. (1959) The comparative bionomics of American *Cryptolestes* (Coleoptera-Cucujidae) that infest stored grain. *Ann. Entomol. Soc. Am.* **52,** 657–65.

BITSCH, J. (1962) Le complex nerveux hypocérébral et les corpora allata des Machilides (Ins. Thysanura). *C. R. Acad. Sci. Paris* **254,** 1501–3.

BLAIS, J. R. (1952) The relationship of the spruce budworm (*Choristoneura fumiferana*, Clem.) to the flowering condition of balsam fir (*Abies balsamea* (L.) Mill.). *Can. J. Zool.* **30,** 1–29.

BLAKE, G. M. (1961) Length of life, fecundity and the oviposition cycle in *Anthrenus verbasci* (L.) (Col., Dermestidae) as affected by adult diet. *Bull. Entomol. Res.* **52,** 459–72.

BLOCH, B., THOMSEN, E. and THOMSEN, M. (1966) The neurosecretory system of the adult *Calliphora erythrocephala.* III. Electron microscopy of the medial neurosecretory cells of the brain and some adjacent cells. *Z. Zellforsch.* **70,** 185–208.

BLUNCK, H. (1912) Das Geschlechtsleben des *Dytiscus marginalis* L. 1. Teil. Die Begattung. *Z. Wiss. Zool.* **102,** 169–248.

BOBB, M. L. (1964) Apparent loss of sex attractiveness by the female of the Virginia-Pine sawfly, *Neodiprion pratti pratti. J. Econ. Entomol.* **57,** 829–30.

BODENHEIMER, F. S. (1929) Studien zur Epidemiologie, Ökologie und Physiologie der afrikanischen Wanderheuschrecke (*Schistocerca gregaria* Forsk.). *Z. Angew. Entomol.* **15,** 435–557.

BODENHEIMER, F. S. (1943) Studies on the life-history and ecology of Coccinellidae: I. The life-history of *Coccinella septempunctata* L. in four different zoogeographical regions. *Bull. Soc. Fouad Ier Entomol.* **27,** 1–28.

BODENHEIMER, F. S. and BEN-NERYA, A. (1937) One-year studies on the biology of the honey-bee in Palestine. *Ann. Appl. Biol.* **24,** 385–403.

BODENHEIMER, F. S. and SWIRSKI, E. (1957) *The Aphidoidea of the Middle East.* The Weizmann Science Press, Jerusalem.

BODENSTEIN, D. (1938) Untersuchungen zum Metamorphoseproblem. III. Über die Entwicklung der Ovarien im thoraxlosen Puppenabdomen. *Biol. Zentr.* **58,** 328–32.

BODENSTEIN, D. (1947) Investigations on the reproductive system of *Drosophila. J. Exp. Zool.* **104,** 101–51.

BODENSTEIN, D. (1953) Studies on the humoral mechanisms in growth and metamorphosis of the cockroach, *Periplaneta americana.* III. Humoral effects on metabolism. *J. Exp. Zool.* **124,** 105–16.

BODENSTEIN, D. and SHAAYA, E. (1968) The function of the accessory sex glands in *Periplaneta americana* (L.). I. A quantitative bioassay for juvenile hormone. *Proc. Natl. Acad. Sci.* **59,** 1223–30.

BODENSTEIN, D. and SPRAGUE, J. B. (1959) The developmental capacities of accessory sex glands in *Periplaneta americana. J. Exp. Zool.* **142,** 177–201.

BODNARYK, R. P. and MORRISON, P. E. (1966) The relationship between nutrition, haemolymph proteins, and ovarian development in *Musca domestica* L. *J. Insect Physiol.* **12,** 963–76.

BOECKH, J., PRIESNER, E., SCHNEIDER, D. and JACOBSON, M. (1963) Olfactory receptor response to the cockroach sexual attractant. *Science* **141,** 716–17.

DE BOISSEZON, P. (1929) Expériences au sujet de la maturation des œufs chez les Culicides. *Bull. Soc. Path. Exotique* **22,** 683–9.

DE BOISSEZON, P. (1933) De l'utilisation des protéines et du fer d'origine végétale dans la maturation des œufs chez *Culex pipiens* L. *C.R. Soc. Biol.* **114,** 487–9.

BOLDYREV, B. T. (1927) Einige Daten über die Spermatophoren-Befruchtung bei den Insekten. *Rev. Russe Entomol.* **21,** 133–6.

BOLDYREV, B. T. (1928a) Biological studies on *Bradyporus multituberculatus* F. W. (Orth., Tettig.). *Eos* **4,** 13–56.

BOLDYREV, B. T. (1928b) Einige Episoden aus dem Geschlechtsleben von *Discoptila fragosoi* Bol. (Orthoptera, Gryllidae). *Rev. Russe Entomol.* **22,** 137–47.

BOMBOSCH, S. (1962) Untersuchung über die Auslösung der Eiablage bei *Syrphus corollae* Fabr. (Dipt. Syrphidae). *Z. Angew. Entomol.* **50,** 81–88.

BONHAG, P. F. (1955) Histochemical studies of the ovarian nurse tissue and oocytes of the milkweed bug, *Oncopeltus fasciatus* (Dallas). I. Cytology, nucleic acids, and carbohydrates. *J. Morphol.* **96,** 381–439.

BONHAG, P. F. (1958) Ovarian structure and vitellogenesis in insects. *Ann. Rev. Entomol.* **3,** 137–60.

BONHAG, P. F. (1959) Histological and histochemical studies on the ovary of the American cockroach *Periplaneta americana* (L.). *Univ. Calif. Publ. Entomol.* **16,** 81–124.

BONNEMAISON, L. (1951) Contribution à l'étude des facteurs provoquant l'apparition des formes ailées et sexuées chez les Aphidinae. *Ann. Epiphyties, Sér. C,* **2,** 1–380.

BONNEMAISON, L. (1958) Facteurs d'apparition des formes sexupares ou sexuées chez puceron cendré du pommier (*Sappaphis plantaginea* Pass.). *Ann. Epiphyties, Sér. C,* **9,** 329–53.

BONNEMAISON, L. (1959) Le puceron cendré du pommier (*Dysaphis plantaginea* Pass.). Morphologie et biologie—méthodes de lutte. *Ann. Epiphyties, Sér. C,* **10,** 257–320.

BONNEMAISON, L. (1961a) Étude de quelques facteurs de la fécondité et de la fertilité chez la Noctuelle du chou (*Mamestra brassicae* L.) (Lep.). II. Influence de la lumière sur les imagos et sur l'accouplement. *Bull. Soc. Entomol. Fr.* **66,** 62–70.

BONNEMAISON, L. (1961b) Étude de quelques facteurs de la fécondité et de la fertilité chez la Noctuelle du chou (*Mamestra brassicae* L.) (Lep.). III. Influence de la diapause. *Bull. Soc. Entomol. Fr.* **66,** 128–33.

Bonnemaison, L. (1964a) Action combinée de la photopériode et de l'effet de groupe sur l'apparition des sexupares ailés de *Dysaphis plantaginea* Pass. (Homoptères, Aphididae). *C.R. Acad. Sci. Paris* **259,** 1663–5.

Bonnemaison, L. (1964b) Action inhibitrice d'une longue photopériode et d'une température élévée sur l'apparition des sexupares ailés de *Dysaphis plantaginea* Pass. (Homoptères, Aphididae). *C.R. Acad. Sci. Paris* **259,** 1768–70.

Bonnemaison, L. (1964c) Observations écologiques sur la Coccinelle à 7 point (*Coccinella septempunctata* L.) dans la région parisienne (Col.). *Bull. Soc. Entomol. Fr.* **69,** 64–83.

Bonnemaison, L. (1966a) Combination de photophases et de scotophases avec des températures élévées ou basses sur la production des sexupares de *Dysaphis plantaginea* Pass. (Homoptères, Aphididae). *C.R. Acad. Sci. Paris* **263,** 177–9.

Bonnemaison, L. (1966b) Action de l'alternance de scotophases et de photophases dans un cycle de 24 h sur la production des sexupares de *Dysaphis plantaginea* Pass. (Homoptères, Aphididae). *C.R. Acad. Sci. Paris* **262,** 2498–501.

Bonnemaison, L. (1966c) Action de l'alternance de scotophases de 12 h et de longues photophases sur la production des sexupares de *Dysaphis plantaginea* Pass. (Homoptères, Aphididae). *C.R. Acad. Sci. Paris* **262,** 2609–11.

Bonnemaison, L. (1967) Action de scotophases de diverses durées dans des rhythmes circadiens, ultra- et infra-circadiens sur la production des gynopares ailés de *Dysaphis plantaginea* Pass. (Homoptères, Aphididae). *C.R. Acad. Sci. Paris* **264,** 2661–3.

Bornemissza, G. F. (1964) Sex attractant of male scorpion flies. *Nature* **203,** 786–7.

Bornemissza, G. F. (1966a) Specificity of male sex attractants in some Australian scorpion flies. *Nature* **209,** 732–3.

Bornemissza, G. F. (1966b) Observations on the hunting and mating behaviour of two species of scorpion flies (Bittacidae: Mecoptera). *Australian J. Zool.* **14,** 371–82.

Börner, C. (1909) Zur Biologie und Systematik der Chermesiden. *Biol. Zentr.* **29,** 118–25, 129–46.

Bounhiol, J. J. (1936) Métamorphose après ablation des corpora allata chez le ver à soie (*Bombyx mori* L.). *C.R. Acad. Sci. Paris* **203,** 388–9.

Bowen, W. R. and Stern, V. M. (1966) Effect of temperature on the production of males and sexual mosaics in a uniparental race of *Trichogramma semifumatum* (Hymenoptera: Trichogrammatidae). *Ann. Entomol. Soc. Am.* **59,** 823–34.

Bowers, B. (1966) Personal communication.

Bowers, B. and Johnson, B. (1966) An electron microscope study of the corpora cardiaca and neurosecretory neurons in the aphid, *Myzus persicae* (Sulz.). *Gen. Comp. Endocrinol.* **6,** 213–30.

Bowers, W. S. and Blickenstaff, C. C. (1966) Hormonal termination of diapause in the alfalfa weevil. *Science* **154,** 1673–4.

Bowers, W. S., Fales, H. M., Thompson, M. J. and Uebel, E. C. (1966) Juvenile hormone: identification of an active compound from balsam fir. *Science* **154,** 1020–21.

Bowers, W. S., Thompson, M. J. and Uebel, E. C. (1965) Juvenile and gonadotropic hormone activity of 10,11-epoxyfarnesenic acid methyl ester. *Life Sci.* **4,** 2323–31.

Bracken, G. K. (1965) Effects of dietary components on fecundity of the parasitoid *Exeristes comstockii* (Cress.) (Hymenoptera: Ichneumonidae). *Can. Entomologist* **97,** 1037–41.

Bracken, G. K. (1966) Role of ten dietary vitamins on fecundity of the parasitoid *Exeristes comstockii* (Cresson) (Hymenoptera: Ichneumonidae). *Can. Entomologist* **98,** 918–22.

Bracken, G. K. and Nair, K. K. (1967) Stimulation of yolk deposition in an Ichneumonid parasitoid by feeding synthetic juvenile hormone. *Nature* **216,** 483–4.

Brandenburg, J. (1953) Der Parasitismus der Gattung *Stylops* an der Sandbiene *Andrena vaga* Pz. *Z. Parasitenk.* **15,** 457–75.

Brandenburg, J. (1956) Das endokrine System des Kopfes von *Andrena vaga* Pz. (Ins. Hymenopt.) und Wirkung der Stylopisation (*Stylops*, Ins. Strepsipt.). *Z. Morphol. Oekol. Tierre* **45,** 343–64.

Brian, M. V. (1951) Caste determination in a Myrmicine ant. *Experientia* **7,** 182–3.

Brian, M. V. (1953a) Oviposition by workers of the ant *Myrmica. Physiol. Comp. Oecol.* **3,** 25–36.

Brian, M. V. (1953b) Brood-rearing in relation to worker number in the ant *Myrmica. Physiol. Zool.* **26,** 355–66.

Brian, M. V. (1954) Studies of caste differentiation in *Myrmica rubra* L. 1. The growth of queens and males. *Insectes Sociaux* **1,** 101–22.

Brian, M. V. (1956) Studies of caste differentiation in *Myrmica rubra* L. 4. Controlled larval nutrition. *Insectes Sociaux* **3,** 369–94.

Brian, M. V. (1957) Caste determination in social insects. *Ann. Rev. Entomol.* **2,** 107–20.

Brian, M. V. (1958) The evolution of queen control in the social Hymenoptera. *Proc. Xth Intern. Congr. Entomol. Montreal* 1956, **2,** 497–502.

Brian, M. V. (1963) Studies of caste differentiation in *Myrmica rubra* L. 6. Factors influencing the course of female development in the early third instar. *Insectes Sociaux* **10,** 91–102.

BRIAN, M. V. (1965a) *Social Insect Populations*. Academic Press, London, New York.
BRIAN, M. V. (1965b) Caste differentiation in social insects. *Symp. Zool. Soc. London* **14,** 13–38.
BRIAN, M. V. and HIBBLE, J. (1964) Studies of caste differentiation in *Myrmica rubra* L. 7. Caste bias, queen age and influence. *Insectes Sociaux* **11,** 223–38.
BRIDGES, C. B. (1916) Non-disjunction as proof of the chromosome theory of heredity. *Genetics* **1,** 1–52.
BRIDGES, C. B. (1921) Triploid intersexes in *Drosophila melanogaster*. *Science* **54,** 252–4.
BRIDGES, C. B. (1922) The origin of variations in sexual and sex-limited characters. *Am. Naturalist* **56,** 51–63.
BRIDGES, C. B. (1932) The genetics of sex in *Drosophila*. In *Sex and Internal Secretion*, E. ALLEN, Ed., pp. 55–93.
BRIEN, P. (1965) Considérations à propos de la reproduction sexuée des invertébrés. *Année Biol., Sér.* 4, **4,** 329–65.
BRINCK, P. (1962) Die Entwicklung der Spermaübertragung der Odonaten. *Proc. XIth Intern. Congr. Entomol. Wien 1960*, **1,** 715–18.
BROODRYK, S. W. and DOUTT, R. L. (1966) The biology of *Coccophagoides utilis* Doutt (Hymenoptera: Aphelinidae). *Hilgardia* **37,** 233–54.
BROUSSE-GAURY, P. (1967a) Mise en évidence, chez les Dictyoptères, d'une quatrième paire de nervi cardiaci. *C.R. Acad. Sci. Paris* **265,** 228–31.
BROUSSE-GAURY, P. (1967b) Généralisation, à divers insectes, de l'innervation deuterocérébrale des corpora cardiaca, et rôle neurosécrétoire des nervi corporis cardiaci IV. *C.R. Acad. Sci. Paris* **265,** 2043–6.
BROUSSE-GAURY, P. (1968) Localisation des noyaux-origines des nerfs paracardiaques de dictyoptères Blaberidae et Blattidae. *C.R. Acad. Sci. Paris* **266,** 1972–5.
BROWER, L. P., BROWER, J. v. Z. and CRANSTON, F. P. (1965) Courtship behavior of the queen butterfly, *Danaus gilippus berenice* (Cramer). *Zoologica* **50,** 1–39.
BROWN, E. H. and KING, R. C. (1964) Studies on the events resulting in the formation of an egg chamber in *Drosophila melangaster*. *Growth* **28,** 41–81.
BROWN, R. G. B. (1964) Courtship behaviour in the *Drosophila obscura* group. I. *D. pseudoobscura*. *Behaviour* **23,** 61–106.
BROWN, R. G. B. (1965) Courtship behaviour in the *Drosophila obscura* group. II. Comparative studies. *Behaviour* **25,** 281–323.
BROWN, S. W. (1959) Lecanoid chromosome behaviour in three more families of the Coccoidea (Homoptera). *Chromosoma* **10,** 278–300.
BROWN, S. W. and MCKENZIE, H. L. (1962) Evolutionary patterns in the armored scale insects and their allies (Homoptera: Coccoidea: Diaspididae, Phoenicococcidae, and Asterolecaniidae). *Hilgardia* **33,** 141–70.
BROWN, S. W. and NELSON-RIES, W. A. (1961) Radiation analysis of a lecanoid genetic system. *Genetics* **46,** 983–1007.
BROWN, S. W. and NUR, U. (1964) Heterochromatic chromosomes in the coccids. *Science* **145,** 130–6.
BRÜNNICH, K. (1922) Graphische Darstellung der Legetätigkeit einer Bienenkönigin. *Arch. Bienenk.* **4,** 137–47.
BRUNSON, M. H. (1937) The influence of the instars of host larvae on the sex of the progeny of *Tiphia popilliavora* Roh. *Science* **86,** 197.
BUCHHOLTZ, C. (1951) Untersuchungen an der Libellen-Gattung *Calopteryx*-Leach unter besonderer Berücksichtigung ethologischer Fragen. *Z. Tierpsychol.* **8,** 273–93.
BUCHHOLTZ, C. (1955) Eine vergleichende Ethologie der orientalischen Calopterygiden (Odonata) als Beitrag zu ihrer systematischen Deutung. *Z. Tierpsychol.* **12,** 364–86.
BUCHHOLTZ, C. (1956) Eine Analyse des Paarungsverhaltens und der dabei wirkenden Auslöser bei den Libellen *Platycnemis pennipes* Pall. und *Pl. dealbata* Klug. *Z. Tierpsychol.* **13,** 13–25.
BUCHLI, H. (1956a) Die Neotenie bei *Reticulitermes*. *Insectes Sociaux* **3,** 131–43.
BUCHLI, H. 1956b) Le cycle de développement des castes chez *Reticulitermes*. *Insectes Sociaux* **3,** 395–401.
BUCHLI, H. (1958) L'origine des castes et les potentialités ontogéniques des termites européens du genre *Reticulitermes* Holmgren. *Ann. Sci. Nat. Zool.*, Sér. 11, **20,** 263–429.
DE BUCK, A. (1935) Beitrag zur Rassenfrage bei *Culex pipiens*. *Z. Angew. Entomol.* **22,** 242–52.
BUCK, J. and BUCK, E. (1966) Biology of synchronous flashing of fireflies. *Nature* **211,** 562–4.
BUCK, J. and BUCK, E. (1968) Mechanism of rhythmic synchronous flashing of fireflies. *Science* **159,** 1319–27.
BUCK, J. B. (1937) Studies on the firefly. II. The signal system and color vision in *Photinus pyralis*. *Physiol. Zool.* **10,** 412–19.
BUCK, J. B. (1948) The anatomy and physiology of the light organ in fireflies. *Ann. N.Y. Acad. Sci.* **49,** 397–482.
BUGNION, E. (1913) *Termitoxenia*. Étude anatomo-histologique. *Ann. Soc. Entomol. Belg.* **57,** 23–44.
BURDICK, D. J. and KARDOS, E. H. (1963) The age structure of fall, winter, and spring populations of *Culex tarsalis* in Kern County, California. *Ann. Entomol. Soc. Am.* **56,** 527–35.
BURGES, H. D. and CAMMELL, M. E. (1964) Effect of temperature and humidity on *Trogoderma anthrenoides* (Sharp) (Coleoptera, Dermestidae) and comparisons with related species. *Bull. Entomol. Res.* **55,** 313–25.

BURISCH, E. (1963) Beiträge zur Genetik von *Megaselia scalaris* Loew (Phoridae). *Z. Vererbungslehre* **94,** 322–30.

BURNETT, G. F. (1951) Observations on the life-history of the red locust, *Nomadacris septemfasciata* (Serv.) in the solitary phase. *Bull. Entomol. Res.* **42,** 473–90.

BUSH, G. L. (1966) Female heterogamety in the family Tephritidae (Acalyptratae, Diptera). *Am. Naturalist* **100,** 119–26.

BUSNEL, M. C. (1963) Caractérisation acoustique de populations d'*Ephippiger* écologiquement voisines. *Ann. Epiphyties* **14,** 25–34.

BUSNEL, M. C. and BUSNEL, R. G. (1954) La directivité acoustique des déplacements de la femelle d'*Oecanthus pellucens* Scop. *C.R. Soc. Biol.* **148,** 830–3.

BUSNEL, R. G., DUMORTIER, B. and BUSNEL, M. C. (1956) Recherches sur le comportement acoustique des Éphippigères (Orthoptères, Tettigoniidae). *Bull. Biol. Fr. Belg.* **90,** 219–86.

BUTENANDT, A., BECKMANN, R., STAMM, D. and HECKER, E. (1959) Über den Sexual-Lockstoff des Seidenspinners *Bombyx mori.* Reindarstellung und Konstitution. *Z. Naturforsch* **14b,** 283–4.

BUTENANDT, A. and HECKER, E. (1961) Synthese des Bombykols, des Sexuallockstoffes des Seidenspinners, und seine geometrischen Isomeren. *Angew. Chem.* **73,** 349–53.

BUTENANDT, A. and REMBOLDT, H. (1958) Über den Weiselzellenfuttersaft der Honigbiene, II. Isolierung von 2-amino-4-hydroxy-6-(1.2-dihydroxy-propyl)-pteridin. *Hoppe-Seyler's Z. Physiol. Chem.* **311,** 79–83.

BUTLER, C. G. (1954) The method and importance of the recognition by a colony of honeybees (*A. mellifera*) of the presence of its queen. *Trans. Roy. Entomol. Soc. London* **105,** 11–29.

BUTLER, C. G. (1956) Some further observations on the nature of the "queen substance" and of its role in the organisation of a honey-bee (*Apis mellifera*) community. *Proc. Roy. Entomol. Soc. London* A, **31,** 12–16.

BUTLER, C. G. (1957a) The control of ovary development in worker honeybees (*Apis mellifera*). *Experientia* **13,** 256–7.

BUTLER, C. G. (1957b) The process of queen supercedure in colonies of honeybees (*Apis mellifera* Linn.). *Insectes Sociaux* **4,** 211–23.

BUTLER, C. G. (1961) The scent of queen honeybees (*A. mellifera* L.) that causes partial inhibition of queen rearing. *J. Insect Physiol.* **7,** 258–64.

BUTLER, C. G. (1967a) Insect pheromones. *Biol. Rev.* **42,** 42–87.

BUTLER, C. G. (1967b) A sex attractant acting as an aphrodisiac in the honeybee (*Apis mellifera* L.). *Proc. Roy. Entomol. Soc. London* A **42,** 71–76.

BUTLER, C. G., CALAM, D. H. and CALLOW, R. K. (1967) Attraction of *Apis mellifera* drones by the odours of the queens of two other species of honeybees. *Nature* **213,** 423–4.

BUTLER, C. G. and CALLOW, R. K. (1968) Pheromones of the honeybee (*Apis mellifera* L.): the "inhibitory scent" of the queen. *Proc. Roy. Entomol. Soc. London* A **43,** 62–65.

BUTLER, C. G., CALLOW, R. K. and CHAPMAN, J. R. (1964) 9-Hydroxydec-trans-2-enoic acid, a pheromone stabilizing honeybee swarms. *Nature* **201,** 733.

BUTLER, C. G., CALLOW, R. K. and JOHNSTON, W. C. (1959) Extraction and purification of "queen substance" from queen bees. *Nature* **184,** 1871.

BUTLER, C. G., CALLOW, R. K. and JOHNSTON, W. C. (1961) The isolation and synthesis of queen substance, 9-oxodec-trans-2-enoic acid, a honeybee pheromone. *Proc. Roy. Soc. London* B, **155,** 417–32.

BUTLER, C. G. and FAIREY, E. M. (1964) Pheromones of the honeybee: biological studies of the mandibular gland secretion of the queen. *J. Apic. Res.* **3,** 65–76.

BUTLER, C. G. and PATON, P. N. (1962) Inhibition of queen rearing by queen honeybees (*Apis mellifera* L.) of different ages. *Proc. Roy. Entomol. Soc. London* A, **37,** 114–16.

BUTLER, C. G. and SIMPSON, J. (1958) The source of the queen substance of the honey-bee (*Apis mellifera* L.). *Proc. Roy. Entomol. Soc. London* A, **33,** 120–2.

BUTLER, C. G. and SIMPSON, J. (1967) Pheromones of the queen honeybee (*Apis mellifera* L.) which enable her workers to follow her when swarming. *Proc. Roy. Entomol. Soc. London* A, **42,** 149–54.

BUTOVITSCH, V. (1939) Zur Kenntnis der Paarung, Eiablage und Ernährung der Cerambyciden. *Entomol. Tidskr.* **60,** 206–58.

BUXTON, P. A. (1930) The biology of a blood-sucking bug, *Rhodnius prolixus. Trans. Roy. Entomol. Soc. London* **78,** 227–36.

BUXTON, P. A. (1948) Experiments with mice and fleas. I. The baby mouse. *Parasitol.* **39,** 119–24.

CADEILHAN, L. (1965a) Stimulation de la ponte d'*Acrolepia assectella* Zell. par la présence de la plante hôte. *C.R. Acad. Sci. Paris* **261,** 1106–9.

CADEILHAN, L. (1965b) Action de la plante-hôte sur l'ovogenèse de l'adulte d'*Acrolepia assectella* Zell. (Insecte Lepidoptère). *C.R. Acad. Sci. Paris* **261,** 1910–13.

CALDWELL, R. L. and DINGLE, H. (1965) Regulation of mating activity in the milkweed bug (*Oncopeltus*) by temperature and photoperiod. *Am. Zoologist* **5,** 685.

CALLAHAN, P. S. (1958) Serial morphology as a technique for determination of reproductive patterns in the corn earworm, *Heliothis zea* (Boddie). *Ann. Entomol. Soc. Am.* **51,** 413–28.

Calori, L. (1848) Sulla generazione vivipara della *Chloe diptera* (Efemera Diptera Linn.). *Nuovi Annali Sci. Natur.* **9**.

Camenzind, R. (1962) Untersuchungen über die bisexuelle Fortpflanzung einer paedogenetischen Gallmücke. *Rev. Suisse Zool.* **69,** 377–84.

Camenzind, R. (1966) Die Zytologie der bisexuellen und parthenogenetischen Fortpflanzung von *Heteropeza pygmaea* Winnertz, eine Gallmücke mit pädogenetischer Vermehrung. *Chromosoma* **18,** 123–52.

Campbell, J. M. (1966) Sexual behavior in the genus *Androchirus. Proc. Entomol. Soc. Washington* **68,** 258–64.

Cancella da Fonseca, J. P. (1965) Oviposition and length of adult life in *Caryedon gonagra* (F.) (Col., Bruchidae). *Bull. Entomol. Res.* **55,** 697–707.

Cantacuzène, A. M. (1967) Histologie des glandes annexes mâles de *Schistocerca gregaria* F. (Orthoptères). Effet de l'allatectomie sur leur structure et leur activité. *C.R. Acad. Sci. Paris* **264,** 93–96.

Cappe de Baillon, P., Favrelle, M. and de Vichet, G. (1937) Parthénogénèse et variation chez les phasmes. III. *Bacillus rossi* Rossi, *Epibacillus lobipes* Luc., *Phobaeticus sinetyi* Bt., *Parasosibia parva* Redt., *Carausius rotundato-lobatus* Br. *Bull. Biol. Fr. Belg.* **71,** 129–89.

Cappe de Baillon, P., Favrelle, M. and de Vichet, G. (1938) Parthénogénese et variation chez les phasmes. IV. Discussion des faits. Conclusions. *Bull. Biol. Fr. Belg.* **72,** 1–47.

Carayon, J. (1952a) Les fécondations hémocoeliennes chez les Hémiptères Nabidés du genre *Alloeorhynchus. C.R. Acad. Sci. Paris* **234,** 751–4.

Carayon, J. (1952b) Les fécondations hémocoeliennes chez les Hémiptères Nabidae du genre *Prostemma. C.R. Acad. Sci. Paris* **234,** 1220–2.

Carayon, J. (1952c) La fécondation hémocoelienne chez *Prostemma guttula* (Hémipt. Nabidae). *C.R. Acad. Sci. Paris* **234,** 1317–19.

Carayon, J. (1952d) Existence chez certains Hémiptères Anthocoridae d'un organe analogue à l'organe de Ribaga. *Bull. Mus. Natl. Hist. Nat.* 2e *Sér.* **24,** 89–97.

Carayon, J. (1953) Organe de Ribaga et fécondation hémocoelienne chez les *Xylocoris* du groupe *galactinus* (Hémipt. Anthocoridae). *C.R. Acad. Sci. Paris* **236,** 1099–1101.

Carayon, J. (1954) Fécondation hémocoelienne chez un Hémiptère Cimicidé dépourvu d'organe de Ribaga. *C.R. Acad. Sci. Paris* **239,** 1542–4.

Carayon, J. (1959) Insemination par "spermalège" et cordon conducteur de spermatozoides chez *Stricticimex brevispinosus* Usinger (Heteroptera, Cimicidae). *Rev. Zool. Bot. Afr.* **60,** 81–104.

Carayon, J. (1962) La viviparité chez les Hétéroptères. *Proc. XIth Intern. Kongr. Entomol. Wien 1960* **1,** 711–14.

Carayon, J. (1964) Un cas d'offrande nuptiale chez les Hétéroptères. *C.R. Acad. Sci. Paris* **259,** 4815–18.

Carayon, J. (1966a) Les inséminations traumatiques accidentelles chez certains Hémiptères Cimicoidea. *C.R. Acad. Sci. Paris* **262,** 2176–9.

Carayon, J. (1966b) Traumatic insemination and the paragenital system. *The Thomas Say Foundation* **7,** 81–166.

Carle, P. (1965) Essai d'analyse expérimentale des facteurs conditionnant la fécondité chez la bruche du haricot (*Acanthoscelides obtectus* Say). *Ann. Epiphyties* **16,** 215–49.

Carlisle, D. B., Ellis, P. E. and Betts, E. (1965) The influence of aromatic shrubs on sexual maturation in the desert locust *Schistocerca gregaria. J. Insect Physiol.* **11,** 1541–58.

Carne, P. B. (1962) The characteristics and behaviour of the saw-fly *Perga affinis affinis* (Hymenoptera). *Australian J. Zool.* **10,** 1–35.

Carr, C. A. H. (1962) Further studies on the influence of the queen in ants of the genus *Myrmica. Insectes Sociaux* **9,** 197–211.

Carson, H. L. (1967) Selection for parthenogenesis in *Drosophila mercatorum, Genetics* **55,** 157–71.

Caspers, H. (1951) Rhythmische Erscheinungen in der Fortpflanzung von *Clunio marinus* (Dipt. Chiron.) und das Problem der lunaren Periodizität bei Organismen. *Arch. Hydrobiol. Suppl.* **18,** 415–594.

Cassagnau, P. and Juberthie, C. (1966) Neurosécrétion et organes endocrines chez *Tomocerus minor* (Collemboles). *C.R. Acad. Sci. Paris* **262,** 793–6.

Cassagnau, P. and Juberthie, C. (1967) Structures nerveuses, neurosécrétion et organes endocrines chez les Collemboles. II. Le complexe cérébral des Entobryomorphes. *Gen. Comp. Endocrinol.* **8,** 489–502.

Cassier, P. (1965a) Interactions des effets du groupement et d'un facteur saisonnier chez *Locusta migratoria migratorioides* (R. & F.). *Bull. Soc. Zool. Fr.* **90,** 39–51.

Cassier, P. (1965b) Contribution à l'étude du comportement phototropique du criquet migrateur (*Locusta migratoria migratorioides* R. et F.). *Ann. Sci. Nat., Zool. Sér.* 12, **7,** 213–358.

Castle, G. B. (1934) The experimental determination of caste differentiation in termites. *Science* **80,** 314.

Castle, G. B. (1946) The damp-wood termites of Western United States, genus *Zootermopsis* (formerly, *Termopsis*). II. An experimental investigation of caste differentiation in *Zootermopsis angusticollis.* In *Termites and Termite Control,* C. A. Kofoid, Ed., pp. 292–310.

Cavanagh, G. C. (1963) The use of the Dadd synthetic diet as a food for adult *Schistocerca gregaria* (Forsk.) and the effects of some additions and modifications to it. *J. Insect Physiol.* **9,** 759–75.

Cazal, P. (1948) Les glandes endocrines rétro-cérébrales des insectes. *Bull. Biol. Fr. Belg. Suppl.* **32,** 1–227.

Chang, T. H. and Riemann, J. G. (1967) H^3-thymidine radioautographic study of spermatogenesis in the boll weevil, *Anthonomus grandis* (Coleoptera, Curculionidae). *Ann. Entomol. Soc. Am.* **60,** 975–9.

Charniaux-Cotton, H. (1965) Hormonal control of sex differentiation in invertebrates. In *Organogenesis*, R. DeHaan and H. Ursprung, Eds., pp. 701–40.

Chauvin, R. (1949) Les stérols dans le régime alimentaire de certains Orthoptères. *C.R. Acad. Sci. Paris* **229,** 902–3.

Chauvin, R. (1952) Déterminisme du polymorphisme social chez les abeilles. *Coll. Intern. Centre Nat. Rech. Sci.* **34,** 117–22.

Chiang, H. C. and Hodson, A. C. (1950) The relation of copulation to fecundity and population growth in *Drosophila melanogaster*. *Ecology* **31,** 255–9.

Chiang, H. C. and Kim, Y. H. (1962) Decapitation-initiated oviposition in crane flies. *Entomol. Exp. Appl.* **5,** 289–90.

Chen, D. H., Robbins, W. E. and Monroe, R. E. (1962) The gonadotropic action of *Cecropia* extracts in allatectomized American cockroaches. *Experientia* **18,** 577–8.

Chen, P. S. and Levenbook, L. (1966) Studies on the haemolymph proteins of the blowfly *Phormia regina*. I. Changes in ontogenetic patterns. *J. Insect Physiol.* **12,** 1595–1609.

Chewyreuv, I. (1913) Le rôle des femelles dans la détermination du sexe du leur descendance dans le groupe des Ichneumonides. *C.R. Soc. Biol.* **74,** 695–9.

Cholodkovsky, N. (1908) Über den weiblichen Geschlechtsapparat einiger viviparen Fliegen. *Zool. Anz.* **33,** 367–76.

Chopard, L. (1914) Sur la vitalité de *Mantis religiosa* L. (Orth. Mantidae); ponte après décapitation. *Bull. Soc. Entomol. Fr.* **19,** 481–2.

Chopard, L. (1948) La parthénogénèse chez les Orthopteroides. *Année Biol. Sér.* 3, **24,** 15–22.

Chopard, L. (1950) Sur l'anatomie et le développement d'une Blatte vivipare. *Proc. VIIIth Intern. Congr. Entomol. Stockholm 1948*, pp. 218–22.

Choudhuri, J. C. B. (1958) Experimental studies on the choice of oviposition sites by two species of *Chorthippus* (Orthoptera: Acrididae). *J. Animal Ecology* **27,** 201–16.

Christophers, S. R. (1915) The male genitalia of *Anopheles*. *Indian J. Med. Res.* **3,** 371–94.

Chumakova, B. M. (1962) Significance of individual food components for the vital activity of mature predatory and parasitic insects. *Vop. Ekol. Kievsk. Univ.* **8,** 133–4.

Clarke, J. M. and Maynard Smith, J. (1955) The genetics and cytology of *Drosophila subobscura*. XI. Hybrid vigour and longevity. *J. Genetics* **53,** 172–80.

Clausen, C. P. (1939) The effect of host size upon the sex ratio of hymenopterous parasites and its relation to methods of rearing and colonization. *J. N.Y. Entomol. Soc.* **47,** 1–9.

Clements, A. N. (1956) Hormonal control of ovary development in mosquitoes. *J. Exp. Biol.* **33,** 211–23.

Clements, A. N. (1963) *The Physiology of Mosquitoes*. Pergamon Press, Oxford, New York.

Coles, G. C. (1964) Some affects of decapitation on metabolism in *Rhodnius prolixus* Stål. *Nature* **203,** 323.

Coles, G. C. (1965a) Haemolymph proteins and yolk formation in *Rhodnius prolixus* Stål. *J. Exp. Biol.* **43,** 425–31.

Coles, G. C. (1965b) Studies on the hormonal control of metabolism in *Rhodnius prolixus* Stål. I. The adult female. *J. Insect Physiol.* **11,** 1325–30.

Colless, D. H. and Chellapah, W. T. (1960) Effects of body weight and size of blood-meal upon egg production in *Aedes aegypti* (Linnaeus) (Diptera, Culicidae). *Ann. Trop. Med. Parasitol.* **54,** 475–82.

Collins, C. W. and Potts, S. F. (1932) Attractants for the flying gypsy moths as an aid in locating new infestations. *U.S. Dept. Agr. Tech. Bull.* No. 336, 1–43.

Colombo, G. (1957) Ricerche citologiche ed istochimiche sull'oogenesi di *Bombyx mori* L. (Lepidoptera). *Arch. Zool. Ital.* **42,** 309–47.

Colombo, G. and Mocellin, E. (1956) Ricerche sulla biologia dell'*Anacridium aegyptium* L. (Orthoptera, Catantopidae). *Redia* **41,** 277–313.

Common, I. F. B. (1954) A study of the ecology of the adult bogong moth, *Agrotis infusa* (Boisd.) (Lepidoptera: Noctuidae), with special reference to its behaviour during migration and aestivation. *Australian J. Zool.* **2,** 223–63.

Comstock, J. H. (1925) *An Introduction to Entomology*. Ithaca, N.Y.

Connin, R. V., Jantz, O. K. and Bowers, W. S. (1967) Termination of diapause in the cereal leaf beetle by hormones. *J. Econ. Entomol.* **60,** 1752–3.

Corbet, P. S. (1957) The life-history of the emperor dragonfly *Anax imperator* Leach (Odonata: Aeshnidae). *J. Animal Ecology* **26,** 1–69.

Corbet, P. S. (1963a) *A Biology of Dragonflies*. Quadrangle Books, Chicago.

Corbet, P. S. (1963b) The oviposition-cycles of certain sylvan culicine mosquitoes (Diptera, Culicidae) in Uganda. *Ann. Trop. Med. Parasitol.* **57,** 371–81.

Corbet, P. S. (1964) Autogeny and oviposition in arctic mosquitoes. *Nature* **203,** 669.

CORBET, P. S. (1965) Reproduction in mosquitoes of the high Arctic. *Proc. XIIth Intern. Congr. Entomol. London 1964*, pp. 817–18.
COUNCE, S. J. and SELMAN, G. G. (1955) The effects of ultrasonic treatment on embryonic development of *Drosophila melanogaster*. *J. Embryol. Exp. Morphol.* **3,** 121–41.
COUSIN, G. (1929) Sur le mode de ponte de *Lucilia sericata* Meig. *C.R. Soc. Biol.* **100,** 731–2.
COUSIN, G. (1933) Étude biologique d'un Chalcidien: *Mormoniella vitripennis* Walk. *Bull. Biol. Fr. Belg.* **67,** 371–400.
COUTURIER, A. (1938) Contribution à l'étude biologique de *Podisus maculiventris* Say, prédateur americain de Doryphore. *Ann. Epiphyties Sér.* 2, **4,** 95–165.
COWAN, F. A. (1932) A study of fertility in the blowfly, *Phormia regina* Meigen. *Ohio J. Sci.* **32,** 389–92.
CRAGG, F. W. (1920) Further observations on the reproductive system of *Cimex*, with special reference to the behaviour of the spermatozoa. *Indian J. Med. Res.* **8,** 32–79.
CRAGG, J. B. (1956) The olfactory behaviour of *Lucilia* species (Diptera) under natural conditions. *Ann. Appl. Biol.* **44,** 467–77.
CRAIG, G. B., JR, (1967) Mosquitoes: female monogamy induced by male accessory gland substance. *Science* **156,** 1499–1501.
CRANDELL, H. A. (1939) The biology of *Pachycrepoideus dubius* Ashmead (Hymenoptera), a pteromalid parasite of *Piophila casei* Linne (Diptera). *Ann. Entomol. Soc. Am.* **32,** 632–54.
CRANE, J. (1955) Imaginal behavior of a Trinidad butterfly, *Heliconius erato hydara* Hewitson, with special reference to the social use of color. *Zoologica* **40,** 167–96.
CRAWLEY, W. C. (1911) Parthenogenesis in worker ants, with special reference to two colonies of *Lasius niger* Linn. *Trans. Roy. Entomol. Soc. London* **59,** 657–63.
CROUSE, H. V. (1960) The nature of the influence of x-translocations on the sex of progeny in *Sciara coprophila*. *Chromosoma* **11,** 146–66.
CRYSTAL, M. M. and MEYNERS, H. H. (1965) Influence of mating on oviposition by screw-worm flies (Diptera: Calliphoridae). *J. Med. Entomol.* **2,** 214–16.
CUMBER, R. A. (1949) The biology of bumble-bees, with special reference to the production of the worker caste. *Trans. Roy. Entomol. Soc. Lond.* **100,** 1–45.
CUMMING, M. E. P. (1959) The biology of *Adelges cooleyi* (Gill.) (Homoptera: Phylloxeridae). *Can. Entomologist* **91,** 601–17.
CUMMING, M. E. P. (1968) The life history and morphology of *Adelges lariciatus* (Homoptera: Phylloxeridae). *Can. Entomologist* **100,** 113–26.
CUMMINGS, M. R. (1968) The cytology of the vitellogenic stages of oogenesis in *Drosophila melanogaster*. Thesis, Northwestern University, Evanston.
CUNHA, A. B. DA and KERR, W. E. (1957) A genetical theory to explain sex determination by arrhenotokous parthenogenesis. *Forma et Functio* **1,** 33–36.
CURTIN, T. J. and JONES, J. C. (1961) The mechanism of ovulation and oviposition in *Aedes aegypti*. *Ann. Entomol. Soc. Am.* **54,** 298–313.
CUTHBERT, F. P., JR. and REID, W. J., JR. (1964) Studies of sex attractant of banded cucumber beetle. *J. Econ. Entomol.* **57,** 247–50.
DADD, R. H. and MITTLER, T. E. (1965) Studies on the artificial feeding of the aphid *Myzus persicae* (Sulzer). III. Some major nutritional requirements. *J. Insect Physiol.* **11,** 717–43.
DAHM, K. H., RÖLLER, H. and TROST, B. M. (1968) The juvenile hormone—IV. Sterochemistry of juvenile hormone and biological activity of some of its isomers and related compounds. *Life Sci.* **7,** 129–37.
DALY, H. V. (1966) Biological studies on *Ceratina dallatorreana*, an alien bee in California which reproduces by parthenogenesis (Hymenoptera: Apoidea). *Ann. Entomol. Soc. Am.* **59,** 1138–54.
DANILEVSKII, A. S. (1965) *Photoperiodism and Seasonal Development of Insects* (English Edition). Oliver & Boyd, Edinburgh and London.
DAVEY, K. G. (1958) The migration of spermatozoa in the female of *Rhodnius prolixus* Stål. *J. Exp. Biol.* **35,** 694–701.
DAVID, J. (1963) Influence de la fécondation de la femelle sur le nombre et la taille des œufs pondus: Étude chez *Drosophila melanogaster* Meig. *J. Insect Physiol.* **9,** 13–24.
DAVID, J. (1964) Influence d'un inhibiteur de l'acide folique sur l'ovogénèse de la Drosophile. I. Étude de la fécondité, du pourcentage d'éclosion et de la taille des œufs. *J. Insect Physiol.* **10,** 805–17.
DAVID, J. (1965) Influence d'un inhibiteur de l'acide folique sur l'ovogénèse de la Drosophile. *Proc. XIIth Intern. Congr. Entomol. London 1964*, 179.
DAVID, W. A. L. and GARDINER, B. O. C. (1961) The mating behaviour of *Pieris brassicae* (L.) in a laboratory culture. *Bull. Entomol. Res.* **52,** 263–80.
DAVIDSON, J. (1929) On the occurrence of the parthenogenetic and sexual forms in *Aphis rumicis* L., with special reference to the influence of environmental factors. *Ann. Appl. Biol.* **16,** 104–34.
DAVIES, D. M., MORRISON, P. E., TAYLOR, G. C. and GOODMAN, T. (1965) Larval nutrient stores, adult diets and reproduction in Diptera. *Proc. XIIth Intern. Congr. Entomol. London 1964*, 174.

Davies, D. M. and Peterson, B. V. (1956) Observations on the mating, feeding, ovarian development, and oviposition of adult black flies (Simuliidae, Diptera). *Can. J. Zool.* **34,** 615–55.

Davies, L. (1961) Ecology of two *Prosimulium* species (Diptera) with references to their ovarian cycles. *Can. Entomologist* **93,** 1113–40.

Davies, L. (1965) On spermatophores in Simuliidae (Diptera). *Proc. Roy. Entomol. Soc. London* A **40,** 30–34.

Davies, R. G. (1961) The postembryonic development of the female reproductive system in *Limothrips cerealium* Haliday (Thysanoptera: Thripidae). *Proc. Zool. Soc. London* **136,** 411–37.

Davis, C. W. C. (1965) A comparative descriptive and experimental study on the embryology of the mosquito *Culex fatigans* Wied. (Diptera: Culicidae) and the sheep fly *Lucilia sericata* Meig. (Diptera: Calliphoridae). Thesis Univ. Sydney, Australia.

Davis, C. W. C. (1967) A comparative study of larval embryogenesis in the mosquito *Culex fatigans* Wiedemann (Diptera: Culicidae) and the sheep-fly *Lucilia sericata* Meigen (Diptera: Calliphoridae). I. Description of embryonic development. *Australian J. Zool.* **15,** 547–79.

Davis, N. T. (1956) The morphology and functional anatomy of the male and female reproductive systems of *Cimex lectularius* L. (Heteroptera, Cimicidae). *Ann. Entomol. Soc. Am.* **49,** 466–93.

Davis, N. T. (1964) Studies of the reproductive physiology of Cimicidae (Hemiptera). I. Fecundation and egg maturation. *J. Insect Physiol.* **10,** 947–63.

Davis, N. T. (1965a) Studies of the reproductive physiology of Cimicidae (Hemiptera). II. Artificial insemination and the function of the seminal fluid. *J. Insect Physiol.* **11,** 355–66.

Davis, N. T. (1965b) Studies of the reproductive physiology of Cimicidae (Hemiptera). III. The seminal stimulus. *J. Insect Physiol.* **11,** 1199–1211.

Day, M. F. (1943) The function of the corpus allatum in muscoid Diptera. *Biol. Bull.* **84,** 127–40.

DeCoursey, J. D. and Webster, A. P. (1952) Effects of insecticides and other substances on oviposition by *Aedes sollicitans. J. Econ. Entomol.* **45,** 1030–4.

Déduit, Y. (1957) Études sur la ponte par autogénèse des Culicides. II. Données numériques sur l'acte de ponte chez la femelle fécondée de *Culex pipiens autogenicus* Roubaud. *C.R. Soc. Biol.* **151,** 974–7.

Déduit, Y. and Callot, J. (1955) Études sur la ponte autogénèse des Culicides. I. Données numériques sur l'acte de ponte chez la femelle vierge isolée de *Culex pipiens autogenicus* Roubaud. *C.R. Soc. Biol.* **149,** 1003–6.

Deegener, P. (1902) Das Duftorgan von *Hepialus hectus* L. *Z. Wiss. Zool.* **71,** 276–95.

Dehn, M. von (1963) Hemmung der Flügelbildung durch Farnesol bei der schwarzen Bohnenlaus, *Dorsalis fabae* Scop. *Naturwiss.* **50,** 578–9.

Dehn, M. von (1967) Über den Photoperiodismus heterogoner Aphiden. Zur Frage der direkten oder indirekten Wirkung der Tageslänge. *J. Insect Physiol.* **13,** 595–612.

Delavault, R. and Royer, M. (1966) Le début d'une seconde spermatogénèse, en fin de vie des individus hermaphrodites, chez *Icerya purchasi* Mask. (Insecte, Homoptère). *C.R. Acad. Sci. Paris* **262,** 2502–5.

Deleurance, E. P. (1948) Le comportement reproducteur est indépendant de la présence des ovaires chez *Polistes* (Hyménoptère-Vespides). *C.R. Acad. Sci. Paris* **227,** 866–7.

Deleurance, E. P. (1950) Sur le mécanisme de la monogynie fonctionnelle chez les *Polistes* (Hyménoptères-Vespides). *C.R. Acad. Sci. Paris* **230,** 782–4.

Deleurance, E. P. (1955) L'influence des ovaires sur l'activité de construction chez les *Polistes* (Hyménoptères Vespides). *C.R. Acad. Sci. Paris* **241,** 1073–5.

Deligne, J. and Pasteels, J. M. (1963) Phénomènes endocrines chez les reines de *Microcerotermes* sp. (Isoptères, Termitidae). *Gen. Comp. Endocrinol.* **3,** 694.

Deoras, P. J. and Bhaskaran, G. (1967) Studies on neuroendocrine system in the housefly *Musca nebulo* (Fabr.). IV. Hormonal control of ovary development. *J. Univ. Bombay* **35,** 73–87.

Depdolla, P. (1928) Die Keimzellenbildung und die Befruchtung bei den Insekten. In Schröder, C., *Handbuch der Entomologie,* **1,** 825–1116.

Detinova, T. S. (1953) Mechanism of gonotrophic harmony in the common malarial mosquito. *Zool. Zh.* **32,** 1178–88.

Detinova, T. S. (1962) Age-grouping methods in Diptera of medical importance. *World Health Org. Monogr.* No. 47, 1–216.

Detinova, T. S. (1968) Age structure of insect populations of medical importance. *Ann. Rev. Entomol.* **13,** 427–50.

Dickson, R. C. (1949) Factors governing the induction of diapause in the oriental fruit moth. *Ann. Entomol. Soc. Am.* **42,** 511–37.

Dimond, J. B., Lea, A. O., Brooks, R. F. and DeLong, D. M. (1955) A preliminary note on some nutritional requirements for reproduction in female *Aedes aegypti. Ohio J. Sci.* **55,** 209–11.

Dimond, J. B., Lea, A. O., Hahnert, W. F., Jr. and DeLong, D. M. (1956) The amino acids required for egg production in *Aedes aegypti* L. *Can. Entomologist* **88,** 57–62.

Dixon, A. F. G. (1963) Reproductive activity of the sycamore aphid, *Drepanosiphum platanoides* (Schr.) (Hemiptera, Aphididae). *J. Animal Ecology* **32,** 33–48.

Doane, J. F. (1961) Movement on the soil surface, of adult *Ctenicera aeripennis destructor* (Brown) and

Hypolithus bicolor Esch. (Coleoptera Elateridae), as indicated by funnel pitfall traps, with notes on captures of other Arthropods. *Can. Entomologust* **93,** 636–44.

Doane, J. F. (1963) Studies on oviposition and fecundity of *Ctenicera destructor* (Brown) (Coleoptera: Elateridae). *Can. Entomologist* **95,** 1145–53.

Doane, W. W. (1960a) Developmental physiology of the mutant female sterile (2) adipose of *Drosophila melanogaster*. I. Adult morphology, longevity, egg production, and egg lethality. *J. Exp. Zool.* **145,** 1–21.

Doane, W. W. (1960b) Developmental physiology of the mutant female sterile (2) adipose of *Drosophila melanogaster*. II. Effects of altered environment and residual genome on its expression. *J. Exp. Zool.* **145,** 23–42.

Dobzhansky, T. (1951) *Genetics and the Origin of Species*, 3rd ed. Columbia Univ. Press, New York.

Dobzhansky, T. and Schultz, J. (1934) The distribution of sex-factors in the x-chromosome of *Drosophila melanogaster*. *J. Genetics* **28,** 349–86.

Dodds, K. S. (1939) Oogenesis in *Neuroterus baccarum* L. *Genetica* **21,** 177–90.

Doncaster, L. (1910) Gametogenesis of the gall-fly, *Neuroterus lenticularis* (*Spathegaster baccarum*). Part I. *Proc. Roy. Soc. London* B, **82,** 88–113.

Doncaster, L. (1911) Gametogenesis of the gall-fly, *Neuroterus lenticularis*. Part II. *Proc. Roy. Soc. London* B, **83,** 476–89.

Doncaster, L. (1916) Gametogenesis and sex-determination in the gall-fly, *Neuroterus lenticularis* (*Spathegaster baccarum*). Part III. *Proc. Roy. Soc. London* B, **89,** 183–200.

Dorman, S. C., Hale, W. C., and Hoskins, W. M. (1938) The laboratory rearing of flesh flies and the relations between temperature, diet and egg production. *J. Econ. Entomol.* **31,** 44–51.

Doucette, C. F. and Eide, P. M. (1955) Influence of sugars on oviposition of narcissus bulk fly. *Ann. Entomol. Soc. Am.* **48,** 343–4.

Doutt, R. L. (1959) The biology of parasitic Hymenoptera. *Ann. Rev. Entomol.* **4,** 161–82.

Downes, J. A. (1955) The food habits and description of *Atrichopogon pollinivorus* sp.n. (Diptera: Ceratopogonidae). *Trans. Roy. Entomol. Soc. London* **106,** 439–53.

Downes, J. A. (1966) Observations on the mating behaviour of the crab hold Mosquito *Deinocerites cancer* (Diptera: Culicidae). *Can. Entomologist* **98,** 1169–77.

Dreischer, H. (1956) Untersuchungen über die Arbeitstätigkeit und Drüsenentwicklung altersbestimmter Bienen im weisellosen Volk. *Zool. Jahrb. Abt. Physiol.* **66,** 429–72.

Drescher, W. and Rothenbuhler, W. C. (1963) Gynandromorph production by egg chilling. Cytological mechanisms in honey bees. *J. Heredity* **54,** 195–201.

Drescher, W. and Rothenbuhler, W. C. (1964) Sex determination in the honey bee. *J. Heredity* **55,** 91–96.

Du Bois, A. M. and Geigy, R. (1935) Beiträge zur Oekologie, Fortpflanzungsbiologie und Metamorphose von *Sialis lutaria* L. *Rev. Suisse Zool.* **42,** 169–248.

Dumortier, B. (1963) Ethological and physiological study of sound emissions in Arthropoda. In *Acoustic Behaviour of Animals*, pp. 583–654.

Dumortier, B., Brieu, S. and Pasquinelly, F. (1957) Facteurs externes contrôlant le rhythme des périodes de chant chez *Ephippiger ephippiger* Fiebig mâle (Orthoptère-Tettigonoidea). *C.R. Acad. Sci. Paris* **244,** 2315–18.

Dunn, J. A. (1959) The survival in the soil apterae of the lettuce root aphid, *Pemphigus bursarius* (L.). *Ann. Appl. Biol.* **47,** 766–71.

Dupont-Raabe, M. (1952) Contribution à l'étude du rôle endocrine du cerveau et notamment de la pars intercérébralis chez les Phasmides. *Arch. Zool. Exp. Gén.* **89,** 128–38.

Dupont-Raabe, M. (1956) Mise en évidence, chez les phasmides, d'une troisième paire de nervi corporis cardiaci, voie possible de cheminement de la substance chromactive tritocérébrales vers les corpora cardiaca. *C.R. Acad. Sci. Paris* **243,** 1240–3.

Dupuis, C. (1955) Les génitalia des Hémiptères Hétéroptères (génitalia externes des deux sexes, voies ectodermiques femelles). Revue de la morphologie. Lexique de la nomenclature. Index bibliographique analytique. *Mem. Mus. Natl. Hist. Nat. Sér. A*, **6,** 183–278.

Dzierzon, J. (1845) *Eichstädter Bienenzeit.*

Earle, N. W., Walker, A. B., Burks, M. L. and Slatten, B. H. (1967) Sparing of cholesterol by cholestanol in the diet of the boll weevil *Anthonomus grandis* (Coleoptera: Curculionidae). *Ann. Entomol. Soc. Am.* **60,** 599–603.

Eastham, L. E. S. and McCully, S. B. (1943) The oviposition responses of *Calandra granaria* Linn. *J. Exp. Biol.* **20,** 35–42.

Eckstein, K. (1911) Beiträge zur Kenntnis des Kiefernspinners *Lasiocampa* (*Gastropacha, Dendrolimus*) *pini* L. *Zool. Jahrb. Abt. Syst.* **31,** 59–164.

Edmunds, L. R. (1954) A study of the biology and life history of *Prosevania punctata* (Brulle) with notes on additional species (Hymenoptera: Evaniidae). *Ann. Entomol. Soc. Am.* **47,** 575–92.

Edwards, F. W. (1926) Extraordinary mating-habits in a mosquito. *Entomol. Mag.* **62,** 23.

Edwards, J. S. (1961) On the reproduction of *Prionoplus reticularis* (Coleoptera, Cerambycidae), with general remarks on reproduction in the Cerambycidae. *Quart. J. Microscop. Sci.* **102,** 519–29.

EDWARDS, R. L. (1954) The host-finding and oviposition behaviour of *Mormoniella vitripennis* (Walker) (Hym., Pteromalidae), a parasite of muscoid flies. *Behaviour* **7,** 88–112.

EHRHARDT, H. J. (1962) Ablage über grosser Eier durch Arbeiterinnen von *Formica polyctena* Förster (Ins. Hym.) in Gegenwart von Königinnen. *Naturwiss.* **49,** 524–5.

EHRHARDT, P. (1968) Die Wirkung verschiedener Spurenelemente auf Wachstum, Reproduktion und Symbionten von *Neomyzus circumflexus* Buckt. (Aphidae, Homoptera, Insecta) bei künstlicher Ernährung. *Z. Vergleich. Physiol.* **58,** 47–75.

EHRMAN, L. (1964) Courtship and mating behavior as a reproductive isolating mechanism in *Drosophila. Am. Zoologist* **4,** 147–53.

EIDMANN, H. (1931) Morphologische und physiologische Untersuchungen am weiblichen Genitalapparat der Lepidopteren. II. Physiologischer Teil. *Z. Angew. Entomol.* **18,** 57–112.

EIDMANN, H. (1935) Zur Kenntnis der Blattschneiderameise *Atta sexdens* L., insbesondere ihrer Ökologie. *Z. Angew. Entomol.* **22,** 385–436.

EITER, K., TRUSCHEIT, E. and BONESS, M. (1967) Neuere Ergebnisse der Chemie von Insektensexuallockstoffen. Synthesen von D,L-10-Acetoxy-hexadecen-(7-cis)-ol-(1),12-Acetoxy-octadecen-(9-cis)-ol-(1) ("Gyplure") und 1-Acetoxy-10-propyl-tridecadien-(5-trans.9). *Ann. Chem.* **709,** 29–45.

ELDRIDGE, B. F. (1962) The influence of daily photoperiod on blood-feeding activity of *Culex tritaeniorhynchus* Giles. *Am. J. Hyg.* **77,** 49–53.

ELDRIDGE, B. F. (1966) Environmental control of ovarian development in mosquitoes of the *Culex pipiens* complex. *Science* **151,** 826–8.

ELDRIDGE, B. F. (1968) The effect of temperature and photoperiod on blood-feeding and ovarian development in mosquitoes of the *Culex pipiens* complex. *Am. J. Trop. Med. Hyg.* **17,** 133–40.

ELLIS, P. E., CARLISLE, D. B. and OSBORNE, D. J. (1965) Desert locusts: sexual maturation delayed by feeding on senescent vegetation. *Science* **149,** 546–7.

ELMHIRST, R. (1912) Some observations on the glowworm (*Lampyris noctiluca*, L.). *Zoologist*, Ser. 4, **16,** 190–2.

EL-TIGANI, M. EL-AMIN (1962) Der Einfluss der Mineraldüngung der Pflanzen auf Entwicklung und Vermehrung von Blattläusen. *Wiss. Z. Univ. Rostock* **11,** 307–24.

ELTRINGHAM, H. (1925) On the source of the sphragidal fluid in *Parnassius apollo* (Lepidoptera). *Trans. Entomol. Soc. London* **73,** 11–15.

ELTRINGHAM, H. (1928) On the production of silk by species of the genus *Hilara* Meig. (Diptera). *Proc. Roy. Soc. London* B, **102,** 327–38.

EMERY, C. (1896) Le polymorphisme des fourmis et la castration alimentaire. *III. Intern. Congr. Zool. 1895*, 395–410.

EMMERICH, H. and BARTH, R. H., JR. (1968) Effect of farnesylmethylether on reproductive physiology in the cockroach, *Byrsotria fumigata* (Guérin). *Z. Naturforsch.* **23b,** 119–20.

ENDERS, E. (1956) Die hormonale Steuerung rhythmischer Bewegungen von Insekten-Ovidukten. *Verhandl. Deut. Zool. Ges. Erlangen* 1955, 113–16.

ENGEL, H. (1935) Biologie und Ökologie von *Cassida viridis* L. *Z. Morphol. Oekol. Tiere* **30,** 41–96.

ENGELHARDT, V. VON (1914) Ueber die Hancockschen Drüsen von *Oecanthus pellucens* Scop. *Zool. Anz.* **44,** 219–27.

ENGELHARDT, V. VON (1915) On the structure of the alluring gland of *Isophya acuminata* Br.-W. *Bull. Soc. Entomol. Moscow* **1,** 58–63.

ENGELMANN, F. (1957a) Die Steuerung der Ovarfunktion bei der ovoviviparen Schabe *Leucophaea maderae* (Fabr.). *J. Insect Physiol.* **1,** 257–78.

ENGELMANN, F. (1957b) Bau und Funktion des weiblichen Geschlechtsapparates bei der ovoviviparen Schabe *Leucophaea maderae* (Fabr.) (Orthoptera) und einige Beobachtungen über die Entwicklung. *Biol. Zentr.* **76,** 722–40.

ENGELMANN, F. (1958) The stimulation of the corpora allata in *Diploptera punctata* (Blattaria). *Anat. Rec.* **132,** 432–3.

ENGELMANN, F. (1959a) The control of reproduction in *Diploptera punctata* (Blattaria). *Biol. Bull.* **116,** 406–19.

ENGELMANN, F. (1959b) Ueber die Wirkung implantierter Prothoraxdrüsen im adulten Weibchen von *Leucophaea maderae* (Blattaria). *Z. Vergleich. Physiol.* **41,** 456–70.

ENGELMANN, F. (1960a) Hormonal control of mating behavior in an insect. *Experientia* **16,** 69–70.

ENGELMANN, F. (1960b) Mechanisms controlling reproduction in two viviparous cockroaches (Blattaria). *Ann. N.Y. Acad. Sci.* **89,** 516–36.

ENGELMANN, F. (1962) Further experiments on the regulation of the sexual cycle in females of *Leucophaea maderae* (Blattaria). *Gen. Comp. Endocrinol.* **2,** 183–92.

ENGELMANN, F. (1963) Die Innervation der Genital- und Postgenitalsegmente bei Weibchen der Schabe *Leucophaea maderae. Zool. Jahrb. Abt. Anat.* **81,** 1–16.

ENGELMANN, F. (1964) Inhibition of egg maturation in a pregnant viviparous cockroach. *Nature* **202,** 724–5.

ENGELMANN, F. (1965a) The mode of regulation of the corpus allatum in adult insects. *Arch. Anat. Microscop. Morphol. Exp.* **54,** 387–404.

ENGELMANN, F. (1965b) Endocrine activation of protein metabolism during egg maturation in an insect. *Am. Zoologist* **5,** 673.

ENGELMANN, F. (1966) Corpus allatum controlled protein biosynthesis in *Leucophaea maderae. Symp. on Insect Endocrines*, Brno.

ENGELMANN, F. and BARTH, R. H., JR. (1968) Endocrine control of female receptivity in *Leucophaea maderae* (Blattaria). *Ann. Entomol. Soc. Am.* **61,** 503–5.

ENGELMANN, F. and LÜSCHER, M. (1957) Die hemmende Wirkung des Gehirns auf die Corpora allata bei *Leucophaea maderae* (Orthoptera). *Verhandl. Deut. Zool. Ges. Hamburg 1956*, 215–20.

ENGELMANN, F. and MÜLLER, H. P. (1966) Fat body respiration as influenced by previously isolated corpora cardiaca. *Naturwiss.* **53,** 388–9.

ENGELMANN, F. and PENNEY, D. (1966) Studies on the endocrine control of metabolism in *Leucophaea maderae* (Blattaria). I. The hemolymph proteins during egg maturation. *Gen. Comp. Endocrinol.* **7,** 314–25.

ENGELMANN, F. and RAU, I. (1965) A correlation between feeding and the sexual cycle in *Leucophaea maderae* (Blattaria). *J. Insect Physiol.* **11,** 53–64.

ENGELS, W. (1966) Der zeitliche Ablauf von Protein- und Kohlenhydratsynthesen in der Oogenese bei *Apis mellifica* L. *Verhandl. Deut. Zool. Ges. Jena 1965*, 243–51.

ENGELS, W. and BIER, K. (1967) Zur Glykogenspeicherung während der Oogenese und ihrer vorzeitigen Auslösung durch Blockierung der RNS-Versorgung (Untersuchungen an *Musca domestica* L.). *Arch. Enticklungsmech. Organ.* **158,** 64–88.

ENGELS, W. and DRESCHER, W. (1964) Einbau von H^3-D-Glucose während der Oogenese bei *Apis mellifica* L. *Experientia* **20,** 445–7.

ENTWISTLE, P. F. (1964) Inbreeding and arrhenotoky in the ambrosia beetle *Xyleborus compactus* (Eichh.) (Coleoptera: Scolytidae). *Proc. Roy. Entomol. Soc. London* A **39,** 83–88.

EVANS, D. (1967) The bisexual and agamic generations of *Besbicus mirabilis* (Hymenoptera: Cynipidae), and their associated insects. *Can. Entomologist* **99,** 187–96.

EVANS, H. E. (1958) The evolution of social life in wasps. *Proc. Xth Intern. Congr. Entomol. Montreal 1956*, **2,** 449–57.

EVERS, A. M. J. (1948) Over de aanhangsels aan de uiteinden der dekschilden bij het ♂ van *Axinotarsus pulicarius* F. *Tijdschr. Entomol.* **89,** 149–54.

EVERS, A. M. J. (1956) Über die Funktion der Excitatoren beim Liebesspiel der Malachiidae. *Entomol. Bl.* **52,** 165–9.

EWEN, A. B. (1966) The corpus allatum and oocyte maturation in *Adelphocoris lineolatus* (Goeze) (Hemiptera: Miridae). *Can. J. Zool.* **44,** 719–27.

EWER, D. W. and EWER, R. F. (1942) The biology and behaviour of *Ptinus tectus* Boie. (Coleoptera, Ptinidae), a pest of stored products. III. The effect of temperature and humidity on oviposition, feeding and duration of life cycle. *J. Exp. Biol.* **18,** 290–305.

EWING, A. W. (1964) The influence of wing area on the courtship behaviour of *Drosophila melanogaster. Animal Behaviour* **12,** 316–20.

EYLES, A. C. (1963) Fecundity and oviposition rhythms in *Nysius huttoni* White (Heteroptera: Lygaeidae). *New Zealand J. Sci.* **6,** 186–207.

EZHIKOV, T. (1934) Individual variability and dimorphism of social insects. *Am. Naturalist* **68,** 333–44.

FABER, W. (1949) Biologische Untersuchungen zur Diapause des Kartoffelkäfers (*Leptinotarsa decemlineata* Say). *Pflanzenschutzb.* **3,** 65–94.

FALK, G. J. and KING, R. C. (1964) Studies on the developmental genetics of the mutant tiny of *Drosophila melanogaster. Growth* **28,** 291–324.

FALKENHAN, H. H. (1932) Biologische Beobachtungen an *Sminthurides aquaticus* (Collembola). *Z. Wiss. Zool.* **141,** 525–80.

FAVARD-SÉRÉNO, C. and DURAND, M. (1963) L'utilisation de nucléosides dans l'ovaire du Grillon et ses variations au cours de l'ovogénèse. II. Incorporation dans l'ADN. *Develop. Biol.* **6,** 206–18.

FELDMAN-MUHSAM, B. (1944) Studies on the ecology of the levant house fly (*Musca domestica vicina* Macq.). *Bull. Entomol. Res.* **35,** 53–67.

FÉRON, M. (1959) Attraction chimique du mâle de *Ceratitis capitata* Wied. (Dipt., Trypetidae) pour la femelle. *C.R. Acad. Sci. Paris* **248,** 2403–4.

FEUERBORN, H. J. (1922) Das Hypopygium "inversum" und "circumversum" der Dipteren. *Zool. Anz.* **55,** 189–212.

FIEDLER, H. G. (1950) Studien zur Biologie und Bekämpfung der Kaffeewanzen (*Antestia lineaticollis* Stal. und *A. faceta* Germ.) in Ostafrika. *Z. Angew. Entomol.* **31,** 473–99.

FIELDING, J. W. (1919) Notes on the bionomics of *Stegomyia fasciata*, Fabr. *Ann. Trop. Med. Parasitol.* **13,** 259–96.

FINKE, C. (1968) Lautäusserungen und Verhalten von *Sigara striata* und *Callicorixa praeusta* (Corixidae Leach., Hydrocorisae Latr.). *Z. Vergleich. Physiol.* **58,** 398–422.

FISCHER, L. H. (1853) *Orthoptera europaea.* Leipzig.

FLANDERS, S. E. (1939) Environmental control of sex in hymenopterous insects. *Ann. Entomol. Soc. Am.* **32,** 11–26.

FLANDERS, S. E. (1942) Oosorption and ovulation in relation to oviposition in the parasitic Hymenoptera. *Ann. Entomol. Soc. Am.* **35,** 251–66.

FLANDERS, S. E. (1945) The role of the spermatophore in the mass propagation of *Macrocentrus ancylivorus* Roh. *J. Econ. Entomol.* **38,** 323–7.

FLANDERS, S. E. (1950) Regulation of ovulation and egg disposal in the parasitic Hymenoptera. *Can. Entomologist* **82,** 134–40.

FLANDERS, S. E. (1957) Ovigenic-ovisorptive cycle in the economy of the honey bee. *Sci. Monthly* **85,** 176–7.

FLORENCE, L. (1921) The hog louse, *Haematopinus suis* Linné: its biology, anatomy, and histology. *Cornell Univ. Agr. Exp. Stat. Mem.* **51,** 641–743.

FLUITER, H. J. DE (1950) De invloed van daglengte en temperatuur op het opterden van de geslachtsdieren bij *Aphis fabae* Scop., de zwarte bonenluis. *Tijdschr. Plantenziekten* **56,** 265–85.

FOLLIOT, R. (1964) Contribution à l'étude de la biologie des *Cynipides gallicoles* (Hyménoptères, Cynipoidea). *Ann. Sci. Nat., Zool. Sér.* 12, **6,** 407–564.

FOOTE, R. H. and THOMAS, C. A. (1959) *Mycophila fungicola* Felt: a redescription and review of its biology (Diptera, Itonididae). *Ann. Entomol. Soc. Am.* **52,** 331–4.

FOREL, A. (1921) *Le monde social des fourmis.* Génève.

FOSTER, W. (1967) Hormone-mediated nutrional control of sexual behavior in male dung flies. *Science* **158,** 1596–7.

FRAENKEL, G. S., FRIEDMAN, S., HINTON, T., LASZLO, S. and NOLAND, J. L. (1955) The effect of substituting carnitine for choline in the nutrition of several organisms. *Arch. Biochem. Biophys.* **54,** 432–9.

FRANZ, J. (1940) Der Tannentriebwickler *Cacoecia murinana* Hb. Beiträge zur Bionomie und Oekologie. *Z. Angew Entomol.* **27,** 345–407.

FRASER, F. C. (1939) The evolution of the copulatory process in the order Odonata. *Proc. Roy. Entomol. Soc. London* A **14,** 125–9.

FRÉON, G. (1964a) Recherches histophysiologiques sur la neurosécrétion dans la chaine nerveuse ventrale du criquet migrateur, *Locusta migratoria. C. R. Acad. Sci. Paris* **259,** 1565–8.

FRÉON, G. (1964b) Contribution à l'étude de la neurosécrétion dans la chaîne nerveuse ventrale du criquet migrateur, *Locusta migratoria* L. *Bull. Soc. Zool. Fr.* **89,** 819–30.

FRINGS, H. and FRINGS, M. (1958) Uses of sounds by insects. *Ann. Rev. Entomol.* **3,** 87–106.

FULTON, B. B. (1915) The tree crickets of New York: life history and bionomics. *Tech. Bull. N.Y. Agr. Exp. Stat.* **42,** 3–47.

FULTON, B. B. (1931) A study of the genus *Nemobius* (Orthoptera, Gryllidae). *Ann. Entomol. Soc. Am.* **24,** 205–37.

FULTON, B. B. (1933) Inheritance of song in hybrids of two subspecies of *Nemobius fasciatus* (Orthoptera). *Ann. Entomol. Soc. Am.* **26,** 368–76.

FUNKE, W. (1957) Zur Biologie und Ethologie einheimischer Lamiinen (Cerambycidae, Coleoptera). *Zool. Jahrb. Abt. Syst.* **85,** 73–176.

GABBUTT, P. D. (1954) Notes on the mating behaviour of *Nemobius sylvestris* (Bosc) (Orth., Gryllidae). *Brit. J. Animal Behaviour* **2,** 84–88.

GABRITSCHEVSKY, E. (1928) Sénescence embryonnaire, rajeunissement et déterminisme des formes larvaires de *Miastor metraloas* (Cecidomyidae. Diptera). Étude expérimentale. *Bull. Biol. Fr. Belg.* **62,** 478–524.

GABRITSCHEVSKY, E. (1930) Der umkehrbare Entwicklungscyclus bei *Miastor metraloas. Arch. Entwicklungsmech. Organ.* **121,** 450–65.

GANAGARAJAH, M. (1965) The neuro-endocrine complex of adult *Nebria brevicollis* (F.) and its relation to reproduction. *J. Insect Physiol.* **11,** 1377–87.

GANAGARAJAH, M. (1966) Seasonal changes in food reserves of *Nebria brevicollis* (Carabidae, Coleoptera). *Entomol. Exp. Appl.* **9,** 314–22.

GARCIA-BELLIDO, A. (1964a) Beziehungen zwischen Vermehrungswachstum und Differenzierung von männlichen Keimzellen von *Drosophila melanogaster. Arch. Entwicklungsmech. Organ.* **155,** 594–610.

GARCIA-BELLIDO, A. (1964b) Analyse der physiologischen Bedingungen des Vermehrungswachstums männlicher Keimzellen von *Drosophila melanogaster. Arch. Entwicklungsmech. Organ.* **155,** 611–31.

GARCIA-BELLIDO, A. (1964c) Das Sekret der Paragonien als Stimulus der Fekundität bei Weibchen von *Drosophila melanogaster. Z. Naturforsch.* **19b,** 491–5.

GARDINER, P. (1953) The morphology and biology of *Ernobius mollis* L. (Coleoptera-Anobiidae). *Trans. Roy. Entomol. Soc. London* **104,** 1–24.

GARY, N. E. (1962) Chemical mating attractant in the queen honey bee. *Science* **136,** 773–4.

GASSNER, G., III and BRELAND, O. P. (1967) A phase-contrast study of the spermatozoa of the tiger beetle (Coleoptera: Cicindelidae). *Ann. Entomol. Soc. Am.* **60,** 244–8.

GAST, R. (1967) Untersuchungen über den Einfluss der Königinnensubstanz auf die Entwicklung der endokrinen Drüsen bei der Arbeiterin der Honigbiene (*Apis mellifica*). *Insectes Sociaux* **14,** 1–12.

GEER, B. W. (1966) Choline activity in the reproduction of *Drosophila melanogaster. Am. Zoologist* **6,** 509.

GEERING, Q. A. and COAKER, T. H. (1960) The effects of different plant foods on the fecundity, fertility and development of a cotton stainer, *Dysdercus superstitiosus* (F.). *Bull. Entomol. Res.* **51,** 61–76.

GEIGY, R. (1931) Action de l'ultra-violet sur le pôle germinal dans l'œuf de *Drosophila melanogaster* (Castration et mutabilité). *Rev. Suisse Zool.* **38,** 187–288.

GELDIAY, S. (1965) Hormonal control of adult diapause in the Egyptian grasshopper, *Anacridium aegyptium* L. *Gen. Comp. Endocrinol.* **5,** 680–1.

GELDIAY, S. (1966) Influence of photoperiod on imaginal diapause in *Anacridium aegyptium* L. *Sci. Reports Fac. Sci. Ege University* No. 40, 1–19.

GELDIAY, S. (1967) Hormonal control of adult reproductive diapause in the Egyptian grasshopper, *Anacridium aegyptium* L. *J. Endocrinol.* **37,** 63–71.

GEORGE, J. A. and HOWARD, M. G. (1968) Insemination without spermatophores in the oriental fruit moth, *Grapholitha molesta* (Lepidoptera: Tortricidae). *Can. Entomologist* **100,** 190–2.

GERHARDT, U. (1913) Copulation und Spermatophoren von Grylliden und Locustiden. *Zool. Jahrb. Abt. Syst.* **35,** 415–532.

GERHARDT, U. (1914) Copulation und Spermatophoren von Grylliden und Locustiden. *Zool. Jahrb. Abt. Syst.* **37,** 1–64.

GERHARDT, U. (1921) Neue Studien über Copulation und Spermatophoren von Grylliden und Locustiden. *Acta Zool.* **2,** 293–327.

GERSHENSON, S. (1928) A new sex-ratio abnormality in *Drosophila obscura. Genetics* **13,** 488–507.

GERVET, J. (1962) Étude de l'effet de groupe sur la ponte dans la société polygyne de *Polistes gallicus* L. (Hymén. Vesp.). *Insectes Sociaux* **9,** 231–63.

GERVET, J. (1964a) Essai d'analyse élémentaire du comportement de ponte chez la guêpe poliste *P. gallicus* L. (Hymén. Vesp.). *Insectes Sociaux* **11,** 21–40.

GERVET, J. (1964b) La ponte et sa régulation dans la société polygyne de *Polistes gallicus* L. (Hyménoptère Vespidé). *Ann. Sci. Nat., Zool. Ser.* 12, **6,** 601–778.

GERVET, J. (1964c) Le comportement d'oophagie différentielle chez *Polistes gallicus* L. (Hymén. Vesp.). *Insectes Sociaux* **11,** 343–82.

GERVET, J. and STRAMBI, A. (1965) Dynamique de la fonction ovarienne chez les *Polistes* (Hymén. Vesp.). Cas de l'ouvrière. *C.R. Acad. Sci. Paris* **260,** 4599–601.

GEYER-DUSZYŃSKA, I. (1959) Experimental research on chromosome elimination in Cecidomyidae (Diptera). *J. Exp. Zool.* **141,** 391–447.

GHILAROV, M. S. (1958) Evolution of insemination character in terrestrial Arthropods. *Zool. Zh.* **37,** 707–35.

GHOURI, A. S. K. and MCFARLANE, J. E. (1957) Reproductive isolation in the house cricket (Orthoptera: Gryllidae). *Psyche* **64,** 30–36.

GHOURI, A. S. K. and MCFARLANE, J. E. (1958) Observations on the development of crickets. *Can. Entomologist* **90,** 158–65.

GIARD, A. (1905) La poecilogonie. *Bull. Sci. Fr. Belg.* **39,** 153–87.

GILBERT, L. I. (1967) Changes in lipid content during the reproductive cycle of *Leucophaea maderae* and effects of the juvenile hormone on lipid metabolism *in vitro. Comp. Biochem. Physiol.* **21,** 237–57.

GILCHRIST, B. M. and HALDANE, J. B. S. (1947) Sex linkage and sex determination in a mosquito, *Culex molestus. Hereditas* **33,** 175–90.

GILLETT, J. D. (1956) Initiation and promotion of ovarian development in the mosquito *Aedes* (*Stegomyia*) *aegypti* (Linnaeus). *Ann. Trop. Med. Parasitol.* **50,** 375–80.

GILLETT, J. D., HADDOW, A. J. and CORBET, P. S. (1959) Observations on the oviposition-cycle of *Aëdes* (*Stegomyia*) *aegypti* (Linnaeus), II. *Ann. Trop. Med. Parasitol.* **53,** 35–41.

GILLETTE, C. P. (1904) Copulation and ovulation in *Anabrus simplex* Hald. *Entomol. News* **15,** 321–4.

GILLIES, M. T. (1956) A new character for the recognition of nulliparous females of *Anopheles gambiae. Bull. World Health Org.* **15,** 451–9.

GIRARDIE, A. (1962) Étude biométrique de la croissance ovarienne après ablation et implantation de corpora allata chez *Periplaneta americana. J. Insect Physiol.* **8,** 199–204.

GIRARDIE, A. (1964) Action de la pars intercérébralis sur le développement de *Locusta migratoria* L. *J. Insect Physiol.* **10,** 599–609.

GIRARDIE, A. (1966) Contrôle de l'activité génitale chez *Locusta migratoria.* Mise en évidence d'un facteur gonadotrope et d'un facteur allatotrope dans la pars intercérébralis. *Bull. Soc. Zool. Fr.* **91,** 423–39.

GIRARDIE, A. and GIRARDIE, J. (1967) Étude histologique, histochimique et ultrastructurale de la pars intercérébralis chez *Locusta migratoria* L. (Orthoptère). *Z. Zellforsch.* **78,** 54–75.

GIRARDIE, A. and VOGEL, A. (1966) Étude du contrôle neuro-humoral de l'activité sexuelle mâle de *Locusta migratoria* (L.). *C.R. Acad. Sci. Paris* **263,** 543–6.

GLASER, R. W. (1923) The effect of food on longevity and reproduction in flies. *J. Exp. Zool.* **38,** 383–412.

GOETSCH, W. (1953) *Vergleichende Biologie der Insekten-Staaten.* 2nd ed. Geest & Portig, Leipzig.

GOETSCH, W. and KÄTHNER, B. (1937) Die Koloniegründung der Formicinen und ihre experimentelle Beeinflussung. *Z. Morphol. Oekol. Tiere* **33,** 202–60.

GOLDSCHMIDT, R. (1911) Über die Vererbung der sekundären Geschlechtscharaktere. *Münch. Med. Wochenschr.* **58,** 2642–3.

GOLDSCHMIDT, R. (1920) Untersuchungen über Intersexualität. *Z. Induktive Abstammungs. Vererbungslehre* **23,** 1–199.

GOLDSCHMIDT, R. (1931) Analysis of intersexuality in the gipsy-moth. *Quart. Rev. Biol.* **6,** 125–42.

GOLDSCHMIDT, R. B. (1949) The interpretation of the triploid intersexes of *Solenobia. Experientia* **5,** 417–25.

GOLDSCHMIDT, R. B. (1955) *Theoretical Genetics.* Univ. Calif. Press, Berkeley.

GOLDSCHMIDT, R. and KATSUKI, K. (1928) Cytologie des erblichen Gynandromorphismus von *Bombyx mori* L. *Biol. Zentr.* **48,** 685–99.

GONTARSKI, H. (1949) Mikrochemische Futtersaftuntersuchungen und die Frage der Königinnenentstehung. *Hessische Biene* **85,** 89–92.

GOOD, N. E. (1933) Biology of the flour beetles, *Tribolium confusum* Duv. and *T. ferrugineum* Fab. *J. Agr. Res.* **46,** 327–34.

GOODCHILD, A. J. P. (1955) Some observations on growth and egg production of the blood-sucking Reduviids, *Rhodnius prolixus* and *Triatoma infestans. Proc. Roy. Entomol. Soc. London* A **30,** 137–44.

GOODWIN, J. A. and MADSEN, H. F. (1964) The mating and oviposition behavior of the navel orangeworm, *Paramyelois transitella* (Walker). *Hilgardia* **35,** 507–25.

GORDON, H. T. (1959) Minimal nutritional requirements of the German roach, *Blattella germanica* L. *Ann. N.Y. Acad. Sci.* **77,** 290–351.

GORDON, R. M. (1922) Notes on the bionomics of *Stegomyia calopus*, Meigen, in Brazil. *Ann. Trop. Med. Parasitol.* **16,** 425–39.

GÖRNITZ, K. (1949) Anlockversuche mit dem weiblichen Sexualduftstoff des Schwammspinners (*Lymantria dispar*) und der Nonne (*Lymantria monacha*). *Anz. Schädlingsk.* **22,** 145–9.

GÖSSWALD, K. (1951) Über den Lebensablauf von Kolonien der roten Waldameise. *Zool. Jahrb. Abt. Syst.* **80,** 27–63.

GÖTZ, B. (1939) Über weitere Versuche zur Bekämpfung der Traubenwickler mit Hilfe des Sexualduftstoffes. *Anz. Schädlingsk.* **15,** 109–14.

GÖTZ, B. (1941a) Der Sexualduftstoff als Bekämpfungsmittel gegen die Traubenwickler im Freiland. *Wein und Rebe* **23,** 75–89.

GÖTZ, B. (1941b) Beiträge zur Analyse des Mottenfluges bei den Traubenwicklern *Clysia ambiguella* und *Polychrosis botrana. Wein und Rebe* **23,** 207–28.

GÖTZ, B. (1951) Die Sexualduftstoffe an Lepidopteren. *Experientia* **7,** 406–18.

GOUGH, L. H. (1928) Apistischer Brief von Südafrika. *Bienenvater* **60,** 30–32.

GOUIN, F. J. (1963) Anatomie, Histologie und Entwicklungsgeschichte der Insekten und der Myriapoden. Das Abdomen der Insekten. *Fortschr. Zool.* **15,** 337–53.

GOWEN, J. W. (1942) On the genetic basis for hermaphroditism. *Anat. Rec.* **84,** 458.

GOWEN, J. W. (1952) Hybrid vigor in *Drosophila.* In *Heterosis*, J. W. GOWEN, Ed., pp. 474–93.

GOWEN, J. W. (1961) Genetic and cytological foundations for sex. In *Sex and Internal Secretion*, YOUNG, Ed., pp. 3–75.

GOWEN, J. W. and FUNG, S. T. C. (1957) Determination of sex through genes in a major sex locus in *Drosophila melanogaster. Heredity* **11,** 397–402.

GOWEN, J. W. and JOHNSON, L. E. (1946) On the mechanism of heterosis. I. Metabolic capacity of different races of *Drosophila melanogaster* for egg production. *Am. Naturalist* **80,** 149–79.

GRANDI, M. (1941) Contributi allo studio degli Efemerotteri italiani. III. *Cloëon dipterum* L. *Bull. Ist Entomol. Univ. Bologna* **13,** 29–71.

GRASSÉ, P. P. (1922) Étude biologique sur le criquet égyptien *Orthacanthacris aegyptia* (L.). *Bull. Biol. Fr. Belg.* **56,** 545–78.

GRASSÉ, P. P. (1949) Ordre des isoptères ou termites. *Traité de Zoologie* **9,** 408–544.

GRASSÉ, P. P. and NOIROT, C. (1946) La production des sexués néoténiques chez le termite à cou jaune (*Calotermes flavicollis* F.): inhibition germinale et inhibition somatique. *C.R. Acad. Sci. Paris* **223,** 869–71.

GRASSÉ, P. P. and NOIROT, C. (1947) Le polymorphisme social du termite à cou jaune (*Calotermes flavicollis* F.). Les faux-ouvriers ou pseudergates et les mues régressives. *C. R. Acad. Sci. Paris* **224,** 219–21.

GRASSÉ, P. P. and NOIROT, C. (1951) La sociotomie: migration et fragmentation de la termitière chez les *Anoplotermes* et les *Trinervitermes. Behaviour* **3,** 146–66.

GRASSÉ, P. P. and NOIROT, C. (1960) Rôle respectif des mâles et des femelles dans la formation des sexués néoténiques chez *Calotermes flavicollis. Insectes Sociaux* **7,** 109–23.

GRASSI, B. and SANDIAS, A. (1896) The constitution and development of the society of termites: observations on their habits; with appendices on the parasitic protozoa of Termitidae, and on the Embiidae. *Quart. J. Microscop. Sci.* **39,** 245–322 and **40,** 1–75.

Green, C. D. (1964) The life history and fecundity of *Folsomia candida* (Willem) var. *distincta* (Bagnall) (Collembola: Isotomidae). *Proc. Roy. Entomol. Soc. London* A **39,** 125–8.

Green, N., Jacobson, M., Henneberry, T. J. and Kishaba, A. N. (1967) Insect sex attractants. VI. 7-dodecen-l-ol acetates and congeners. *J. Med. Chem.* **10,** 533–5.

Greenberg, J. (1951) Some nutritional requirements of adult mosquitoes (*Aedes aegypti*) for oviposition. *J. Nutr.* **43,** 27–35.

Gregory, G. E. (1965) The formation and fate of the spermatophore in the African migratory locust, *Locusta migratoria migratorioides* Reiche und Fairmaire. *Trans. Roy. Entomol. Soc. London* **117,** 33–66.

Gresson, R. A. R. (1929) Yolk-formation in certain Tenthredinidae. *Quart. J. Miscrocop. Sci.* **73,** 345–64.

Gresson, R. A. R. and Threadgold, L. T. (1962) Extrusion of nuclear material during oogenesis in *Blatta orientalis*. *Quart. J. Microscop. Sci.* **103,** 141–5.

Griffiths, J. T., Jr. and Tauber, O. E. (1942) Fecundity, longevity, and parthenogenesis of the American roach, *Periplaneta americana* L. *Physiol. Zool.* **15,** 196–209.

Grison, P. (1944) Inhibition de l'ovogénèse chez le Doryphore (*Leptinotarsa decemlineata* Say) nourri avec des feuilles sénescentes de pomme de terre. *C.R. Acad. Sci. Paris* **219,** 295–6.

Grison, P. (1948) Action des lécithines sur la fécondité du Doryphore. *C.R. Acad. Sci. Paris* **227,** 1172–4.

Grison, P. (1952) Relations entre l'état physiologique de la plante-hôte, *Solanum tuberosum* et la fécondité du Doryphore, *Leptinotarsa decemlineata* Say. *Trans. IXth Intern. Congr. Entomol. Amsterdam 1951,* **1,** 331–7.

Grison, P. (1957) Les facteurs alimentaires de la fécondité chez le doryphore (*Leptinotarsa decemlineata* Say) (Col. Chrysomelidae). *Ann. Epiphyties Sér. C,* **8,** 305–81.

Grison, P. (1958) L'influence de la plante-hôte sur la fécondité de l'insecte phytophage. *Entomol. Exp. Appl.* **1,** 73–93.

Grison, P. and Ritter, R. (1961) Effets du groupement sur l'activité et la ponte du doryphore *Leptinotarsa decemlineata* Say (Col. Chrysomelidae). *Insectes Sociaux* **8,** 109–23.

de Groot, A. P. and Voogd, S. (1954) On the ovary development in queenless worker bees (*Apis mellifica* L.). *Experientia* **10,** 384–5.

Guerra, A. A. and Bishop, J. L. (1962) The effect of aestivation on sexual maturation in the female alfalfa weevil (*Hypera postica*). *J. Econ. Entomol.* **55,** 747–9.

Günther, K. G. (1961) Funktionell-anatomische Untersuchung des männlichen Kopulationsapparates der Flöhe unter besonderer Berücksichtigung seiner postembryonalen Entwicklung (Siphonaptera). *Deut. Entomol. Z.* N.F. **8,** 258–349.

Guppy, J. C. (1961) Life-history, behaviour, and ecology of the clover seed midge, *Dasyneura leguminicola* (Lint.) (Diptera: Cecidomyiidae), in Eastern Ontario. *Can. Entomologist* **93,** 59–73.

Guyénot, E. (1913a) Études biologiques sur une mouche, *Drosophila ampelophila* Löw. IV. Nutrition des larves et fécondité. *C.R. Soc. Biol.* **74,** 270–2.

Guyénot, E. (1913b) Études biologiques sur une mouche, *Drosophila ampelophila* Löw. V. Nutrition des adultes et fécondité. *C.R. Soc. Biol.* **74,** 332–4.

Haddow, A. J., Corbet, P. S. and Gillett, J. D. (1960) Laboratory observations on the oviposition-cycle in the mosquito *Aedes* (*Stegomyia*) *apicoargenteus* Theobald. *Ann. Trop. Med. Parasitol.* **54,** 392–6.

Haddow, A. J. and Gillett, J. D. (1957) Observations on the oviposition-cycle of *Aedes* (*Stegomyia*) *aegypti* (Linnaeus). *Ann. Trop. Med. Parasitol.* **51,** 159–69.

Haddow, A. J. and Gillett, J. D. (1958) Laboratory observations on the oviposition-cycle in the mosquito *Taeniorhynchus* (*Coquillettidia*) *fuscopennatus* Theobald. *Ann. Trop. Med. Parasitol.* **52,** 320–5.

Haddow, A. J., Gillett, J. D. and Corbet, P. S. (1961) Observations on the oviposition-cycle of *Aedes* (*Stegomyia*) *aegypti* (Linnaeus). V. *Ann. Trop. Med. Parasitol.* **55,** 343–56.

Hadorn, E., Remensberger, P. and Tobler, H. (1964) Autonomie in der Hodenentwicklung und Dissoziation von Chemogenese und Histogenese bei *Drosophila melanogaster*. *Rev. Suisse Zool.* **71,** 583–92.

Hadorn, E. and Zeller, H. (1943) Fertilitätsstudien an *Drosophila melanogaster*. I. Untersuchungen zum altersbedingten Fertilitätsabfall. *Arch. Entwicklungsmech. Organ.* **142,** 276–300.

Haeger, J. S. and Provost, M. W. (1965) Colonization and biology of *Opifex fuscus*. *Trans. Roy. Soc. New Zealand, Zool.* **6,** 21–31.

Hafez, M. (1947) The biology and life-history of *Apanteles ruficrus* Hal. *Bull. Soc. Fouad Ier, Entomol.* **31,** 225–49.

Hagan, H. R. (1931) The embryogeny of the polyctenid, *Hesperoctenes fumarius* Westwood, with reference to viviparity in insects. *J. Morphol. Physiol.* **51,** 1–117.

Hagan, H. R. (1951) *Embryology of the Viviparous Insects*. The Ronald Press Company, New York.

Hagan, H. R. (1954a) The reproductive system of the army-ant queen, *Eciton* (*Eciton*). Part 1. General anatomy. *Am. Mus. Novitates No.* 1663, 1–12.

Hagan, H. R. (1954b) The reproductive system of the army-ant queen, *Eciton* (*Eciton*). Part 3. The oocyte cycle. *Am. Mus. Novitates No.* 1665, 1–20.

Hagen, K. S. (1950) Fecundity of *Chrysopa californica* as affected by synthetic foods. *J. Econ. Entomol.* **43,** 101–4.

Hagen, K. S. (1958) Honeydew as an adult fruit fly diet affecting reproduction. *Proc. Xth Intern. Congr. Entomol. Montreal 1956*, **3,** 25–30.
Hagen, K. S. (1962) Biology and ecology of predaceous Coccinellidae. *Ann. Rev. Entomol.* **7,** 289–326.
Hagen, K. S. and Tassan, R. L. (1966) The influence of protein hydrolysates of yeasts and chemically defines upon the fecundity of *Chrysopa carnea* Stephens (Neuroptera). *Acta Soc. Zool. Bohemoslov.* **30,** 219–27.
Hamilton, A. G. (1936) The relation of humidity and temperature to the development of three species of African locusts—*Locusta migratoria migratorioides* (R. & F.), *Schistocerca gregaria* (Forsk.), *Nomadacris septemfasciata* (Serv.). *Trans. Roy. Entomol. Soc. London* **85,** 1–60.
Hamilton, A. G. (1955) Parthenogenesis in the desert locust (*Schistocerca gregaria* Forsk.) and its possible effect on the maintenance of the species. *Proc. Roy. Entomol. Soc. London* A **30,** 103–14.
Hamilton, W. D. (1964) The genetical evolution of social behaviour I, II. *J. Theoret. Biol.* **7,** 1–52.
Hamilton, W. D. (1967) Extraordinary sex ratios. *Science* **156,** 477–88.
Hamm, A. H. (1908) Observations on *Empis livida*, L. *Entomol. Mo. Mag.* **44,** 181–4.
Hamm, A. H. (1909a) Observations on *Empis opaca*, F. *Entomol. Mo. Mag.* **45,** 132–4.
Hamm, A. H. (1909b) Further observations on the Empinae. *Entomol. Mo. Mag.* **45,** 157–62.
Hamon, J. (1963) Étude de l'âge physiologique des femelles d'anophèles dans les zones traitées au DDT, et non traitées, de la région de Bobo-Dioulasso, Haute-Volta. *Bull. World Health Org.* **28,** 83–109.
Hampton, U. M. (1952) Reproduction in the housefly (*Musca domestica* L.). *Proc. Roy. Entomol. Soc. London* A **27,** 29–32.
Hancock, J. L. (1905) The habits of the striped meadow cricket (*Oecanthus fasciatus* Fitch). *Am. Naturalist* **39,** 1–11.
Handlirsch, A. (1889) Beitrag zur Kenntnis des Gespinstes von *Hilara sartrix* Becker. *Verhandl. Zool. Bot. Ges. Wien* **39,** 623–6.
Hanna, A. D. (1947) Studies on the Mediterranean fruit-fly *Ceratitis capitata* Wied. (Diptera-Trypaneidae). *Bull. Soc. Fouad Ier, Entomol.* **31,** 251–285.
Hannah-Alava, A. (1964) The brood-pattern of X-ray-induced mutational damage in the germ cells of *Drosophila melanogaster* males. *Mutation Res.* **1,** 414–36.
Hannah-Alava, A. (1965) The premeiotic stages of spermatogenesis. *Advan. Genet.* **13,** 157–26.
Hannah-Alava, A. and Stern, C. (1957) The sexcombs in males and intersexes of *Drosophila melanogaster*. *J. Exp. Zool.* **134,** 533–56.
Hansen, H. J. (1894) On the structure and habits of *Hemimerus talpoides* Walk. *Entomol. Tidskr.* **15,** 65–93.
Hanström, B. (1938) Zwei Probleme betreffs der hormonalen Lokalisation im Insektenkopf. *Acta Univ. Lund*, N.F. **39,** 1–17.
Hanström, B. (1942) Die Corpora cardiaca und Corpora allata der Insekten. *Biol. Gen.* **15,** 485–531.
Hardenberg, J. D. F. (1929) Beiträge zur Kenntnis der Pupiparen. *Zool. Jahrb. Abt. Anat.* **50,** 497–570.
Harlow, P. M. (1956) A study of ovarial development and its relation to adult nutrition in the blowfly *Protophormia terrae-novae* (R.D.). *J. Exp. Biol.* **33,** 777–97.
Harper, A. M. (1959) Gall aphids on poplar in Alberta. II. Periods of emergence from galls, reproductive capacities, and predators of aphids in galls. *Can. Entomologist* **91,** 680–5.
Harris, R. G. (1924) Sex of adult Cecidomyidae (*Oligarces* sp.) arising from larvae produced by paedogenesis. *Psyche* **31,** 148–54.
Harris, W. V. (1958) Colony formation in the Isoptera. *Proc. Xth Intern. Congr. Entomol. Montreal 1956*, **2,** 435–9.
Harvey, E. N. (1952) *Bioluminescence*. Academic Press, New York.
Harwood, R. F. and Halfhill, E. (1964) The effect of photoperiod on fat body and ovarian development of *Culex tarsalis* (Diptera: Culicidae). *Ann. Entomol. Soc. Am.* **57,** 596–600.
Hase, A. (1918) Beobachtungen über den Kopulationsvorgang bei der Bettwanze (*Cimex lectularius* L.). *Sitzungsb. Ges. Naturforsch. Freunde Berlin 1918*, 311–22.
Hase, A. (1934) Zur Fortpflanzungsphysiologie der blutsaugenden Wanze *Rhodnius pictipes* (Hemipt. Heteropt.). Beiträge zur experimentellen Parasitologie. *Z. Parasitenk.* **6,** 129–44.
Hashimoto, H. (1965) Discovery of *Clunio takahashi* Tokunaga from Japan. *Japan J. Zool.* **15,** 13–29.
Haskell, P. T. (1953) The stridulation behaviour of the domestic cricket. *Brit. J. Animal Behaviour* **1,** 120–1.
Haskell, P. T. (1956) Hearing in certain Orthoptera. II. The nature of the response of certain receptors to natural and imitation stridulation. *J. Exp. Biol.* **33,** 767–76.
Haskell, P. T. (1957) Stridulation and its analysis in certain Geocorisae (Hemiptera, Heteroptera). *Proc. Zool. Soc. London* **129,** 351–8.
Haskell, P. T. (1958) Stridulation and associated behaviour in certain Orthoptera. 2. Stridulation of females and their behaviour with males. *Animal Behaviour* **6,** 27–42.
Haskell, P. T. (1960) Stridulation and associated behaviour in certain Orthoptera. 3. The influence of the gonads. *Animal Behaviour* **8,** 76–81.
Haskell, P. T. (1961) *Insect Sounds*. H. F. & G. Witherby, London.
Haskins, C. P. and Enzmann, E. V. (1945) On the occurrence of impaternate females in the Formicidae. *J. N.Y. Entomol. Soc.* **53,** 263–77.

Hassan, A. I. (1939) The biology of some British Delphacidae (Homopt.) and their parasites with special reference to the Strepsiptera. *Trans. Roy. Entomol. Soc. London* **89,** 345–84.

Hathaway, D. S. and Selman, G. G. (1961) Certain aspects of cell lineage and morphogenesis studies in embryos of *Drosophila melanogaster* with ultra-violet micro-beam. *J. Embryol. Exp. Morphol.* **9,** 310–25.

Hauschteck, E. (1962) Die Cytologie der Pädogenese und der Geschlechtsbestimmung einer heterogonen Gallmücke. *Chromosoma* **13,** 163–82.

Haydak, M. H. (1943) Larval food and development of castes in the honeybee. *J. Econ. Entomol.* **36,** 778–92.

Heath, H. (1927) Caste formation in the termite genus *Termopsis. J. Morphol.* **43,** 387–425.

Heberdey, R. F. (1931) Zur Entwicklungsgeschichte, vergleichenden Anatomie und Physiologie der weiblichen Geschlechtsausführwege der Insekten. *Z. Morphol. Oekol. Tiere* **22,** 416–586.

Hecht, O. (1933a) Experimentelle Beiträge zur Biologie der Stechmücken. III. Die Blutverdauung und Eireifung bei *Anopheles maculipennis* am Ende der Überwinterung. *Z. Angew. Entomol.* **20,** 126–35.

Hecht, O. (1933b) Experimentelle Beiträge zur Biologie der Steckmücken. IV. *Arch. Schiffs. Tropenhyg.* **37,** 256–71.

Hecht, O. (1933c) Die Blutnahrung, die Erzeugung der Eier und die Überwinterung der Stechmückenweibchen. *Arch. Schiffs. Tropenhyg.* 37, Beiheft No. 3, 1–87.

Heed, W. B. and Kircher, H. W. (1965) Unique sterol in the ecology and nutrition of *Drosophila pachea. Science* **149,** 758–761.

Hegner, R. W. (1908) Effects of removing the germ-cell determinants from the eggs of some Chrysomelid beetles. Preliminary report. *Biol. Bull.* **16,** 19–26.

Hegner, R. W. (1909) The origin and early history of the germ cells in some Chrysomelid beetles. *J. Morphol.* **20,** 231–96.

Hegner, R. W. (1911) Experiments with Chrysomelid beetles. III. The effects of killing parts of the eggs of *Leptinotarsa decemlineata. Biol. Bull.* **20,** 237–51.

Hegner, R. W. (1914) Studies on germ cells. *J. Morphol.* **25,** 375–509.

Hegner, R. W. and Russell, C. P. (1916) Differential mitoses in the germ-cell cycle of *Dineutes nigrior. Proc. Natl. Acad. Sci. U.S.* **2,** 356–60.

Heldmann, G. (1936) Über das Leben auf Waben mit mehreren überwinterten Weibchen von *Polistes gallica* L. *Biol. Zentr.* **56,** 389–400.

Henslee, E. D. (1966) Sexual isolation in a parthenogenetic strain of *Drosophila mercatorum. Am. Naturalist* **100,** 191–7.

Heron, R. J. (1955) Studies on the starvation of last-instar larvae of the larch sawfly, *Pristiphora erichsonii* (Htg.) (Hymenoptera: Tenthredinidae). *Can. Entomologist* **87,** 417–27.

Heron, R. J. (1966) The reproductive capacity of the larch sawfly and some factors of concern in its measurement. *Can. Entomologist* **98,** 561–578.

Herron, J. C. (1953) Biology of the sweet clover weevil and notes on the biology of the clover root curculio. *Ohio J. Sci.* **53,** 105–12.

Herter, K. (1943) Zur Fortpflanzungbiologie eines lebendgebärenden Ohrwurmes (*Prolabia arachidis* Yersin). *Z. Morphol. Oekol. Tiere* **40,** 158–80.

Herter, K. (1965) Vergleichende Beobachtungen und Betrachtungen über die Fortpflanzungsbiologie der Ohrwürmer. *Z. Naturforsch.* **20b,** 365–75.

Hess, G. (1942) Über den Einfluss der Weisellosigkeit und des Fruchtbarkeitsvitamins E auf die Ovarien der Bienenarbeiterin. *Schweiz. Bienen.* Beiheft **1,** 33–110.

Hess, W. N. (1920) Notes on the biology of some common Lampyridae. *Biol. Bull.* **38,** 39–76.

Hewer, H. R. (1934) Studies in *Zygaena* (Lepidoptera). Part II. The mechanism of copulation and the passage of the sperm in the female. *Proc. Zool. Soc. London* 1934, 513–27.

Hewitt, C. G. (1906) Some observations on the reproduction of the Hemiptera-Cryptocerata. *Trans. Entomol. Soc. London* **54,** 86–90.

Heymer, A. (1967) Contribution à l'étude du comportement de ponte du genre *Platycnemis* Burmeister, 1839 (Odonata; Zygoptera). *Z. Tierpsychol.* **24,** 645–50.

Heymons, R. (1909) Eine Plazenta bei einem Insekt (*Hemimerus*). *Verhandl. Deut. Zool. Ges. Frankfurt 1909,* 97–107.

Heymons, R. (1912) Über den Genitalapparat und die Entwicklung von *Hemimerus talpoides* Walk. *Zool. Jahrb. Suppl.* **15,** 2, 141–84.

Heymons, R. (1929) Die Zahl der Eiröhren bei den Coprini (Coleoptera). *Zool. Anz.* **85,** 35–38.

Highnam, K. C. (1961) Induced changes in the amounts of material in the neurosecretory system of the desert locust. *Nature* **191,** 199–200.

Highnam, K. C. (1962a) Neurosecretory control of ovarian development in *Schistocerca gregaria. Quart. J. Microscop. Sci.* **103,** 57–72.

Highnam, K. C. (1962b) Neurosecretory control of ovarian development in the desert locust. *Mem. Soc. Endocrinol. No.* 12, 379–90.

Highnam, K. C. (1964) Endocrine relationships in insect reproduction. *Roy. Entomol. Soc. London Symp.* **2,** 26–42.

HIGHNAM, K. C. (1965) Some aspects of neurosecretion in arthropods. *Zool. Jahrb. Abt. Physiol.* **71**, 558–82.

HIGHNAM, K. C. and HASKELL, P. T. (1964) The endocrine systems of isolated and crowded *Locusta* and *Schistocerca* in relation to oocyte growth, and the effects of flying upon maturation. *J. Insect Physiol.* **10**, 849–64.

HIGHNAM, K. C., HILL, L. and MORDUE, W. (1965) The effects of starvation and removal of the frontal ganglion upon the endocrine control of oocyte development in the desert locust. *Gen. Comp. Endocrinol.* **5**, 685–6.

HIGHNAM, K. C. and LUSIS, O. (1962) The influence of mature males on the neurosecretory control of ovarian development in the desert locust. *Quart. J. Microscop. Sci.* **103**, 73–83.

HIGHNAM, K. C., LUSIS, O. and HILL, L. (1963a) The role of the corpora allata during oocyte growth in the desert locust, *Schistocerca gregaria* Forsk. *J. Insect Physiol.* **9**, 587–96.

HIGHNAM, K. C., LUSIS, O. and HILL, L. (1963b) Factors affecting oocyte resorption in the desert locust *Schistocerca gregaria* (Forskal). *J. Insect Physiol.* **9**, 827–37.

HILL, L. (1962) Neurosecretory control of haemolymph protein concentration during ovarian development in the desert locust. *J. Insect Physiol.* **8**, 609–19.

HILL, L. (1965) The incorporation of C^{14}-glycine into the proteins of the fat body of the desert locust during ovarian development. *J. Insect Physiol.* **11**, 1605–15.

HILL, L., MORDUE, W. and HIGHNAM, K. C. (1966) The endocrine system, frontal ganglion, and feeding during maturation in the female desert locust. *J. Insect Physiol.* **12**, 1197–1208.

HILL, R. E. (1946) Influence of food plants on fecundity, larval development and abundance of the tuber flea beetle in Nebraska. *Nebraska Agr. Exp. Sta. Res. Bull.* No. 143.

HILLE RIS LAMBERS, D. (1960) Some notes on morph determination in aphids. *Entomol. Ber. Amsterdam* **20**, 110–13.

HILLE RIS LAMBERS, D. (1966) Polymorphism in Aphididae. *Ann. Rev. Entomol.* **11**, 47–78.

HINTON, H. E. (1964) Sperm transfer in insects and the evolution of haemocoelic insemination. *Roy. Entomol. Soc. London Symp.* **2**, 95–107.

HIROYOSHI, T. (1964) Sex-limited inheritance and abnormal sex ratio in strains of the housefly. *Genetics* **50**, 373–85.

HOBSON, R. P. (1938) Sheep blow-fly investigations. VII. Observations on the development of eggs and oviposition in the sheep blow-fly, *Lucilia sericata* Mg. *Ann. Appl. Biol.* **25**, 573–82.

HODEK, I. (1962) Experimental influencing of the imaginal diapause in *Coccinella septempunctata* L. (Col., Coccinellidae), 2nd part. *Acta Soc. Entomol. Čechslov.* **59**, 297–313.

HOFENEDER, K. (1923) Stylops in copula. *Verhandl. Zool. Bot. Ges. Wien* **73**, 128–34.

HOFFMANN, R. (1954) Zur Fortpflanzungsbiologie und zur intrauterinen Entwicklung von *Glossina palpalis*. *Acta Tropica* **11**, 1–57.

HOHORST, W. (1936) Die Begattungsbiologie der Grille *Oecanthus pellucens* Scopoli. *Z. Morphol. Oekol. Tiere* **32**, 227–75.

HÖLLDOBLER, B. (1962) Zur Frage der Oligogynie bei *Camponotus ligniperda* Latr. und *Camponotus herculeanus* L. (Hym. Formicidae). *Z. Angew. Entomol.* **49**, 337–52.

HOLMGREN, N. (1904) Uber vivipare Insekten. *Zool. Jahrb. Abt. Syst.* **19**, 431–68.

HÖRMANN-HECK, S. VON (1957) Untersuchungen über den Erbgang einiger Verhaltensweisen bei Grillenbastarden (*Gryllus campestris* L. und *Gryllus bimaculatus* De Geer). *Z. Tierpsychol.* **14**, 137–83.

HORSFALL, W. R. and ANDERSON, J. F. (1964) Thermal stress and anomalous development of mosquitoes (Diptera: Culicidae). II. Effect of alternating temperatures on dimorphism of adults of *Aedes stimulans*. *J. Exp. Zool.* **156**, 61–89.

HOSOI, T. (1954) Egg production in *Culex pipiens pallens* Coquillett. II. Influence of light and temperature on activity of females. *Japan. J. Med. Sci. Biol.* **7**, 75–81.

HOUGHTON, C. O. (1909) Observations on the mating habits of *Oecanthus*. *Entomol. News* **20**, 274–9.

HOUSE, H. L. (1966) Effects of vitamins E and A on growth and development, and the necessity of vitamin E for reproduction in the parasitoid *Agria affinis* (Fallén) (Diptera, Sarcophagidae). *J. Insect Physiol.* **12**, 409–17.

HOUSE, H. L. and BARLOW, J. S. (1960) Effects of oleic acid and other fatty acids on the growth of *Agria affinis* (Fall.) (Diptera: Sarcophagidae). *J. Nutr.* **72**, 409–14.

HOWE, R. W. (1951) Studies on beetles of the family Ptinidae. V. The oviposition rate of *Ptinus tectus* Boield. under natural conditions. *Bull. Entomol. Res.* **42**, 445–53.

HOWE, R. W. (1952) The biology of the rice weevil, *Calandra oryzae* (L.). *Ann. Appl. Biol.* **39**, 168–80.

HOWLETT, M. (1907) Notes on the coupling of *Empis borealis*. *Entomol. Mo. Mag.* **43**, 229–32.

HSIAO, T. H. and FRAENKEL, G. (1968) Selection and specificity of the Colorado potato beetle for solanaceous and nonsolanaceous plants. *Ann. Entomol. Soc. Am.* **61**, 493–503.

HSU, W. S. (1952) The history of the cytoplasmic elements during vitellogenesis in *Drosophila melanogaster*. *Quart. J. Microscop. Sci.* **93**, 191–206.

HSU, W. S. (1953) The origin of proteid yolk in *Drosophila melanogaster*. *Quart. J. Microscop. Sci.* **94**, 23–28

HUBER, F. (1955) Sitz und Bedeutung nervöser Zentren für Instinkthandlungen beim Männchen von *Gryllus campestris* L. *Z. Tierpsychol.* **12,** 12–48.
HUDSON, A. and MCLINTOCK, J. (1967) A chemical factor that stimulates oviposition by *Culex tarsalis* Coquillet (Diptera, Culicidae). *Animal Behaviour* **15,** 336–41.
HUDSON, B. N. A. (1956) The behaviour of the female mosquito in selecting water for oviposition. *J. Exp. Biol.* **33,** 478–92.
HUET, C. and LENDER, T. (1962) Étude du développement de l'appareil génital femelle de *Tenbrio molitor*. *Bull. Soc. Zool. Fr.* **87,** 36–40.
HUFF, C. G. (1929) Ovulation requirements of *Culex pipiens* Linn. *Biol. Bull.* **56,** 347–50.
HUGGANS, J. L. and BLICKENSTAFF, C. C. (1964) Effects of photoperiod on sexual development in the alfalfa weevil. *J. Econ. Entomol.* **57,** 167–8.
HUGHES-SCHRADER, S. (1924) Reproduction in *Acroschismus wheeleri* Pierce. *J. Morphol.* **39,** 157–205.
HUGHES-SCHRADER, S. (1925) Cytology of hermaphroditism in *Icerya purchasi* (Coccidae). *Z. Zellforsch.* **2,** 264–92.
HUGHES-SCHRADER, S. (1926) Spermatogenesis in *Icerya purchasi*—A correction. *Science* **63,** 500–1.
HUGHES-SCHRADER, S. (1928) Origin and differentiation of the male and female germ cells in the hermaphrodite of *Icerya purchasi* (Coccidae). *Z. Zellforsch.* **6,** 509–40.
HUGHES-SCHRADER, S. (1944) A primitive coccid chromosome cycle in *Puto* sp. *Biol. Bull.* **87,** 167–76.
HUGHES-SCHRADER, S. (1948) Cytology of coccids (Coccoidea-Homoptera). *Advan. Genet.* **2,** 127–203.
HUGHES-SCHRADER, S. (1963) Hermaphroditism in an African Coccid, with notes on other Margarodids (Coccoidea-Homoptera). *J. Morphol.* **113,** 173–84.
HUGHES-SCHRADER, S. and MONAHAN, D. F. (1966) Hermaphroditism in *Icerya zeteki* Cockerell, and the mechanism of gonial reduction in Iceryine Coccids (Coccoidea: Margarodidae Morrison). *Chromosoma* **20,** 15–31.
HUIGNARD, J. (1964) Recherches histophysiologiques sur le contrôle hormonal de l'ovogénèse chez *Gryllus domesticus* L. *C.R. Acad. Sci. Paris* **259,** 1557–60.
HUSAIN, M. A. and MATHUR, C. B. (1945) Studies on *Schistocerca gregaria* Forsk. XIII. Sexual life. *Indian J. Entomol.* **7,** 89–101.
HUSSON, R. and PALÉVODY, C. (1967) La parthénogénèse chez les collemboles. *Ann. Soc. Entomol. Fr.* N.S. **3,** 631–3.
IBRAHIM, M. M. (1955) Studies on *Coccinella undecimpunctata aegyptiaca* Rche. II. Biology and life-history (Coleoptera: Coccinellidae). *Bull. Soc. Entomol. Egypt* **39,** 395–423.
IKEDA, H. (1965) Interspecific transfer of the "sex-ratio" agent of *Drosophila willistoni* in *Drosophila bifasciata* and *Drosophila melanogaster*. *Science* **147,** 1147–8.
ILSE, D. (1937) New observations on responses to colours in egglaying butterflies. *Nature* **140,** 544–5.
ISHAY, J. (1964) Observations sur la biologie de la guêpe orientale *Vespa orientalis* F. *Insectes Sociaux* **11,** 193–206.
ITO, Y. (1960) Territorialism and residentiality in a dragonfly, *Orthetrum albistylum speciosum* Uhler (Odonata: Anisoptera). *Ann. Entomol. Soc. Am.* **53,** 851–3.
IWANOFF, P. P. and MESTSCHERSKAJA, K. A. (1935) Die physiologischen Besonderheiten der geschlechtlich unreifen Insektenovarien und die zyklischen Veränderungen ihrer Eigenschaften. *Zool. Jahrb., Abt. Physiol.* **55,** 281–348.
IWANOWA, S. A. (1926) Zur Frage über die Spermatophorenbefruchtung bei den Acridoidea (*Locusta migratoria* L.). *Zool. Anz.* **65,** 75–86.
JACK, R. W. (1916) Parthenogenesis amongst the workers of the Cape Honey-Bee. *Trans. Roy, Entomol. Soc. London* **64,** 396–403.
JACKSON, C. H. N. (1948) The eclosion of tsetse (*Glossina*) larvae (Diptera). *Proc. Roy. Entomol. Soc. London* A **23,** 36–38.
JACKSON, D. J. (1958) Observations on the biology of *Caraphractus cinctus* Walker (Hymenoptera: Mymaridae), a parasitoid of the eggs of Dytiscidae. *Trans. Roy. Entomol. Soc. London* **110,** 533–54.
JACOB, F. H. (1964) A new species of *Thuleaphis* H.R.L. (Homoptera: Aphidoidea) from Wales, Scotland and Iceland. *Proc. Roy. Entomol. Soc. London* B **33,** 111–16.
JACOB, J. and SIRLIN, J. L. (1959) Cell function in the ovary of *Drosophila*. I. DNA classes in the nurse cell nuclei as determined by autoradiography. *Chromosoma* **10,** 210–28.
JACOBI, E. F. (1939) Über Lebensweise, Auffinden des Wirtes und Regulierung der Individuenzahl von *Mormoniella vitripennis* Walker. *Arch. Neerl. Zool.* **3,** 197–282.
JACOBS, W. (1950) Vergleichende Verhaltensstudien an Feldheuschrecken. *Z. Tierpsychol.* **7,** 169–216.
JACOBS, W. (1953) Verhaltensbiologische Studien an Feldheuschrecken. *Z. Tierpsychol. Suppl.* 1, p. 228.
JACOBSON, L. A. (1965) Mating and oviposition of the pale Western cutworm, *Agrotis orthogonia* Morrison (Lepidoptera: Noctudidae), in the laboratory. *Can. Entomologist* **97,** 994–1000.
JACOBSON, M. (1965) *Insect Sex Attractants*. Interscience Publishers, New York.
JACOBSON, M. and BEROZA, M. (1963) Chemical insect attractants. *Science* **140,** 1367–73.
JACOBSON, M. and BEROZA, M. (1965) American cockroach sex attractant. *Science* **147,** 748–9.

Jacobson, M., Beroza, M. and Jones, W. A. (1960) Isolation, identification, and synthesis of the sex attractant of gypsy moth. *Science* **132,** 1011–12.

Jacobson, M., Beroza, M. and Jones, W. A. (1961) Insect sex attractants. I. The isolation, identification, and synthesis of the sex attractant of the gypsy moth. *J. Am. Chem. Soc.* **83,** 4819–24.

Jacobson, M., Beroza, M. and Yamamoto, R. T. (1962) Isolation and identification of the sex attractant of the American cockroach. *Science* **139,** 48–49.

Jacobson, M. and Jones, W. A. (1962) Insect sex attractants. II. The synthesis of a highly potent gypsy moth sex attractant and some related compounds. *J. Org. Chem.* **27,** 2523–4.

Jacobson, M., Lilly, C. E. and Harding, C. (1968) Sex attractant of sugar beet wireworm: identification and biological activity. *Science* **159,** 208–210.

Janet, C. (1903) *Observations sur les guêpes.* Paris.

Jay, S. C. (1968) Factors influencing ovary development of worker honeybees under natural conditions. *Can. J. Zool.* **46,** 345–7.

Jazdowska-Zagrodzinska, B. (1966) Experimental studies on the role of "polar granules" in the segregation of pole cells in *Drosophila melanogaster. J. Embryol. Exp. Morphol.* **16,** 391–9.

Jefferson, R. N., Shorey, H. H. and Rubin, R. E. (1968) Sex pheromones of noctuid moths. XVI. The morphology of the female sex pheromone glands of eight species. *Ann. Entomol. Soc. Am.* **61,** 861–5.

Jensen, J. P. (1909) Courting and mating of *Oecanthus fasciatus* Harris. *Can. Entomologist* **41,** 25–27.

Jentsch, S. (1936) Ovoviviparie bei einer einheimischen Copeognathenart (*Hyperetes guestphalicus*). *Zool. Anz.* **116,** 287–9.

Johansson, A. S. (1954) Corpus allatum and egg production in starved milkweed bugs. *Nature* **174,** 89.

Johansson, A. S. (1955) The relationship between corpora allata and reproductive organs in starved female *Leucophaea maderae* (Blattaria). *Biol. Bull.* **108,** 40–44.

Johansson, A. S. (1958) Relation of nutrition to endocrine-reproductive functions in the milkweed bug *Oncopeltus fasciatus* (Dallas) (Heteroptera: Lygaeidae). *Nytt Mag. Zool.* **7,** 3–132.

Johansson, A. S. (1964) Feeding and nutrition in reproductive processes in insects. *Roy. Entomol. Soc. London Symp.* **2,** 43–55.

John, O. (1923) Fakultative Viviparität bei Thysanopteren. *Entomol. Mitteil.* **12,** 227–32.

Johnson, C. (1962a) Breeding behavior and oviposition in *Calopteryx maculatum* (Beauvais) (Odonata: Calopterygidae). *Am. Midland Naturalist* **68,** 242–7.

Johnson, C. (1962b) Reproductive isolation in damselflies and dragonflies (Order Odonata). *Texas J. Sci.* **14,** 297–304.

Johnson, C. (1965) Mating and oviposition of damselflies in the laboratory. *Can. Entomologist* **97,** 321–6.

Jolicoeur, H. and Topsent, E. (1892) Études sur l'écrivain ou gribouri (*Adoxus vitis* Kirby). *Mem. Soc. Zool. Fr.* **5,** 723–30.

Joly, L. (1960) Fonctions des corpora allata chez *Locusta migratoria* (L.). Strasbourg, 1960, 1–103.

Joly, L. (1964) Contrôle du fonctionnement ovarien chez *Locusta migratoria* L. I. Effets de castrations totales et de ligatures unilatérales de l'oviducte. *J. Insect Physiol.* **10,** 437–42.

Joly, P. (1942) Sur le rôle des corpora allata dans la ponte des Dytiscidés. *C.R. Acad. Sci. Paris* **214,** 807–9.

Joly, P. (1945) La fonction ovarienne et son contrôle humoral chez les Dytiscidés. *Arch. Zool. Exp. Gén.* **84,** 49–164.

Joly, P. (1950) Fonctionnement ovarien des Carabes. *C.R. Soc. Biol.* **144,** 1217–20.

Jones, J. C. and Wheeler, R. E. (1965a) Studies on spermathecal filling in *Aedes aegypti* (Linnaeus). I. Description. *Biol. Bull.* **129,** 134–50.

Jones, J. C. and Wheeler, R. E. (1965b) Studies on spermathecal filling in *Aedes aegypti* (Linnaeus). II. Experimental. *Biol. Bull.* **129,** 532–45.

Jones, W. A., Jacobson, M. and Martin, D. F. (1966) Sex attractant of the pink bollworm moth: isolation, identification and synthesis. *Science* **152,** 1516–17.

De Jong, J. K. (1938) The influence of the quality of the food on the egg-production in some insects. *Treubia* **16,** 445–68.

Judge, F. D. (1968) Polymorphism in a subterranean aphid, *Pemphigus bursarius.* I. Factors affecting the development of sexuparae. *Ann. Entomol. Soc. Am.* **61,** 819–27.

Judson, C. L. (1968) Physiology of feeding and oviposition behavior in *Aedes aegypti* (L.). Experimental dissociation of feeding and oogenesis. *J. Med. Entomol.* **5,** 21–23.

Junker, H. (1923) Cytologische Untersuchungen an den Geschlechtsorganen der halbzwitterigen Steinfliege *Perla marginata* (Panzer). *Arch. Zellforsch.* **17,** 185–259.

Jura, C. (1964) Cytological and experimental observations on the origin and fate of the pole cells in *Drosophila virilis* Sturt. Part II. Experimental analysis. *Acta Biol. Cracoviensis* **7,** 89–103.

Kahle, W. (1908) Die Paedogenesis der Cecidomyiden. *Zoologica* (Stuttgart) **21,** 1–80.

Kahn, M. C., Celestin, W. and Offenhauser, W. (1945) Recording of sounds produced by certain disease-carrying mosquitoes. *Science* **101,** 335–6.

Kahn, M. C. and Offenhauser, W. (1949) The first field tests of recorded mosquito sounds used for mosquito destruction. *Am. J. Trop. Med.* **29,** 811–25.

KAISER, P. (1949) Histologische Untersuchungen über die Corpora allata und Prothoraxdrüsen der Lepidopteren in Bezug auf ihre Funktion. *Arch. Entwicklungsmech. Organ.* **144,** 99–131.

KAISER, P. (1956) Die Bedeutung der Hormonalorgane für die Kastendifferenzierung der Termiten. *Verhandl. Deut. Zool. Ges. Erlangen 1955*, 101–6.

KAISSLING, K. E. and RENNER, M. (1968) Antennale Rezeptoren für die Queen substance und Sterzelduft bei der Honigbiene. *Z. Vergleich. Physiol.* **59,** 357–61.

KAMAL, A. S. (1954) Ecological and nutritional studies on the cherry fruit fly. *J. Econ. Entomol.* **47,** 959–65.

KARLINSKY, A. (1963) Effets de l'ablation des corpora allata imaginaux sur le développement ovarien de *Pieris brassicae* L. (Lépidoptère). *C.R. Acad. Sci. Paris* **256,** 4101–3.

KARLINSKY, A. (1967a) Reprise de la vitellogénèse après implantation de corpora allata chez *Pieris brassicae* L. (Lépidoptère). *C.R. Acad. Sci. Paris* **264,** 1735–8.

KARLINSKY, A. (1967b) Influence des corpora allata sur le fonctionnement ovarien en milieu mâle de *Pieris brassicae* L. (Lépidoptère). *C.R. Acad. Sci. Paris* **265,** 2040–2.

KARLINSKY, A. (1967c) Corpora allata et vitellogénèse chez les Lépidoptères. *Gen. Comp. Endocrinol.* **9,** 511–12.

KARLSON, P. and BUTENANDT, A. (1959) Pheromones (Ectohormones) in insects. *Ann. Rev. Entomol.* **4,** 39–58.

KARLSON, P. and LÜSCHER, M. (1959) "Pheromones": a new term for a class of biologically active substances. *Nature* **183,** 55–56.

KATSUKI, K. (1935) Weitere Versuche über erbliche Mosaikbildung und Gynandromorphismus bei *Bombyx mori* L. *Biol. Zentr.* **55,** 361–83.

KAUFMANN, T. (1965) Biological studies on some Bavarian Acridoidea (Orthoptera), with special reference to their feeding habits. *Ann. Entomol. Soc. Am.* **58,** 791–801.

KAYE, J. S. (1962) Acrosome formation in the house cricket. *J. Cell Biol.* **12,** 411–31.

KEILIN, D. (1916) Sur la viviparité chez les Diptères et sur les larves de Diptères vivipares. *Arch. Zool. Exp. Gén.* **55,** 393–415.

KELLER, J. C., MITCHELL, E. B., MCKIBBEN, G. and DAVICH, T. B. (1964) A sex attractant for female boll weevils from males. *J. Econ. Entomol.* **57,** 609–10.

KELLOGG, V. L. (1907) Some silkworm moth reflexes. *Biol. Bull.* **12,** 152–4.

KENNEDY, J. S. (1942) On water-finding and oviposition by captive mosquitoes. *Bull. Entomol. Res.* **32,** 279–301.

KENNEDY, J. S. (1949) A preliminary analysis of oviposition behavior by *Locusta* (Orthoptera, Acrididae) in relation to moisture. *Proc. Roy. Entomol. Soc. London* A **24,** 83–89.

KENNEDY, J. S. and STROYAN, H. J. G. (1959) Biology of aphids. *Ann. Rev. Entomol.* **4,** 139–60.

KENTEN, J. (1955) The effect of photoperiod and temperature on reproduction in *Acyrthosiphon pisum* (Harris) and on the forms produced. *Bull. Entomol. Res.* **46,** 599–624.

KERN, P. (1912) Über die Fortpflanzung und Eibildung bei einigen Caraben. *Zool. Anz.* **40,** 345–51.

KERR, W. E. (1950a) Genetic determination of castes in the genus *Melipona. Genetics* **35,** 143–152.

KERR, W. E. (1950b) Evolution of the mechanism of caste determination in the genus *Melipona. Evolution* **4,** 7–13.

KERR, W. E. (1962) Genetics of sex determination. *Ann. Rev. Entomol.* **7,** 157–76.

KERR, W. E. and NIELSEN, R. A. (1967) Sex determination in bees (Apinae). *J. Apicult. Res.* **6,** 3–9.

KESSEL, E. L. (1955) The mating activities of balloon flies. *Syst. Zool.* **4,** 97–104.

KESSEL, E. L. (1959) Introducing *Hilara wheeleri* Melander as a balloon maker, and notes on other North American balloon flies (Diptera: Empididae). *Wasmann J. Biol.* **17,** 221–30.

KESSEL, E. L. and KARABINOS, J. V. (1947) *Empimorpha geneatis* Melander, a balloon fly from California, with a chemical examination of its balloons. *Pan-Pacific Entomologist* **23,** 181–92.

KESSEL, E. L. and KESSEL, B. B. (1951) A new species of balloon-bearing *Empis* and an account of its mating activities (Diptera: Empididae). *Wasmann J. Biol.* **9,** 137–46.

KESSEL, E. L. and KESSEL, B. B. (1961) Observations on the mating behavior of *Platypezina pacifica* Kessel (Diptera: Platypezidae). *Wasmann J. Biol.* **19,** 295–9.

KESSEL, R. G. and BEAMS, H. W. (1963) Micropinocytosis and yolk formation in oocytes of the small milkweed bug. *Exp. Cell Res.* **30,** 440–3.

KHALIFA, A. (1949a) The mechanism of insemination and the mode of action of the spermatophore in *Gryllus domesticus. Quart. J. Microscop. Sci.* **90,** 281–92.

KHALIFA, A. (1949b) Spermatophore production in *Trichoptera* and some other insects. *Trans. Roy. Entomol. Soc. London* **100,** 449–71.

KHALIFA, A. (1950a) Sexual behaviour in *Gryllus domesticus* L. *Behaviour* **2,** 264–74.

KHALIFA, A. (1950b) Spermatophore production in *Galleria mellonella* L. (Lepidoptera). *Proc. Roy. Entomol. Soc. London* A **25,** 33–42.

KHALIFA, A. (1952) A contribution to the study of reproduction in the bed-bug (*Cimex lectularius* L.) (Hemiptera-Heteroptera). *Bull. Soc. Fouad* 1[er] *Entomol.* **36,** 311–36.

KING, P. E. and RICHARDS, J. G. (1968) Oosorption in *Nasonia vitripennis* (Hymenoptera: Pteromalidae). *J. Zool.* **154,** 495–516.

KING, R. C. (1964a) Studies on early stages of insect oogenesis. *Roy. Entomol. Soc. London Symp.* **2,** 13–25.

KING, R. C. (1964b) Further information concerning the envelopes surrounding dipteran eggs. *Quart. J. Microscop. Sci.* **105,** 209–11.

KING, R. C. and AGGARWAL, S. K. (1965) Oogenesis in *Hyalophora cecropia. Growth* **29,** 17–83.

KING, R. C., AGGARWAL, S. K. and AGGARWAL, U. (1968) The development of the female *Drosophila* reproductive system. *J. Morphol.* **124,** 143–66.

KING, R. C., AGGARWAL, S. K. and BODENSTEIN, D. (1966) The comparative submicroscopic cytology of the corpus allatum–corpus cardiacum complex of wild type and fes adult female *Drosophila melanogaster. J. Exp. Zool.* **161,** 151–76.

KING, R. C., BENTLEY, R. M. and AGGARWAL, S. K. (1966) Some of the properties of the components of *Drosophila* ooplasm. *Am. Naturalist* **100,** 365–7.

KING, R. C. and BURNETT, R. G. (1959) Autoradiographic study of uptake of tritiated glycine, thymidine, and uridine by fruit fly ovaries. *Science* **129,** 1674–5.

KING, R. C. and FALK, G. J. (1960) *In vitro* incorporation of uridine-H^3 into developing fruit fly oocytes. *J. Biophys. Biochem. Cytol.* **8,** 550–3.

KING, R. C. and KOCH, E. A. (1963) Studies on the ovarian follicle cells of *Drosophila. Quart. J. Microscop. Sci.* **104,** 297–320.

KING, R. C. and MILLS, R. P. (1962) Oogenesis in adult *Drosophila.* XI. Studies of some organelles of the nutrient stream in egg chambers of *D. melanogaster* and *D. willistoni. Growth* **26,** 235–53.

KING, R. C., RUBINSON, A. C. and SMITH, R. F. (1956) Oogenesis in adult *Drosophila melanogaster. Growth* **20,** 121–57.

KING, R. C. and SANG, J. R. (1959) Oogenesis in adult *Drosophila melanogaster.* VIII. The role of folic acid in oogenesis. *Growth* **23,** 37–53.

KIRCHER, H. W., HEED, W. B., RUSSELL, J. S. and GROVE, J. (1967) Senita cactus alkaloids: their significance to Sonoran desert *Drosophila* ecology. *J. Insect Physiol.* **13,** 1869–74.

KIRK, H. B. (1922) Notes on the mating-habits and early life-history of Culicid *Opifex fuscus* Hutton. *Trans. New Zealand Inst.* **54,** 400–6.

KIRKPATRICK, T. W. (1937) Studies on the ecology of coffee plantations in East Africa. II. The autecology of *Antestia spp.* (Pentatomidae) with a particular account of a strepsipterous parasite. *Trans. Roy. Entomol. Soc. London* **86,** 247–343.

KITZMILLER, J. B. (1953) Mosquito genetics and cytogenetics. *Rev. Brasil. Malariol.* **5,** 285–359.

KITZMILLER, J. B. (1959) Parthenogenesis in *Culex fatigans. Science* **129,** 837–8.

KLATT, B. (1913) Experimentelle Untersuchungen über die Beziehungen zwischen Kopulation und Eiablage beim Schwammspinner. *Biol. Zentr.* **33,** 620–8, 629–38.

KLATT, B. (1920) Beiträge zur Sexualphysiologie des Schwammspinners. *Biol. Zentr.* **40,** 539–58.

KLOBOUCEK, V. (1913) Das Liebesleben der *Musca domestica* L. (Stubenfliege). *Intern. Entomol. Z.* **7,** 138–9, 142–3.

KLOMP, H. and GRUYS, P. (1965) The analysis of factors affecting reproduction and mortality in a natural population of the pine looper, *Bupalus piniarius* L. *Proc. XIIth Intern. Congr. Entomol. London* 1964 369–372.

KLUG, W. S. (1968) Oogenesis in the su^2-Hw mutant of *Drosophila melanogaster.* Thesis, Northwestern University, Evanston.

KNERER, G. and PLATEAUX-QUÉNU, C. (1966) Sur le polymorphisme des femelles chez quelques Halictinae (Insectes Hymenoptères) paléarctiques. *C.R. Acad. Sci. Paris* **263,** 1759–61.

KOCH, E. A. and KING, R. C. (1966) The origin and early differentiation of the egg chamber of *Drosophila melanogaster. J. Morphol.* **119,** 283–304.

KOCH, E. A. and KING, R. C. (1968) Oocyte determination in *Drosophila melanogaster. J. Cell Biol.* **39,** 74A.

KOCH, E. A., SMITH, P. A. and KING, R. C. (1967) The division and differentiation of *Drosophila* cystocytes. *J. Morphol.* **121,** 55–70.

KOPEĆ, S. (1912) Untersuchungen über Kastration und Transplantation bei Schmetterlingen. *Arch. Entwicklungsmech. Organ.* **33,** 1–116.

KOREF-SANTIBAÑEZ, S. (1964) Reproductive isolation between the sibling species *Drosophila pavani* and *Drosophila gaucha. Evolution* **18,** 245–51.

KORNHAUSER, S. I. (1919) The sexual characteristics of the Membracid, *Thelia bimaculata* (Fabr.). I. External changes induced by *Aphelopus theliae* (Gahan). *J. Morphol.* **32,** 531–636.

KOVAČEVIĆ, Ž. (1956) Die Nahrungswahl und das Auftreten der Pflanzenschädlinge. *Anz. Schädlingsk.* **29,** 97–101.

KOZHANTSHIKOV, I. W. (1938) Carbohydrate and fat metabolism in adult Lepidoptera. *Bull. Entomol. Res.* **29,** 103–14.

KRACZKIEWICZ, Z. (1950) Recherches cytologiques sur les chromosomes de *Lasioptera rubi* Heeg. (Cecidomyidae). *Zool. Polon.* **5,** 73–117.

Kraińska, M. (1961) A morphological and histochemical study of oogenesis in the gall-fly *Cynips folii*. *Quart. J. Microscop. Sci.* **102,** 119–29.

Krassilstschik, J. (1893) Zur Entwicklungsgeschichte der Phytophthires. *Zool. Anz.* **16,** 69–76.

Kühn, A. (1919) *Die Orientierung der Tiere im Raum.* Jena.

Kullenberg, B. (1947) Über Morphologie und Funktion des Kopulationsapparates der Capsiden und Nabiden. *Zool. Bidrag Uppsala* **24,** 217–418.

Kummer, H. (1960) Experimentelle Untersuchungen zur Wirkung von Fortpflanzungs-Faktoren auf die Lebensdauer von *Drosophila melanogaster*-Weibchen. *Z. Vergleich. Physiol.* **43,** 642–79.

Kunz, W. (1966) Zur Chromosomenstruktur in den Oocyten der Heuschrecke *Locusta migratoria* L. *Naturwiss.* **53,** 23.

Labeyrie, V. (1959) Sur la fécondité des femelles vierges de *Diadromus pulchellus* Wsm. (Hym. Ichneumonidae). *Bull. Soc. Entomol. Fr.* **64,** 58–60.

Labeyrie, V. (1960a) Contribution à l'étude de la dynamique des populations d'insectes. I. Influence stimulatrice de l'hôte *Acrolepia assectella* Z. sur la multiplication d'un Hyménoptère Ichneumonidae (*Diadromus* sp.). *Entomophaga Mém. No.* 1, 1–193.

Labeyrie, V. (1960b) Action de la présence des grains de haricot sur l'ovogénèse d'*Acanthoscelides obtectus* Say (Coléoptère, Bruchidae). *C.R. Acad. Sci. Paris* **250,** 2626–8.

Labeyrie, V. (1962) Stimulation de l'ovogénèse de *Diadromus pulchellus* Wsm. provoqué par la présence des cocons de son hôte *Acrolepia assectella* Z. *Proc. XIth Intern. Congr. Entomol. Wien* 1960, **1,** 722–7.

Labeyrie, V. (1964) Action sélective de la fréquence de l'hôte utilisable (*Acrolepia assectella* Zel.) sur *Diadromus pulchellus* Wsm. (Hymenoptera Ichneumonidae): la variabilité de la fécondité en fonction de l'intensité de la stimulation. *C.R. Acad. Sci. Paris* **259,** 3644–7.

Labine, P. A. (1964) Population biology of the butterfly, *Euphydryas editha*. I. Barriers to multiple inseminations. *Evolution* **18,** 335–6.

LaBrecque, G. C., Meifert, D. W. and Smith, C. N. (1962) Mating competitiveness of chemosterilized and normal house flies. *Science* **136,** 388–9.

LaChance, L. E. and Bruns, S. B. (1963) Oogenesis and radiosensitivity in *Cochliomyia hominivorax* (Diptera: Calliphoridae). *Biol. Bull.* **124,** 65–83.

Lal, R. and Haque, E. (1955) Effect of nutrition under controlled conditions of temperature and humidity on longevity and fecundity of *Sphaerophoria scuttellaris* (Fabr.) (Syrphidae: Diptera). Efficacy of its maggots as aphid predators. *Indian J. Entomol.* **17,** 317–25.

Lampel, G. (1962) Formen und Steuermechanismen des Generationswechsels bei Insekten. *Zool. Anz.* **168,** 1–26.

Lampel, G. (1965a) Neue Aspekte in der Terminologie des Aphidoidea-Generations- und Wirtswechsels. *Proc. XIIth Intern. Congr. Entomol. London 1964*, 115–17.

Lampel, G. (1965b) Die Erscheinungsformen des Blattlaus-Generations- und Wirtswechsels (Homoptera, Aphidoidea). *Rev. Suisse Zool.* **72,** 609–18.

Lampel, G. (1968) *Die Biologie des Blattlaus-Generationswechsels.* Fischer Verlag, Jena.

Lamy, M. (1967) Mise en évidence, par électrophorèse sur acétate de cellulose, d'une proteine vitellogène dans l'hémolymphe de l'imago femelles de la Piéride du Chou (*Pieris brassicae* L.). *C.R. Acad. Sci. Paris* **265,** 990–3.

Landa, V. (1960) Origin, development and function of the spermatophore in cockchafer (*Melolontha melolontha* L.). *Acta Soc. Entomol. Czechoslov.* **57,** 297–316.

Landa, V. (1961) Experiments with an artificial spermatophore in cockchafer (*Melolontha melolontha* L.). *Acta Soc. Entomol. Czechoslov.* **58,** 296–301.

Lang, C. A. (1956) The influence of mating on egg production by *Aedes aegypti*. *Am. J. Trop. Med. Hyg.* **5,** 909–14.

Langley, P. A. (1968) The effect of host pregnancy on the reproductive capability of the tsetse fly, *Glossina morsitans*, in captivity. *J. Insect Physiol.* **14,** 121–33.

Lanier, G. N. (1966) Interspecific mating and cytological studies of closely related species of *Ips* DeGeer and *Orthotomicus* Ferrari (Coleoptera: Scolytidae). *Can. Entomologist* **98,** 175–98.

Lanier, G. N. and Oliver, J. H., Jr. (1966) "Sex-ratio" condition: unusual mechanisms in bark beetles. *Science* **153,** 208–9.

Larsen, J. R. (1958) Hormone-induced ovarian development in mosquitoes. *Science* **127,** 587.

Larsen, J. R. and Bodenstein, D. (1959) The humoral control of egg maturation in the mosquito. *J. Exp. Zool.* **140,** 343–81.

Larsen, J. R., Peadt, R. E. and Peterson, L. G. (1966) Olfactory and oviposition responses of the house fly to domestic manures, with notes on an autogenous strain. *J. Econ. Entomol.* **59,** 610–15.

Larson, A. O. and Fisher, C. K. (1924) Longevity and fecundity of *Bruchus quadrimaculatus* Fab. as influenced by different foods. *J. Agr. Res.* **29,** 297–305.

Laugé, G. (1962) Influence de la température d'élevage sur l'expression des caractères sexuels externes et internes des intersexués triploides de *Drosophila melanogaster*. *C.R. Acad. Sci. Paris* **255,** 1798–1800.

Laugé, G. (1964) Féminisation des génitalia externes des intersexués triploides de *Drosophila melanogaster*

Meig. sous l'action d'une température élevée. Mise en évidence d'une phase sensible au cours de développement. *C.R. Acad. Sci. Paris* **259**, 4156–9.

LAUGÉ, G. (1966) Étude comparative des effets d'un traitement thermiques sur le développement des gonades et de divers caractères sexuels primaires chez les intersexués triploides de *Drosophila melanogaster* Meig. *Bull. Soc. Zool. Fr.* **91**, 661–86.

LAUGÉ, G. (1967) Utilisation d'une température léthale dans l'étude des effets de traitements thermiques sur le développement de divers caractères sexuels primaires chez les intersexués triploides de *Drosophila melanogaster* Meig. *C.R. Soc. Biol.* **161**, 16–21.

LAURENCE, B. R. (1964) Autogeny in *Aedes* (*Finlaya*) *togoi* Theobald (Diptera, Culicidae). *J. Insect Physiol.* **10**, 319–31.

LAUTERBACH, G. (1953) Begattung und Geburt bei den Strepsipteren. *Naturwiss.* **40**, 516.

LAUTERBACH, G. (1954) Begattung und Larvengeburt bei den Strepsipteren. Zugleich ein Beitrag zur Anatomie der Stylops-Weibchen. *Z. Parasitenk.* **16**, 255–97.

LAUVERJAT, S. (1964) Caractères histochimiques de la sécrétion dans les voies génitales femelles et données expérimentales sur l'origine de l'oothéque chez *Locusta migratoria* R. et F. (Insecte, Orthoptère). *C.R. Acad. Sci. Paris* **258**, 4348–51.

LAUVERJAT, S. (1967) Rôle de corpora allata dans l'acquisition des caractères sexuels du corps gras chez *Locusta migratoria migratorioides* R. et F. (Insecte Orthoptère). *Bull. Soc. Zool. Fr.* **92**, 39–50.

LAVEN, H. (1951) Untersuchungen und Deutungen zum *Culex-pipiens*-Komplex. *Trans. Roy. Entomol. Soc. London* **102**, 365–8.

LAVERDURE, A. M. (1967a) Mode d'action des corpora allata au cours de la vitellogénèse chez *Tenebrio molitor* (Coleoptère). *C.R. Acad. Sci. Paris* **265**, 145–6.

LAVERDURE, A. M. (1967b) Culture *in vitro* des ovaires de *Tenebrio molitor* (Coléoptère). Importance de la composition du milieu sur la survie, la croissance et la vitellogénèse. *C.R. Acad. Sci. Paris* **265**, 505–7.

LEA, A. O. (1963) Some relationships between environment, corpora allata, and egg maturation in Aedine mosquitoes. *J. Insect Physiol.* **9**, 793–809.

LEA, A. O. (1964a) Selection for autogeny in *Aedes aegypti* (Diptera: Culicidae). *Ann. Entomol. Soc. Am.* **57**, 656–7.

LEA, A. O. (1964b) Studies on the dietary and endocrine regulation of autogenous reproduction in *Aedes taeniorhynchus* (Wied.). *J. Med. Entomol.* **1**, 40–44.

LEA, A. O. (1967) The medial neurosecretory cells and egg maturation in mosquitoes. *J. Insect Physiol.* **13**, 419–29.

LEA, A. O. and THOMSEN, E. (1962) Cycles in the synthetic activity of the medial neurosecretory cells of *Calliphora erythrocephala* and their regulation. *Mem. Soc. Endocrinol.* **12**, 345–7.

LEAHY, M. G. (1966) Egg deposition in *D. melanogaster* increased by transplant of male paragonia. *Drosophila Inf. Serv.* **41**, 145–6.

LEAHY, M. G. (1967) Non-specificity of the male factor enhancing egg-laying in Diptera. *J. Insect Physiol.* **13**, 1283–92.

LEAHY, M. G. and CRAIG, G. B., JR. (1965) Accessory gland substance as a stimulant for oviposition in *Aedes aegypti* and *A. albopictus*. *Mosquito News* **25**, 448–52.

LEAHY, M. G. and LOWE, M. L. (1967) Purification of the male factor increasing egg deposition in *D. melanogaster*. *Life Sci.* **6**, 151–6.

LE BERRE, J. R. (1965) Quelques considérations d'ordre écologique et physiologique sur la diapause du doryphore *Leptinotarsa decemlineata* Say. *C.R. Soc. Biol.* **159**, 2131–5.

LEBRUN, D. (1967a) Hormone juvénile et formation des soldats chez le termite à cou jaune, *Calotermes flavicollis* Fabr. *C.R. Acad. Sci. Paris* **265**, 996–7.

LEBRUN, D. (1967b) Nouvelles recherches sur le déterminisme endocrinien du polymorphisme de *Calotermes flavicollis*. *Ann. Soc. Entomol. Fr.* N.S. **3**, 867–71.

LEDERER, G. (1960) Verhaltensweisen der Imagines und der Entwicklungsstadien von *Limenitis camilla camilla* L. (Lep. Nymphalidae). *Z. Tierpsychol.* **17**, 521–46.

LEDOUX, A. (1950) Recherche sur la biologie de la fourmi fileuse (*Oecophylla longinoda* Latr.). *Ann. Sci. Nat., Zool. Sér.* 11, **12**, 314–461.

LEDOUX, A. (1954) Recherches sur le cycle chromosomique de la fourmi fileuse *Oecophylla longinoda* Latr. (Hyménoptère Formicoidea). *Insectes Sociaux* **1**, 149–75.

LEE, R. D. (1954) Oviposition by the poultry bug. *J. Econ. Entomol.* **47**, 224–6.

LEE, R. D. (1955) The biology of the Mexican chicken bug, *Haematosiphon inodorus* (Dugés). *Pan-Pacific Entomologist* **31**, 47–61.

LEE, V. H. (1968) Parthenogenesis and autogeny in *Culicoides bambusicola* Lutz (Ceratopogonidae, Diptera). *J. Med. Entomol.* **5**, 91–3.

LEES, A. D. (1955) *The Physiology of Diapause in Arthropods*. Cambridge Univ. Press, Cambridge.

LEES, A. D. (1959) The role of photoperiod and temperature in the determination of parthenogenetic and sexual forms in the aphid *Megoura viciae* Buckton. I. The influence of these factors on apterous virginoparae and their progeny. *J. Insect Physiol.* **3**, 92–117.

LEES, A. D. (1960a) The role of photoperiod and temperature in the determination of parthenogenic and sexual forms in the aphid *Megoura viciae* Buckton. II. The operation of the "interval timer" in young clones. *J. Insect Physiol.* **4,** 154–75.

LEES, A. D. (1960b) Some aspects of animal photoperiodism. *Cold Spring Harbor Symp. Quant. Biol.* **25,** 261–8.

LEES, A. D. (1961) Clonal polymorphism in aphids. *Roy. Entomol. Soc. London Symp.* **1,** 68–79.

LEES, A. D. (1963) The role of photoperiod and temperature in the determination of parthenogenetic and sexual forms in the aphid *Megoura viciae* Buckton. III. Further properties of the maternal switching mechanism in apterous aphids. *J. Insect Physiol.* **9,** 153–64.

LEES, A. D. (1964) The location of the photoperiodic receptors in the aphid *Megoura viciae* Buckton. *J. Exp. Biol.* **41,** 119–33.

LEES, A. D. (1966) The control of polymorphism in aphids. *Advan. Insect Physiol.* **3,** 207–77.

LEES, A. D. (1967a) The production of the apterous and alate forms in the aphid *Megoura viciae* Buckton, with special reference to the role of crowding. *J. Insect Physiol.* **13,** 289–318.

LEES, A. D. (1967b) Direct and indirect effects of day length on the aphid *Megoura viciae* Buckton. *J. Insect Physiol.* **13,** 1781–5.

LEIGH, T. F. (1966) A reproductive diapause in *Lygus hesperus* Knight. *J. Econ. Entomol.* **59,** 1280–1.

LENDER, T. and LAVERDURE, A. M. (1964a) La vitellogénèse et la sécrétion des corpora allata de *Tenebrio molitor* (Coléoptère). *C.R. Acad. Sci. Paris* **258,** 1086–8.

LENDER, T. and LAVERDURE, A. M. (1964b) L'évolution de l'ovaire. La vitellogénèse et la sécrétion des corpora allata chez la nymphe et l'adulte de *Tenebrio molitor* (Coléoptère). *Bull. Soc. Zool. Fr.* **89,** 495–509.

LENDER, T. and LAVERDURE, A. M. (1967) Culture *in vitro* des ovaires de *Tenebrio molitor* (Coléoptère). Croissance et vitellogénèse. *C.R. Acad. Sci. Paris* **265,** 451–4.

LEONG, J. K. L. and OATMAN, E. R. (1968) The biology of *Campoplex haywardi* (Hymenoptera: Ichneumonidae), a primary parasite of the potato tuberworm. *Ann. Entomol. Soc. Am.* **61,** 26–36.

LERNER, I. M. (1954) *Genetic Homeostasis.* John Wiley & Sons, New York.

LEROY, Y. (1964) Les caractères sexuels et le comportement acoustique des males d'*Homoeogryllus reticulatus* Fabricius. *Bull. Soc. Entomol. Fr.* **69,** 7–14.

LESPÉS, M.C. (1855) Mémoire sur les spermatophores des grillons. *Ann. Sci. Nat., Zool., Sér.* 4, **3,** 366–77.

LESPÉS, M. C. (1856) Recherches sur l'organisation et les mœurs du termite lucifuge. *Ann. Sci. Nat., Zool., Sér.* 4, **5,** 227–82.

LEVENTHAL, E. A. (1965) A study of the "sex ratio" condition in *Drosophila bifasciata. Am. Zoologist* **5,** 649.

L'HÉLIAS, C. (1953) Étude comparée de l'azote total et de l'azote non protéinique chez le phasme *Dixippus morosus* après ablation des corpora allata. *C.R. Acad. Sci. Paris* **236,** 2439–41.

LHOSTE, J. and ROCHE, A. (1960) Organes odoroférants des mâles de *Ceratitis capitata* (Dipt. Trypetidae). *Bull. Soc. Entomol. Fr.* **65,** 206–10.

LIGHT, S. F. (1942) The determination of the castes of social insects. *Quart. Rev. Biol.* **17,** 312–26; **18,** 46–63.

LIGHT, S. F. (1944a) Parthenogenesis in termites of the genus *Zootermopsis. Univ. Calif. Publ. Zool.* **43,** 405–12.

LIGHT, S. F. (1944b) Experimental studies on ectohormonal control of the development of supplementary reproductives in the termite genus *Zootermopsis* (formerly *Termopsis*). *Univ. Calif. Publ. Zool.* **43,** 413–54.

LIGHT, S. F. and WEESNER, F. M. (1951) Further studies on the production of supplementary reproductives in *Zootermopsis* (Isoptera). *J. Exp. Zool.* **117,** 397–414.

LILLY, C. E. and MCGINNIS, A. J. (1965) Reactions of male click beetles in the laboratory to olfactory pheromones. *Can. Entomologist* **97,** 317–21.

LINDBERG, H. (1939) Der Parasitismus der auf *Chloriona*-Arten (Homoptera Cicadina) lebenden Strepsiptere *Elenchinus chlorionae* n.sp. sowie die Einwirkung derselben auf ihren Wirt. *Acta. Zool. Fenn.* **22,** 1–179.

LLOYD, D. C. (1966) Some aspects of egg resorption in *Leptomastix dactylopii* How. (Hymenoptera, Encyrtidae). *Entomophaga* **11,** 365–81.

LLOYD, J. E. (1964) Notes on flash communication in the firefly *Pyractomena dispersa* (Coleoptera: Lampyridae). *Ann. Entomol. Soc. Am.* **57,** 260–1.

LOAN, C. (1961) *Pygostolus falcatus* (Nees) (Hymenoptera, Braconidae), a parasite of *Sitona* species (Coleoptera, Curculionidae). *Bull. Entomol. Res.* **52,** 473–88.

LOAN, C. and HOLDAWAY, F. G. (1961) *Microctonus aethiops* (Nees) auctt. and *Perilitus rutilus* (Nees) (Hymenoptera: Braconidae), European parasites of *Sitona* weevils (Coleoptera: Curculionidae). *Can. Entomologist* **93,** 1057–79.

LOHER, W. (1959) Contribution to the study of sexual behaviour of *Schistocerca gregaria* Forskal (Orthoptera: Acrididae). *Proc. Roy. Entomol. Soc. London* A **34,** 49–56.

LOHER, W. (1960) The chemical acceleration of the maturation process and its hormonal control in the male of the desert locust. *Proc. Roy. Soc. London* B **153,** 380–97.

LOHER, W. (1962) Die Kontrolle des Weibchengesangs von *Gomphocerus rufus* L. (Acridinae) durch die Corpora allata. *Naturwiss.* **49,** 406.

LOHER, W. (1965) Hormonale Kontrolle der Oocytenentwicklung bei der Heuschrecke *Gomphocerus rufus* L. (Acridinae). *Zool. Jahrb. Abt. Physiol.* **71**, 677–84.

LOHER, W. (1966a) Nervöse und hormonale Kontrolle des Sexualverhaltens beim Weibchen der Heuschrecke *Gomphocerus rufus* L. *Verhandl. Deut. Zool. Ges. Jena 1965*, 386–91.

LOHER, W. (1966b) Die Steuerung sexueller Verhaltensweisen und der Oocytenentwicklung bei *Gomphocerus rufus* L. *Z. Vergleich. Physiol.* **53**, 277–316.

LOHER, W. and BROUGHTON, W. B. (1954) Études sur le comportement acoustique de *Chorthippus bicolor* (Charp.) avec quelques notes comparatives sur des espèces voisines (Acrididae). In *Coll. L'Acoustique des Orthoptères*, R. G. BUSNEL, Ed., pp. 248–77.

LOHER, W. and HUBER, F. (1964) Experimentelle Untersuchungen am Sexualverhalten des Weibchens der Heuschrecke *Gomphocerus rufus* L. (Acridinae). *J. Insect Physiol.* **10**, 13–36.

LOHER, W. and HUBER, F. (1966) Nervous and endocrine control of sexual behaviour in a grasshopper (*Gomphocerus rufus* L., Acridinae). *Symp. Soc. Exp. Biol.* **20**, 381–400.

LOIBL, E. (1958) Zur Ethologie und Biologie der deutschen Lestiden (Odonata). *Z. Tierpsychol.* **15**, 54–81.

LOZINSKY, V. A. (1961) On the correlation existing between the weight of pupae and the number and weight of eggs of *Lymantria dispar* L. *Zool. Zh.* **40**, 1571–3.

LUBBOCK, J. (1884) *Ants, Bees, and Wasps.* New York.

LUKEFAHR, M. J. and MARTIN, D. F. (1964) The effect of various larval and adult diets on the fecundity and longevity of the bollworm, tobacco budworm, and cotton leafworm. *J. Econ. Entomol.* **57**, 233–5.

LUKOSCHUS, F. (1956) Untersuchungen zur Entwicklung der Kastenmerkmale bei der Honigbiene (*Apis mellifica* L.). *Z. Morphol. Oekol. Tiere* **45**, 157–97.

LUM, P. T. M. (1961) The reproductive system of some Florida mosquitoes. II. The male accessory glands and their roles. *Ann. Entomol. Soc. Am.* **54**, 430–3.

LUND, H. O. (1938) Studies on longevity and productivity in *Trichogramma evanescens*. *J. Agr. Res.* **56**, 421–39.

LÜSCHER, M. (1952a) Untersuchungen über das individuelle Wachstum bei der Termite *Kalotermes flavicollis* Fabr. (Ein Beitrag zum Kastenbildungsproblem.) *Biol. Zentr.* **71**, 529–43.

LÜSCHER, M. (1952b) Die Produktion und Elimination von Ersatzgeschlechtstieren bei der Termite *Kalotermes flavicollis* Fabr. *Z. Vergleich. Physiol.* **34**, 123–41.

LÜSCHER, M. (1956a) Die Entstehung von Ersatzgeschlechtstieren bei der Termite *Kalotermes flavicollis* Fabr. *Insectes Sociaux* **3**, 119–28.

LÜSCHER, M. (1956b) Hemmende und fördernde Faktoren bei der Entstehung der Ersatzgeschlechtstiere bei der Termite *Kalotermes flavicollis* Fabr. *Rev. Suisse Zool.* **63**, 261–7.

LÜSCHER, M. (1960) Hormonal control of caste differentiation in termites. *Ann. N.Y. Acad. Sci.* **89**, 549–63.

LÜSCHER, M. (1961) Social control of polymorphism in termites. *Roy. Entomol. Soc. London Symp.* **1**, 57–67.

LÜSCHER, M. (1968) Oocyte protection—a function of a corpus cardiacum hormone in the cockroach *Nauphoeta cinerea*. *J. Insect Physiol.* **14**, 685–8.

LÜSCHER, M. and ENGELMANN, F. (1955) Über die Steuerung der Corpora allata-Funktion bei der Schabe *Leucophaea maderae*. *Rev. Suisse Zool.* **62**, 649–57.

LÜSCHER, M. and SPRINGHETTI, A. (1960) Untersuchungen über die Bedeutung der Corpora allata für die Differenzierung der Kasten bei der Termite *Kalotermes flavicollis* F. *J. Insect Physiol.* **5**, 190–212.

LÜSCHER, M. and WALKER, I. (1963) Zur Frage der Wirkungsweise der Königinnenpheromone bei der Honigbiene. *Rev. Suisse Zool.* **70**, 304–11.

MACDONALD, W. W. (1956) *Aedes aegypti* in Malaya. II. Larval and adult biology. *Ann. Trop. Med. Parasitol.* **50**, 399–414.

MACFIE, J. M. S. (1915) Observations on the bionomics of *Stegomyia fasciata*. *Bull. Entomol. Res.* **6**, 205–29.

MACGILLIVRAY, M. E. and ANDERSON, G. B. (1964) The effect of photoperiod and temperature on the production of gamic and agamic forms in *Macrosiphum euphorbiae* (Thomas). *Can. J. Zool.* **42**, 491–510.

MACGREGOR, M. E. (1932) The occurrence of Roubaud's "Race autogène" in a German strain of *Culex pipiens* in England: with notes on rearing and bionomics. *Trans. Roy. Soc. Trop. Med. Hyg.* **26**, 307–14.

MACKENSEN, O. (1943) The occurrence of parthenogenetic females in some strains of honeybees. *J. Econ. Entomol.* **36**, 465–7.

MACKENSEN, O. (1951) Viability and sex determination in the honey bee (*Apis mellifera* L.). *Genetics* **36**, 500–9.

MACKERRAS, M. J. (1933) Observations on the life-histories, nutritional requirements and fecundity of blowflies. *Bull. Entomol. Res.* **24**, 353–62.

MAELZER, D. A. (1960) The behaviour of the adult of *Aphodius tasmaniae* Hope (Col., Scarabaeidae) in South Australia. *Bull. Entomol. Res.* **51**, 643–70.

MAGNUS, D. (1950) Beobachtungen zur Balz und Eiablage des Kaisermantels *Argynnis paphia* L. (Lep., Nymphalidae). *Z. Tierpsychol.* **7**, 435–49.

MAGNUS, D. (1953) Über optische "Schlüsselreize" beim Paarungsverhalten des Kaisermantels *Argynnis paphia* L. (Lep. Nymph.). *Naturwiss.* **40**, 610–11.

MAGNUS, D. (1955) Zum Problem der "überoptimalen' Schlüsselreize. *Verhandl. Deut. Zool. Ges. Tübingen 1954*, 317–25.

MAGNUS, D. B. E. (1958a) Experimental analysis of some "overoptimal" sign-stimuli in the mating-behaviour of the fritillary butterfly *Argynnis paphia* L. (Lepidoptera: Nymphalidae). *Proc. Xth Intern. Congr. Entomol. Montreal 1956*, **2,** 405–18.

MAGNUS, D. (1958b) Experimentelle Untersuchungen zur Bionomie und Ethologie des Kaisermantels *Argynnis paphia* L. (Lep. Nymph.). I. Über optische Auslöser von Anfliegreaktionen und ihre Bedeutung für das Sichfinden der Geschlechter. *Z. Tierpsychol.* **15,** 397–426.

MAINX, F. (1959) Die Geschlechtsverhältnisse der Phoride *Megaselia scalaris* und das Problem einer alternativen Geschlechtsbestimmung. *Z. Vererbungslehre* **90,** 251–6.

MAINX, F. (1962) Ein neuer Modus der genotypischen Geschlechtsbestimmung. *Biol. Zentr.* **81,** 335–40.

MAINX, F. (1964a) Neuere Befunde zur Geschlechtsbestimmung der Organismen. *Verhandl. Deut. Zool. Ges. München 1963*, 60–67.

MAINX, F. (1964b) The genetics of *Megaselia scalaris* Loew (Phoridae): A new type of sex determination in Diptera. *Am. Naturalist* **98,** 415–30.

MAINX, F. (1966) Die Geschlechtsbestimmung bei *Megaselia scalaris* Loew (Phoridae). *Z. Vererbungslehre* **98,** 49–60.

MAKIELSKI, S. K. (1966) The structure and maturation of the spermatozoa in *Sciara coprophila*. *J. Morphol.* **119,** 11–42.

MAMPE, C. D. and NEUNZIG, H. H. (1966) Function of the stridulating organs of *Conotrachelus nenuphar* (Coleoptera: Curculionidae). *Ann. Entomol. Soc. Am.* **59,** 614–15.

MAMSCH, E. (1965) Regulation der Fruchtbarkeit von Ameisenarbeiterinnen ohne Königin und ohne "Königinsubstanz". *Naturwiss.* **52,** 168.

MAMSCH, E. (1967) Quantitative Untersuchungen zur Regulation der Fertilität im Ameisenstaat durch Arbeiterinnen, Larven und Königin. *Z. Vergleich. Physiol.* **55,** 1–25.

MANGLITZ, G. R. (1958) Aestivation of the alfalfa weevil. *J. Econ. Entomol.* **51,** 506–8.

MANNING, A. (1966a) Corpus allatum and sexual receptivity in female *Drosophila melanogaster*. *Nature* **211,** 1321–2.

MANNING, A. (1966b) Sexual behaviour. *Roy. Entomol. Soc. London Symp.* **3,** 59–68.

MANNING, A. (1967a) The control of sexual receptivity in female *Drosophila*. *Animal Behaviour* **15,** 239–50.

MANNNG, A. (1967b) Antennae and sexual receptivity in *Drosophila melanogaster* females. *Science* **158,** 136–7.

MARCHAL, P. (1896) La reproduction et l'évolution des guêpes sociales. *Arch. Zool. Exp. Gén.* 3e *Sér.* **4,** 1–100.

MARCHAL, P. (1933) Les aphides de l'orme et leurs migrations. *Ann. Epiphyties Sér.* 1, **19,** 207–329.

MARCOVITCH, S. (1923) Plant lice and light exposure. *Science* **58,** 537–8.

MARCOVITCH, S. (1924) The migration of the Aphididae and the appearance of the sexual forms as affected by the relative length of daily light exposure. *J. Agr. Res.* **27,** 513–22.

MARKKULA, M. and ROIVAINEN, S. (1961) The effect of temperature, plant food, and starvation on the oviposition of some *Sitona* (Col., Curculionidae) species. *Ann. Entomol. Fenn.* **27,** 30–45.

MARSHALL, J. F. and STALEY, J. (1936) Exhibition of "autogenous" and "stenogamous" characteristics by *Theobaldia subochrea*, Edwards (Diptera, Culicidae). *Nature* **137,** 580–1.

MARTIN, J. (1966) Female heterogamety in *Polypedilum nubifer* (Diptera: Nematocera). *Am. Naturalist* **100,** 157–9.

MARTOJA, R. (1964) Un type particulier d'appareil génital femelle chez les insectes: les ovarioles adénomorphes du Coléoptère "*Steraspis speciosa*" (Heterogastra, Buprestidae). *Bull. Soc. Zool. Fr.* **89,** 614–41.

MAST, S. O. (1912) Behavior of fire-flies (*Photinus pyralis*)? with special reference to the problem of orientation. *J. Animal Behavior* **2,** 256–72.

MATHIS, M. (1938) Influence de la nutrition larvaire sur la fécondité du *Stegomyia* (*Aedes aegypti* L.). *Bull. Soc. Pathol. Exotique* **31,** 640–6.

MATSUDA, R. (1958) On the origin of the external genitalia of insects. *Ann. Entomol. Soc. Am.* **51,** 84–94.

MATTHES, D. (1959) Das Paarungsverhalten (Paarungsspiel und Kopulation) des Malachiiden *Troglops albicans* L. *Zool. Anz.* **163,** 153–60.

MATTHES, D. (1960) Sozialsekrete und ihre Rolle im sexualbiologischen Geschehen der Insekten. *Naturwiss. Rundschau* **13,** 299–301.

MATTHES, D. (1962a) Excitatoren und Paarungsverhalten mitteleuropäischer Malachiiden (Coleopt., Malacodermata). *Z. Morphol. Oekol. Tiere* **51,** 375–546.

MATTHES, D. (1962b) Zur Sexualbiologie von *Malachius viridis* F. (Coleopt., Malacodermata). *Entomol. Bl.* **58,** 162–7.

MATTHES, E. (1951) Der Einfluss der Fortpflanzung auf die Lebensdauer eines Schmetterlings (*Fumea crassiorella*). *Z. Vergleich. Physiol.* **33,** 1–13.

MATTHEY, R. (1941) Étude biologique et cytologique de *Saga pedo* Pallas (Orthoptères-Tettigoniidae). *Rev. Suisse Zool.* **48,** 92–142.

MATTHEY, R. (1945) Cytologie de la parthénogénèse chez *Pycnoscelus surinamensis* L. *Rev. Suisse Zool.* **52,** Suppl. 1, 1–109.

MATTHEY, R. (1946) Démonstration du caractère géographique de la parthénogénèse de *Saga pedo* Pallas et de sa polyploidie, par comparaison avec les espèces bisexuées *S. ephippigera* Fisch. et *S. gracilipes* Uvar. *Experientia* **2,** 260–1.

MATTHEY, R. (1948) La formule chromosomiale de *Pycnoscelus surinamensis* L. Race bisexuée et race parthénogénétique. Existence probable d'une parthénogénèse diploïde facultative. *Arch. Julius Klaus-Stift. Vererbungsforsch.* **23,** 517–20.

MAYER, H. (1957) Zur Biologie und Ethologie einheimischer Collembolen. *Zool. Jahrb. Abt. Syst.* **85,** 501–70.

MAYER, M. S. and BRAZZEL, J. R. (1963) The mating behavior of the boll weevil, *Anthonomus grandis*. *J. Econ. Entomol.* **56,** 605–9.

MAYNARD SMITH, J. (1956) Fertility, mating behaviour and sexual selection in *Drosophila subobscura*. *J. Genetics* **54,** 261–79.

MAYR, E. (1950) The role of the antennae in the mating behavior of female *Drosophila*. *Evolution* **4,** 149–54.

MAYR, E. (1963) *Animal Species and Evolution*. Harvard Univ. Press, Cambridge.

MCCLUNG, C. E. (1902) The accessory chromosome-sex determinant. *Biol. Bull.* **3,** 43–84.

MCDERMOTT, F. A. (1911) Some further observations on the light-emission of American Lampyridae: the photogenic function as a mating adaptation in the Photinini. *Can. Entomologist* **43,** 399–406.

MCMULLEN, R. D. (1967) A field study of diapause in *Coccinella novemnotata* (Coleoptera: Coccinellidae). *Can. Entomologist* **99,** 42–49.

MEAD-BRIGGS, A. R. (1964) The reproductive biology of the rabbit flea *Spilopsyllus cuniculi* (Dale) and the dependence of this species upon the breeding of its hosts. *J. Exp. Biol.* **41,** 371–402.

MEAD-BRIGGS, A. R. and RUDGE, A. J. B. (1960) Breeding of the rabbit flea, *Spilopsyllus cuniculi* (Dale): requirement of a "factor" from a pregnant rabbit for ovarian maturation. *Nature* **187,** 1136–7.

MECZNIKOFF, C. (1865) Ueber die Entwicklung der Cecidomyienlarve aus dem Pseudovum. *Arch. Naturgesch.* **31,** 304–10.

MEDNIKOVA, M. V. (1952) Endocrine glands corpora allata and corpora cardiaca of mosquitoes (Fam. Culicidae). (In Russian.) *Zool. Zh.* **31,** 676–85.

MEIKLE, J. E. S. and MCFARLANE, J. E. (1965) The role of lipid in the nutrition of the house cricket, *Acheta domesticus* L. (Orthoptera: Gryllidae). *Can. J. Zool.* **43,** 87–98.

DE MEILLON, B. and GOLBERG, L. (1946) Nutritional studies on bloodsucking arthropods. *Nature* **158,** 269–70.

MEINERT, F. (1864) Weitere Erläuterungen über die von Prof. Nic. Wagner beschriebene Insectenlarve, welche sich durch Sprossenbildung vermehrt. *Z. Wiss. Zool.* **14,** 394–9.

MEINWALD, J., MEINWALD, Y. C., WHEELER, J. W., EISNER, T. and BROWER, L. P. (1966) Major components in the exocrine secretion of a male butterfly (*Lycorea*). *Science* **151,** 583–5.

MELANDER, A. L. (1940) *Hilara granditarsis*, a balloon-maker. *Psyche* **47,** 55–56.

MELLANBY, K. (1939) Fertilization and egg production in the bed-bug, *Cimex lectularius* L. *Parasitol.* **31,** 193–211.

MENON, M. (1966) Endocrine influence on yolk deposition in insects. *J. Animal Morphol. Physiol.* **12,** 76–80.

MENUSAN, H. (1935) Effects of constant light, temperature and humidity on the rate of total amount of oviposition of the bean weevil, *Bruchus obtectus* Say. *J. Econ. Entomol.* **28,** 448–53.

MERCIER, L. (1914) Caractère sexuel secondaire chez les panorpes. Le rôle des glandes salivaires des mâles. *Arch. Zool. Exp. Gén.* **55,** 1–5.

MERCIER, L. (1920) Les glandes salivaires des panorpes sont-elles sous la dépendance des glandes génitales? *C.R. Soc. Biol.* **83,** 470–1.

MERLE, J. and DAVID, J. (1967) Fonctionnement ovarien et ponte chez les femelles vierges et les femelles fécondées de *Drosophila melanogaster* soumises à une alimentation exclusivement glucidiques. *C.R. Acad. Sci. Paris* **264,** 2028–30.

MERTON, L. F. H. (1959) Studies in the ecology of the Moroccan locust (*Dociostaurus maroccanus* Thunberg) in Cyprus. *Anti-Locust Bull.* **34,** 1–123.

METZ, C. W. (1938) Chromosome behavior, inheritance and sex determination in *Sciara*. *Am. Naturalist* **72,** 485–520.

MEYER, G. F. (1961) Interzelluläre Brücken (Fusome) im Hoden und im Ei-Nährzellverband von *Drosophila melanogaster*. *Z. Zellforsch.* **54,** 238–51.

MICHELSEN, A. (1963) Observations on the sexual behaviour of some longicorn beetles, subfamily Lepturinae (Coleoptera, Cerambycidae). *Behaviour* **22,** 152–66.

MICHELSEN, A. (1965) The sexual behaviour of longhorned beetles (Cerambycidae: Coleoptera). *Proc. XIIth Intern. Congr. Entomol. London 1964*, 334–5.

MICHENER, C. D. (1958) The evolution of social behavior in bees. *Proc. Xth Intern. Congr. Entomol. Montreal 1956*, **2,** 441–7.

MICHENER, C. D. (1964) Reproductive efficiency in relation to colony size in hymenopterous societies. *Insectes Sociaux* **11,** 317–42.

MICHENER, C. D. and LANGE, R. B. (1958) Observations on the behaviour of the Brasilian Halictid bees. V, *Chloralictus*. *Insectes Sociaux* **5,** 379–407.

MICHENER, C. D. and WILLIE, A. (1961) The bionomics of a primitively social bee, *Lasioglossum inconspicuum*. *Kansas Univ. Sci. Bull.* **42,** 1123–1202.

MIK, J. (1894) Ein Beitrag zur Biologie einiger Dipteren. *Wien. Entomol. Z.* **13,** 261–84.

MIKA, G. (1959) Über das Paarungsverhalten der Wanderheuschrecke *Locusta migratoria* R. und F. und deren Abhängigkeit vom Zustand der inneren Geschlechtsorgane. *Zool. Beitr.* N.F. **4,** 153–203.

MILLER, D. (1939) Blow-flies (Calliphoridae) and their associates in New Zealand. *Cawthron Inst. Monogr. No.* 2, pp. 68.

MILLER, T., JEFFERSON, R. N. and THOMSON, W. W. (1967) Sex pheromone of noctuid moths. XI. The ultrastructure of the apical region of cells of the female sex pheromone gland of *Trichoplusia ni*. *Ann. Entomol. Soc. Am.* **60,** 707–8.

MILLS, R. R., GREENSLADE, F. C. and COUCH, E. F. (1966) Studies on vitellogenesis in the American cockroach. *J. Insect Physiol.* **12,** 767–79.

MINCHIN, E. A. (1905) Report on the anatomy of the tsetse-fly (*Glossina palpalis*). *Proc. Roy. Soc. London* B **76,** 531–47.

MINKS, A. K. (1967) Biochemical aspects of juvenile hormone action in the adult *Locusta migratoria*. *Arch. Neerl. Zool.* **17,** 175–258.

MISSONNIER, J. (1961) Contribution à l'étude des facteurs qui conditionnent la fécondité des adultes et les pullulations de *Pegomyia betae* Curt. (Diptera Muscidae). *C.R. Acad. Sci. Paris* **253,** 907–9.

MISSIONNIER, J. and STENGEL, M. (1966) Étude des facteurs de fécondité des adultes de *Chortophila brassicae* B., *Hylemyia antiqua* Meig. et *Pegomyia betae* Curt. *Ann. Epiphyties* **17,** 5–41.

MOKIA, G. G. (1941) Contribution to the study of hormones in insects. *C.R. Acad. Sci. U.R.S.S.* **30,** 371–3.

MÖLLRING, F. K. (1956) Autogene und anautogene Eibildung bei *Culex* L. Zugleich ein Beitrag zur Frage der Unterscheidung autogener und anautogener Weibchen an Hand von Eiröhrenzahl und Flügellänge. *Z. Tropenmed. Parasitol.* **7,** 15–48.

MONRO, J. (1953) Stridulation in the Queensland fruit fly *Dacus* (*Strumeta*) *tryoni* Frogg. *Australian J. Sci.* **16,** 60–62.

MONROE, R. E. (1959) Role of cholesterol in housefly reproduction. *Nature* **184,** 1513.

MONROE, R. E. (1960) Effect of dietary cholesterol on the house fly reproduction. *Ann. Entomol. Soc. Am.* **53,** 821–4.

MONROE, R. E., HOPKINS, T. L. and VALDER, S. A. (1967) Metabolism and utilization of cholesterol-4-C^{14} for growth and reproduction of aseptically reared houseflies, *Musca domestica* L. *J. Insect Physiol.* **13,** 219–33.

MONROE, R. E., KAPLANIS, J. N. and ROBBINS, W. R. (1961) Sterol storage and reproduction in the house fly. *Ann. Entomol. Soc. Am.* **54,** 537–9.

MONROE, R. E. and LAMB, N. J. (1968) Effect of commercial proteins on house fly reproduction. *Ann. Entomol. Soc. Am.* **61,** 456–9.

MONTAGNER, H. (1966) Sur l'origine des mâles dans les sociétés de Guêpes du genre *Vespa*. *C.R. Acad. Sci. Paris* **263,** 785–7.

MOORE, B. P., WOODROFFE, G. E. and SANDERSON, A. R. (1956) Polymorphism and parthenogenesis in a ptinid beetle. *Nature* **177,** 847–8.

MOORE, C. G. (1963) Seasonal variation in autogeny in *Culex tarsalis* Coq. in northern California. *Mosquito News* **23,** 238–41.

MORDUE, W. (1965a) Studies on oocyte production and associated histological changes in the neuro-endocrine system in *Tenebrio molitor* L. *J. Insect Physiol.* **11,** 493–503.

MORDUE, W. (1965b) The neuro-endocrine control of oocyte development in *Tenebrio molitor* L. *J. Insect Physiol.* **11,** 505–11.

MORDUE, W. (1965c) Neuro-endocrine factors in the control of oocyte production in *Tenebrio molitor* L. *J. Insect Physiol.* **11,** 617–29.

MORDUE, W. (1967) The influence of feeding upon the activity of the neuroendocrine system during oocyte growth in *Tenebrio molitor*. *Gen. Comp. Endocrinol.* **9,** 406–15.

MORDVILKO, A. (1928) The evolution of cycles and the origin of heteroecy (migrations) in plant-lice. *Ann. Mag. Nat. Hist.* 10*th Ser.* **2,** 570–82.

MORDVILKO, A. (1935) Die Blattläuse mit unvollständigem Generationszyklus und ihre Entstehung. *Ergeb. Fortschr. Zool.* **8,** 38–328.

MORÈRE, J. L. and LE BERRE, J. R. (1967) Étude au laboratoire du développement de la Pyrale *Plodia interpunctella* (Hübner) (Lép. Phycitidae). *Bull. Soc. Entomol. Fr.* **72,** 157–66.

MORSE, R. H., GARY, N. E. and JOHANSSON, T. S. K. (1962) Mating of virgin queen honey bees (*Apis mellifera* L.) following mandibular gland extirpation. *Nature* **194,** 605.

MOURSI, A. A. (1946) The effect of temperature on development and reproduction of *Mormoniella vitripennis* (Walker). *Bull. Soc. Fouad Ier*, *Entomol.* **30,** 39–61.

MUCKENTHALER, F. A. (1964) Autoradiographic study of nucleic acid synthesis during spermatogenesis in the grasshopper, *Melanoplus differentialis*. *Exp. Cell Res.* **35,** 531–47.

MÜLLER, F. (1873a) Beiträge zur Kenntnis der Termiten. I. Die Geschlechtsteile der Soldaten von *Calotermes*. *Jena. Z. Med. Naturwiss.* **7,** 333–40.

MÜLLER, F. (1873b) Beiträge zur Kenntnis der Termiten. III. Die "Nymphen mit kurzen Flügelscheiden" (Hagen). *Jena. Z. Med. Naturwiss.* **7,** 451–63.

MÜLLER, H. J. (1954) Der Saisondimorphismus bei Zikaden der Gattung *Euscelis* Brullé. *Beitr. Entomol.* **4,** 1–56.

MÜLLER, H. J. (1962) Über die Abhängigkeit der Oogenese von *Stenocranus minutus* Fabr. (Hom. Auchenorrhyncha) von Dauer und Art der täglichen Beleuchtung. *Proc. XI. Intern. Kongr. Entomol. Wien 1960*, **1,** 678–89.

MÜLLER, H. P. (1965a) Der Einfluss der Corpora allata auf Paarungsverhalten und Eireifung bei *Euthystria brachyptera* Ocsk. (Acrididae). *Naturwiss.* **52,** 93–94.

MÜLLER, H. P. (1965b) Zur Frage der Steuerung des Paarungsverhaltens und der Eireifung bei der Feldheuschrecke *Euthystira brachyptera* Ocsk. unter besonderer Berücksichtigung der Rolle der Corpora allata. *Z. Vergleich. Physiol.* **50,** 447–97.

MÜLLER, O. (1957) Biologische Studien über den frühen Kastanienwickler *Pammene juliana* (Stephens) (Lep: Tortricidae) und seine wirtschaftliche Bedeutung für den Kanton Tessin. *Z. Angew. Entomol.* **41,** 73–111.

MURDY, W. H. and CARSON, H. L. (1959) Parthogenesis in *Drosophila mangabeirai* Malog. *Am. Naturalist* **93,** 355–63.

MURRAY, M. D. (1956) Observations on the biology of Calliphorids in New Zealand. *New Zealand J. Sci. Tech.* A **38,** 193–208.

MUSPRATT, J. (1951) The bionomics of an African *Megarhinus* (Dipt., Culicidae) and its possible use in biological control. *Bull. Entomol. Res.* **42,** 355–70.

MÜSSBICHLER, A. (1952) Die Bedeutung äusserer Einflüsse und der Corpora allata bei der Afterweiselentstehung von *Apis mellifica*. *Z. Vergleich. Physiol.* **34,** 207–21.

MYERS, K. (1952) Oviposition and mating behaviour of the Queensland fruit-fly (*Dacus* (*Strumeta*) *tryoni* (*Frogg.*)) and the solanum fruit-fly (*Dacus* (*Strumeta*) *cacuminatus* (Hering)). *Australian J. Sci. Res.* B **5,** 264–81.

NACHTSHEIM, H. (1913) Cytologische Studien über die Geschlechtsbestimmung bei der Honigbiene (*Apis mellifica* L.). *Arch. Zellforsch.* **11,** 169–241.

NAISSE, J. (1963a) Détermination sexuelle chez *Lampyris noctiluca* L. (Insecte Coléoptère Malacoderme). *C.R. Acad. Sci. Paris* **256,** 799–800.

NAISSE, J. (1963b) Cellules neurosécrétrices chez *Lampyris noctiluca* L. (Insecte Coléoptère Malacoderme). *C.R. Acad. Sci. Paris* **256,** 3895–7.

NAISSE, J. (1965) Contrôle endocrinien de la différenciation sexuelle chez les insectes. *Arch. Anat. Microscop. Morphol. Exp.* **54,** 417–28.

NAISSE, J. (1966a) Contrôle endocrinien de la différenciation sexuelle chez l'insecte *Lampyris noctulica* (Coléoptère Malacoderme Lampyride). I. Role androgène des testicules. *Arch. Biol.* **77,** 139–201.

NAISSE, J. (1966b) Contrôle endocrinien de la différenciation sexuelle chez *Lampyris noctiluca* (Coléoptère Lampyridae). II. Phénomènes neurosécrétoires et endocrines au cours du développement postembryonnaire chez le mâle et la femelle. *Gen. Comp. Endocrinol.* **7,** 85–104.

NAISSE, J. (1966c) Contrôle endocrinien de la différenciation sexuelle chez *Lampyris noctiluca* (Coléoptère Lampyridae). III. Influence des hormones de la pars intercerebralis. *Gen. Comp. Endocrinol.* **7,** 105–10.

NARBEL-HOFSTETTER, M. (1955) La pseudogamie chez *Luffia lapidella* Goeze (Lepid. Psychide). *Rev. Suisse Zool.* **62,** 224–9.

NARBEL-HOFSTETTER, M. (1961) Cytologie comparée de l'espèce parthénogénétique *Luffia ferchaultella* Steph. et de l'espèce bisexuée, *L. lapidella* Goeze (Lepidoptera, Psychidae). *Chromosoma* **12,** 505–52.

NARBEL-HOFSTETTER, M. (1963) Cytologie de la pseudogamie chez *Luffia lapidella* Goeze (Lepidoptera, Psychidae). *Chromosoma* **13,** 623–45.

NASH, T. A. M., JORDAN, A. M. and BOYLE, J. A. (1966) Effect of host pregnancy and pupal production by the tsetse fly. *Nature* **212,** 1581–2.

NATH, V. (1924) Egg-follicle of *Culex*. *Quart J. Microscop. Sci.* **69,** 151–75.

NATH, V. (1956) Cytology of spermatogenesis. *Intern. Rev. Cytology* **5,** 395–453.

NATH, V., GUPTA, B. L. and MITTAL, L. C. (1960) Position of the proximal centriole in flagellate spermatozoa. *Nature* **186,** 899–900.

NAYAR, K. K. (1953) Corpus allatum in *Iphita limbata* Stal. *Current Sci.* **22,** 241–2.

NAYAR, K. K. (1956) The structure of the corpus allatum of *Iphita limbata* (Hemiptera). *Quart J. Microscop. Sci.* **97,** 83–88.

NAYAR, K. K. (1958a) Studies on the neurosecretory system of *Iphita limbata* Stal. Part V. Probable endocrine basis of oviposition in the female insect. *Proc. Indian Acad. Sci.* **47,** 233–51.

NAYAR, K. K. (1958b) Probable endocrine mechanism controlling oviposition in the insect *Iphita limbata* Stal. *2nd Int. Symp. Neurosecretion 1957*, 102–4.

NEEDHAM, J. G., TRAVER, J. R. and HSU, Y. C. (1935) *The Biology of Mayflies*. Comstock Publishing Company, Ithaca, N.Y.

NEUGEBAUER, W. (1961) Wirkungen der Exstirpation und Transplantation der Corpora allata auf den Sauerstoffverbrauch, die Eibildung und den Fettkörper von *Carausius* (*Dixippus*) *morosus* Br. et Redt. *Arch. Entwicklungsmech. Organ.* **153,** 314–52.

NEUMANN, D. (1963) Über die Steuerung der lunaren Schwärmperiodik der Mücke *Clunio marinus*. *Verhandl. Deut. Zool. Ges. Wien 1962*, 275–85.

NEUMANN, D. (1965) Photoperiodische Steuerung der 15-tägigen lunaren Metamorphose-Periodik von *Clunio*-Populationen (Diptera: Chironomidae). *Z. Naturforsch.* **20b,** 818–19.

NEUMANN, D. (1966a) Die intraspezifische Variabilität der lunaren und täglichen Schlüpfzeiten von *Clunio marinus* (Diptera: Chironomidae). *Verhandl. Deut. Zool. Ges. Jena 1965*, 223–33.

NEUMANN, D. (1966b) Die lunare und tägliche Schlüpfperiodik der Mücke *Clunio*. Steuerung und Abstimmung auf die Gezeitenperiodik. *Z. Vergleich. Physiol.* **53,** 1–61.

NEUMANN, H. (1958) Der Bau und die Funktion der männlichen Genitalapparate von *Trichocera annulata* Meig. und *Tipula paludosa* Meig. *Deut. Entomol. Z.* N.F. 5, 235–98.

NEWSON, H. D. and BLAKESLEE, T. E. (1957) Observations of a laboratory colony of the mosquito *Culex tritaeniorhynchus* Giles. *Mosquito News* **17,** 308–11.

NICHOLSON, A. J. (1921) The development of the ovary and ovarian eggs of a Mosquito, *Anopheles maculipenis*, Meig. *Quart. J. Microscop. Sci.* **65,** 395–448.

NICKLAS, R. B. (1960) The chromosome cycle of a primitive cecidomyiid—*Mycophila speyeri*. *Chromosoma* **11,** 402–18.

NIELSEN, E. T. (1959) Copulation of *Glyptotendipes* (*Phytotendipes*) *paripes* Edwards. *Nature* **184,** 1252–3.

NIELSEN, E. T. and HAEGER, J. S. (1960) Swarming and mating in mosquitoes. *Misc. Publ. Entomol. Soc. Am.* **1,** 72–95.

NIKOLEI, E. (1958) Untersuchungen über den Generationswechsel pädogenetischer Gallmücken. *Rev. Suisse Zool.* **65,** 390–6.

NIKOLEI, E. (1961) Vergleichende Untersuchungen zur Fortpflanzung heterogoner Gallmücken unter experimentellen Bedingungen. *Z. Morphol. Oekol. Tiere* **50,** 281–329.

NOIROT, C. (1954) Le polymorphisme des termites supérieurs. *Année Biol.* **30,** 461–74.

NOLAN, W. J. (1925) The brood-rearing cycle of the honeybee. *U.S. Dept. Agr. Bull. No.* 1349.

NOÑIDEZ, J. F. (1920) The internal phenomena of reproduction in *Drosophila*. *Biol. Bull.* **39,** 207–30.

NORMANN, T. C. (1965) The neurosecretory system of the adult *Calliphora erythrocephala*. I. The fine structure of the corpus cardiacum with some observations on adjacent organs. *Z. Zellforsch.* **67,** 461–501.

NORRIS, M. J. (1933) Contributions towards the study of insect fertility. II. Experiments on the factors influencing fertility in *Ephestia kühniella* Z. (Lepidoptera, Phycitidae). *Proc. Zool. Soc. London 1933*, 903–34.

NORRIS, M. J. (1934) Contributions towards the study of insect fertility. III. Adult nutrition, fecundity, and longevity in the genus *Ephestia* (Lepidoptera, Phycitidae). *Proc. Zool. Soc. London 1934*, 333–60.

NORRIS, M. J. (1936) Experiments on some factors affecting fertility in *Trogoderma versicolor* Creutz. (Coleoptera, Dermestidae). *J. Animal Ecology* **5,** 19–22.

NORRIS, M. J. (1950) Reproduction in the African migratory locust (*Locusta migratoria migratorioides* R. & F.) in relation to density and phase. *Anti-Locust Bull.* **6,** 1–50.

NORRIS, M. J. (1952) Reproduction in the desert locust (*Schistocerca gregaria* Forsk.) in relation to density and phase. *Anti-Locust Bull.* **13,** 1–49.

NORRIS, M. J. (1954) Sexual maturation in the desert locust (*Schistocerca gregaria* Forsk.) with special reference to the effects of grouping. *Anti-Locust Bull.* **18,** 1–44.

NORRIS, M. J. (1957) Factors affecting the rate of sexual maturation of the desert locust (*Schistocerca gregaria* Forsk.) in the laboratory. *Anti-Locust Bull.* **28,** 1–26.

NORRIS, M. J. (1959a) Reproduction in the red locust (*Nomadacris septemfasciata* Serville) in the laboratory. *Anti-Locust Bull.* **36,** 1–46.

NORRIS, M. J. (1959b) The influence of day-length on imaginal diapause in the red locust, *Nomadacris septemfasciata* (Serv.). *Entomol. Exp. Appl.* **2,** 154–68.

NORRIS, M. J. (1960) Group effects of feeding in adult males of the desert locust, *Schistocerca gregaria* (Forsk.), in relation to sexual maturation. *Bull. Entomol. Res.* **51,** 731–53.

NORRIS, M. J. (1962a) Diapause induced by photoperiod in a tropical locust, *Nomadacris septemfasciata* (Serv.). *Ann. Appl. Biol.* **50,** 600–3.

NORRIS, M. J. (1962b) The effects of density and grouping on sexual maturation, feeding and activity in caged *Schistocerca gregaria*. *Coll. Intern. Centre Natl. Rech. Sci.* **114,** 23–35.

NORRIS, M. J. (1964) Environmental control of sexual maturation in insects. *Roy. Entomol. Soc. London Symp.* **2,** 56–65.

NOVAK, K. and SEHNAL, F. (1965) Imaginaldiapause bei den in periodischen Gewässern lebenden Trichopteren. *Proc. XIIth Intern. Congr. Entomol. London 1964*, 434.

NOVITSKI, E., PEACOCK, W. J. and ENGEL, J. (1965) Cytological basis of "sex ratio" in *Drosophila pseudoobscura. Science* **148,** 516–17.

NUR, U. (1963) Meiotic parthenogenesis and heterochromatization in a soft scale, *Pulvinaria hydrangeae* (Coccoidea: Homoptera). *Chromosoma* **14,** 123–39.

ODHIAMBO, T. R. (1966a) Corpus allatum hormone at the cellular level in *Schistocerca gregaria. Acta Tropica* **23,** 264–71.

ODHIAMBO, T. R. (1966b) Growth and the hormonal control of sexual maturation in the male desert locust, *Schistocerca gregaria* (Forsk.). *Trans. Roy. Entomol. Soc. London* **118,** 393–412.

ODHIAMBO, T. R. (1966c) The metabolic effects of the corpus allatum hormone in the male desert locust. I. Lipid metabolism. *J. Exp. Biol.* **45,** 45–50.

ODHIAMBO, T. R. (1966d) The fine structure of the corpus allatum of the sexually mature male of the desert locust. *J. Insect Physiol.* **12,** 819–28.

OELHAFEN, F. (1961) Zur Embryogenese von *Culex pipiens*: Markierungen und Exstirpationen mit UV-Strahlenstich. *Arch. Entwicklungsmech. Organ.* **153,** 120–57.

OESER, R. (1961) Vergleichend-morphologische Untersuchungen über den Ovipositor der Hymenopteren. *Mitt. Zool. Mus. Berlin* **37,** 3–119.

OKA, H. (1930) Morphologie und Ökologie von *Clunio pacificus* Edwards (Diptera, Chironomidae). *Zool. Jahrb. Abt. Syst.* **59,** 253–80.

OKA, H. and HASHIMOTO, H. (1959) Lunare Perodizität in der Fortpflanzung einer pazifischen Art von *Clunio* (Diptera, Chironomidae). *Biol. Zentr.* **78,** 545–59.

OLBERG, G. (1959) *Das Verhalten der solitären Wespen Mitteleuropas* (*Vespidae, Pompilidae, Sphecidae*). VEB Deutscher Verlag der Wissenschaften, Berlin.

OLDIGES, H. (1959) Der Einfluss der Temperatur auf Stoffwechsel und Eiproduktion von Lepidopteren. *Z. Angew. Entomol.* **44,** 115–66.

OLIVIERI, G. and OLIVIERI, A. (1965) Autoradiographic study of nucleic acid synthesis during spermatogenesis in *Drosophila melanogaster. Mutation Res.* **2,** 366–80.

ORR, C. W. M. (1964a) The influence of nutritional and hormonal factors on egg development in the blowfly *Phormia regina* (Meig.). *J. Insect Physiol.* **10,** 53–64.

ORR, C. W. M. (1964b) The influence of nutritional and hormonal factors on the chemistry of the fat body, blood, and ovaries of the blowfly *Phormia regina* Meig. *J. Insect Physiol.* **10,** 103–19.

OSSIANNILSSON, F. (1949) Insect drummers: a study of the morphology and function of the sound-producing organ of Swedish Homoptera *Auchenorrhyncha* with notes on their sound-production. *Opusc. Entomol. Suppl.* **10,** 1–145.

OSTEN-SACKEN, C. R. (1877) A singular habit of *Hilara. Entomol. Mo. Mag.* **14,** 126–7.

OTTO, D. (1960) Zur Erscheinung der Arbeiterinnenfertilität und Parthenogenese bei der kahlrückigen roten Waldameise (*Formica polyctena* Först.) (Hym.). *Deut. Entomol. Z.* N.F. **7,** 1–9.

OUDEMANS, J. T. (1899) Falter aus castrierten Raupen. *Zool. Jahrb. Abt. Syst.* **12,** 71–88.

OZEKI, K. (1949) Mechanism of ovarian maturation in the earwig. *Zool. Mag. Tokyo* **58,** 232–6.

PAIN, J. (1954) Sur l'ectohormone des reines d'abeilles. *C.R. Acad. Sci. Paris* **239,** 1869–70.

PAIN, J. (1960) De l'influence du nombre des abeilles encagées sur la formation des œufs dans les ovaires de l'ouvrière. *C.R. Acad. Sci. Paris* **250,** 2629–31.

PAIN, J. (1961a) Sur quelques facteurs alimentaires, accélérateurs du développement des œufs dans les ovaires des ouvrières d'abeilles (*Apis m.* L.). *Insectes Sociaux* **8,** 31–93.

PAIN, J. (1961b) Absence du pouvoir d'inhibition de la phéromone I sur le développement ovarien des jeunes ouvrières d'abeilles. *C.R. Acad. Sci. Paris* **252,** 2316–17.

PAIN, J. (1961c) Sur le phéromone des reines d'abeilles et ses effets physiologiques. *Ann. Abeille* **4,** 73–152.

PAIN, J. and BARBIER, M. (1963) Structures chimiques et propriétés biologiques de quelques substances identifiées chez l'abeille. *Insectes Sociaux* **10,** 129–42.

PALM, N. B. (1948) Normal and pathological histology of the ovaries in *Bombus* Latr. (Hymenopt.). *Opuscula Entomol. Suppl.* VII, 1–101.

PALM, N. B. (1949) Sexual differences in size and structure of the corpora allata in some insects. *Kungl. Svens. Vetenskapsakad. Handl. Ser.* 4, 1, No. 6, 1–24.

PALMEN, J. A. (1884) *Über paarige Ausführgänge der Geschlechtsorgane bei Insekten.* Helsingfors.

PANELIUS, S. (1968) Germ line and oogenesis during paedogenetic reproduction in *Heteropeza pygmaea* Winnertz (Diptera: Cecidomyiidae). *Chromosoma* **23,** 333–45.

PAPILLON, M. (1960) Étude préliminaire de la répercussion du groupement des parents sur les larves nouveau-nées de *Schistocerca gregaria* Forsk. *Bull. Biol. Fr. Belg.* **94,** 203–63.

PARDI, L. (1942) Richerche sui Polistini. V. La poliginia iniziale di *Polistes gallicus* (L.). *Boll. 1st Entomol. Univ. Bologna* **14,** 1–106.

PARDI, L. (1944) Ricerche sui Polistini. VII. La "dominazione" e il ciclo ovarico annuale in *Polistes gallicus* (L.). *Boll. 1st Entomol. Univ. Bologna* **15,** 25–84.

PARDI, L. (1948a) Beobachtungen über das interindividuelle Verhalten bei *Polistes gallicus*. *Behaviour* **1**, 138–72.

PARDI, L. (1948b) Dominance order in *Polistes* wasps. *Physiol. Zool.* **21**, 1–13.

PARDI, L. (1950) Recenti ricerche sulla divisione di lavoro negli Imenotteri sociali. *Boll. Zool. Suppl.* **17**, 17–66.

PARKER, H. L. and THOMPSON, W. R. (1928) Contribution à la biologie des *Chalcidiens entomophages*. *Ann. Soc. Entomol. Fr.* **97**, 425–65.

PARKER, J. R. (1914) The life history of the sugar-beet root-louse (*Pemphigus betae* Doane). *J. Econ. Entomol.* **7**, 136–41.

PASCHKE, J. D. (1959) Production of the agamic alate form of the spotted alfalfa aphid, *Therioaphis maculata* (Buckton) (Homoptera: Aphidae). *Univ. Calif. Publ. Entomol.* **16**, 125–80.

PASSERA, L. (1966) La ponte des ouvrières de la fourmi *Plagiolepis pygmaea* Latr. (Hymén. Formicidae): Œufs reproducteurs et œufs alimentaires. *C.R. Acad. Sci. Paris* **263**, 1095–8.

PATTERSON, J. T. (1946) A new type of isolating mechanism in *Drosophila*. *Proc. Natl. Acad. Sci.* **32**, 202–8.

PATTERSON, J. T. (1947) The insemination reaction and its bearing on the problem of speciation in the mulleri subgroup. *Univ. Texas Publ. No.* 4720, 41–77.

PATTON, W. S. and CRAGG, F. W. (1913) *A Textbook of Medical Entomology*. London.

PEACOCK, A. D. and SANDERSON, A. R. (1939) The cytology of the thelytokous parthenogenetic sawfly *Thrinax macula*. *Trans. Roy. Soc. Edinburgh* **59**, 647–60.

PEARL, R. (1932) The influence of density of population upon egg production in *Drosophila melanogaster*. *J. Exp. Zool.* **63**, 57–84.

PENER, M. P. (1965) On the influence of corpora allata on maturation and sexual behaviour of *Schistocerca gregaria*. *J. Zool.* **147**, 119–36.

PENER, M. P. (1967a) Comparative studies on reciprocal interchange of the corpora allata between males and females of adult *Schistocerca gregaria* (Forskal) (Orthoptera: Acrididae). *Proc. Roy. Entomol. Soc. London* A, **42**, 139–48.

PENER, M. P. (1967b) Effects of allatectomy and sectioning of the nerves of the corpora allata on oocyte growth, male sexual behaviour, and colour change in adults of *Schistocerca gregaria*. *J. Insect Physiol.* **13**, 665–84.

PENER, M. P. (1968) The effect of corpora allata on sexual behaviour and "adult diapause" in males of the red locust. *Entomol. Exp. Appl.* **11**, 94–100.

PERDECK, A. C. (1957) The isolating value of specific song patterns in two sibling species of grasshoppers (*Chorthippus brunneus* Thunb. and *C. biguttulus* L.). *Behaviour* **12**, 1–75.

PERKINS, R. C. L. (1918) The pairing of Stylops and "assembling" of the males. *Proc. Roy. Entomol. Soc. London* **66**, lxx–lxxv.

PERRY, A. S. and MILLER, S. (1965) The essential role of folic acid and the effect of antimetabolites on growth and metamorphosis of housefly larvae *Musca domestica* L. *J. Insect Physiol.* **11**, 1277–87.

PESSON, P. (1941) Description du mâle de *Pulvinaria mesembryanthemi* Vallot et observations biologiques sur cette espèce (Hemipt. Coccidae). *Ann. Soc. Entomol. Fr.* **110**, 71–77.

PESSON, P. (1950) Sur un phénomène de phorésie des spermatozoides par des cellules oviductaires, chez *Aspidiotus ostreaeformis* Curt. (Hemiptera–Homoptera–Coccoidea). *Proc. VIIIth Intern. Congr. Entomol. Stockholm 1948*, 566–70.

PETERSEN, B. and TENOW, O. (1954) Studien am Rapsweissling und Bergweissling (*Pieris napi* L. und *Pieris bryoniae* O.). Isolation und Paarungsbiologie. *Zool. Bidrag Uppsala* **30**, 169–98.

PETERSEN, B., TÖRNBLOM, O. and BODIN, N. O. (1951) Verhaltensstudien am Rapsweissling und Bergweissling (*Pieris napi* L. und *Pieris bryoniae* Ochs.). *Behaviour* **4**, 67–84.

PETERSEN, W. (1928) Über die Sphragis und das Spermatophragma der Tagfaltergattung *Parnassius* (Lep.). *Deut. Entomol. Z.* **1928**, 407–13.

PFEIFFER, I. W. (1939) Experimental study of the function of the corpora allata in the grasshopper, *Melanoplus differentialis*. *J. Exp. Zool.* **82**, 439–61.

PFEIFFER, I. W. (1940) Further studies on the function of the corpora allata in relation to the ovaries and oviducts of *Melanoplus differentialis*. *Anat. Rec. Suppl.* **78**, 39–40.

PFEIFFER, I. W. (1945) Effect of the corpora allata on the metabolism of adult female grasshoppers. *J. Exp. Zool.* **99**, 183–233.

PFLUGFELDER, O. (1937a) Untersuchungen über die Funktion der Corpora allata der Insekten. *Verhandl. Deut. Zool. Ges. Bremen 1937*, 121–9.

PFLUGFELDER, O. (1937b) Bau, Entwicklung und Funktion der Corpora allata und cardiaca von *Dixippus morosus* Br. *Z. Wiss. Zool.* **149**, 477–512.

PFLUGFELDER, O. (1938) Untersuchungen über die histologischen Veränderungen und das Kernwachstum der Corpora allata von Termiten. *Z. Wiss. Zool.* **150**, 451–67.

PFLUGFELDER, O. (1948) Volumetrische Untersuchungen an den Corpora allata der Honigbiene, *Apis mellifica* L. *Biol. Zentr.* **67**, 223–41.

PHILLIPS, D. M. (1966a) Observations on spermiogenesis in the fungus gnat *Sciara coprophila*. *J. Cell Biol.* **30**, 477–97.

PHILLIPS, D. M. (1966b) Fine structure of *Sciara coprophila* sperm. *J. Cell Biol.* **30**, 499–517.

PHIPPS, J. (1949) The structure and maturation of the ovaries in British Acrididae (Orthoptera). *Trans. Roy. Entomol. Soc. London* **100**, 233–47.

PHIPPS, J. (1966) Ovulation and oocyte resorption in Acridoidea (Orthoptera). *Proc. Roy. Entomol. Soc. London* A, **41**, 78–86.

PICARD, F. (1926) Recherches sur la biologie de l'altise de la vigne (*Haltica ampelophaga* Guér.). *Ann. Epiphyties Sér.* 1, **12**, 177–96.

PICKFORD, R. (1958) Observations on the reproductive potential of *Melanoplus bilituratus* (Wlk.) (Orthoptera, Acrididae) reared on different food plants in the laboratory. *Can. Entomologist* **90**, 483–5.

PICKFORD, R. (1966) Development, survival and reproduction of *Camnula pellucida* (Scudder) (Orthoptera: Acrididae) in relation to climate conditions. *Can. Entomologist* **98**, 158–69.

PIERANTONI, U. (1911) Larven-Hermaphroditismus von *Icerya purchasi*. *Z. Wiss. Insektenbiol.* **7**, 322–3.

PIERANTONI, U. (1914) Studii sullo sviluppo d'*Icerya purchasi* Mask. II. Origine ed evoluzione degli organi sessuali maschili. Ermafroditismo. *Arch. Zool. Ital.* **7**, 27–49.

PIERCE, F. N. (1911) Viviparity in Lepidoptera. *Entomologist* **44**, 309–10.

PIJNACKER, L. P. (1966a) The development of follicle cells in the ovarioles of *Carausius morosus* Br. (Orthoptera, Phasmidae). *Experientia* **22**, 158–9.

PIJNACKER, L. P. (1966b) The maturation divisions of the parthenogenetic stick insect *Carausius morosus* Br. (Orthoptera, Phasmidae). *Chromosoma* **19**, 99–112.

PIPKIN, S. B. (1940) Multiple sex genes in the X chromosome of *Drosophila melanogaster*. *Univ. Texas Publ. No.* 4032, 126–56.

PIPKIN, S. B. (1947) A search for sex genes in the second chromosome of *Drosophila melanogaster* using the triploid method. *Genetics* **32**, 592–607.

PLATEAUX-QUÉNU, L. (1961) Les sexués de remplacement chez les insectes sociaux. *Année Biol. Sér.* 3, **37**, 177–216.

POINTING, P. J. (1961) The biology and behaviour of the European pine shoot moth, *Rhyacionia buoliana* (Schiff.), in Southern Ontario. I. Adult. *Can. Entomologist* **93**, 1098–1112.

POPE, P. (1953) Studies of the life histories of some Queensland Blattidae (Orthoptera). *Proc. Roy. Soc. Queensland* **63**, 23–59.

POPOV, G. B. (1954) Notes on the behaviour of swarms of the desert locust (*Schistocerca gregaria* Forskal) during oviposition in Iran. *Trans. Roy. Entomol. Soc. London* **105**, 65–77.

PORTCHINSKY, J. A. (1885) Comparative biology of the necrophagous and coprophagous larvae. *Horae Soc. Entomol. Ross.* **19**, 210–44.

PORTCHINSKY, J. A. (1910) *Recherches biologiques sur le* Stomoxys calcitrans *L. et biologie comparée des mouches coprophagues*. St. Petersburg.

POSSOMPÈS, B. (1955) Corpus allatum et développement ovarien chez *Calliphora erythrocephala* Meig. (Diptère). *C.R. Acad. Sci. Paris* **241**, 2001–4.

POSSOMPÈS, B. (1956) Développement ovarien après ablation du corpus allatum juvénile chez *Calliphora erythrocephala* Meig. (Diptère), et chez *Sipyloidea sipylus* W. (Phasmoptère). *Ann. Sci. Nat., Zool. Sér.* 11, **18**, 313–14.

POULSON, D. F. and WATERHOUSE, D. F. (1960) Experimental studies on pole cells and midgut differentiation in Diptera. *Australian J. Biol. Sci.* **13**, 541–67.

POULTON, E. B. (1913) Empidae and their prey in relation to courtship. *Entomol. Mo. Mag.* **49**, 177–80.

POULTON, E. B. (1921) The courtship of the cicada, *Monometapa insignis* Dist. (Tibicinae), observed in Tanganyika Territory. *Proc. Roy. Entomol. Soc. London* **69**, 63–66.

POWELL, E. (1938) The biology of *Cephalonomia tarsalis* (Asch.), a vespoid wasp (Bethylidae: Hymenoptera) parasitic on the sawtoothed grain beetle. *Ann. Entomol. Soc. Am.* **31**, 44–49.

PRATT, H. S. (1899) The anatomy of the female genital tract of the Pupipara as observed in *Melophagus ovinus*. *Z. Wiss. Zool.* **66**, 16–42.

PRICE, R. D. (1958) Notes on the biology and laboratory colonization of *Wyeomyia smithii* (Coquillett) (Diptera: Culicidae). *Can. Entomologist* **90**, 473–8.

PRINGLE, J. A. (1938) A contribution to the knowledge of *Micromalthus debilis* Lec. (Coleoptera). *Trans. Roy. Entomol. Soc. London* **87**, 271–86.

PURO, J. (1964) Temporal distribution of X-ray induced recessive lethals and recombinants in the post-sterile broods of *Drosophila melanogaster* males. *Mutation Res.* **1**, 268–78.

PUTNAM, P. and SHANNON, R. C. (1934) The biology of *Stegomyia* under laboratory conditions. *Proc. Entomol. Soc. Washington* **36**, 217–42.

PUTTARUDRIAH, M. and BASAVANNA, G. P. C. (1953) Beneficial coccinellids of Mysore. *Indian J. Entomol.* **15**, 87–96.

PUTTLER, B. (1963) Notes on the biology of *Hemiteles graculus* (Hymenoptera: Ichneumonidae) parasitizing the alfalfa weevil, *Hypera postica*. *Ann. Entomol. Soc. Am.* **56**, 857–9.

QUEDNAU, F. W. (1967) Notes on mating behavior and oviposition of *Chrysocharis laricinellae* (Hymenoptera: Eulophidae), a parasite of the larch casebearer (*Coleophora laricella*). *Can. Entomologist* **99,** 326–31.

QUÉNU, C. (1958a) Sur l'existence de caste chez *Halictus marginatus* (Brullé) (Insecte Hyménoptère). *C.R. Acad. Sci. Paris* **246,** 1294–6.

QUÉNU, C. (1958b) Sur la possibilité d'une inhibition des ouvrières par la reine chez *Halictus marginatus* (Brullé) (Insecte Hyménoptère). *C.R. Acad. Sci. Paris* **246,** 1102–4.

QUO, F. (1959) Studies on the reproduction of the oriental migratory locust: the physiological effects of castration and copulation. *Acta Entomol. Sinica* **9,** 464–76.

RAABE, M. (1964) Nouvelles recherches sur la neurosécrétion chez les insectes. *Ann. Endocrinol.* **25,** 107–12.

RAHN, R. (1968) Rôle de la plante-hôte sur l'attractivité sexuelle chez *Acrolepia assectella* Zeller (Lep. Plutellidae). *C.R. Acad. Sci. Paris* **266,** 2004–6.

RAIGNIER, A. and BOVEN, J. VAN (1955) Études taxonomique, biologique et biometrique des *Dorylus* du sous-genre *Anomma* (Hymenoptera Formicidae). *Ann. Mus. Roy. Congo Belge,* N.S. *Sci. Zool.* **2,** 9–359.

RAMAMURTY, P. S. (1963) Über die Herkunft der Ribonukleinsäure in den wachsenden Eizellen der Skorpionsfliege *Panorpa communis* (Insecta, Mecoptera). *Naturwiss.* **50,** 383–4.

RAMAMURTY, P. S. (1964) On the contribution of the follicle epithelium to the deposition of yolk in the oocyte of *Panorpa communis* (Mecoptera). *Exp. Cell Res.* **33,** 601–5.

RASSO, S. C. and FRAENKEL, G. (1954) The food requirements of the adult female blow-fly, *Phormia regina* (Meigen), in relation to ovarian development. *Ann. Entomol. Soc. Am.* **47,** 636–45.

RAU, P. and RAU, N. (1913) The biology of *Stagmomantis carolina. Trans. Acad. Sci. St. Louis* **22,** 1–58.

RAU, P. and RAU, N. (1914) Longevity in saturnid moths and its relation to the function of reproduction. *Trans. Acad. Sci. St. Louis* **23,** 1–78.

RAU, P. and RAU, N. L. (1929) The sex attraction and rhythmic periodicity in giant saturniid moths. *Trans. Acad. Sci. St. Louis* **26,** 82–221.

REGEN, J. (1909) Kastration und ihre Folgeerscheinungen bei *Gryllus campestris* L. *Zool. Anz.* **34,** 477–8.

REGEN, J. (1910) Kastration und ihre Folgeerscheinungen bei *Gryllus campestris* L. *Zool. Anz.* **35,** 427–32.

REGEN, J. (1912) Experimentelle Untersuchungen über das Gehör von *Liogryllus campestris* L. *Zool. Anz.* **40,** 305–16.

REGEN, J. (1913) Über die Anlockung des Weibchens von *Gryllus campestris* L. durch telephonisch übertragene Stridulationslaute des Männchens. *Pflügers Arch. Ges. Physiol.* **155,** 193–200.

REGEN, J. (1923) Über die Orientierung das Weibchens von *Liogryllus campestris* L. nach dem Stridulationsschall des Männchens. Ein Beitrag zur Physiologie des tympanalen Sinnesorgans. *Sitzungsb. Akad. Wissenschaft. Wien* **132,** 81–88.

REICHENBACH, H. (1902) Über Parthenogenese bei Ameisen und andere Beobachtungen an Ameisenkolonien in künstlichen Nestern. *Biol. Zentr.* **22,** 461–5.

REMBOLD, H. and HANSER, G. (1964) Über den Weiselzellfuttersaft der Honigbiene. VIII. Nachweis des determinierenden Prinzips im Futtersaft der Königinnenlarven. *Hoppe-Seyler's Z. Physiol. Chem.* **339,** 251–4.

RENNER, M. (1952) Analyse der Kopulationsbereitschaft des Weibchens der Feldheuschrecke *Euthystira brachyptera* Ocsk. in ihrer Abhängigkeit vom Zustand des Geschlechtsapparates. *Z. Tierpsychol.* **9,** 122–54.

RETHFELDT, C. (1924) Die Viviparität bei *Chrysomela varians* Schaller. *Zool. Jahrb. Abt. Anat.* **46,** 245–302.

RETTENMEYER, C. W. (1963) Behavioral studies of army ants. *Univ. Kansas Sci. Bull.* **44,** 281–465.

RHOADES, M. M. (1961) Meiosis. In BRACHET, J. and A. E. MIRSKY, Eds., *The Cell* **3,** 1–75.

RIBAGA, C. (1896) Sopra un organo particolare delle cimice dei letti (*Cimex lectularius* L.). *Riv. Pat. Veg.* **5,** 343–53.

RIBBANDS, C. R. (1953) *The Behaviour and Social Life of Honeybees.* Bee Research Ass. Ltd., London.

RICHARDS, O. W. (1927) Sexual selection and allied problems in the insects. *Biol. Rev.* **2,** 298–364.

RICHARDS, O. W. and RICHARDS, M. J. (1951) Observations on the social wasps of South America (Hymenoptera Vespidae). *Trans. Roy. Entomol. Soc. London* **102,** 1–170.

RICHARDS, T. J. (1952) *Nemobius sylvestris* in S.E. Devon. *Entomologist* **85,** 83–87, 108–11, 136–41, 161–6.

RIDDIFORD, L. M. (1967) Trans-2-hexenal: mating stimulant for *polyphemus* moths. *Science* **158,** 139–41.

RIDDIFORD, L. M. and WILLIAMS, C. M. (1967) Volatile principle from oak leaves: role in sex life of the *polyphemus* moth. *Science* **155,** 589–99.

RIEGERT, P. W. (1965) Effects of grouping, pairing, and mating on the bionomics of *Melanoplus bilituratus* (Walker) (Orthoptera: Acrididae). *Can. Entomologist* **97,** 1046–51.

RIEMANN, J. G., MOEN, D. J. and THORSON, B. J. (1967) Female monogamy and its control in houseflies. *J. Insect Physiol.* **13,** 407–18.

RIES, E. (1932) Die Prozesse der Eibildung und des Eiwachstums bei Pediculiden und Mallophagen. *Z. Zellforsch.* **16,** 314–88.

RILEY, C. V. (1890) A viviparous cockroach. *Ins. Life* **3** 443–4.

RITCHOT, C. and MCFARLANE, J. E. (1962) The effects of wheat germ oil and linoleic acid on growth and reproduction of the house cricket. *Can. J. Zool.* **40,** 371–4.

ROBBINS, W. E., KAPLANIS, J. N., THOMPSON, M. J., SHORTINO, T. J., COHEN, C. F. and JOYNER, S. C. (1968) Ecdysones and analogs: effects on development and reproduction of insects. *Science* **161,** 1158–9.

ROBBINS, W. E. and SHORTINO, T. J. (1962) Effect of cholesterol in the larval diet on ovarian development in the adult housefly. *Nature* **194,** 502–3.

ROBERT, P. (1965) Influence de la plante-hôte sur l'activité reproductrice de la teigne de la Betterave *Scrobipalpa* (*Phthorimaea*) *ocellatella* Boyd (Lépidoptère Plutellidae). *Proc. XIIth Intern. Congr. Entomol. London 1964*, 552–3.

ROBERTS, F. H. S. and O'SULLIVAN, P. J. (1948) Studies on the behaviour of adult Australasian Anophelines. *Bull. Entomol. Res.* **39,** 159–78.

ROBERTSON, F. W. and SANG, J. H. (1944) The ecological determinants of population growth in a *Drosophila* culture. I. Fecundity of adult flies. *Proc. Roy. Soc. London* B **132,** 258–77.

ROBERTSON, J. G. (1961) Ovariole number in Coleoptera. *Can. J. Zool.* **39,** 245–63.

ROBERTSON, J. G. (1966) The chromosomes of bisexual and parthenogenetic species of *Calligrapha* (Coleoptera: Chrysomelidae) with notes on sex ratio, abundance and egg number. *Can. J. Genet. Cytol.* **8,** 695–732.

ROEDER, K. D. (1935) An experimental analysis of the sexual behavior of the praying mantis (*Mantis religiosa* L.). *Biol. Bull.* **69,** 203–20.

ROEDER, K. D., TOZIAN, L. and WEIANT, E. A. (1960) Endogenous nerve activity and behaviour in the mantis and cockroach. *J. Insect Physiol.* **4,** 45–62.

ROELOFS, W. L. and ARN, H. (1968) Sex attractant of the red-banded leaf roller moth. *Nature* **219,** 513.

ROELOFS, W. L. and FENG, K. C. (1968) Sex pheromone specificity tests in the Tortricidae—an introductory report. *Ann. Entomol. Soc. Am.* **61,** 312–16.

ROHDENDORF, E. (1965) Der Einfluss der Allatektomie auf adulte Weibchen von *Thermobia domestica* Packard (Lepismatidae, Thysanura). *Zool. Jahrb. Abt. Physiol.* **71,** 685–93.

RÖLLER, H. (1962) Über den Einfluss der Corpora allata auf den Stoffwechsel der Wachsmotte. *Naturwiss.* **49,** 524.

RÖLLER, H., BJERKE, J. S., NORGARD, D. W. and MCSHAN, W. H. (1966) *Proc. Intern. Symp. Insect Endocrinol. Brno 1966.*

RÖLLER, H., DAHM, K. H., SWEELY, C. C. and TROST, B. M. (1967) The structure of the juvenile hormone. *Angew. Chem.* **6,** 179–80.

RÖLLER, H., PIEPHO, H. and HOLZ, I. (1963) Zum Problem der Hormonabhängigkeit des Paarungsverhaltens bei Insekten. Untersuchung an *Galleria mellonella* (L.). *J. Insect Physiol.* **9,** 187–94.

ROSTAND, J. (1950) *La Parthénogénèse Animale.* Paris.

ROTH, L. M. (1948) A study of mosquito behavior. An experimental laboratory study of the sexual behavior. of *Aedes aegypti* (Linnaeus). *Am. Midland Naturalist* **40,** 265–352.

ROTH, L. M. (1952) The tergal gland of the male cockroach, *Supella supellectilium. J. Morphol.* **91,** 469–77.

ROTH, L. M. (1962) Hypersexual activity induced in females of the cockroach *Nauphoeta cinerea. Science* **138,** 1267–9.

ROTH, L. M. (1964a) Control of reproduction in female cockroaches with special reference to *Nauphoeta cinerea.* I. First pre-oviposition period. *J. Insect Physiol.* **10,** 915–45.

ROTH, L. M. (1964b) Control of reproduction in female cockroaches with special reference to *Nauphoeta cinerea.* II. Gestation and postparturition. *Psyche* **71,** 198–243.

ROTH, L. M. (1967) Water changes in cockroach oothecae in relation to the evolution of ovoviviparity and viviparity. *Ann. Entomol. Soc. Am.* **60,** 928–46.

ROTH, L. M. (1968) A teratological specimen of *Pycnoscelus surinamensis. Ann. Entomol. Soc. Am.* **61,** 777–9.

ROTH, L. M. and BARTH, R. H., JR. (1964) The control of sexual receptivity in female cockroaches. *J. Insect Physiol.* **10,** 965–75.

ROTH, L. M. and BARTH, R. H., JR. (1967) The sense organs employed by cockroaches in mating behavior *Behaviour* **28,** 58–94.

ROTH, L. M. and DATEO, G. P., JR. (1964) Uric acid in the reproductive system of males of the cockroach. *Blattella germanica. Science* **146,** 782–4.

ROTH, L. M. and STAY, B. (1959) Control of oocyte development in cockroaches. *Science* **130,** 271–2.

ROTH, L. M. and STAY, B. (1961) Oocyte development in *Diploptera punctata* (Eschscholtz) (Blattaria). *J. Insect Physiol.* **7,** 186–202.

ROTH, L. M. and STAY B. (1962a) Oocyte development in *Blattella germanica* and *Blattella vaga* (Blattaria). *Ann. Entomol. Soc. Am.* **55,** 633–42.

ROTH, L. M. and STAY, B. (1962b) A comparative study of oocyte development in false ovoviviparous cockroaches. *Psyche* **69,** 165–208.

ROTH, L. M. and WILLIS, E. R. (1952a) Method for isolating males and females in laboratory colonies of *Aedes aegypti. J. Econ. Entomol.* **45,** 344.

ROTH, L. M. and WILLIS, E. R. (1952b) A study of cockroach behavior. *Am. Midland Naturalist* **47,** 66–129.

ROTH, L. M. and WILLIS, E. R. (1954) The reproduction of cockroaches. *Smiths Misc. Coll.* **122,** No. 12, 1–49.

ROTH, L. M. and WILLIS, E. R. (1955a) Water relations of cockroach oothecae. *J. Econ. Entomol.* **48,** 33–36.

ROTH, L. M. and WILLIS, E. R. (1955b) Water content of cockroach eggs during embryogenesis in relation to oviposition behavior. *J. Exp. Zool.* **128,** 489–510.

ROTH, L. M. and WILLIS, E. R. (1955c) Intra-uterine nutrition of the "Beetle-Roach" *Diploptera dytiscoides* (Serv.) during embryogenesis, with notes on its biology in the laboratory (Blattaria: Diplopteridae). *Psyche* **62,** 55–68.

ROTH, L. M. and WILLIS, E. R. (1956) Parthenogenesis in cockroaches. *Ann. Entomol. Soc. Am.* **49,** 195–204.

ROTH, L. M. and WILLIS, E. R. (1958) An analysis of oviparity and viviparity in the Blattaria. *Trans. Am. Entomol. Soc.* **83,** 221–38.

ROTH, L. M. and WILLIS, E. R. (1961) A study of bisexual and parthenogenetic strains of *Pycnoscelus surinamensis* (Blattaria: Epilamprinae). *Ann. Entomol. Soc. Am.* **54,** 12–25.

ROTH, T. F. (1966) Changes in the synaptinemal complex during meiotic prophase in mosquito oocytes. *Protoplasma* **61,** 346–86.

ROTH, T. F. and PORTER, K. R. (1962) Specialized sites on the cell surface for protein uptake. *Electron Microscopy*, BREESE, Ed. 2, LL-4.

ROTH, T. F. and PORTER, K. R. (1964) Yolk protein uptake in the oocyte of the mosquito *Aedes aegypti* L. *J. Cell Biol.* **20,** 313–32.

ROTHENBUHLER, W. C. (1957) Diploid male tissues as new evidence on sex determination in honey bees. *J. Heredity* **48,** 160–8.

ROTHENBUHLER, W. C., GOWEN, J. W. and PARK, O. W. (1952) Androgenesis with zygogenesis in gynandromorphic honey bees (*Apis mellifera* L.). *Science* **115,** 637–8.

ROTHSCHILD, M. and FORD, R. (1965a) Reproductive hormones of the host controlling the sexual cycle of the rabbit flea (*Spilopsyllus cuniculi* Dale). *Proc. XIIth Intern. Congr. Entomol. London 1964*, 801–2.

ROTHSCHILD, M. and FORD, R. (1965b) Observations on gravid rabbit fleas (*Spilopsyllus cuniculi* Dale) parasitising the hare (*Lepus europaeus* Pallas), together with further speculations concerning the course of myxomatosis at Ashton, Northants. *Proc. Roy. Entomol. Soc. London* A, **40,** 109–17.

ROTHSCHILD, M. and FORD, R. (1966) Hormones of the vertebrate host controlling ovarian regression and copulation of the rabbit flea. *Nature* **211,** 261–6.

ROUBAUD, E. (1909) Recherches biologiques sur les conditions de viviparité et de vie larvaire de *Glossina palpalis* R. Desv. *C.R. Acad. Sci. Paris* **148,** 195–7.

ROUBAUD, E. (1922) Recherches sur la fécondité et la longévité de la mouche domestique. *Ann. Inst. Pasteur* **36,** 765–83.

ROUBAUD, E. (1929) Cycle autogène d'attente et générations hivernales suractives inapparentes chez le moustique commun, *Culex pipiens* L. *C.R. Acad. Sci. Paris* **188,** 735–8.

ROUBAUD, E. (1930) Sur l'existence des races biologiques génétiquement distinctes chez le moustique commun *Culex pipiens. C.R. Acad. Sci. Paris* **191,** 1386–8.

ROUBAUD, E. (1933) Essai synthétique sur la vie du moustique commun (*Culex pipiens*). *Ann. Sci. Nat., Zool. Sér.* 10, **16,** 5–168.

ROUSSEL, J. P. (1967) Fonctions des corpora allata et contrôle de la pigmentation chez *Gryllus bimaculatus* De Geer. *J. Insect Physiol.* **13,** 113–30.

ROY, D. N. (1936) On the role of blood in ovulation in *Aedes aegypti*, Linn. *Bull. Entomol. Res.* **27,** 423–9.

ROY, D. N. and SIDDONS, L. B. (1939) On the life history and bionomics of *Chrysomyia rufifacies* Macq. (order Diptera, family Calliphoridae). *Parasitol.* **31,** 442–7.

ROYER, M. (1966) Conditions biologiques de l'accouplement et de la ponte de *Gryllus bimaculatus* (Orth. Ensif. Gryllidae). *Ann. Soc. Entomol. Fr.* N.S. **2,** 671–85.

RULE, H. D., GODWIN, P. A. and WATERS, W. E. (1965) Irradiation effects on spermatogenesis in the gypsy moth, *Porthetria dispar* (L.). *J. Insect Physiol.* **11,** 369–78.

RUPPLI, E. and LÜSCHER, M. (1964) Die Elimination überzähliger Ersatzgeschlechtstiere bei der Termite *Kalotermes flavicollis* (Fabr.). *Rev. Suisse Zool.* **71,** 626–32.

RUPPRECHT, R. (1968) Das Trommeln der Plecopteren. *Z. Vergleich. Physiol.* **59,** 38–71.

RUTTNER, F. (1956a) The mating of the honeybee. *Bee World* **37,** 3–15.

RUTTNER, F. (1956b) Zur Frage der Spermaübertragung bei der Bienenkönigin. *Insectes Sociaux* **3,** 351–9.

RUTTNER, F. (1961) Die Innervation der Fortpflanzungsorgane der Honigbiene (*Apis mellifica* L.). *Z. Bienenforsch.* **5,** 253–66.

RUTTNER, F. and KAISSLING, K. E. (1968) Über die interspezifische Wirkung des Sexuallockstoffes von *Apis mellifica* und *Apis cerana. Z. Vergleich. Physiol.* **59,** 362–70.

RYCKMAN, R. E. (1958) Description and biology of *Hesperocimex sonorensis*, new species, an ectoparasite of the purple martin (Hemiptera, Cimicidae). *Ann. Entomol. Soc. Am.* **51,** 33–47.

SADO, T. (1963) Spermatogenesis of the silkworm and its bearing on radiation induced sterility. *J. Fac. Agr. Kyusu Univ.* **12,** 359–404.

SAHRHAGE, D. (1953) Ökologische Untersuchungen an *Thermobia domestica* (Packard) und *Lepisma saccharina* L. *Z. Wiss. Zool.* **157,** 77–168.

SAKAGUCHI, B., OISHI, K. and KOBAYASHI, S. (1965) Interference between "sex-ratio" agents of *Drosophila willistoni* and *Drosophila nebulosa. Science* **147,** 160–2.

SALKELD, E. H. (1959) Notes on anatomy, life-history, and behaviour of *Aphaereta pallipes* Say) (Hymenoptera: Braconidae), a parasite of the onion maggot, *Hylemya antiqua* (Meig.). *Can. Entomologist* **91,** 93–97.

SALT, G. (1938) Further notes on *Trichogramma semblidis. Parasitol.* **30,** 511–22.

SALT, G. (1940) Experimental studies in insect parasitism. VII. The effects of different hosts on the parasite *Trichogramma evanescens* Westw. (Hym. Chalcidoidea). *Proc. Roy. Entomol. Soc. London* A **15,** 81–95.

SAMEOTO, D. D. and MILLER, R. S. (1966) Factors controlling the productivity of *Drosophila melanogaster* and *D. simulans. Ecology* **47,** 695–704.

SANDERSON, A. R. (1960) The cytology of a diploid bisexual spider beetle, *Ptinus clavipes* Panzer and its triploid gynogenetic form *mobilis* Moore. *Proc. Roy. Soc. Edinburgh* B **67,** 333–50.

SANG, J. H. and KING, R. C. (1961) Nutritional requirements of axenically cultured *Drosophila melanogaster* adults. *J. Exp. Biol.* **38,** 793–809.

SATO, H. (1931) Untersuchungen über die künstliche Parthenogenese des Seidenspinners *Bombyx mori.* IV. *Biol. Zentr.* **51,** 382–94.

SAUNDERS, D. S. (1961) Studies on ovarian development in tsetse flies (*Glossina*, Diptera). *Parasitol.* **51,** 545–64.

SAUTER, W. (1956) Morphologie und Systematik der schweizerischen *Solenobia*-Arten (Lep. Psychidae). *Rev. Suisse Zool.* **63,** 451–550.

SCHALLER, F. (1951) Lauterzeugung und Hörvermögen von *Corixa* (*Callicorixa*) *striata* L. *Z. Vergleich. Physiol.* **33,** 476–86.

SCHALLER, F. (1953) Untersuchungen zur Fortpflanzungsbiologie arthropleoner Collembolen. *Z. Morphol. Oekol. Tiere* **41,** 265–77.

SCHALLER, F. (1954) Indirekte Spermatophorenübertragung bei *Campodea* (Apterygota, Diptura). *Naturwiss.* **41,** 406–7.

SCHALLER, F. (1965) Mating behaviour of lower terrestrial arthropods from the phylogenetical point of view. *Proc. XIIth Intern. Congr. Entomol. London 1964*, 297–8.

SCHARRER, B. (1946) The relationship between corpora allata and reproductive organs in adult *Leucophaea maderae* (Orthoptera). *Endocrinology* **38,** 46–55.

SCHARRER, B. (1952) Neurosecretion XI. The effects of nerve section on the intercerebralis-cardiacum-allatum system of the insect *Leucophaea maderae. Biol. Bull.* **102,** 261–72.

SCHARRER, B. (1955) "Castration cells" in the central nervous system of an insect (*Leucophaea maderae* Blattaria). *Trans. N.Y. Acad. Sci.* **17,** 520–5.

SCHARRER, B. (1963) Neurosecretion. XIII. The ultrastructure of the corpus cardiacum of the insect *Leucophaea maderae. Z. Zellforsch.* **60,** 761–96.

SCHARRER, B. (1964) Histophysiological studies on the corpus allatum of *Leucophaea maderae.* IV. Ultrastructure during normal activity cycle. *Z. Zellforsch.* **62,** 125–48.

SCHARRER, B. and VON HARNACK, M. (1958) Histophysiological studies on the corpus allatum of *Leucophaea maderae.* I. Normal life cycle in male and female adults. *Biol. Bull.* **115,** 508–20.

SCHARRER, B. and VON HARNACK, M. (1961) Histophysiological studies on the corpus allatum of *Leucophaea maderae.* III. The effect of castration. *Biol. Bull.* **121,** 193–208.

SCHARRER, E. and SCHARRER, B. (1963) *Neuroendocrinology*. Columbia Univ. Press, New York and London.

SCHEDL, K. E. (1936) Der Schwammspinner (*Porthetria dispar* L.) in Eurasien, Afrika und Neuengland. *Monogr. Angew. Entomol. No.* 12, 1–242.

SCHEURER, R. and LÜSCHER, M. (1966) Die phasenspezifische Eireifungkompetenz der Ovarien von *Leucophaea maderae. Rev. Suisse Zool.* **73,** 511–16.

SCHLIWA, W. and SCHALLER, F. (1963) Die Paarbildung des Springschwanzes *Podura aquatica* (Apterygota (Urinsekten), Collembola). *Naturwiss.* **50,** 698.

SCHLOTTMAN, L. L. and BONHAG, P. F. (1956) Histology of the ovary of the adult mealworm *Tenebrio molitor* L. (Coleoptera, Tenebrionidae). *Univ. Calif. Publ. Entomol.* **11,** 351–94.

SCHMIDT, E. L. and WILLIAMS, C. M. (1953) Physiology of insect diapause. V. Assay of the growth and differentiation hormone of Lepidoptera by the method of tissue culture. *Biol. Bull.* **105,** 174–87.

SCHMIEDER, R. G. (1933) The polymorphic forms of *Melittobia chalybii* Ashmead and the determining factors involved in their production (Hymenoptera: Chalcidoidea, Eulophidae). *Biol. Bull.* **65,** 338–54.

SCHMIEDER, R. G. (1938) The sex ratio in *Melittobia chalybii* Ashmead, gametogenesis and cleavage in females and in haploid males (Hymenoptera: Chalcidoidea). *Biol. Bull.* **74,** 256–66.

SCHNEIDER, D. (1962) Electrophysiological investigation on the olfactory specificity of sexual attracting substances in different species of moths. *J. Insect Physiol.* **8,** 15–30.

SCHNEIDER, D. (1966) Chemical sense communication in insects. *Symp. Soc. Exp. Biol.* **20,** 273–97.

SCHNEIDER, D., BLOCK, B. C. and PRIESNER, E. (1967) Die Reaktion der männlichen Seidenspinner auf Bombykol und seine Isomeren: Elektroantennogramm und Verhalten. *Z. Vergleich. Physiol.* **54,** 192–209.

SCHNEIDER, D. and HECKER, E. (1956) Zur Elektrophysiologie der Antenne des Seidenspinners *Bombyx mori* bei Reizung mit angereicherten Extrakten des Sexuallockstoffes. *Z. Naturforsch.* **11b,** 121–4.

SCHNEIDER, H. (1955) Vergleichende Untersuchungen über Parthenogenese und Entwicklungsrhythmen bei einheimischen Psocopteren. *Biol. Zentr.* **74,** 273–310.

SCHNEIRLA, T. C. (1938) A theory of army-ant behavior based upon the analysis of activities in a representative species. *J. Comp. Psychol.* **25,** 51–90.

SCHNEIRLA, T. C. (1949) Army-ant life and behavior under dry-season conditions. 3. The course of reproduction and colony behavior. *Bull. Am. Mus. Nat. Hist.* **94,** 1–81.

SCHNEIRLA, T. C. (1956) A preliminary survey of colony division and related processes in two species of terrestrial army ants. *Insectes Sociaux* **3,** 49–69.

SCHNEIRLA, T. C. (1957a) Theoretical consideration of cyclic processes in doryline ants. *Proc. Am. Philos. Soc.* **101,** 106–33.

SCHNEIRLA, T. C. (1957b) A comparison of species and genera in the ant subfamily Dorylinae with respect to functional pattern. *Insectes Sociaux* **4,** 259–98.

SCHNEIRLA, T. C. (1958) The behavior and biology of certain Nearctic army ants. Last part of the functional season, Southeastern Arizona. *Insectes Sociaux* **5,** 216–55.

SCHNEIRLA, T. C. (1961) The behavior and biology of certain nearctic Doryline ants. Sexual broods and colony division in *Neivamyrmex nigrescens*. *Z. Tierpsychol.* **18,** 1–32.

SCHNEIRLA, T. C. and BROWN, R. Z. (1952) Sexual broods and the production of young queens in two species of army ants. *Zoologica* **37,** 5–32.

SCHOENEMUND, E. (1912) Zur Biologie und Morphologie einiger *Perla*-Arten. *Zool. Jahrb. Abt. Anat.* **34,** 1–56

SCHOLL, H. (1956) Die Chromosomen parthenogenetischer Mücken. *Naturwiss.* **43,** 91–92.

SCHRADER, F. (1921) The chromosomes of *Pseudococcus nipae*. *Biol. Bull.* **40,** 259–71.

SCHULZE, H. (1926) Über die Eiablage des Schmetterlings *Trochilium apiforme* L. *Zool. Anz.* **68,** 233–8.

SCHUSTER, M. F. (1965) Studies on the biology of *Dusmetia sangwani* (Hymenoptera: Encyrtidae). *Ann. Entomol. Soc. Am.* **58,** 272–5.

SCHUURMANS, S. J. H. (1923) *De bloedzuigende Arthropoda van Nederlandsch Ost-Indie. V. De bestrijding der Luisvliegenplaag.* Buitenzorg.

SCHWALB, H. H. (1961) Beiträge zur Biologie der einheimischen Lampyriden *Lampyris noctiluca* Geoffr. und *Phausis splendidula* Lec. und experimentelle Analyse ihres Beutefang- und Sexualverhaltens. *Zool. Jahrb. Abt. Syst.* **88,** 399–550.

SCHWARTZ, H. (1932) Der Chromosomenzyklus von *Tetraneura ulmi* De Geer. *Z. Zellforsch.* **15,** 645–87.

SCHWARTZ, P. H., JR. and TURNER, R. B. (1966) A dietary study of the adult eye gnat *Hippelates pusio* (Diptera: Chloropidae). *Ann. Entomol. Soc. Am.* **59,** 277–80.

SCHWINCK, I. (1953) Über den Sexualduftstoff der Pyraliden. *Z. Vergleich. Physiol.* **35,** 167–74.

SCHWINCK, I. (1954) Experimentelle Untersuchungen über Geruchssinn und Strömungswahrnehmung in der Orientierung bei Nachtschmetterlingen. *Z. Vergleich. Physiol.* **37,** 19–56.

SCHWINCK, I. (1955a) Weitere Untersuchungen zur Frage der Geruchsorientierung der Nachtschmetterlinge: Partielle Fühleramputation bei Spinnermännchen, insbesondere am Seidenspinner *Bombyx mori* L. *Z. Vergleich. Physiol.* **37,** 439–58.

SCHWINCK, I. (1955b) Freilandversuche zur Frage der Artspezifität des weiblichen Sexualduftstoffes der Nonne (*Lymantria monacha* L.) und des Schwammspinners (*Lymantria dispar* L.). *Z. Angew. Entomol.* **37,** 349–57.

SCHWINCK, I. (1958) A study of olfactory stimuli in the orientation of moths. *Proc. Xth Intern. Congr. Entomol. Montreal 1956*, **2,** 577–82.

SCOTT, A. C. (1936) Haploidy and aberrant spermatogenesis in a coleopteran, *Micromalthus debilis* Leconte. *J. Morphol.* **59,** 485–515.

SCOTT, A. C. (1938) Paedogenesis in the Coleoptera. *Z. Morphol. Oekol. Tiere* **33,** 633–53.

SCOTT, A. C. (1941) Reversal of sex production in *Micromalthus*. *Biol. Bull.* **81,** 420–31.

SCUDDER, G. G. E. (1961a) The functional morphology and interpretation of the insect ovipositor. *Can. Entomologist* **93,** 267–72.

SCUDDER, G. G. E. (1961b) The comparative morphology of the insect ovipositor. *Trans. Roy. Entomol. Soc. London* **113,** 25–40.

SCUDDER, G. G. E. (1964) Further problems in the interpretation and homology of the insect ovipositor. *Can. Entomologist* **96,** 405–17.

SEGUY, E. (1951) Ordre des mallophages. *Traité de Zoologie* **10,** 1341–64.

SEIBERT, A. (1922) Eierlegende Arbeiterinnen trotz vorhandener normaler Königin. *Arch. Bienenk.* **4,** 31–34.

SEIDEL, S. (1963) Experimentelle Untersuchungen über die Grundlage der Sterilität von transformer-(tra) Männchen bei *Drosophila melanogaster*. *Z. Vererbungslehre* **94,** 215–41.

SEILER, J. (1937) Ergebnisse aus der Kreuzung parthenogenetischer und zweigeschlechtlicher Schmetterlinge. V. Die *Solenobia*-Intersexe und die Deutungen des Phänomens der Intersexualität. *Rev. Suisse Zool.* **44,** 283–307.

SEILER, J. (1949) Das Intersexualitätsphänomen. *Experientia* **5,** 425–38.

SEILER, J. (1958) Die Entwicklung des Genitalapparates bei triploiden Intersexen von *Solenobia triquetrella* F.R. (Lepid. Psychidae). Deutung des Intersexualitätsphänomens. *Arch. Entwicklungsmech. Organ.* **150,** 199–372.

SEILER, J. (1959) Untersuchungen über die Entstehung der Parthenogenese bei *Solenobia triquetrella* F.R. (Lepidoptera, Psychidae). I. Die Zytologie der bisexuellen *S. triquetrella*, ihr Verhalten und ihr Sexualverhältnis. *Chromosoma* **10,** 73–114.

SEILER, J. (1961) Untersuchungen über die Entstehung der Parthenogenese bei *Solenobia triquetrella* F.R. (Lepidoptera, Psychidae). III. Die geographische Verbreitung der drei Rassen von *Solenobia triquetrella* (bisexuell, diploid und tetraploid parthenogenetisch) in der Schweiz und in angrenzenden Ländern und die Beziehungen zur Eiszeit. Bemerkungen über die Entstehung der Parthenogenese. *Z. Vererbungslehre* **92,** 261–316.

SEILER, J. (1963) Untersuchungen über die Entstehung der Parthenogenese bei *Solenobia triquetrella* F.R. (Lepidoptera, Psychidae). IV. Wie besamen begattete diploid and tetraploid parthenogenetische Weibchen von *Solenobia triquetrella* ihre Eier? Schicksal der Richtungskörper im besamten und unbesamten Ei. *Z. Vererbungslehre* **94,** 29–66.

SEILER, J. (1964) Untersuchungen über die Entstehung der Parthenogenese bei *Solenobia triquetrella* F.R. (Lepidoptera, Psychidae). V. Biologische und zytologische Beobachtungen zum Übergang von der diploiden zur tetraploiden Parthenogenese. *Chromosoma* **15,** 503–39.

SEILER, J. and PUCHTA, O. (1956) Die Fortpflanzungsbiologie der Solenobien (Lepid. Psychidae). Verhalten bei Artkreuzungen und F_1-Resultate. *Arch. Entwicklungsmech. Organ.* **149,** 115–246.

SEILER, J. and SCHÄFFER, K. (1960) Untersuchungen über die Entstehung der Parthenogenese bei *Solenobia triquetrella* F.R. (Lepidoptera, Psychidae). II. Analyse der diploid parthenogenetischen *S. triquetrella.* Verhalten, Aufzuchtresultate und Zytologie. *Chromosoma* **11,** 29–102.

SEKHAR, P. S. (1957) Mating, oviposition, and discrimination of hosts by *Aphidius testaceipes* (Cresson) and *Praon aguti* Smith, primarily parasites of aphids. *Ann. Entomol. Soc. Am.* **50,** 370–5.

SEKUL, A. A. and SPARKS, A. N. (1967) Sex pheromone of the fall army-worm moth: isolation, identification and synthesis. *J. Econ. Entomol.* **60,** 1270–2.

SELANDER, R. B. (1964) Sexual behavior in blister beetles (Coleoptera: Meloidae). I. The genus *Pyrota. Can. Entomologist* **96,** 1037–82.

SEN, S. K. (1917) Preliminary note on role of blood in ovulation in Culicidae. *Indian J. Med. Res.* **4,** 729–53.

SENGEL, P. and BULLIÈRE, D. (1966) Ponte naturelle et provoquée chez *Blabera craniifer* Burm. (Insectes Dictyoptère). *C.R. Acad. Sci. Paris* **262,** 1286–88.

SETHI, S. L. and SWENSON, K. G. (1967) Formation of sexuparae in the aphid *Eriosoma pyricola*, on pear roots. *Entomol. Exp. Appl.* **10,** 97–102.

SHANNON, R. C. and HADJINICALAO, J. (1941) Egg production of Greek Anophelines in nature. *J. Econ. Entomol.* **34,** 300–5.

SHAUMAR, N. (1966) Anatomie du système nerveux et analyse des facteurs externes pouvant intervenir dans le déterminisme du sexe chez les Ichneumonidae Pimplinae. *Ann. Sci. Nat., Zool. Sér.* 12, **8,** 391–494.

SHEN, S. K. and BERRYMAN, A. A. (1967) The male reproductive system and spermatogenesis of the European pine shoot moth, *Rhyacionia buoliana* (Lepidoptera: Olethreutidae), with observations on the effects of gamma irradiation. *Ann. Entomol. Soc. Am.* **60,** 767–74.

SHOREY. H. H. (1963) The biology of *Trichoplusia ni* (Lepidoptera: Noctuidae). II. Factors affecting adult fecundity and longevity. *Ann. Entomol. Soc. Am.* **56,** 476–80.

SHOREY, H. H. (1964) Sex pheromones of noctuid moths. II. Mating behavior of *Trichoplusia ni* (Lepidoptera: Noctuidae) with special reference to the role of the sex pheromone. *Ann. Entomol. Soc. Am.* **57,** 371–7.

SHOREY, H. H. and GASTON, L. K. (1964) Sex pheromones of noctuid moths. III. Inhibition of male responses to the sex pheromone in *Trichoplusia ni* (Lepidoptera: Noctuidae). *Ann. Entomol. Soc. Am.* **57,** 775–9.

SHOREY, H. H. and GASTON, L. K. (1965) Sex pheromones of noctuid moths. VII. Quantitative aspects of the production and release of pheromone by females of *Trichoplusia ni* (Lepidoptera: Noctuidae). *Ann. Entomol. Soc. Am.* **58,** 604–8.

SHOREY, H. H., GASTON, L. K. and FUKUTO, T. R. (1964) Sex pheromones of noctuid moths. I. A quantitative bioassay for the sex pheromone of *Trichoplusia ni* (Lepidoptera: Noctuidae). *J. Econ. Entomol.* **57,** 252–4.

SHOREY, H. H., GASTON, L. K. and ROBERTS, J. S. (1965) Sex pheromones of noctuid moths. VI. Absence of behavioral specificity for the females sex pheromones of *Trichoplusia ni* versus *Autographa californica*, and *Heliothis zea* versus *H. virescens* (Lepidoptera: Noctuidae). *Ann. Entomol. Soc. Am.* **58,** 600–3.

SHOREY, H. H., GASTON, L. K. and SAARIO, C. H. (1967) Sex pheromone of noctuid moths. XIV. Feasibility of behavioral control by disrupting pheromone communication in cabbage looper. *J. Econ. Entomol.* **60,** 1541–5.

SHOREY, H. H., MCFARLAND, S. U. and GASTON, L. K. (1968) Sex pheromones of noctuid moths. XIII. Changes in pheromone quantity as related to reproductive age and mating history, in females of seven species of Noctuidae (Lepidoptera). *Ann. Entomol. Soc. Am.* **61,** 372–6.

SHOREY, H. H., MORIN, K. L. and GASTON, L. K. (1968) Sex pheromones of noctuid moths. XV. Timing or development of pheromone-responsiveness and other indicators of reproductive age in males of eight species. *Ann. Entomol. Soc. Am.* **61,** 857–61.

SHUEL, R. W. and DIXON, S. E. (1960) The early establishment of dimorphism in the female honeybee, *Apis mellifera* L. *Insectes Sociaux* **7,** 265–82.

SHULL, A. F. (1928) Duration of light and the wings of the aphid *Macrosiphum solanifolii. Arch. Entwicklungsmech. Organ.* **113,** 210–39.

SHULL, A. F. (1929) The effect of intensity and duration of light and of duration of darkness, partly modified by temperature, upon wing production in aphids. *Arch. Entwicklungsmech. Organ.* **115,** 825–51.

SIEBOLD, C. T. VON (1837) Die Spermatozoen der wirbellosen Tiere. 4. Teil. Die Spermatozoen in den befruchteten Insektenweibchen. *Arch. Anat. Physiol.* **1837,** 392–439.

SIEW, Y. C. (1965a) The endocrine control of adult reproductive diapause in the Chrysomelid beetle, *Galeruca tanaceti* (L.). I. *J. Insect Physiol.* **11,** 1–10.

SIEW, Y. C. (1965b) The endocrine control of adult reproductive diapause in the Chrysomelid beetle *Galeruca tanaceti* (L.). II. *J. Insect Physiol.* **11,** 463–79.

SIEW, Y. C. (1965c) The endocrine control of adult reproductive diapause in the Chrysomelid beetle, *Galeruca tanaceti* (L.). III. *J. Insect Physiol.* **11,** 973–81.

SIEW, Y. C. (1966) Some physiological aspects of adult reproductive diapause in *Galeruca tanaceti* (L.) (Coleoptera: Chrysomelidae). *Trans. Roy. Entomol. Soc. London* **118,** 359–74.

SILVERSTEIN, R. M., BROWNLEE, R. G., BELLAS, T. E., WOOD, D. L. and BROWNE, L. E. (1968) Brevicomin: principal sex attractant in the frass of the female western pine beetle. *Science* **159,** 889–91.

SILVERSTEIN, R. M., RODIN, J. O., BURKHOLDER, W. E. and GORMAN, J. E. (1967) Sex attractant of the black carpet beetle. *Science* **157,** 85–87.

SILVESTRI, F. (1941) Studi sugli "Strepsiptera" (Insecta). II. Descrizione, biologia e sviluppo postembrionale dell'*Halictophagus tettigometrae* Silv. *Boll. Lab. Zool. Portici* **32,** 11–48.

SILVESTRI, F. (1942) Studi sugli "Strepsiptera" (Insecta). III. Descrizione e biologia di 6 specie italiene di *Mengenilla. Boll. Lab. Zool. Portici* **32,** 197–282.

SIMMONDS, F. J. (1947) The biology of the parasites of *Loxostege sticticalis,* L., in North America—*Bracon vulgaris* (Cress.) (Braconidae, Agathinae). *Bull. Entomol. Res.* **38,** 145–55.

SIMMONDS, F. J. (1953) Observations on the biology and mass-breeding of *Spalangia drosophilae* Ashm. (Hymenoptera, Spalangiidae), a parasite of the frit-fly, *Oscinella frit* (L.). *Bull. Entomol. Res.* **44,** 773–8.

SINGH, K. R. P. and BROWN, A. W. A. (1957) Nutritional requirements of *Aedes aegypti* L. *J. Insect Physiol.* **1,** 199–220.

SLÁMA, K. (1964a) Hormonal control of respiratory metabolism during growth, reproduction, and diapause in female adults of *Pyrrhocoris apterus* L. (Hemiptera). *J. Insect Physiol.* **10,** 283–303.

SLÁMA, K. (1964b) Hormonal control of haemolymph protein concentration in the adults of *Pyrrhocoris apterus* L. (Hemiptera). *J. Insect Physiol.* **10,** 773–82.

SLIFER, E. H. (1937) The origin and fate of the membranes surrounding the grasshopper egg together with some experiments on the source of the hatching enzyme. *Quart. J. Microscop. Sci.* **79,** 493–506.

SMITH, P. A. and KING, R. C. (1968) Genetic control of synaptonemal complexes in *Drosophila melanogaster. Genetics* **60,** 335–51.

SMITH, S. G. (1960) Cytogenetics of insects. *Ann. Rev. Entomol.* **5,** 69–84.

SNODGRASS, R. E. (1935) *Principles of Insect Morphology.* McGraw-Hill, New York, London.

SNODGRASS, R. E. (1956) *Anatomy of the Honey Bee.* Comstock Publishing Associates, Ithaca, N.Y.

SNODGRASS, R. E. (1957) A revised interpretation of the external reproductive organs of male insects. *Smiths. Misc. Coll.* 135, No. 6.

SNOW, J. W. and CALLAHAN, P. S. (1967) Laboratory mating studies of the corn earworm, *Heliothis zea* (Lepidoptera: Noctuidae). *Ann. Entomol. Soc. Am.* **60,** 1066–71.

SNOW, S. J. (1928) Effect of ovulation upon seasonal history in the alfalfa weevil. *J. Econ. Entomol.* **21,** 752–61.

SOLOMON, J. D. (1967) Carpenterworm oviposition. *J. Econ. Entomol.* **60,** 309.

SOLOMON, J. D. and MORRIS, R. C. (1966) Sex attraction of the carpenterworm moth. *J. Econ. Entomol.* **59,** 1534–5.

SONDHI, K. C. (1967) Studies in aging. III. The physiological effects of injecting hemolymph from outbred donors into inbred hosts in *Drosophila melanogaster. Proc. Natl. Acad. Sci.* **57,** 965–71.

SOULIÉ, J. (1964) Le contrôle par les ouvrières de la monogynie des colonies chez *Sphaerocrema striatula* (Myrmicidae Cremastogastrini). *Insectes Sociaux* **11,** 383–8.

SOULIÉ, J. and DAN DICKO, L. (1966) La monogynie chez certaines colonies de fourmis de la tribu des Cremastogastrini (Hymenoptera, Formicoidea, Myrmicidae). *Insectes Sociaux* **13,** 139–44.

SPEICHER, B. R. (1937) Oogenesis in a thelytokous wasp, *Nemeritis canescens* (Grav.). *J. Morphol.* **61**, 453–72.

SPENCER, G. J. (1930) The fire brat, *Thermobia domestica* Packard (Lepismidae) in Canada. *Can. Entomologist* **62**, 1–2.

SPIELMAN, A. (1957) The inheritance of autogeny in the *Culex pipiens* complex of mosquitoes. *Am. J. Hyg.* **65**, 404–35.

SPIETH, H. T. (1947) Sexual behavior and isolation in *Drosophila*. I. The mating behavior of species of the Willistoni group. *Evolution* **1**, 17–31.

SPIETH, H. T. (1949) Sexual behavior and isolation in *Drosophila*. II. The interspecific mating behavior of species of the Willistoni group. *Evolution* **3**, 67–81.

SPIETH, H. T. (1951) Mating behavior and sexual isolation in the *Drosophila virilis* species group. *Behaviour* **3**, 105–45.

SPIETH, H. T. (1952) Mating behavior within the genus *Drosophila* (Diptera). *Bull. Am. Mus. Nat. Hist.* **99**, 399–474.

SPIETH, H. T. and HSU, T. C. (1950) The influence of light on the mating behavior of seven species of the *Drosophila melanogaster* species group. *Evolution* **4**, 316–25.

SPRADBERY, J. P. (1965) The social organization of wasp communities. *Symp. Zool. Soc. London* **14**, 61–96.

SPRINGER, F. (1917) Über den Polymorphismus bei den Larven von *Miastor metraloas*. *Zool. Jahrb. Abt. Syst.* **40**, 57–118.

SPRINGHETTI, A. (1962) Sul controllo dell'attivita dell'ovario in *Nauphoeta cinerea*. *Boll. Zool.* **29**, 805–20.

STAAL, G. B. (1961) *Studies on the Physiology of Phase Induction in* Locusta migratoria migratorioides *R. and F.* Wageningen.

STADLER, H. (1926) Drohnenbrütigkeit bei Wespen. *Zool. Anz.* **66**, 92–96.

STALKER, H. D. (1956) On the evolution of parthenogenesis in *Lonchoptera* (Diptera). *Evolution* **10**, 345–59.

STAVRAKI-PAULOPOULOU, H. G. (1966) Contribution à l'étude de la capacité reproductrice et de la fécondité réelle d'*Opius consolor* SZEPL (Hymenoptera-Braconidae). *Ann. Epiphyties* **17**, 391–435.

STAY, B. (1965) Protein uptake in the oocytes of the *Cecropia* moth. *J. Cell Biol.* **26**, 49–62.

STAY, B. and GELPERIN, A. (1966) Physiological basis of ovipositional behaviour in the false ovoviviparous cockroach, *Pycnoscelus surinamensis* (L.). *J. Insect Physiol.* **12**, 1217–26.

STAY, B. and ROTH, L. M. (1958) The reproductive behavior of *Diploptera punctata* (Blattaria: Diplopteridae). *Proc. Xth Intern. Congr. Entomol. Montreal 1956*, **2**, 547–52.

STEFFAN, A. W. (1961) Die Stammes- und Siedlungsgeschichte des Artenkreises *Sacchiphantes viridis* (Ratzeburg 1843) (Adelgidae, Aphidoidea). *Zoologica* (Stuttgart) **109**, 1–112.

STEFFAN, A. W. (1962a) Die Artenkreise der Gattung *Sacchiphantes* (Adelgidae, Aphidoidea). *Proc. XIth Intern. Congr. Entomol. Wien 1960*, **1**, 57–63.

STEFFAN, A. W. (1962b) *Sacchiphantes-abietis*-Befall an *Picea abies* in Montenegro. *Z. Angew. Zool.* **49**, 281–95.

STEFFAN, A. W. (1964) Problems of evolution and speciation in Adelgidae (Homoptera: Aphidoidea). *Can. Entomologist* **96**, 155–7.

STEFFAN, A. W. (1968a) Zum Generations- und Chromosomenzyklus der Adelgidae (Homoptera, Aphidoidea). *Verhandl. Deut. Zool. Ges. Heidelberg 1967*, 762–73.

STEFFAN, A. W. (1968b) Evolution und Systematik der Adelgidae (Homoptera: Aphidina). Eine Verwandtschaftsanalyse auf vorwiegend ethologischer, zytologischer und karyologischer Grundlage. *Zoologica* (Stuttgart) **115**, 1–139.

STEIN, F. (1847) *Vergleichende Anatomie und Physiologie der Insekten.* I. *Monographie. Die weiblichen Geschlechtsorgane der Käfer.* Berlin.

STEINBRECHT, R. A. (1964a) Feinstruktur und Histochemie der Sexualduftdrüse des Seidenspinners *Bombyx mori* L. *Z. Zellforsch.* **64**, 227–61.

STEINBRECHT, R. A. (1964b) Die Abhängigkeit der Lockwirkung des Sexualduftorgans weiblicher Seidenspinner (*Bombyx mori*) von Alter und Kopulation. *Z. Vergleich. Physiol.* **48**, 341–56.

STEINBRECHT, R. A. and SCHNEIDER, D. (1964) Die Faltung der äusseren Zellmembran in den Sexuallockstoff-Drüsenzellen des Seidenspinners. *Naturwiss.* **51**, 41.

STEINER, A. (1932) Die Arbeitsteilung der Feldwespe *Polistes dubia* K. *Z. Vergleich. Physiol.* **17**, 101–52.

STEINER, P. (1930) Studien an *Panorpa communis* L. I. Zur Biologie, *Z. Morphol. Oekol. Tiere* **17**, 1–25.

STERN, V. M. and BOWEN, W. R. (1968) Further evidence of a uniparental race of *Trichogramma semifumatum* at Bishop, California. *Ann. Entomol. Soc. Am.* **61**, 1032–3.

STERN, V. M. and SMITH, R. F. (1960) Factors affecting egg production and oviposition in populations of *Colias philodice eurytheme* Boisduval (Lepidoptera: Pieridae). *Hilgardia* **29**, 411–54.

STEWART, J. W., WHITCOMB, W. H. and BELL, K. O. (1967) Estivation studies of the convergent lady beetle in Arkansas. *J. Econ. Entomol.* **60**, 1730–5.

STICH, H. F. (1963) An experimental analysis of the courtship pattern of *Tipula oleracea* (Diptera). *Can. J. Zool.* **41**, 99–109.

STRAMBI, A. (1965) Influence du parasite *Xenos vesparum* Rossi (Strepsiptère) sur la neurosécrétion des individus du sexe femelle de *Polistes gallicus* L. (Hyménoptère, Vespide). *C.R. Acad. Sci. Paris* **260,** 3768–9.

STRAMBI, A. (1967) Quelques effets de la castration sur la neurosécrétion protocérébrale des femelles de *Polistes* (Hyménoptère Vespidés). *C.R. Acad. Sci. Paris* **264,** 2031–4.

STRANGWAYS-DIXON, J. (1961a) The relationship between nutrition, hormones, and reproduction in the blowfly *Calliphora erythrocephala* (Meig.). I. Selective feeding in relation to the reproductive cycle, the corpus allatum volume and fertilization. *J. Exp. Biol.* **38,** 225–35.

STRANGWAYS-DIXON, J. (1961b) The relationships between nutrition, hormones, and reproduction in the blowfly *Calliphora erythrocephala* (Meig.). II. The effect of removing the ovaries, the corpus allatum and the median neurosecretory cells upon selective feeding, and the demonstration of the corpus allatum cycle. *J. Exp. Biol.* **38,** 637–46.

STRANGWAYS-DIXON, J. (1962) The relationship between nutrition, hormones and reproduction in the blowfly *Calliphora erythrocephala* (Meig.). III. The corpus allatum in relation to nutrition, the ovaries, innervation and the corpus cardiacum. *J. Exp. Biol.* **39,** 293–306.

STRIDE, G. O. (1956) On the courtship behaviour of *Hypolimnas misippus* L. (Lepidoptera, Nymphalidae), with notes on the mimetic association with *Danaus chrysippus* L. (Lepidoptera, Danaidae). *Brit. J. Animal Behaviour* **4,** 52–68.

STRIDE, G. O. (1958a) On the courtship behaviour of a tropical mimetic butterfly, *Hypolimnas misippus* L. (Nymphalidae). *Proc. Xth Intern. Congr. Entomol. Montreal 1956*, **2,** 419–24.

STRIDE, G. O. (1958b) Further studies on the courtship behaviour of African mimetic butterflies. *Animal Behaviour* **6,** 224–30.

STRONG, L. (1965a) The relationships between the brain, corpora allata, and oocyte growth in the Central American locust, *Schistocerca* sp. I. The cerebral neurosecretory system, the corpora allata, and oocyte growth. *J. Insect Physiol.* **11,** 135–46.

STRONG, L. (1965b) The relationships between the brain, corpora allata, and oocyte growth in the Central American locust, *Schistocerca* sp. II. The innervation of the corpora allata, the lateral neurosecretory complex, and oocyte growth. *J. Insect Physiol.* **11,** 271–80.

STRONG, L. (1966) Endocrinology of imaginal diapause in the female red locust, *Nomadacris septemfasciata* (Serv.). *Nature* **212,** 1276–8.

STRÜBING, H. (1958) Lautäusserung—der entscheidende Faktor für das Zusammenfinden der Geschlechter bei Kleinzikaden (Homoptera-Auchenorrhyncha). *Zool. Beitr.* N.F. **4,** 15–21.

STRÜBING, H. (1960) Eiablage und photoperiodischbedingte Generationsfolge von *Chloriona smaragdula* Stal und *Euidella speciosa* Boh. (Homoptera-Auchenorrhyncha). *Zool. Beitr.* N.F. **5,** 301–32.

STÜRCKOW, B. (1965) The electroantennogram (EAG) as an assay for the reception of odours by the gypsy moth. *J. Insect Physiol.* **11,** 1573–84.

STURM, H. (1955) Beiträge zur Ethologie einiger mitteldeutscher Machiliden. *Z. Tierpsychol.* **12,** 337–63.

STURM, H. (1956) Die Paarung beim Silberfischchen *Lepisma saccharina. Z. Tierpsychol.* **13,** 1–12.

STURTEVANT, A. H. (1945) A gene in *Drosophila melanogaster* that transforms females into males. *Genetics* **30,** 297–99.

STURTEVANT, A. H. and DOBZHANSKY, T. (1936) Geographical distribution and cytology of "sex ratio" in *Drosophila pseudoobscura* and related species. *Genetics* **21,** 473–90.

SUBBA RAO, B. R. and GOPINATH, K. (1961) The effects of temperature and humidity on the reproductive potential of *Apanteles angaleti* Muesebeck (Braconidae: Hymenoptera). *Entomol. Exp. Appl.* **4,** 119–22.

SUNDRY, R. A. (1966) A comparative study of the efficiency of three predatory insects, *Coccinella septempunctata* L. (Coleoptera, Coccinellidae), *Chrysopa carnea* St. (Neuroptera, Chrysopidae), and *Syrphus bibesii* L. (Diptera, Syrphidae) at two different temperatures. *Entomophaga* **11,** 395–404.

SUOMALAINEN, E. (1940) Polyploidy in parthenogenetic Curculionidae. *Hereditas* **26,** 51–64.

SUOMALAINEN, E. (1947) Parthenogenese und Polyploidie bei Rüsselkäfern (Curculionidae). *Hereditas* **33,** 425–56.

SUOMALAINEN, E. (1950) Parthenogenesis in animals. *Advan. Genet.* **3,** 193–253.

SUOMALAINEN, E. (1954) Zur Zytologie der parthenogenetischen Curculioniden der Schweiz. *Chromosoma* **6,** 627–55.

SUOMALAINEN, E. (1962) Significance of parthenogenesis in the evolution of insects. *Ann. Rev. Entomol.* **7,** 349–66.

SUOMALAINEN, E. (1965) Die Polyploidie bei dem parthenogenetischen Blattkäfer *Adoxus obscurus* L. (Coleoptera, Chrysomelidae). *Zool. Jahrb. Abt. Syst.* **92,** 183–92.

SURTEES, G. (1964) Observations on some effects of temperature and isolation on fecundity of female weevils, *Sitophilus granarius* (L.) (Coleoptera, Curculionidae). *Entomol. Exp. Appl.* **7,** 249–52.

SWEETMAN, H. L. (1938) Physical ecology of the firebrat, *Thermobia domestica* (Packard). *Ecol. Monogr.* **8,** 287–311.

SWELLENGREBEL, N. H. (1929) La dissociation des fonctions sexuelles et nutritives (dissociation gono-trophique) d'*Anopheles maculipennis*. *Ann. Inst. Pasteur* **43,** 1370–89.

SZÉKESSY, W. (1937) Über Parthenogenese bei Koleopteren. *Biol. Gen.* **12,** 577–90.

TAKENOUCHI, Y. (1957) Polyploidy in some parthenogenetic weevils. (A preliminary report.) *Annot. Zool. Japan* **30,** 38–41.

TANAKA, Y. (1953) Genetics of the silkworm, *Bombyx mori. Advan. Genet.* **5,** 239–317.

TASHIRO, H. and CHAMBERS, D. L. (1967) Reproduction in the California red scale, *Aonidiella aurantii* (Homoptera: Diaspididae). I. Discovery and extraction of a female sex pheromone. *Ann. Entomol. Soc. Amer.* **60,** 1166–70.

TATE, P. and VINCENT, M. (1936) The biology of autogenous and anautogenous races of *Culex pipiens* L. (Diptera: Culicidae). *Parasitol.* **28,** 115–45.

TAZIMA Y. (1944) Studies on the chromosomal aberrations in the silkworm. II. Translocation involving second and W-chromosomes. *Bull. Seric. Exp. Sta. Japan* **12,** 109–81.

TAZIMA, Y. (1964) *The Genetics of the Silkworm.* Logos Press, London.

TELFER, W. H. (1954) Immunological studies of insect metamorphosis. II. The role of a sex-limited blood protein in egg formation by the *Cecropia* silkworm. *J. Gen. Physiol.* **37,** 539–58.

TELFER, W. H. (1960) The selective accumulation of blood proteins by the oocytes of saturniid moths. *Biol. Bull.* **118,** 338–51.

TELFER, W. H. (1961) The route of entry and localization of blood proteins in the oocytes of saturniid moths. *J. Biophys. Biochem. Cytol.* **9,** 747–59.

TELFER, W. H. (1965) The mechanism and control of yolk formation. *Ann. Rev. Entomol.* **10,** 161–84.

TELFER, W. H. and ANDERSON, L. M. (1968) Functional transformation accompanying the initiation of a terminal growth phase in the *Cecropia* moth oocyte. *Develop. Biol.* **17,** 512–35.

TELFER, W. H. and MELIUS, M. E., JR. (1963) The mechanism of blood protein uptake by insect oocytes. *Am. Zoologist* **3,** 185–91.

TELFER, W. H. and RUTBERG, L. D. (1960) The effects of blood protein depletion on the growth of the oocytes in the *Cecropia* moth. *Biol. Bull.* **118,** 352–66.

THIELE, H. U. (1966) Einflüsse der Photoperiode auf die Diapause von Carabiden. *Z. Angew. Entomol.* **58,** 143–9.

THIELE, H. U. (1968) Formen der Diapausesteuerung bei Carabiden. *Verhandl. Deut. Zool. Ges. Heidelberg 1967*, 358–64.

THOMAS, H. T. (1950) Field notes on the mating habits of *Sarcophaga* Meigen (Diptera). *Proc. Roy. Entomol. Soc. London* A **25,** 93–98.

THOMAS, K. K. and NATION J. L. (1966) Control of a sex-limited haemolymph protein by corpora allata during ovarian development in *Periplaneta americana* (L.). *Biol. Bull.* **130,** 254–64.

THOMSEN, E. (1940) Relation between corpus allatum and ovaries in adult flies (Muscidae). *Nature* **145,** 28–29.

THOMSEN, E. (1943) An experimental and anatomical study of the corpus allatum in the blow-fly, *Calliphora erythrocephala* Meig. *Vidensk Medd. Naturh. Foren. Kbh.* **106,** 320–405.

THOMSEN, E. (1948a) The gonadotropic hormones in the Diptera. *Bull. Biol. Fr. Belg. Suppl.* **33,** 68–80.

THOMSEN, E. (1948b) Effect of removal of neurosecretory cells in the brain of adult *Calliphora erythrocephala* Meig. *Nature* **161,** 439–40.

THOMSEN, E. (1952) Functional significance of the neurosecretory brain cells and the corpus cardiacum in the female blow-fly, *Calliphora erythrocephala* Meig. *J. Exp. Biol.* **29,** 137–72.

THOMSEN, E. (1954) Studies on the transport of neurosecretory material in *Calliphora erythrocephala* by means of ligaturing experiments. *J. Exp. Biol.* **31,** 322–30.

THORPE, W. H. and CAUDLE, H. B. (1938) A study of the olfactory responses of insect parasites to the food plant of their host. *Parasitol.* **30,** 523–8.

TINBERGEN, N. (1951) *The Study of Instinct.* Clarendon Press, Oxford.

TINBERGEN, N., MEEUSE, B. J. D., BOEREMA, L. K. and VAROSSIEAU, W. W. (1942) Die Balz des Samtfalters, *Eumenis* (*Satyrus*) *semele* (L.). *Z. Tierpsychol.* **5,** 182–226.

TISCHNER, H. and SCHIEF, A. (1955) Fluggeräusch und Schallwahrnehmung bei *Aedes aegypti* L. (Culicidae). *Verhandl. Deut. Zool. Ges. Tübingen 1954*, 543–60.

TITSCHACK, E. (1930) Untersuchungen über das Wachstum, den Nahrungsverbrauch und die Eierzeugung. III. *Cimex lectularius* L. *Z. Morphol. Oekol. Tiere* **17,** 471–551.

TOBA, H. H., KISHABA, A. N. and WOLF, W. W. (1968) Bioassay of the synthetic female sex pheromone of the cabbage looper. *J. Econ. Entomol.* **61,** 812–6.

TOKUNAGA, M. (1935) Chironomidae from Japan (Diptera). V. Supplementary report on the Clunioninae. *Mushi* **8,** 1–20.

TOMBES, A. S. (1964a) Respiratory and compositional study on the aestivating insect, *Hypera postica* (Gyll.) (Curculionidae). *J. Insect Physiol.* **10,** 997–1003.

TOMBES, A. S. (1964b) Seasonal changes in the reproductive organs of the alfalfa weevil, *Hypera postica* (Coleoptera: Curculionidae), in South Carolina. *Ann. Entomol. Soc. Am.* **57,** 422–6.

TOMBES, A. S. (1966) Aestivation (summer diapause) in *Hypera postica* Coleoptera: Curculionidae). I. Effect of aestivation, photoperiods, and diet on total fatty acids. *Ann. Entomol. Soc. Am.* **59,** 376–80.

TOMBES, A. S. and BODENSTEIN, D. (1967) The neuroendocrine system of the adult alfalfa weevil, *Hypera postica. Am. Zoologist* **7,** 722.

TOMBES, A. S. and DUNIPACE, A. J. (1967) Neuroendocrine-lipid relationships in the prediapause female *Hypera postica* (Coleoptera: Curculionidae). *J. Gen. Physiol.* **50,** 2503–4.

TOMBES, A. S. and SMITH, D. S. (1966) Ultrastructural studies on the corpora cardiaca-allata complex of the adult alfalfa weevil, *Hypera postica. Am. Zoologist* **6,** 575–6.

TOSCHI, C. A. (1965) The taxonomy, life histories, and mating behavior of the green lacewings of Strawberry Canyon (Neuroptera: Chrysopidae). *Hilgardia* **36,** 391–431.

TRAVIS, B. V. (1939) Habits of the June beetle, *Phyllophaga lanceolata* (Say) in Iowa. *J. Econ. Entomol.* **32,** 690–3.

TREHEN, P. (1965) À propos de l'offrande nuptiale chez *Hilara maura* Fab. et *Hilara pilosa* Zett. (Insectes Diptères Empididae). *C.R. Acad. Sci. Paris* **260,** 2603–5.

TROUVELOT, B. and GRISON, P. (1935) Variations de fécondité du *Leptinotarsa decemlineata* Say avec les Solanum tuberifères consommés par l'insecte. *C.R. Acad. Sci. Paris* **201,** 1053–5.

TSCHINKEL, W., WILLSON, C. and BERN, H. A. (1967) Sex pheromone of the mealworm beetle (*Tenebrio molitor*). *J. Exp. Zool.* **164,** 81–85.

TSIU-NGUN, W. and QUO, F. (1963) The endocrine role of the corpora allata on the egg maturation in the armyworm *Leucania separata* Walker (Lepidoptera). *Acta Entomol. Sinica* **12,** 411–21.

TURNER, N. (1960) The effect of inbreeding and crossbreeding on numbers of insects. *Ann. Entomol. Soc. Am.* **53,** 686–8.

TUXEN, S. L. (1956) *Taxonomist's Glossary of Genitalia in Insects.* Ejnar Munksgaard, Copenhagen.

TWOHY, D. W. and ROZEBOOM, L. E. (1957) A comparison of food reserves in autogenous and anautogenous *Culex pipiens* populations. *Am. J. Hyg.* **65,** 316–24.

UESHIMA, N. (1966) Cytology and cytogenetics. *The Thomas Say Foundation,* **7,** 183–237.

ULLERICH, F. H. (1958) Monogene Fortpflanzung bei der Fliege *Chrysomyia albiceps. Z. Naturforsch.* **13b,** 473–4.

ULLERICH, F. H. (1961) Geschlechtsbestimmung bei der Fliege *Phormia regina. Naturwiss.* **48,** 559–60.

ULLERICH, F. H. (1963) Geschlechtschromosomen und Geschlechtsbestimmung bei einigen Calliphorinen (Calliphoridae, Diptera). *Chromosoma* **14,** 45–110.

ULLERICH, F. H., BAUER, H. and DIETZ, R. (1964) Geschlechtsbestimmung bei Tipuliden (Nematocera, Diptera). *Chromosoma* **15,** 591–605.

ULRICH, H. (1934) Experimentelle Untersuchungen über den Generationswechsel einer pädogenetischen Gallmücke. *Rev. Suisse Zool.* **41,** 423–8.

ULRICH, H. (1936) Experimentelle Untersuchungen über den Generationswechsel der heterogonen Cecidomyide *Oligarces paradoxus. Z. Induktive Abstammungs. Vererbungslehre* **71,** 1–60.

ULRICH, H. (1963a) Generationswechsel und Geschlechtsbestimmung einer Gallmücke mit viviparen Larven. *Verhandl. Deut. Zool. Ges. Wien 1962,* 139–52.

ULRICH, H. (1963b) Vergleichend-histologische und zyklische Untersuchungen an den weiblichen Geschlechtsorganen und den innersekretorischen Drüsen adulter Hippobosciden (Diptera Pupipara). *Deut. Entomol. Z.* N.F. **10,** 28–71.

URBAHN, E. (1913) Abdominale Duftorgane bei weiblichen Schmetterlingen. *Jena Z. Naturwiss.* **50,** 277–358.

UVAROV, B. P. (1928) *Locusts and Grasshoppers.* London.

VALENTINE, J. M. (1931) The olfactory sense of the adult mealworm beetle *Tenebrio molitor* (Linn.). *J. Exp. Zool.* **58,** 165–227.

VANDEL, A. (1931) *La Parthénogénèse.* Paris.

VANDERBERG, J. P. (1963) Synthesis and transfer of DNA, RNA, and protein during vitellogenesis in *Rhodnius prolixus* (Hemiptera). *Biol. Bull.* **125,** 556–75.

VANDERZANT, E. S. (1963) Nutrition of the adult boll weevil: oviposition on defined diets and amino acid requirements. *J. Insect Physiol.* **9,** 683–91.

VANDERZANT, E. S., POOL, M. C. and RICHARDSON, C. D. (1962) The role of ascorbic acid in the nutrition of three cotton insects. *J. Insect Physiol.* **8,** 287–97.

VAN EMDEN, H. F. (1966) Studies on the relations of insect and host plants. III. A comparison of the reproduction of *Brevicoryne brassicae* and *Myzus persicae* (Hemiptera: Aphididae) on Brussels sprout plants supplied with different rates of nitrogen and potassium. *Entomol. Exp. Appl.* **9,** 444–60.

VAN ERP, A. (1960) Mode of action of the inhibitory substance of the honeybee queen. *Insectes Sociaux* **7,** 207–11.

VELTHUIS, H. H. W., VERHEIJEN, F. J. and GOTTENBOS, H. J. (1965) Laying worker honey bee: similarities to the queen. *Nature* **207,** 1314.

VERMEIL, C. (1953) De la reproduction par autogénèse chez *Aedes* (O.) *detritus* Haliday. *Bull. Soc. Pathol. Exotique* **46,** 971–3.

VODJDANI, S. (1954) Contribution à l'étude des punaises des céréales et en particulier d'*Eurygaster integriceps* Put. (Hemiptera, Pentatomidae, Scutellerinae). *Ann. Epiphyties Sér. C,* **5,** 105–60.

VOGEL, G. (1954) Das optische Weibchenschema bei *Musca domestica. Naturwiss.* **41,** 482–3.

VOGEL, G. (1957) Verhaltensphysiologische Untersuchungen über die den Weibchensprung des Stubenfliegen-Mänrchens (*Musca domestica*) auslösenden optischen Faktoren. *Z. Tierpsychol.* **14,** 309–23.

VOGEL, G. (1958) Supernormale Auslösereize bei *Sarcophaga carnaria* (Dipt.). *Zool. Beitr.* N.F. **4**, 69–76.

VOGEL, W. (1950) Eibildung und Embryonalentwicklung von *Melolontha vulgaris* F. und ihre Auswertung für die chemische Maikäferbekämpfung. *Z. Angew. Entomol.* **31**, 537–82.

VOGT, M. (1940) Die Förderung der Eireifung innerhalb heteroplastisch transplantierter Ovarien von *Drosophila* durch die gleichzeitige Implantation der arteigenen Ringdrüse. *Biol. Zentr.* **60**, 479–84.

VOGT, M. (1942) Weiteres zur Frage der Artspezifität gonadotroper Hormone. Untersuchungen an *Drosophila*-Arten, *Arch. Entwicklungsmech. Organ.* **141**, 424–54.

VOGT, M. (1943) Zur Produktion gonadotropen Hormones durch Ringdrüsen des ersten Larvenstadiums bei *Drosophila. Biol. Zentr.* **63**, 467–70.

VÖHRINGER, K. (1934) Zur Biologie der grossen Wachsmotte (*Galleria mellonella* Lin.). III. Morphologische und biologische Untersuchungen am Falter der grossen Wachsmotte (*Galleria mellonella* Lin.). *Zool. Jahrb. Abt. Anat.* **58**, 275–302.

VOLK, S. (1964) Untersuchungen zur Eiablage von *Syrphus corollae* Fabr. (Diptera: Syrphidae). *Z. Angew. Entomol.* **54**, 365–86.

VOLOZINA, N. V. (1967) The effect of the amount of blood taken and additional carbohydrate nutrition on oogenesis in females of blood-sucking mosquitoes of the genus *Aedes* (Diptera, Culicidae) of various weights and ages. *Entomol. Rev. USSR* **46**, 27–32.

VOOGD, S. (1955) Inhibition of ovary development in worker bees by extraction fluid of the queen. *Experientia* **11**, 181–2.

VOOGD, S. (1956) The influence of a queen on the ovary development in worker bees. *Experientia* **12**, 199–201.

VOUKASSOVITCH, P. (1949) Facteurs conditionnels de la ponte chez *Acanthoscelides obtectus. Bull. Mus. Hist. Nat., Belgrade* B 224–34.

VROMAN, H. E., KAPLANIS, J. N. and ROBBINS, W. E. (1965) Effect of allatectomy on lipid biosynthesis and turnover in the female American cockroach, *Periplanta americana* (L.). *J. Insect Physiol.* **11**, 897–904.

WAGNER, N. (1863) Beitrage zur Lehre von der Fortpflanzung der Insectenlarven. *Z. Wiss. Zool.* **13**, 513–27.

WAGNER, N. (1865) Über die viviparen Gallmückenlarven. *Z. Wiss. Zool.* **15**, 106–17.

WALDBAUER, G. P. (1962) The growth and reproduction of maxillectomized tobacco hornworms feeding on normally rejected non-solanaceous plants. *Entomol. Exp. Appl.* **5**, 147–58.

WALDRON, I. (1964) Courtship sound production in two sympatric sibling *Drosophila* species. *Science* **144**, 191–3.

WALKER, M. F. (1966) Some observations on the behaviour and life-history of the Jersey tiger moth, *Euplagia quadripunctaria* Poda (Lep., Arctiidae), in the "Valley of the Butterflies", Rhodes. *Entomologist* **99**, 1–24.

WALLIS, R. C. (1954) A study of oviposition activity of mosquitoes. *Am. J. Hyg.* **60**, 135–68.

WALOFF, N. (1957) The effect of the number of queens of the ant *Lasius flavus* (Fab.) (Hym., Formicidae) on their survival and on the rate of development of the first brood. *Insectes Sociaux* **4**, 391–408.

WALOFF, N. and RICHARDS, O. W. (1958) The biology of the Chrysomelid beetle, *Phytodecta olivacea* (Forster) (Coleoptera: Chrysomelidae). *Trans. Roy. Entomol. Soc. London* **110**, 99–116.

WANG, C. M. (1964) Laboratory observations on the life history and habits of the face fly, *Musca autumnalis* (Diptera: Muscidae). *Ann. Entomol. Soc. Am.* **57**, 563–9.

WARTHEN, D. (1968) Synthesis of cis-9-tetradecen-1-ol acetate, the sex pheromone of the fall armyworm. *J. Med. Chem.* **11**, 371–3.

WASMANN, E. (1901) *Termitoxenia*, ein neues flügelloses, physogastres Dipterengenus aus Termitennestern. *Z. Wiss. Zool.* **70**, 289–98.

WASMANN, E. (1910) Nachträge zum sozialen Parasitismus und der Sklaverei bei den Ameisen. *Biol. Zentr.* **30**, 453–64, 475–96, 515–24.

WATSON, J. A. L. (1964) Moulting and reproduction in the adult firebrat, *Thermobia domestica* (Packard) (Thysanura, Lepismatidae). II. The reproductive cycles. *J. Insect Physiol.* **10**, 399–408.

WATSON, J. A. L. (1967) Reproduction, feeding activity, and growth in the adult firebrat, *Lepismodes inquilinus* Newman (Thysanura, Lepismatidae). *J. Insect Physiol.* **13**, 1689–98.

WAY, M. J. and BANKS, C. J. (1967) Intra-specific mechanisms in relation to the natural regulation of numbers of *Aphis fabae* Scop. *Ann. Appl. Biol.* **59**, 189–205.

WEBBER, L. G. (1955) The relationship between larval and adult size of the Australian sheep blowfly *Lucilia cuprina* (Wied.). *Australian J. Zool.* **3**, 346–53.

WEBBER, L. G. (1958) Nutrition and reproduction in the Australian sheep blowfly *Lucilia cuprina. Australian J. Zool.* **6**, 139–44.

WEBER, H. (1930) *Biologie der Hemipteren.* Springer, Berlin.

WEBER, H. (1954) *Grundriss der Insektenkunde.* Gustav Fischer Verlag, Stuttgart.

WEED, I. G. (1936) Removal of corpora allata on egg production in the grasshopper, *Melanoplus differentialis. Proc. Soc. Exp. Biol. Med.* **34**, 883–5.

WEESNER, F. M. (1960) Evolution and biology of the termites. *Ann. Rev. Entomol.* **5**, 153–70.

WEIDNER, H. (1934) Beiträge zur Morphologie und Physiologie des Genitalapparates der weiblichen Lepidopteren. *Z. Angew. Entomol.* **21**, 240–90.

WEIH, A. S. (1951) Untersuchungen über das Wechselsingen (Anaphonie) und über das angeborene Lautschema einiger Feldheuschrecken. *Z. Tierpsychol.* **8,** 1–41.

WEIRICH, G. (1963) Zur Frage der hormonalen Regulation der Eireifung bei Insekten. Diss. Univ. München.

WENK, P. (1965a) Über die Biologie blutsaugender Simuliiden (Diptera). III. Kopulation, Blutsaugen und Eiablage von *Boophthora erythrocephala* de Geer im Laboratorium. *Z. Tropenmed. Parasitol.* **16,** 207–26.

WENK, P. (1965b) Über die Biologie blutsaugender Simuliiden (Diptera). II. Schwarmverhalten, Geschlechterfindung und Kopulation. *Z. Morphol. Oekol. Tiere* **55,** 671–713.

WESENBERG-LUND, C. (1913) Odonaten-Studien. *Intern. Rev. Hydrobiol. Hydrograph* **6,** 155–228.

WESSON, L. G. (1940) An experimental study on caste determination in ants. *Psyche* **47,** 105–11.

WEST, M. J. (1967) Foundress associations in polistine wasps: dominance hierarchies and the evolution of social behavior. *Science* **157,** 1584–5.

WEYER, F. (1934) Der Einfluss der Larvenernährung auf die Fortpflanzungsphysiologie verschiedener Stechmücken. *Arch. Schiffs. Tropenhyg.* **38,** 394–8.

WEYER, F. (1935) Die Rassenfrage bei *Culex pipiens* in Deutschland. *Z. Parasitenk.* **8,** 104–15.

WHARTON, D. R. A., BLACK, E. D. and MERRITT, C., JR. (1963) Sex attractant of the American cockroach. *Science* **142,** 1257.

WHARTON, D. R. A., MILLER, G. L. and WHARTON, M. L. (1954a) The odorous attractant of the American cockroach, *Periplaneta americana* (L.). I. Quantitative aspects of the response to the attractant. *J. Gen. Physiol.* **37,** 461–9.

WHARTON, D. R. A., MILLER, G. L. and WHARTON, M. L. (1954b) The odorous attractant of the American cockroach, *Periplaneta americana* (L.). II. A bioassay method for the attractant. *J. Gen. Physiol.* **37,** 471–81.

WHARTON, M. L. and WHARTON, D. R. A. (1957) The production of sex attractant substance and of oothecae by the normal and irradiated American cockroach, *Periplaneta americana* L. *J. Insect Physiol.* **1,** 229–39.

WHEELER, M. R. (1947) The insemination reactions in intraspecific matings of *Drosophila. Univ. Texas Publ. No.* 4720, 78–115.

WHEELER, W. M .(1910) *Ants, their Structure, Development, and Behavior.* Columbia Univ. Press, New York.

WHEELER, W. M. (1937) *Mosaics and Other Anomalies Among Ants.* Harvard Univ. Press, Cambridge.

WHITE, D. (1965) Changes in size of the corpus allatum in a polymorphic insect. *Nature* **208,** 807.

WHITE, D. F. (1968) Cabbage aphid: effect of isolation on form and on endocrine activity. *Science* **159,** 218–219.

WHITE, M. J. D. (1948) The chromosomes of the parthenogenetic mantid *Brunneria borealis. Evolution* **2,** 90–93.

WHITE, M. J. D. (1954) *Animal Cytology and Evolution,* 2nd ed. Cambridge Univ. Press, Cambridge.

WHITE, M. J. D. (1955) Patterns of spermatogenesis in grasshoppers. *Australian J. Zool.* **3,** 222–6.

WHITE, M. J. D. (1964) Cytogenetic mechanisms in insect reproduction. *Roy. Entomol. Soc. London Symp.* **2,** 1–12.

WHITING, A. R. (1961) Genetics of *Habrobracon. Advan. Genet.* **10,** 295–348.

WHITING, P. W. (1918) Sex determination and biology of a parasite wasp, *Habrobracon brevicornis* (Wesmael). *Biol. Bull.* **34,** 250–6.

WHITING, P. W. (1935) Sex determination in bees and wasps. *J. Heredity* **26,** 263–78.

WHITING, P. W. (1938) Anomalies and caste determination in ants. *J. Heredity* **29,** 189–93.

WHITING, P. W. (1943) Multiple alleles in complementary sex determination of *Habrobracon. Genetics* **28,** 365–82.

WHITING, P. W. (1945) The evolution of male haploidy. *Quart. Rev. Biol.* **20,** 231–60.

WICK, J. R. and BONHAG, P. F. (1955) Postembryonic development of the ovaries of *Oncopeltus fasciatus* (Dallas). *J. Morphol.* **96,** 31–60.

WIESE, L. (1960) Die diplogenotypische Geschlechtsbestimmung. *Fortschr. Zool.* **12,** 295–335.

WIESE, L. (1966) Geschlechtsbestimmung. *Fortschr. Zool.* **18,** 139–206.

WIGGLESWORTH, V. B. (1934) The physiology of ecdysis in *Rhodnius prolixus* (Hemiptera). II. Factors controlling moulting and metamorphosis. *Quart. J. Microscop. Sci.* **77,** 191–222.

WIGGLESWORTH, V. B. (1936) The function of the corpus allatum in the growth and reproduction of *Rhodnius prolixus* (Hemiptera). *Quart. J. Microscop. Sci.* **79,** 91–121.

WIGGLESWORTH, V. B. (1943) The fate of haemoglobin in *Rhodnius prolixus* (Hemiptera) and other blood-sucking arthropods. *Proc. Roy. Soc. London* B **131,** 313–39.

WIGGLESWORTH, V. B. (1948) The functions of the corpus allatum in *Rhodnius prolixus* (Hemiptera). *J. Exp. Biol.* **25,** 1–14.

WIGGLESWORTH, V. B. (1952) The thoracic gland in *Rhodnius prolixus* (Hemiptera) and its role in moulting. *J. Exp. Biol.* **29,** 561–70.

WIGGLESWORTH, V. B. (1961) Some observations on the juvenile hormone effect of farnesol in *Rhodnius prolixus* Sta. (Hemiptera). *J. Insect Physiol.* **7,** 73–78.

DE WILDE, J. (1954) Aspects of diapause in adult insects with special regard to the Colorado beetle, *Leptinotarsa decemlineata* Say. *Arch. Neerl. Zool.* **10,** 375–85.

De Wilde, J. (1958) Perception of the photoperiod by the Colorado potato beetle (*Leptinotarsa decemlineata* Say). *Proc. Xth Intern. Congr. Entomol. Montreal 1956*, **2,** 213–18.

De Wilde, J. and De Boer, J. A. (1961) Physiology of diapause in the adult Colorado beetle. II. Diapause as a case of pseudo-allatectomy. *J. Insect Physiol.* **6,** 152–61.

De Wilde, J. and Bonga, H. (1958) Observations on threshold intensity and sensitivity to different wave lengths of photoperiodic responses in the Colorado beetle (*Leptinotarsa decemlineata* Say) *Entomol. Exp. Appl.* **1,** 301–7.

De Wilde, J., Duintjer, C. S. and Mook, L. (1959) Physiology of diapause in the adult Colorado beetle (*Leptinotarsa decemlineata* Say). I. The photoperiod as a controlling factor. *J. Insect Physiol.* **3,** 75–85.

De Wilde, J. and Stegwee, D. (1958) Two major effects of the corpus allatum in the adult Colorado beetle (*Leptinotarsa decemlineata* Say). *Arch. Neerl. Zool.* **13,** 277–89.

Wilkens, J. L. (1965) Nutrition and endocrine function in egg maturation of *Sarcophaga bullata. Am. Zoologist* **5,** 673.

Wilkens, J. L. (1967a) Reproduction in the fleshfly *Sarcophaga bullata* and its endocrine control. Ph.D. Thesis, Univ. California, Los Angeles.

Wilkens, J. L. (1967b) The control of egg maturation in *Sarcophaga bullata* (Diptera). *Am. Zoologist* **7,** 723–4.

Wilkens, J. L. (1968) The endocrine and nutritional control of egg maturation in the fleshfly *Sarcophaga bullata. J. Insect Physiol.* **14,** 927–43.

Wilkes, A. (1964) Inherited male-producing factor in an insect that produces its males from unfertilized eggs. *Science* **144,** 305–7.

Wilkes, A. (1965) Sperm transfer and utilization by the arrhenotokous wasp *Dahlbominus fuscipennis* (Zett.) (Hymenoptera: Eulophidae). *Can. Entomologist* **97,** 647–57.

Willey, R. B. (1961) The morphology of the stomodeal nervous system in *Periplaneta americana* (L.) and other Blattaria. *J. Morphol.* **108,** 219–62.

Williams, C. M. (1952) Physiology of insect diapause. IV. The brain and prothoracic glands as an endocrine system in the *Cecropia* silkworm. *Biol. Bull.* **103,** 120–38.

Williams, J. R. (1951) The factors which promote and influence the oviposition of *Nemeritis canescens* Grav. (Ichneumonidae, Ophioninae). *Proc. Roy. Entomol. Soc. London* A, **26,** 49–58.

Williams, R. W. (1961) Parthenogenesis and autogeny in *Culicoides bermudensis* Williams. *Mosquito News* **21,** 116–17.

Willis, E. R., Riser, G. R. and Roth, L. M. (1958) Observation on reproduction and development in cockroaches. *Ann. Entomol. Soc. Am.* **51,** 53–69.

Wilson, E. O. (1953) The origin and evolution of polymorphism in ants. *Quart. Rev. Biol.* **28,** 136–56.

Wilson, E. O. (1965) Chemical communication in the social insects. *Science* **149,** 1064–71.

Wilson, E. O. (1966) Behaviour of social insects. *Roy. Entomol. Soc. London Symp.* **3,** 81–96.

Wilson, E. O. and Bossert, W. H. (1963) Chemical communication among animals. *Recent Progress in Hormone Research*, pp. 673–715.

Wilson, F. (1938) Some experiments on the influence of environment upon the forms of *Aphis chloris* Koch (Aphididae). *Trans. Roy. Entomol. Soc. London* **87,** 165–80.

Wilson, F. and Woolcock, L. T. (1960) Temperature determination of sex in a parthenogenetic parasite, *Ooencyrtus submetallicus* (Howard) (Hymenoptera: Encyrtidae). *Australian J. Zool.* **8,** 153–69.

Winge, Ö. (1937) Goldschmidt's theory of sex determination in *Lymantria. J. Genetics* **34,** 81–89.

Withycombe, C. L. (1922) Notes on the biology of some British Neuroptera (Planipennia). *Trans. Entomol. Soc. London* **70,** 501–94.

Woke, P. A. (1937a) Comparative effects of the blood of different species of vertebrates on egg-production of *Aedes aegypti* Linn. *Am. J. Trop. Med.* **17,** 729–45.

Woke, P. A. (1937b) Comparative effects of the blood of man and of canary on egg-production of *Culex pipiens* Linn. *J. Parasitol.* **23,** 311–13.

Woke, P. A., Ally, M. S. and Rosenberger, C. R. (1956) The numbers of eggs developed related to the quantities of human blood ingested in *Aedes aegypti* (L.) (Diptera: Culicidae). *Ann. Entomol. Soc. Am.* **49,** 435–41.

Wong, S. K. and Thornton, I. W. B. (1968) The internal morphology of the reproductive system of some psocid species. *Proc. Roy. Entomol. Soc. London* A, **43,** 1–12.

Woodhill, A. R. (1941) The oviposition responses of three species of mosquitoes (*Aedes* (*Stegomyia*) *aegypti* Linnaeus, *Culex* (*Culex*) *fatigans* Wiedmann, *Aedes* (*Pseudoskusea*) *concolor* Taylor), in relation to the salinity of the water. *Proc. Linn. Soc. New South Wales* **66,** 267–92.

Woodroffe, G. E. (1958) The mode of reproduction of *Ptinus clavipes* Panzer form *mobilis* Moore (= *P. latro* Auct.) (Coleoptera: Ptinidae). *Proc. Roy. Entomol. Soc. London* A **33,** 25–30.

Woodrow, D. F. (1965a) The responses of the African migratory locust *Locusta migratoria migratorioides* R. & F. to the chemical composition of the soil at oviposition. *Animal Behaviour* **13,** 348–56.

Woodrow, D. F. (1965b) Laboratory analysis of oviposition behaviour in the red locust, *Nomadacris septemfasciata* (Serv.). *Bull. Entomol. Res.* **55,** 733–45.

WOYKE, J. (1963) Rearing and viability of diploid drone larvae. *J. Apicult. Res.* **2,** 77–84.
WOYKE, J. (1965) Rearing diploid drone larvae in queen cells in a colony. *J. Apicult. Res.* **4,** 143–8.
WOYKE, J. and KNYTEL, A. (1966) The chromosome number as proof that drones can arise from fertilized eggs of the honeybee. *J. Apicult. Res.* **5,** 149–54.
WÜLKER, W. (1961) Untersuchungen über die Intersexualität der Chironomiden (Dipt.) nach *Paramermis*-Infektion. *Arch. Hydrobiol. Suppl.* 25, 127–81.
WÜLKER, W. (1962) Parasitäre und nichtparasitäre geschlechtliche Aberrationen bei Chironomiden (Dipt.). *Verhandl. Deut. Zool. Ges. Saarbrücken 1961*, 132–9.
WYATT, I. J. (1961) Pupal paedogenesis in the Cecidomyiidae (Diptera). *Proc. Roy. Entomol. Soc. London* A, **36,** 133–43.
WYLIE, H. G. (1966) Some mechanisms that affect the sex ratio of *Nasonia vitripennis* (Walk.) (Hymenoptera: Pteromalidae) reared from superparasitized housefly pupae. *Can. Entomologist* **98,** 645–53.
YAMAMOTO, R. T. and FRAENKEL, G. S. (1960) The specificity of the tobacco hornworm, *Protoparce sexta*, to solanaceous plants. *Ann. Entomol. Soc. Am.* **53,** 503–7.
YAMASHITA, Y., TANI, K. and KOBAYASHI, M. (1961) The effects of allatectomy on the number of eggs in the silkworm, *Bombyx mori*. *Acta Sericol. No.* 39, 12–15.
YOELI, M. and MER, G. G. (1938) The relation of blood feeds to the maturation of ova in *Anopheles elutus*. *Trans. Roy. Soc. Trop. Med. Hyg.* **31,** 437–44.
YOSHIKAWA, K. (1962) Introductory studies on the life economy of polistine wasps. VII. Comparative considerations and phylogeny. *J. Biol. Osaka City Univ.* **13,** 45–64.
ZALOKAR, M. (1960) Sites of ribonucleic acid and protein synthesis in *Drosophila*. *Exp. Cell Res.* **19,** 184–6.
ZANDER, E. (1925) Die Königinnenzucht im Lichte der Beckerschen Untersuchungen. *Erlanger Jahrb. Bienenk.* **3,** 224–46.
ZELUETA, J. DE (1950) Comparative oviposition experiments with caged mosquitoes. *Am. J. Hyg.* **52,** 133–42.
ZIPPELIUS, H. M. (1949) Die Paarungsbiologie einiger Orthopteren-Arten. *Z. Tierpsychol.* **6,** 372–90.
ZWÖLFER, W. (1930) Beiträge zur Kenntnis der Schädlingsfauna Kleinasiens. I. Untersuchungen zur Epidemiologie der Getreidewanze *Eurygaster integriceps* Put. (Hemip. Het.). *Z. Angew. Entomol.* **17,** 227–52.
ZWÖLFER, W. (1931) Studien zur Ökologie und Epidemiologie der Insekten. I. Die Kiefereule, *Panolis flammea* Schiff. *Z. Angew Entomol.* **17,** 475–562.
ZWÖLFER, W. (1933) Studien zur Ökologie, insbesondere zur Bevölkerungslehre der Nonne, *Lymantria monacha* L. *Z. Angew. Entomol.* **20,** 1–50.

Literature to the Addendum on page 189

ENGELMANN, F. (1969) Female specific protein: biosynthesis controlled by corpus allatum in *Leucophaea maderae*. *Science* **165,** 407–9.
JOLY, P. (1969) Résultats d'injections de fortes doses d'hormone juvénile à *Locusta migratoria* en phase grégaire. *C.R. Acad. Sci. Paris* **268,** 1634–5.
MEYER, A. S., SCHNEIDERMAN, H. A., HANZMANN, E. and KO, J. H. (1968) The two juvenile hormones from the *Cecropia* silk moth. *Proc. Natl. Acad. Sci.* **60,** 853–60.
RÖLLER, H., BJERKE, J. S., HOLTHAUS, L. M., NORGARD, D. W., and MCSHAN, W. H. (1969) Isolation and biological properties of the juvenile hormone. *J. Insect Physiol.* **15,** 379–89.
RÖLLER, H. and DAHM, K. H. (1968) The chemistry and biology of juvenile hormone. *Rec. Prog. Hormone Res.* **24,** 651–80.

INDEX

OTHER TITLES IN THE ZOOLOGY DIVISION

General Editor: G. A. KERKUT

Vol. 1. RAVEN—*An Outline of Developmental Physiology*
Vol. 2. RAVEN—*Morphogenesis: The Analysis of Molluscan Development*
Vol. 3. SAVORY—*Instinctive Living*
Vol. 4. KERKUT—*Implications of Evolution*
Vol. 5. TARTAR—*The Biology of Stentor*
Vol. 6. JENKIN—*Animal Hormones—A Comparative Survey*
Vol. 7. CORLISS—*The Ciliated Protozoa*
Vol. 8. GEORGE—*The Brain as a Computer*
Vol. 9. ARTHUR—*Ticks and Disease*
Vol. 10. RAVEN—*Oogenesis*
Vol. 11. MANN—*Leeches (Hirudinea)*
Vol. 12. SLEIGH—*The Biology of Cilia and Flagella*
Vol. 13. PITELKA—*Electron-microscopic Structure of Protozoa*
Vol. 14. FINGERMAN—*The Control of Chromatophores*
Vol. 15. LAVERACK—*The Physiology of Earthworms*
Vol. 16. HADZI—*The Evolution of the Metazoa*
Vol. 17. CLEMENTS—*The Physiology of Mosquitoes*
Vol. 18. RAYMONT—*Plankton and Productivity in the Oceans*
Vol. 19. POTTS AND PARRY—*Osmotic and Ionic Regulation in Animals*
Vol. 20. GLASGOW—*The Distribution and Abundance of Tsetse*
Vol. 21. PANTELOURIS—*The Common Liver Fluke*
Vol. 22. VANDEL—*Biospeleology—The Biology of Cavernicolous Animals*
Vol. 23. MUNDAY—*Studies in Comparative Biochemistry*
Vol. 24. ROBINSON—*Genetics of the Norway Rat*
Vol. 25. NEEDHAM—*The Uniqueness of Biological Materials*
Vol. 26. BACCI—*Sex Determination*
Vol. 27. JØRGENSEN—*Biology of Suspension Feeding*
Vol. 28. GABE—*Neurosecretion*
Vol. 29. APTER—*Cybernetics and Development*
Vol. 30. SHAROV—*Basic Arthropodan Stock*
Vol. 31. BENNETT—*The Aetiology of Compressed Air Intoxication and Inert Gas Narcosis*
Vol. 32. PANTELOURIS—*Introduction to Animal Physiology and Physiological Genetics*
Vol. 33. HAHN and KOLDOVSKÝ—*Utilization of Nutrients during Postnatal Development*
Vol. 34. TROSHIN—*The Cell and Environmental Temperature*
Vol. 35. DUNCAN—*The Molecular Properties and Evolution of Excitable Cells*
Vol. 36. JOHNSTON and ROOTS—*Nerve Membranes*
Vol. 37. THREADGOLD—*The Ultrastructure of the Animal Cell*
Vol. 38. GRIFFITHS—*Echidnas*
Vol. 39. RYBAK—*Principles of Zoophysiology*
Vol. 40. PURCHON—*The Biology of the Mollusca*
Vol. 41. MAUPIN—*Blood Platelets in Man and Animals, Volumes 1 and 2*
Vol. 42. BRØNDSTED—*Planarian Regeneration*
Vol. 43. PHILLIS—*The Pharmacology of Synapses*
Vol. 44. ENGELMANN—*The Physiology of Insect Reproduction*

OTHER DIVISIONS IN THE SERIES IN PURE AND APPLIED BIOLOGY

BIOCHEMISTRY

BOTANY

MODERN TRENDS
IN PHYSIOLOGICAL SCIENCES

PLANT PHYSIOLOGY